ARTIFICIAL INTELLIGENCE AND STRUCTURAL ENGINEERING

ARTIFICIAL INTELLIGENCE
and
STRUCTURAL ENGINEERING

Edited by
B. H. V. TOPPING

CIVIL-COMP PRESS

CIVIL-COMP PRESS

10 Saxe-Coburg Place
Edinburgh

British Library Cataloguing in Publication Data
A catalogue record for this book is available from the British Library

ISBN 0-948749-15-6

Printed in Scotland
by
MEIGLE PRINTERS
Market Street, Galashiels

∘ CONTENTS ∘

PREFACE

The papers included in this volume and its companion "Artificial Intelligence and Civil Engineering" (Civil-Comp Press, ISBN 0-948749-14-8, 1991) were presented at the Second International Conference on the Application of Artificial Intelligence to Civil and Structural Engineering, held at Exeter College, Oxford from 3rd to 5th September 1991.

The proceedings of the first conference in this series were published as "Artificial Intelligence Techniques and Applications for Civil and Structural Engineers" (Civil-Comp Press, ISBN 0-948749-13-X, 1989) and this volume included thirty six papers. Growing interest in the application of artificial intelligence, particularly neural networks and machine learning, to an increasing number of domains is reflected in the large number of papers in the two volumes.

I should like to thank all authors for their contributions and in particular those who travelled to Oxford to present their papers at the Conference.

I should also like to thank the members of the Conference Advisory Board for their help and assistance. The Board Members were: Professor H Adeli, Ohio State University, Columbus, Ohio, United States of America; Dr M Alshawi, University of Salford, United Kingdom; Dr T Arciszewski, Wayne State University, Detroit, Michigan, United States of America; Dr B Balachandran, University of Sydney, Australia; Dr E Balagurusamy, National Centre for Expert Systems, Osmania University Campus, Hyderabad, India; Professor D Blockley, Bristol University, United Kingdom; Dr G Cameron, University of Waterloo, Waterloo, Ontario, Canada; Dr W T Chan, National University of Singapore; Professor J Christian, University of New Brunswick, Fredericton, Canada; Dr I E G Davey-Wilson, Oxford Polytechnic, Headington, Oxford; E William East, US Army Construction Engineering Research Laboratory, Champaign, United States of America; Dr M Eisenberger, Technion, Israel Institute of Technology, Technion City, Haifa, Israel; Professor K Fazio, Center for Building Studies, Concordia University, Montreal, Quebec, Canada; Professor S J Fenves, Carnegie-Mellon University, Pittsburgh, Pennsylvania, United States of America; Dr Renate Fruchter,Stanford University, United States of America; Dr H Furuta, Kyoto University, Japan; Professor James Garrett, Carnegie-Mellon University, Pittsburgh, Pennsylvania, United States of America; Professor J S Gero, Department of Architectural Science, University of Sydney, Australia; Professor D E Grierson, University of Waterloo, Waterloo, Ontario, Canada; Professor P Hajela, Rensselaer Polytechnic Institute, Troy, New York, United States of America; Professor K C Hover, Cornell University, Ithaca, United States of America; A T Humphrey, GEC Marconi Research Centre, Chelmsford, Essex, United Kingdom; Professor C W Ibbs, University of California, Berkeley, United States of America; Dr D G Jamieson, Thames Water, Reading, Berkshire, United Kingdom; Professor P W Jowitt, Heriot-Watt University, Riccarton, Edinburgh, United Kingdom; Dr S Jozwiak, Institute for Fundamental Technological Research, Polish Academy of Sciences, Warsaw, Poland; K Kahkonen, Technical Research Centre of Finland, The Laboratory of Urban Planning and Building Design, Espoo, Finland; Dr Krishnaamoorthy, Indian Institute of Technology, Madras, India; Dr V K Koumousis, National Technical University, Athens, Greece; Dr B Kumar, Strathclyde University, Glasgow, United Kingdom; John Lansdown, Centre for Advanced Studies in Computer Aided Art and Design, Middlesex Polytechnic, Barnet, United Kingdom; Professor K. Lawrence, University of Texas at Arlington, Arlington, United States of America; Professor Liu Xihui, Institute of Computer Science and Knowledge Engineering, China Academy of Electronics and Information Technology, Beijing, PR China; Professor I MacLeod, Strathclyde University, Glasgow, United Kingdom; Dr I May, University of Bradford, Bradford, United Kingdom; Dr J.C. Miles, School of Engineering, University of Wales, Cardiff, United Kingdom; Dr J Oliphant, Heriot-Watt University, Riccarton, Edinburgh, United Kingdom; Dr G Powell, University of California, Berkeley, California, United States of America; Dr D Rehak, Carnegie-Mellon University, Pittsburgh, Pennsylvania, United States of America; Dr S G Ritchie, University of California,

Irvine, United States of America; Dr M A Rosenman, Department of Architectural Science, University of Sydney, Australia; Dr S Rowlinson, University of Hong Kong, Hong Kong; Professor G Roy, The University of Western Australia, Nedlands, Australia; Professor H C Shah, Stanford University, California, United States of America; Dr W J Spencer, Swimburne Institute of Technology, Hawthorn, Victoria, Australia; Professor D Sriram, Massachusetts Institute of Technology, Cambridge, Massachusetts, United States of America; Dr Chee-Kiong Soh, Department of Civil Engineering, Nanyang Technological Institute, Singapore; D Taffs, Ove Arup & Partners, London, United Kingdom; Dr A Watson, Leeds University, Leeds, United Kingdom; and Professor G Yagawa, University of Tokyo, Japan.

Finally I should like to acknowledge the contributions made by my research associates and students during the preparation of this volume. In particular, the contributions made by Bimal Kumar, Mansour Jadid, Asad Khan, Nick Bitoulas and János Sziveri are gratefully acknowledged.

Barry H.V. Topping
Heriot-Watt University
Edinburgh

September 1991

1 STRUCTURAL MODELLING, REPRESENTATION AND ANALYSIS

FORM, FUNCTION AND BEHAVIOR IN STRUCTURAL ENGINEERING KNOWLEDGE REPRESENTATION

J.A. Abdalla, D.H.D. Phan and H.C. Howard
Center for Integrated Facility Engineering,
Stanford University, Stanford, United States of America

Typical structural engineering objects are complex entities that include large, yet closely related data and knowledge. A great deal of attention should be given to the representation of the elements that make up these complex objects. In order to reduce their inherent complexity, these objects can be decomposed into several simpler primitive objects which are related to themselves and also to their composite object by various relationships. A natural way of decomposing these objects is along the fundamental aspects of *form*, *function* and *behavior* of the design object which also correspond to the sources for data and knowledge of the object. *Form* (or *structure*) describes the physical characteristics of the object, *function* depicts the object role and its intended use, and *behavior* describes the object's manifestation of its functions, when subject to the environment stimuli. The structural engineer reasons about these fundamental aspects during the design process and across various stages of the object lifecycle. Consequently, form, function and behavior should be incorporated into the representation of the design object.

This paper addresses form, function, and behavior representation of structural engineering design objects. It then presents several sample hierarchies of primitive object classes that are characterized according to specific form, function, or behavior criteria. To illustrate the concept, the paper shows a detailed example of a base plate for a transmission pole. This work is part of a current effort to develop the Primitive-Composite data model for structural engineering that will also be briefly presented.

Key Words

Form, Function, Behavior, Knowledge Representation, Primitive Object, Composite Object, Characterization Hierarchy, Relationship, Data Model, Product Model.

INTRODUCTION

The concepts of *form*, *function*, and *behavior* were introduced, with some variations, in various fields of applications. These fields include: qualitative physics [Ref. 2, Ref. 9], function-based representation [Ref. 15, Ref. 16, Ref. 10], engineering design [Ref. 3, Ref. 11, Ref. 17, Ref. 12], design knowledge capture [Ref. 7, Ref. 4], etc. In this research, we are exploring the use of form, function and behavior in Data Modelling and Knowledge Representation. Form, function and behavior are used here as criteria for characterizing primitive object classes into several class hierarchies (or *primitive characterization hierarchies*), each of which relates to a single concept. This will result in the representation of simpler primitive object classes and in more homogeneous primitive characterization hierarchies. Composite object classes can then be defined in terms of these primitive classes from several characterization hierarchies. This will enhance the atomicity, modularity and cleanness desired in the representation of complex structural engineering objects. To accurately represent a structural engineering object, it is essential to represent the ingredient primitive objects and their relationships. The composite object can then be formed by combining these primitive objects using the appropriate relationships. Consequently, much attention should be given to the representation of the elements that make up the composite objects, namely the primitive objects and relationship types.

This paper outlines the major characterization hierarchies of primitive object classes where form, function and behavior are clearly separated. The paper also describes how primitive objects from these hierarchies can be used to define composite objects. First, an overview of the Primitive-Composite data model (which is the main goal of this work) is presented. Form, function and behavior are then introduced along with several primitive characterization hierarchies. To illustrate the concept, an integrated example of a base plate for a transmission pole is presented. Finally, the benefits of this modelling approach are stated.

OVERVIEW OF THE PRIMITIVE-COMPOSITE DATA MODEL

This research is part of an ongoing effort at the Center for Integrated Facility Engineering (CIFE) at Stanford University to develop the *Primitive-Composite (P-C) data model* [Ref. 8]. The long-term objectives of the data model are to help integrating structural engineering activities and to contribute to the integration of facility engineering. The P-C data model is based on the object-oriented data model, and therefore incorporates key object-oriented concepts of *class* (object class), *instance* (or object), *attribute*, and *methods*. In addition, the data model incorporates a number of extensions to the core object-oriented data model. The motivation for these extensions is three-fold: to represent complex structural engineering objects, to offer maximum flexibility in modeling structural engineering data and knowledge, and to facilitate data exchange among different applications.

The following items are extensions to the object-oriented data model in the Primitive-Composite data model. Items 1 to 4 are extensions to describe how objects are organized to represent independent concepts and how these objects can be combined to create complex abstractions that suit the user views. Items 5 to 8 are extensions to support the data exchange using the Primitive-Composite data model.

1. **Primitive Class and Primitive Instance** — A primitive class represents an atomic entity that deals with one simple concept (e.g., shape, material, or function). A primitive class is non-decomposable with respect to the particular domain of modeling. A primitive instance is an instance of a primitive class.

2. **Primitive Class Hierarchy** — A primitive class hierarchy groups primitive classes that represent increasing specializations of a single concept of form, function, or behavior. As an example, two-dimensional shapes can be specialized into 3-sided shapes, 4-sided shapes, 5-sided shapes, while 4-sided shapes can be further specialized into squares, rectangles, trapezoids.

3. **Composite Class** — A composite class consists of a heterogeneous set of related primitive classes that represent a complex concept in the modeling domain or in an application. It is used to describe a common combination of domain concepts that may be found in many structural engineering objects. For example, a "Beams" composite class describes the concept of beam as a combination of specific forms, functions, and behaviors, and provides users with the abstraction of the beam object that they are accustomed to.

4. **Composite Instance** — A composite instance is an instance of a composite class. For example, a "beam23" combines physical form objects that represent its location and shape, a function object that describes its load resisting role, and a behavior object that defines the beam as a flexural analysis element.

5. **Primitive data base schema** — A primitive data base schema is a set of primitive characterization hierarchies that define the basic concepts of a domain. These concepts can be used directly or indirectly by the domain specialists for different applications sharing the same primitive schema. The latter may include some composite classes that are commonly seen by the domain specialist, but not defined specifically for any particular application.

6. **Primitive data base** — A primitive data base contains instances of primitive classes from a primitive schema. As described later, a primitive data base can be used as a medium of data exchange between different application systems.

7. **Composite data base schema** — A composite data base schema includes a subset of the primitive schema, and a set of composite classes that defines an application view of the domain data and suits the particular needs of specialized domain tasks. Composite classes, in this case, provide a convenient framework for formalizing higher level abstractions of complex structural engineering objects, in particular in applications that make use of the primitive schema.

8. **Composite data base** — A composite data base is a data base that contains instances of composite and primitive classes from a composite schema.

FORM, FUNCTION AND BEHAVIOR

Form, function, and behavior are the basic conceptual building blocks for defining characterization hierarchies of primitive object classes. Using the Primitive-Composite approach, we can define structural design objects as composite objects in terms of form, function, and behavior primitives. These primitive objects can be selected from different characterization hierarchies. This section identifies the elements of form, function, and behavior that is needed to represent structural design objects. The detailed development of the P-C data model that incorporates form, function, and behavior representation of structural engineering objects is the subject of our ongoing research [Ref. 1, Ref. 13].

Form

Form describes the physical characteristics of an object. There are many types of form description: spatial, geometric, topological, material, fabrication, etc. This section covers those aspects of form necessary to define structural engineering design objects.

- **Spatial Form:** This form describes where and how a physical design object is located, oriented, and realized in three-dimensional space. Such a description includes the spatial envelope of the object, and its location and orientation with respect to certain global reference data or relative to other objects in its environment. The spatial envelope of a physical object can be defined in terms of a local coordinate system and the dimensions (length, width, and height) of its spatial enclosure. For example, a beam object in a typical structure can be located and oriented in terms of its local coordinate system or relative to its floor or frame objects. Figure 1 below illustrates two characterization hierarchies for coordinate system and spatial enclosure forms.

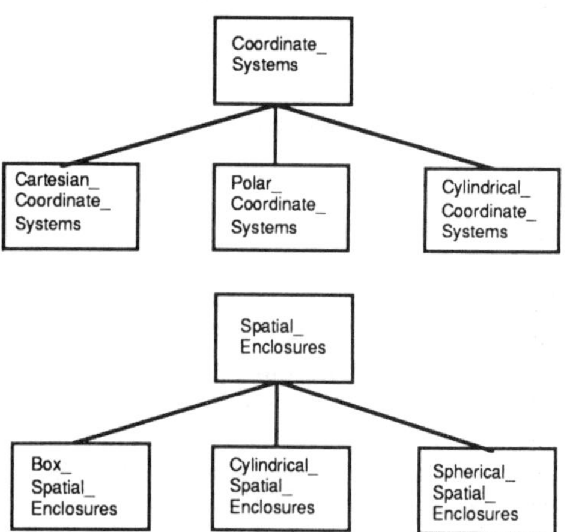

Figure 1: Sample Spatial Form Characterization Hierarchies

- **Geometric Form:** This form defines the geometric shape (and dimensioning) of the object in terms of geometric elements such as points, lines, curves, surfaces, etc. Physical objects are three-dimensional, but their shapes can be represented in many ways by different geometric forms. For example, a rectangular column object can be represented as three 1-Dimensional (1-D) geometric forms that are its width, depth and height. The same column can be represented as one 1-D geometric form (height) and one 2-D geometric form (section-shape), or it can be represented as a solid 3-D geometric form (parallelopiped). Many geometric entities are defined in the PDES/STEP Integrated Product Information Model (IPIM) [Ref. 18]. Figure 2 below illustrates a characterization hierarchy for two-dimensional rectilinear shapes.

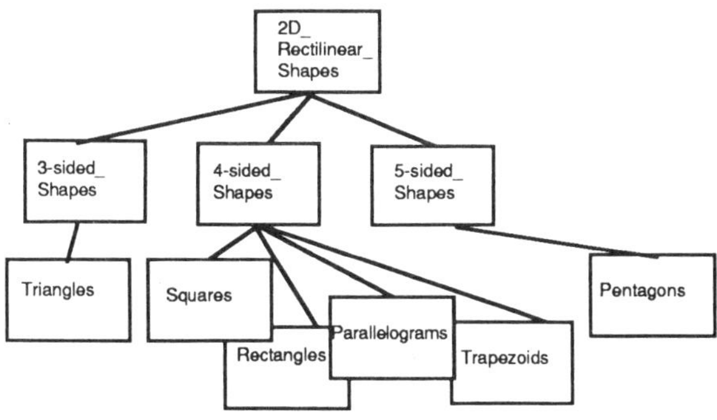

Figure 2: A Sample Geometry Form Characterization Hierarchy

- **Topological Form:** This form defines the connectivity of the object, in the constructed environment, in terms of topological elements. In structural engineering, a wire frame model of the structure is commonly constructed to define the topology of the structure. Such a wire frame model is analogous to a finite element model used for structural analysis purpose. As shown in Figure 3 below, *vertices* (of dimensionality 0), *edges* (of dimensionality 1), *faces* (of dimensionality 2), and *volumes* (of dimensionality 3) are topological primitive entities defined in the PDES/STEP IPIM [Ref. 18] and the GARM model [Ref. 6].

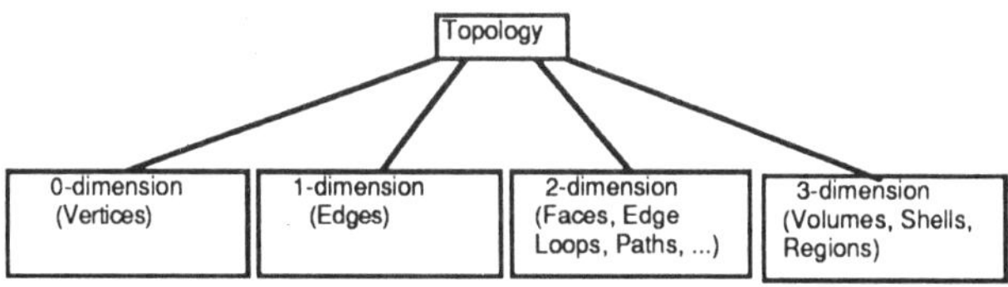

Figure 3: A Sample Topology Characterization Hierarchy

- **Cross-Section Properties Form:** This form includes the cross-sectional properties of the object such as section dimensions, section modulus, area, moments of inertia, torsional rigidity, etc. These properties are derived from the section shape forms of the object. There are several AISC standard shape primitives for structural steel members that are categorized as Rolled Sections, Plates, Bars, etc.

- **Material Form:** This form describes the material type and material properties of the object. The material types used in civil engineering include steel, reinforced concrete, asphalt, mortar, wood, etc. According to the PDES IPMP [Ref. 18], material properties can be classified into groups such as physical, structural, thermal, and thermal expansion. In [Ref. 14] details of material representations for finite element applications have been provided. Figure 4 shows characterization hierarchies for homogeneous material properties and material structural properties. The primitive classes in these hierarchies can be used to represent common material in structural engineering such as steel, aluminum, wood, concrete, etc.

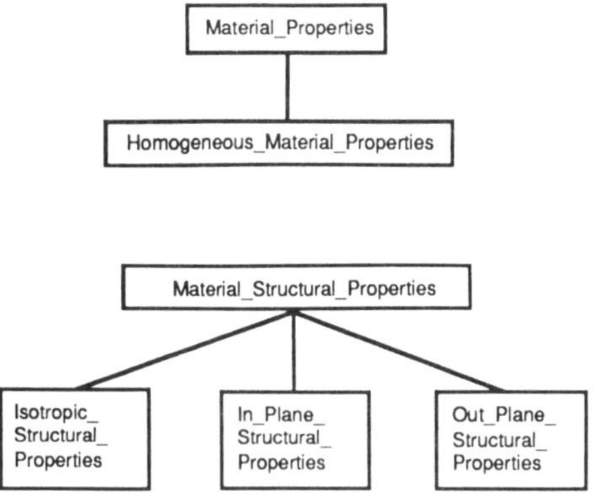

Figure 4: Sample Material Form Characterization Hierarchies

- **Hierarchical Aggregation Form:** This form is used to describe the hierarchical framework of structural design systems and their component objects. Structural members such as beam, column, girder, etc., are designed not independently of one another, but from some preconceived load resisting systems to which they belong. Common entities of this form include "Systems", "Assemblies", "Arrangements", "Members", "Parts", and "Segments".

- **(Part) Fabrication Form:** This form includes fabrication features of an engineering part prescribed by the designer. There is a large set of standard fabrication features such as taper, bend, cut out hole, edge clipping, edge preparation, NC mark, etc. Standard fabrication features are defined in the PDES Integrated Product Information Model [Ref. 18] and the NIDDESC Ship Structural Model [Ref. 5].

- **Load Form:** This form includes loading description features such load intensity, loading shape, and other factors that describe a load. Figure 5 shows a simple applied load form characterization hierarchy.

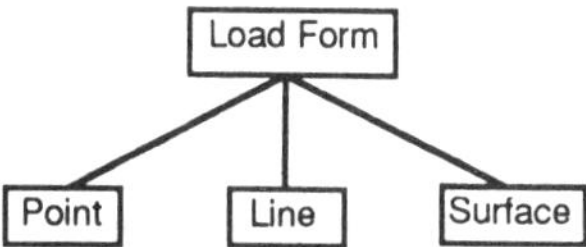

Figure 5: A Sample Load Form Characterization Hierarchy

Function

The function description of an object reflects the intended role or purpose of the object in its constructed environment. An object may serve several functions. For example, structural functions of the building elements relate to some aspects of load carrying, load transferring, part connecting, member supporting roles, etc. These functions come from the specialized relationships among the building elements, or between the building elements and other entities such as loads, load cases, etc. In the Primitive-Composite data model, a number of key structural engineering function objects (primitive) are identified. As illustrated in Figure 6, these primitive objects include: load resisting function, load transmitting function, object supporting function, object connecting function, and object bracing function.

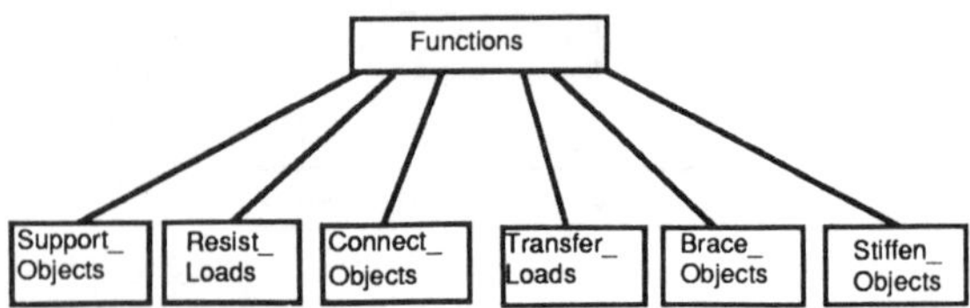

Figure 6: A Sample Function Characterization Hierarchy

EXAMPLE Figure 7 below provides an example of a flat slab, and a column with dropped panel and column capital, and a footing. This example shows the key functions of structural engineering objects such as load resisting function, object supporting function, object connecting function, etc. The *load resisting function* of an object is to resist loads that are applied directly or are transmitted to it from another object. The column object in Figure 7 resists its own weight, the weight of the column-capital, the weight of the dropped-panel, the weight of the flat-slab, the concentrated load directly applied to it, and the distributed loads transmitted to it from the flat slab. The *load transmitting function* of an object is to transfer the loading it carries to other objects in the load path. For example, the flat slab transmits the load it carries to the column through the dropped panel and the column capital. The column object then transmits the load to the footing. The *object supporting function* is to support another object. This function enables the transferring of loads from the supported object, to the supporting object, down to the next object in the load path. The column, for instance, supports the column capital. The column capital supports the dropped panel, which in turn supports the flat slab. The *object connecting function* connects two or more objects together. For example, a column-to-footing connection object connects the column object to the footing object. The *object bracing function* is to brace another object.

In addition, there are several other structural engineering functions that can be included. For instance, the dropped-panel reduces the shear stress in the slab within the area of the dropped-panel to avoid punching shear. The column-capital has a primary function to increase the stiffness of the column object, and thus to reduce the bending moment carried by the flat slab object.

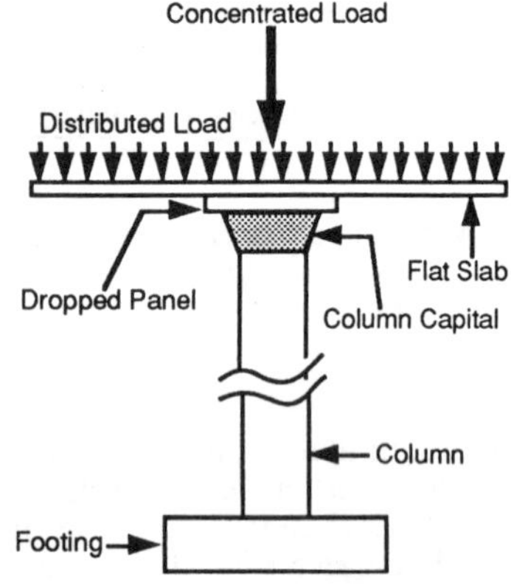

Figure 7: Example of a Column and Flat Slab

Behavior

The behavior of a design object is the way the object responds to the environmental stimuli in carrying out a certain function. Since an object may perform several functions, it follows that it may exhibit different behaviors, each of which corresponds to a particular function or combination of functions. For example, in resisting gravity loads, a column object develops internal axial stresses; in resisting lateral loads, it exhibits shear and bending stresses. In structural engineering, the behavior of a structural component under the influence of loading is manifested in terms of internal forces, stresses, deflection, deformation, vibration, etc. There are also well defined behavior models of structural elements such as flexural bending, torsion, axial buckling, etc.

The design of structural elements includes criteria that impose certain limits on their behavior. These design criteria ensure the acceptable performance of the design object according to professional standards with regard to the following aspects: strength (stresses and internal forces), serviceability (deflection, vibration, cracking, etc.), ductility, stability, and reliability. As shown in Figure 8, these design aspects are used as criteria to characterize the behavior primitive classes.

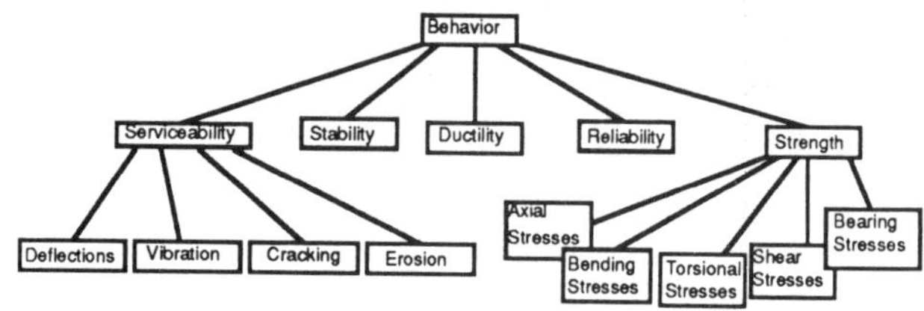

Figure 8: A Sample Behavior Hierarchy

Summary of Form, Function and Behavior

The major aspects of form, function, and behavior are summarized in Figure 9 below. They represent three orthogonal planes. A composite object class can be located within the region defined by these three planes, where it makes use of certain primitive classes from each of these planes. Although not shown in the figure, there are some underlying mappings and relationships among primitive objects of the different planes.

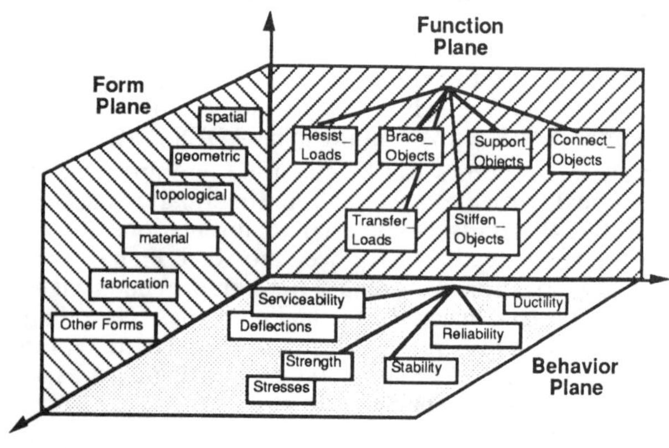

Figure 9: Form, Function, and Behavior in Structural Engineering

AN INTEGRATED EXAMPLE: TRANSMISSION POLE BASE PLATE

In this section, we show an integrated example that demonstrates the use of form, function, and behavior primitive objects in defining a composite object of a base plate for a transmission pole. The example also shows the effects of such a representation on different tasks in the design process.

Typical transmission steel poles consist of steel shafts of various lengths (up to forty feet). These shafts are tapered, hollow, and cylindrical; their cross section can be hexagonal, octagonal, or twelve-sided. The size at the pole base shaft is determined from structural analysis/design of the pole structure for various loading conditions. The reaction loads at the base of the pole are used to design the base plate and anchor bolts assembly (as shown in Figure 10) whose functions are to carry and to transmit these loads to the foundation. Each load case of the pole structure yields a combined axial load P, a resultant shear force V, and an overturning moment M as the applied loads on the base plate.

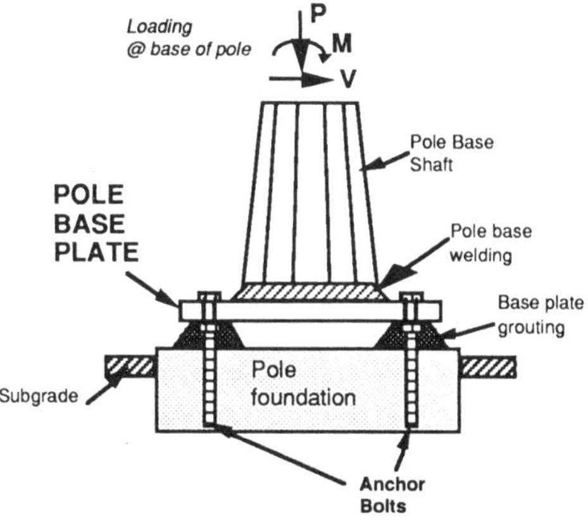

Figure 10: Transmission Pole Base Plate and Anchor Bolts Assembly

A sample base plate with a 4-bolt-hole pattern is illustrated in Figure 11 below. In general, the pole base plate is made of normal strength (grade A-36) or high strength (grade A-572 or other) steel material.

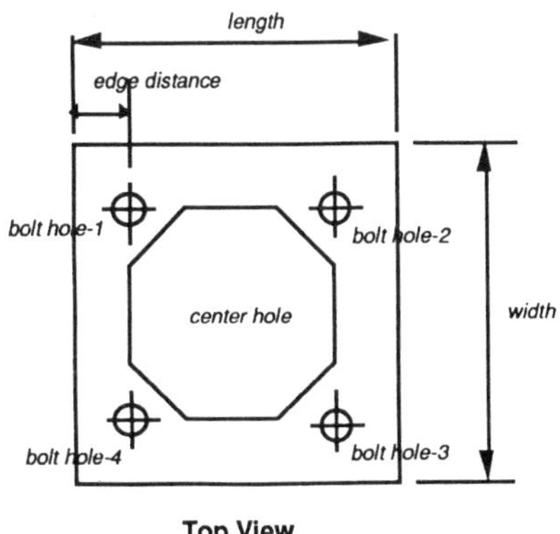

Figure 11: A Sample Base Plate with 4-Bolt-Hole pattern

FORM REPRESENTATION OF THE BASE PLATE COMPOSITE OBJECT The physical description of the base plate object (or instance) used in this example are represented in terms of its form primitive objects. The latter are instances of object classes from disjoint characterization hierarchies that correspond to the form characterization criteria as presented earlier in Section 3.1. Figures 12, 13, 14, 15, 16, and 17 below illustrate the base plate composite object and its spatial, geometric, topological, cross section properties, material, and fabrication form primitives respectively. These figures also show the corresponding primitive object classes from which these primitives are instantiated. Some of these classes are based on entities defined in PDES/STEP IPMP [Ref. 18] and in NIDDESC [Ref. 5].

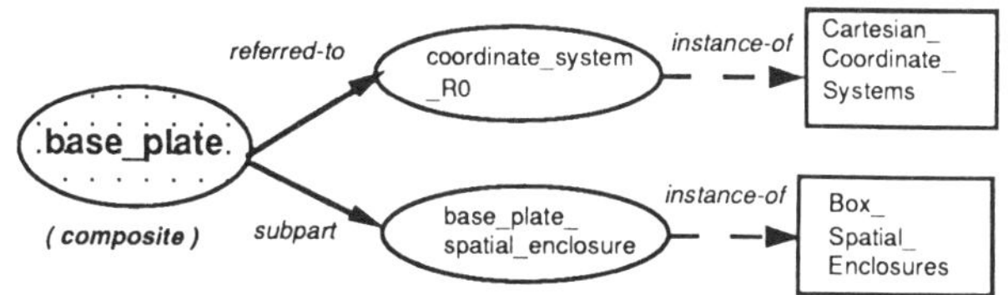

Figure 12: The Base Plate Composite Object and its Spatial Form Primitives

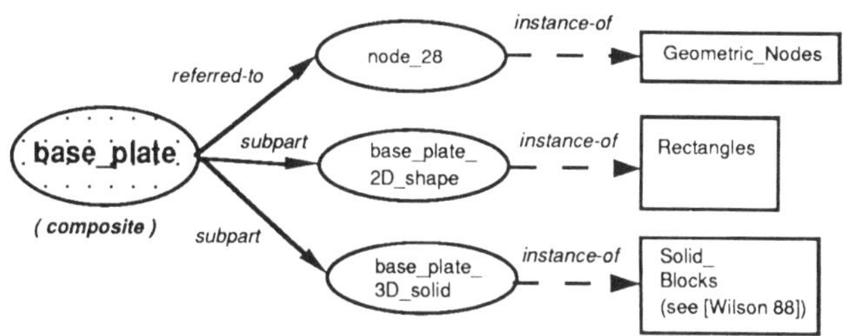

Figure 13: The Base Plate Composite Object and its Geometric Form Primitives

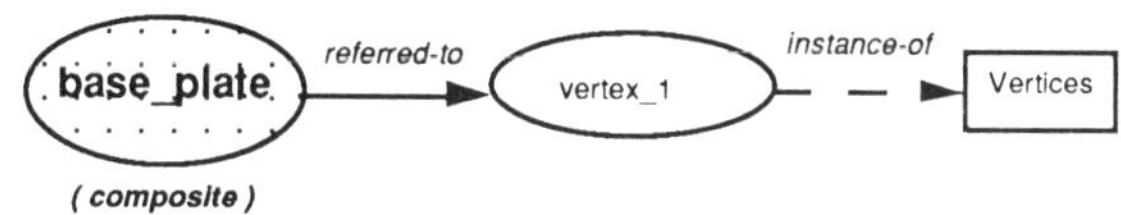

Figure 14: The Base Plate Composite Object and its Topological Form Primitives

Figure 15: The Base Plate Composite Object and its Cross Section Properties Form Primitives

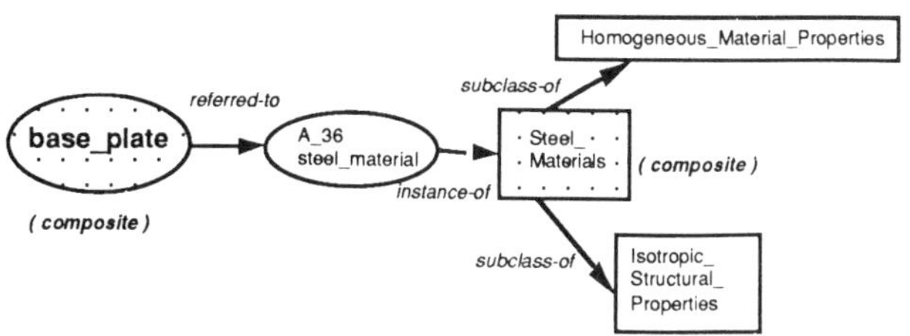

Figure 16: The Base Plate Composite Object and its Material Properties Form Primitives

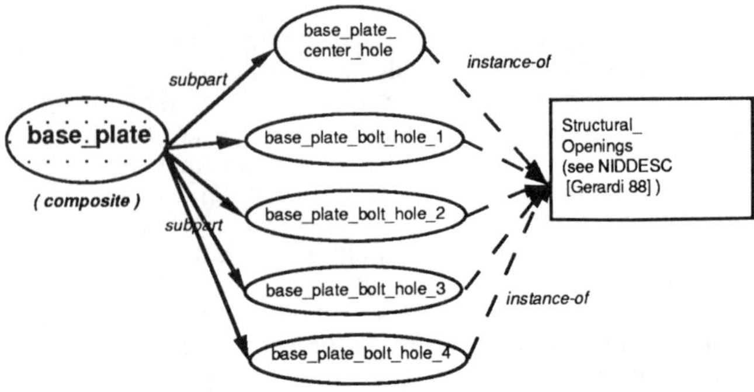

Figure 17: The Base Plate Composite Object and its Fabrication Form Primitives

FUNCTION AND BEHAVIOR REPRESENTATION OF THE BASE PLATE COMPOSITE OBJECT The first function of the base plate is to transfer the reaction loads P, V, and M at the base of the pole to the anchor bolts (which in turn transfer them to the foundation). These external loads transform into axial loads (Tb on the tension side and Cb on the compression side), and shear loads (Vb) of the anchor bolts. The combined loading effect on the set of anchor bolts, however, applies the axial bolt loads (i.e., Tb and Cb) on the base plate whose second function is to resist these loads. Figure 18 illustrates the analysis model used to analyze and design the base plate. In this model, the base plate is subject to both shear and bending stresses.

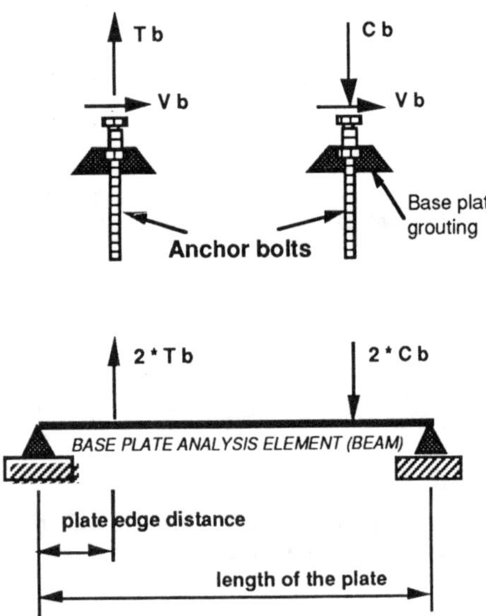

Figure 18: Base Plate Analysis Model

Figures 19 and 20 shows the base plate composite object and its function and behavior primitive objects respectively. Again, these primitive objects are instances of object classes from different characterization hierarchies that correspond to specific function or behavior characterization criteria as presented earlier in Sections 3.2 and 3.3. These figures also show the corresponding primitive object classes from which these primitive objects are instantiated.

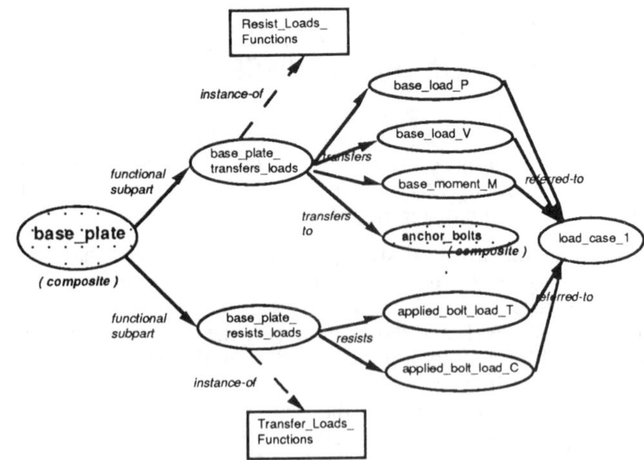

Figure 19: The Base Plate Composite Object and its Function Primitives

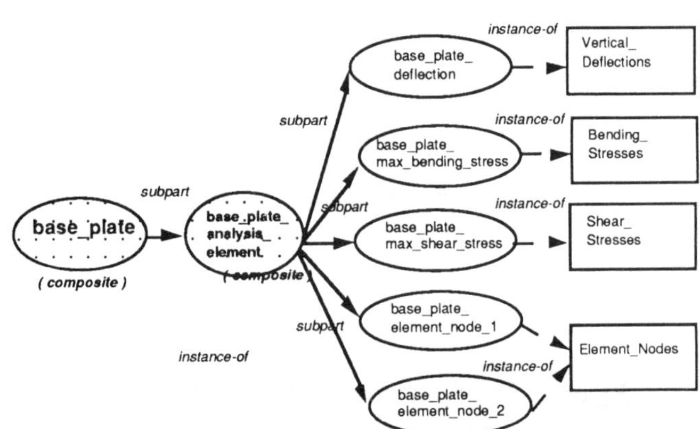

Figure 20: Base Plate Instance and its Behavior Form Primitives

THE BASE PLATE COMPOSITE OBJECT As shown previously, the base plate composite object is represented in terms of a number of form, function, and behavior primitive objects. The disjoint primitive characterization hierarchies provide the primitive object classes from which these primitive objects are instantiated. Each of these hierarchies corresponds to a specific characterization criterion, and therefore describes a single concept of form, function, and behavior. As illustrated in Figure 21, the complete base plate composite object definition is done by selecting the proper primitive classes from various characterization hierarchies (all of which are not shown here due to the limited scope of the paper), and thereby creating the appropriate primitive instances.

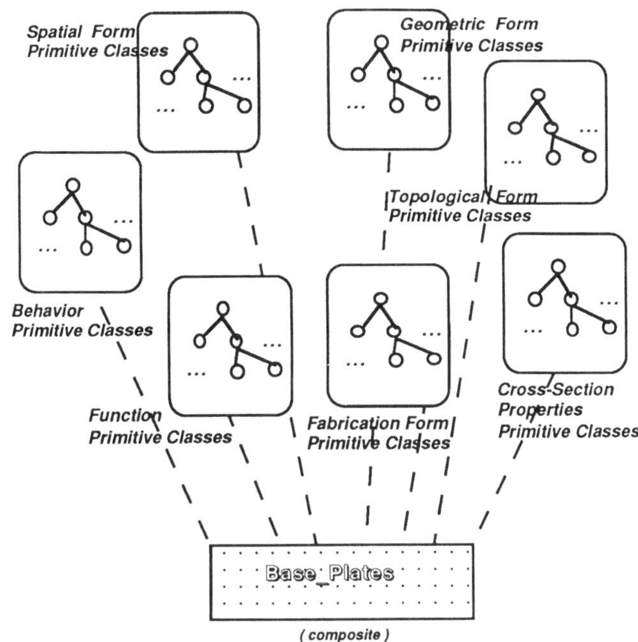

Figure 21: The Base Plate Composite Objects in Terms of Primitive Classes

EFFECT OF THE PRIMITIVE-COMPOSITE REPRESENTATION OF THE BASE PLATE OBJECT The Primitive-Composite approach as illustrated previously enhances the modularity, reusability, and cleanness in representing complex structural engineering objects such as the transmission pole base plate. Instead of a single large object that includes all the data and knowledge about the base plate, the latter is decomposed into smaller primitive objects. Each of these primitive objects describes one particular aspect of form, function, or behavior of the design object.

To illustrate this concept, consider four design tasks: pole geometry configuration, base plate preliminary design, base plate analysis, and base plate detailed design. In the first task, only the spatial configuration and positioning of the base plate are considered. In the second task, the designer defines the functional specifications (i.e., the load transferring and resisting mechanisms) of the base plate and its anchor bolt connectors. In the third task, the designer works out an analysis model of the base plate (similar to the one shown in Figure 18), and computes the maximum bending and shear stresses of the base plate. In the fourth task, the designer determines the final sizing of the base plate, and specifies its fabrication features. In all of these tasks, the composite base plate object can include only the primitive objects that are actually needed for the task. Table 1 below shows the actual primitive objects that are used to represent the base plate in the four tasks. In short, a representation of this nature enables different project participants to concentrate on certain aspects of the overall composite object. It also allows the design object to gradually evolve, and facilitate the exchange (at the level of primitive objects) among different design applications of only the data that they actually need.

	Pole Geometry Configu-ration	Base Plate Preliminary Design	Base Plate Analysis	Base Plate Detailed Design
Spatial Form Primitives				
coordinate_system_R0	X		X	X
base_plate_spatial_enclosure	X			
Geometric Form Primitives				
node_28	X	X	X	X
base_plate_2D_shape			X	X
base_plate_3D_solid			X	X
Topological Form Primitives				
vertex_1	X	X		X
Material Form Primitives				
A_36_steel_material			X	X
Cross-Section Properties Form Primitives				
base_plate_X-section_properties			X	X

Fabrication Form Primitives				
base_plate_center_hole				X
base_plate_bolt_hole_1				X
base_plate_bolt_hole_2				X
base_plate_bolt_hole_3				X
base_plate_bolt_hole_4				X
Function Primitives				
base_plate_transfers_loads		X	X	X
base_plate_resists_loads		X	X	X
Behavior Primitives				
base_plate_deflection			X	X
base_plate_max_bending_stress			X	X
base_plate_max_shear_stress			X	X
base_plate_element_node_1			X	X
base_plate_element_node_2			X	X

TABLE 1: **Different Design Tasks vs. Primitives of the Composite Base Plate Object.**

SUMMARY AND CONCLUSION

The Primitive-Composite data model for structural engineering uses form, function, and behavior criteria to define characterization hierarchies of primitive object classes. Composite object classes that represent complex structural engineering objects can then be defined in terms of these primitive classes. Generalization, aggregation, and association relationship types are allowed in the definition of a composite object class. Having primitive classes of form, function and behavior provides the users with a common base vocabulary (or language) for the domain data and knowledge. This common base of data and knowledge can be shared among different project participants, design phases, and application programs. This modelling approach leads to a clean and efficient representation of complex structural engineering objects due to the following reasons:

1. **Clean primitive views are maintained.** In this approach, the elements of form, function, and behavior are separated in disjoint primitive characterization hierarchies. Each of these hierarchies uses only a single criterion to define the primitive classes, and thus provides a clean view about a specific aspect of the complex design object description. This characterization method when fully utilized enables the developer (or different developers) to concentrate on one particular view of the object at a time.

2. **Various project participants (i.e., architects, designers, fabricators, etc.) can easily create their own customized views of the design object.** By having the available primitive classes of form, function, and behavior, several customized views of the design object can easily be assembled. This is done simply by combining primitive objects from different form, function and behavior hierarchies to form the desired composite object class.

3. **This approach supports the representation of evolving design objects.** Since composite objects are indeed built from smaller primitive objects, more primitive classes can be added to the schema, and consequently, more composite objects can be defined to reflect the current stage of the design object's life cycle.

4. **Data can be efficiently and easily exchanged among different project participants, various stages of the design life cycle, or different design applications.** This data exchange can be accomplished at the level of primitive objects where several project participants, design stages, or application programs can share a similar set of primitive objects of form, function, and behavior, and therefore have a common vocabulary and speak the same language.

This is a continuing research in the area of data modelling for structural engineering. Our immediate objectives are to formalize the Primitive-Composite data model, to define a comprehensive primitive schema for structural steel, to develop software tools that automate the primitive and composite schema design, and to study how the data exchange among different application can be done using the Primitive-Composite approach.

ACKNOWLEDGMENT

This work has been supported by a grant from the Center for Integrated Facility Engineering (CIFE) at Stanford University. The authors extend their thanks to Ajay Lavakare who has contributed ideas to our ongoing work in the area of data modelling.

REFERENCES

1. Abdalla, Jamal A., Phan, H. D., and Howard, H. C., *Theoretical Development of the Primitive-Composite Data Model for Structural Engineering,* Forthcoming Report, Center for Integrated Facility Engineering, Stanford University, Stanford, CA, 1991.

2. De Kleer, J. and Brown, J. S., "A Qualitative Physics Based on Confluences," *Artificial Intelligence,* Vol. 24, No. 3, pp. 7-83, 1984.

3. Jain, D., Luth, G. P., Krawinkler, H., and Law, K. H., "A Formal Approach to Automating Conceptual Structural Design," *Technical Report No. 31,* Center for Integrated Facility Engineering, Stanford University, Stanford, CA, 1990.

4. Garcia, A. C. B. and Howard, H. C., "Building a Model for Augmented Design Documentation," *Artificial Intelligence in Design,* Edinburgh, UK, 1991.

5. Gerardi, M., et al., NIDDESC: The Navy/Industry Digital Data Exchange Standards Committee, *Reference Model For Ship Structural Systems,* Version 3.0, ISO TC184/SC4/WG1 Document 3.2.2.5, October 1988.

6. Gielingh, W., "General AEC Reference Model (GARM): an Aid for the Integration of Application Specific Product Definition Models," Conceptual Modelling of Buildings, *CIB Proceedings,* Publication 126, edited by P. Christiansson and H. Karlsson, The Swedish Building Centre, October 1988.

7. Gruber, T. R. and Russel, D. M., *Design Knowledge and Design Rationale: A Framework for Representation, Capture, and Use,* Technical Report KSL 90-45, Knowledge Systems Laboratory, Stanford University, Stanford, CA, 1990.

8. Howard, H. Craig, Abdalla, Jamal A., and Phan, H. D., "The Primitive-Composite Approach to Structural Data Modelling," To appear in the *ASCE Journal of Computing in Civil Engineering,* Special Issue on Databases, 1992.

9. Iwasaki, Y., "Qualitative Physics," In *The Handbook of Artificial Intelligence,* Vol. IV, edited by A. Barr, P. R. Cohen, and E. A. Feigenbaum, pp. 325-413, Addison-Wesley, 1989.

10. Keuneke, A. M., "Device Representation — The Significance of Functional Knowledge," *IEEE Expert,* pp. 22-25, April 1991.

11. La Rota, J. L., Biswas, G., and Basu, P. K., "A Model-Based Approach to Structural Design." *Applications of Artificial Intelligence in Engineering V, Proceedings of the Fifth International Conference,* Vol. 1, edited by J.S. Gero, Boston, MA, pp. 3-22, 1990.

12. Luth, G. P., *Representation and Reasoning for Integrated Structural Design of High-Rise Commercial Office Buildings.* Forthcoming unpublished Ph.D. Thesis, Department of Civil Engineering, Stanford University, Stanford, CA, 1991.

13. Phan, D. H., *Conceptual Development and Data Base Applications of the Primitive-Composite Data Model for Structural Engineering.* Unpublished Ph.D. Thesis Proposal, Department of Civil Engineering, Stanford University, Stanford, CA, 1991.

14. Rumble, J., *Exchange of Product Model Data - Part 45: Materials,* ISO CD 10303-45. NIST, A323 Physics Bldg, Gaithersburg, MD 20899, 1990.

15. Sembugamoorthy, V. and Chandrasekaran, B., "Functional Representation of Devices and Compilation of Diagnostic Problem-Solving Systems," *Experience, Memory, and Reasoning,* edited by J. L. Kolodner and C. K. Riesbeck, pp. 47-73, 1986.

16. Sticklen, J. and Bond, W. E., "Functional Reasoning and Functional Modeling," *IEEE Expert,* pp. 20-21, April 1991.

17. Umeda, Y., Takeda, H., Tomiyama, T., and Yoshikawa, H., "Function, Behavior, and Structure." *Applications of Artificial Intelligence in Engineering V, Proceedings of the Fifth International Conference,* Vol. 1, edited by J.S. Gero, Boston, MA, 177-194, 1990.

18. Wilson, P.R., and Kennicott, P.R., PDES/STEP Integrated Product Information Model (IPIM), ISO TC 184/SC4/WG1, Doc. 4.1.2., Working Draft, September 1988.

COMPUTATIONAL ARTIFICIAL INTELLIGENCE: A HYBRID COMPUTATIONAL-HEURISTIC KNOWLEDGE-BASED SYSTEM FOR SETTING TIMESTEPS IN DYNAMIC FINITE ELEMENT PROGRAMS

M.R. Ramirez
Department of Civil Engineering, The John Hopkins University, Baltimore, United States of America

ABSTRACT

An expert system, **ETUDES** - **E**xpert **T**ime integration control **U**sing **DE**ep and **S**urface Knowledge System, which addresses the determination of the timestep for time integration of linear structural dynamic equations, is described. This timestep may be also be applicable for a moderately nonlinear simulation of the same structure. The system also determines whether an explicit or implicit method is most efficient for the particular simulation. A production rule programming system written in OPS5 is used for the implementation of this prototype expert system. Issues relating to the expert system architecture for this application, such as knowledge representation, reasoning mechanisms, and structure are discussed. Two types of knowledge representations are used, one to hold information which uniquely identifies a simulation (scenario) and another type which encodes the causal knowledge used to obtain the scenario. The reasoning mechanism is a hybrid of computational and heuristic processes. The domain of this expert system has the characteristic of being rich in data but sparse in knowledge therefore it is found in the knowledge acquisition process that there exists no golden rule that can be applied for every simulation and that each simulation has unique attributes which require special attention when setting a timestep and its corresponding integrator. The domain is thus subdivided into subdomains such that within a restricted neighborhood of an abstract simulation space, sets of parameters may be safely interpolated so that the result is robust and reproducible. Since the system uses deep knowledge from the domain of structural dynamics, relationships from that domain are used to structure the knowledge so that different simulations can be compared and arbitrary simulations can be controlled with equivalent degrees of accuracy. The surface component interpolates the data within a restricted subdomain by utilizing a heuristic knowledge structure defined as a scenario. A scenario is defined as a discretized spectrum of n bands whereby a band is defined to have homogeneous dynamic characteristics and/or similar importance to the overall result within the band. Although the setting of the timestep is overall not a procedural process, the subtasks needed to navigate in the domain are procedural and thus the computational process forms part of the reasoning mechanism. Thus, the concept of *computational artificial intelligence* is introduced. Essentially, heuristic measures are used to estimate the effects of factors affecting the response and thus the error. These factors are; the load distribution, the stiffness distribution, and the load-structure interaction distribution. The key feature is that only distributions are needed since all parameters are referenced to the energy of the system defined in terms of the response. Thus, absolute quantities which would be known only apostiori are not needed.

The domain knowledge is comprised of the analysis of error in time integration and the behavior of structural dynamic systems. Error arising in numerical simulations is categorized into method-dependent (arising from the discrete aspect of the integrator) and method-independent (arising from implementation of algorithm on a finite precision machine and the spatial discretization error). For the purposes of analysis, method-independent errors are assumed and verified to be negligible. The various types of method-dependent time integration errors; frequency distortion (a nonlinear compression/expansion of the temporal axis) and amplitude attenuation (a numerical damping which can differ from the actual system damping), and methods for measuring their effects are examined. These measures vary in terms of their domain of measure; some measure error over the entire spatial domain others only at a point in space. The same can be said for the temporal domain. From these four combinations the measure over the time domain for the entire simulation length for one point in space is chosen as being the best tradeoff between accuracy and computational complexity and is called the cumulative squared error. The point in space is ideally the the point at which the spectral displacement occurs. Two types of approximations are introduced: a truncated spectrum model and a reduced band spectrum. The truncated model references estimates of the first n modes of the system whereas the reduced band spectrum approximates between the modes of the system, thus corresponding to a reduction in the degrees of freedom in the frequency domain. Mathematical inequalities on the aggregate effect of each component are then used to bound the error. Error estimation using apriori information is considered and described within the framework of the discrete spectrum models. The conditions governing the definition of an appropriate scenario are derived from the standpoint of structural dynamics, in particular modal analysis. Considerations which effect the setting of error tolerances are introduced such as response spectrum content, spectrum information content, and external error masking. Two general types of algorithms are proposed for the setting of the timestep once the scenario is heuristically defined; - an iterative model which accounts for the current charactersitics of the simulation and a single-iterate model which produces a pareto-optimal result based on variations of a typical simulation. The prototype is evaluated by measuring its performance in various benchmark model problems. Currently available methods are used as a basis of reference to show the expert system's sensitivity to various parameters which affect the error in a simulation. The results prove to be robust for reasonable subdomains. The limitations of the system and future research directions are given with regard to increasing the amount of knowledge extraction that can be effected on the user input and to the degree of applicability for this system for more nonlinear systems.

INTRODUCTION

A consequence of the advances in finite element technology has been an increasingly large burden on the user as to the choice of input parameters and methods. These parameters are intended to give the computer program more flexibility in handling diverse simulations. However, they prove to be a double-edged sword since all too often the user is at a loss as to how to rationally pick these parameters so as to achieve reasonable results without gross misuse of computer resources or unacceptably large errors. Flexibility without control results in confusion and ultimately presents an inherent limitation in the use of complex finite element programs.

In many cases this issue is resolved by consulting with the resident expert(s) or alternatively by trial and error. In the case of expert consultation, the user gets enough advice to solve the immediate problem and yet remains ignorant of how those decisions were made and how they would vary under different conditions. In view of the amount of time that a typical "guru" has available, this approach is justifiable but the end result is that most users guess values or constantly use the same values regardless of the characteristics of the problem at hand. When the parameter values are obtained by trial and error, errors may be unacceptably large or the simulation may require excessive time. Worst of all, in some cases, the results may be worthless.

Most finite element program user manuals do not provide much advice regarding acceptable timestep values for time integration of the dynamic equations. This is understandable since the choice of timestep is a function of the type of simulation, loadings, structural properties, and algorithms used. The typical software developer cannot possibly foresee all of the simulations that may be attempted with the software. Even a resident expert is limited by the types of simulations that he/she has performed. The problem of determining one aspect of this space of input variables, the time step and integration method, is addressed by this expert system. Simple formulas for the selection of a timestep based on the highest frequency of interest have been developed, see for example [Ref. 22]. However, these formulas, as shown in the examples in this paper can be quite conservative because they do not reflect the influence of other important factors: the frequency spectrum of the load, the spectral character of the structure, the interaction between these two, and the relative importance of the various parts of the response spectrum to the analyst. Incorporating the influence of these factors exactly is quite difficult. In most cases the user is at a loss as to how to estimate their effects and their complex relationships. This situation renders it an excellent candidate for expert system application. The type of knowledge involved in setting a time step is heuristic in nature, yet in order to have some advantage over basic knowledge, these heuristics must be based on rational foundations. In addition, this knowledge is characterized by an ill-structured character and empiricism. As is often the case in expert system development, the knowledge acquisition phase proves to be the bottleneck in the process. To facilitate this phase a convenient framework for the knowledge base must be set up to allow the expert to define knowledge in an organized manner. Furthermore, the knowledge base must be easily expandable and modular so that incremental development will be possible.

A key feature in this expert system was the development of a band scenario concept which describes the various aspects of a simulation in terms of bands in the frequency domain (Fig.1). The load, structure, and load-structure interaction are estimated in the frequency domain to provide a consistent method for estimating the time integration error. Several partial scenarios have been incorporated in the system and can serve as a framework for describing more specific simulations.

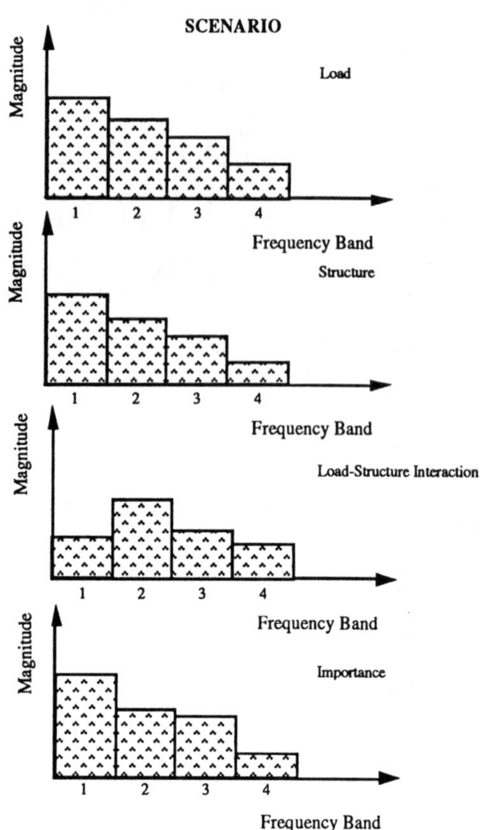

Fig. 1. Scenario definition for four factors affecting error control

The objectives of this expert system were:

- to allow the user to employ as much knowledge as is available about the current problem;
- to arrive at a solution, which although not necessarily optimal, should be near optimal (optimality would require a more rigorous mathematical treatment involving optimization over a large number of variables which would probably be computationally unfeasible);
- to provide the user an awareness of how the selection was made and how sensitive the parameters are to various conditions;
- to have the capability of communicating with the finite element program;
- and finally to permit a level of performance equal to or greater than that obtained from methods currently available.

ETUDES-INITIAL is implemented as an intelligent front end to a finite element program in order to accomplish these objectives.

EXPERT SYSTEMS IN ENGINEERING

Expert systems seek to embody the knowledge of an expert in a form that can be readily used by non-experts. Some of the main motivations for creating expert systems are to:

1. facilitate the efficient dissemination of knowledge.
2. enhance the understanding of users of complex solution strategies.
3. reduce the search space required for a solution.
4. synthesize numerous considerations and satisfy multiple goals.
5. augment available resources by assisting in complicated processes.

At the core of these motivations is the availability of the powerful heuristics tool to reduce the search space of possible solutions. Heuristics simplify the decision process by abstracting a rule(s) that is generally applicable given a set of circumstances.

Exhaustive searches are eliminated through the judicious use of heuristics for decision tree pruning. That is, heuristics essentially remove possible decision paths and thus, make the search more efficient. This accounts for some of the success of expert systems. Examples of expert systems are identification systems such as PROSPECTOR [Refs. 10,11] which locates mineral deposits using Bayesian decision theory to arrive at recommendations, or classification systems such as MOLGEN [Refs. 36,37] which determines strains of bacterial infection.

The limitation of such heuristics is that they may lead to erroneous and inconsistent results since the abstraction may not be necessarily applicable for the current situation presented to the system. These inconsistencies are as crucial in engineering as they are in other fields. However, the fact that engineering has an abundance of laws which lend consistency and order leads to a special environment not found in other fields. To achieve the most benefit from working with AI tools in an engineering setting, heuristics need to be coalesced with established engineering principles in a manner so that neither efficiency nor accuracy is lost.

The present expert system can be categorized as a classification and control expert system since it classifies simulation types and uses knowledge which is simulation-type dependent to control the simulation's accuracy by setting the timestep and selecting the time integrator.

Previous researchers have shown the benefits of applying expert system technology to the problem of parameter selection for finite element programs. SACON [Ref. 5] is an intelligent front-end to the MARC finite element analysis code, which guides the user as to the most optimal options for analyzing a particular structure - i.e. numerical methods, elements, and material properties. [Ref. 8] wrote PLASHTRAN, an intelligent consultant for the NASTRAN code to aid users in selecting parameters and options for two-dimensional finite element analysis. The approach taken in PLASHTRAN is of a tutorial nature such that the program gives advice to the user only if it is requested. Even then, the user can override the assistant's recommendation.

The problem of knowledge acquisition is quite crucial in these types of systems since it can determine the system's feasibility [Ref. 21]. The basic characteristic of parameter selection problems which causes knowledge acquisition to be inefficient is that they are data-intensive but knowledge-sparse. That is, there is a lot of data from which to choose but it proves to be an overwhelming task to try to establish order. One method of addressing this characteristic is to categorize the data into subsets of the subdomain which have a particular set of typical parameters. This is analogous to the development path that expert system technology has taken. At the onset, the General Problem Solver [Ref. 27] attempted to solve arbitrary problems without limitations in scope. This proved to be quite inefficient as the amount of knowledge required to solve each additional problem proved overwhelming. The reduction in scope to a limited subdomain allows for a knowledge base to be built with domain-specific knowledge. The difference in the present problem is that the subdomain itself, timestep selection for finite element analysis, has a considerable size. Thus, the subdomain has to be subdivided into sub-sub-domains which will be called subsets and take the form of scenarios in this expert system. This will be described in more detail later.

With human experts, the abstractions that are made are generally applicable only for a subset of the domain. Thus, the knowledge base loses robustness as one approaches the boundary of these subsets. To circumvent this subdomain limitation, [Ref. 18] used a clustering algorithm to recognize patterns in output data from numerous runs given different input parameters. By grouping these clusters of data they were able to establish ranges of values for parameters that produced acceptable results. An example of these parameters is the number of integration points through the thickness for beam elements. The knowledge acquisition process in their system was automated by the application of cluster analysis to the program output, whereas, SACON [Ref. 5] used the more conventional knowledge acquisition approach of interviewing experts. The conventional approach required two man-months to get the knowledge and an additional two man-months to enter into the rulebase, while the cluster analysis approach can take many runs of considerable CPU consumption to obtain sufficient robustness of the recommendations. In another context, the fact that the knowledge acquisition process is dependent on an expert raises the problem of conflict resolution should multiple experts be used [Ref. 26]. This issue has been addressed in the literature by weighing recommendations, and establishing rules for maintaining consistency. The advantage of using human experts is that the reasoning mechanism is similar to the General Problem Solver in that the expert(s) is bringing to bear on the problem a wealth of experience, judgment, and concepts which are difficult to attribute to a particular subdomain.

The approach presented here is to establish a structure and methodology for setting up a knowledge base for an arbitrary problem in a subset which possesses the gestalt of the simulation with regard to time integration. The main goal of the system is to establish meta-rules - rules to make rules, so that this arbitrary subset can be addressed in a consistent manner. The expert system can then take advantage of both the human ability to abstract by allowing for judgment in what constitutes a different simulation type as well as the computer's ability to capture detail by using heuristics and procedural routines to arrive at pareto-optimal solutions.

Expert System Architecture

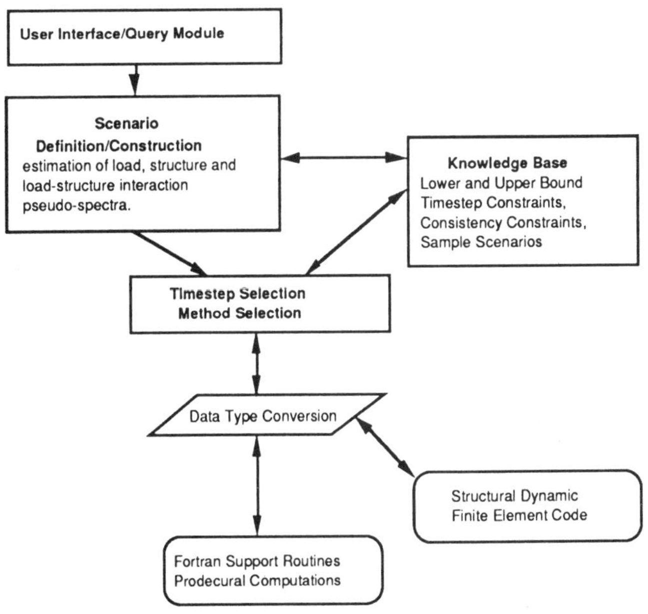

Fig. 2 Expert System Architecture

The architecture of the expert system, Fig. 2, consists of the expert system running as a host to the finite element program with data type conversion in the communication link since the expert system uses LISP variables and the FE program is written in FORTRAN. The expert system also communicates with FORTRAN support routines which do the procedural computations described in subsequent sections. The main function of the artificial intelligence component is to intelligently query the user so as to obtain as much information as is available and pertinent to the problem. The user interface uses the the knowledge base to build/define scenarios and then passes control to the solution component which selects the timestep using a combination of heuristic and procedural approaches. The prototype, ETUDES, is really a hybrid of conventional Boolean and mathematical logic. Viewed from a different perspective this expert system uses knowledge both in the deep sense - the finite element program being controlled and in the surface sense - the heuristic and experiential knowledge which is used to guide the process of optimizing the temporal integration operator. The benefits of using these two levels of knowledge have been discussed [Ref. 17] and implemented [Refs. 1,41].

KNOWLEDGE REPRESENTATION AND STRUCTURE

The knowledge representation for the knowledge base to a large extent predetermines and limits the type of knowledge that can be incorporated in the expert system. The choices available for knowledge representation are essentially divided into two types: declarative forms which are used to represent facts or assertions and procedural forms which are used to capture actions or procedures. Instances of declarative representation include frames, semantic networks and logic, while procedural representations include subroutines and production rules. Details of these knowledge representations are available in a number of AI texts [Refs. 29,34,43].

Viewed from a different perspective, knowledge representations can be differentiated by the "grainsize" of the knowledge entity. That is, semantic networks are coarser in grain than frames which in turn are coarser than production rules. The character of the knowledge that can be represented with each knowledge entity is intrinsically related to the structure of the entity. Semantic networks incorporate distributed knowledge about the relations between nodal knowledge components, while a rule can only make a relation between the antecedent and the consequent. Thus, a rule is better suited for connecting two distinct states which have a particular set of characteristics. A network can provide a wealth of information about a group of distinct but related states.

In spite of these differences, all of these knowledge forms can be used to portray virtually any knowledge base. However, certain knowledge representations are better suited for knowledge bases with certain characteristics: knowledge bases that are composed of primarily fine-grained knowledge such as isolated states will be better represented by rules; for knowledge bases that are composed of many related states with intrinsic dependencies and a hierarchical structure, networks, frames, trees, or lists are more appropriate. The appropriateness of knowledge entities is also affected by factors related to the implementations of these structures on a particular hardware platform.

Based on an initial review of the characteristics of the knowledge to be encoded in the present knowledge base, we found that it required a fine-grained knowledge representation to capture heuristic knowledge which is very specific in scope i.e. certain input variables affect only a limited subset of output parameters. Therefore, a production rule format was chosen for the heuristic component. In addition, another type of knowledge structure was needed to facilitate the establishment of relationships found during knowledge acquisition. The coarser grain of a frame format was chosen as ideal in providing a suitable structure in which knowledge could be imbued with the qualities of inheritance, hierarchy, and intra-frame relationships so that.knowledge about a certain type of simulation could be stored in a manner which enables the expert system to concentrate on the factors which are critical to that specific simulation .

Another facet of knowledge classification is the depth of reasoning that is associated with a given knowledge entity. Production rules and frames embody surface knowledge which essentially consists of static dependencies between knowledge entities. For production rules, the condition-action pair represents a fixed relationship between two endpoints of a very complex reasoning mechanism involving many other intrinsic relation ships and factors. The production rule is an abstraction which ignores this intervening detail in an effort to simplify an expert's reasoning. At the other extreme are causal networks which embody deep knowledge in the sense that the interdependencies between knowledge entities is dynamic and a restricted subset of the reasoning mechanism used by the expert is incorporated. This subset includes knowledge about the underlying basic physical and causal processes. For the present expert system, it was deemed necessary to incorporate knowledge about the underlying structural dynamics and computational aspects into the knowledge base.

The coding of mathematical and physical relationships for a pure form (symbolic) of a causal network is very difficult. It would also require a complex dedicated inferencing engine consisting of artificial reasoning capabilities operating in a causal framework. As a viable compromise, it was decided to embody deep knowledge in the form of a mathematical model. Since the majority of this knowledge is mathematical in nature, it is necessary for the tool to have mathematical capabilities.

In summary, the essential requirements for the expert system development tool were:

- to allow both a rule-based knowledge base and a frame knowledge representation;
- to permit deep knowledge to be incorporated in the form of mathematical models and furthermore, since the finite element program is a source of deep knowledge, the tool must be able to communicate with it.

The finite element program provides feedback on the fulfillment of goals for a given simulation using the timestep and time integration method recommended by the expert system.

A survey of the available expert system development tools [Ref. 40], was made to find one which had these capabilities and allowed for the project objectives to be met while still providing reasonably fast execution time. Among the tools examined were EMYCIN [Ref. 39], KAS [Ref. 10], EXPERT [Ref. 42], OPS5 [Ref. 13], ROSIE [Refs. 14,15,20], RLL [Ref. 16], HEARSAY-III [Ref. 2], AGE [Ref. 28], EXSYS [Ref. 23], RuleMaster [Ref. 25]. As is the case in finite element programs, the tradeoff of flexibility versus control is intrinsic in expert system development tools. Some tools provide so much flexibility that it is difficult to ascertain the correct combination of methodologies unless one has a sufficient amount of experience working with that particular tool. These tools are not quite as attractive when one considers one of the primary characteristics of expert system development: incomplete initial program specification. That is, most expert systems are developed first as a prototype based on an initial estimate of software specifications. This prototype is constantly revised through the process of incremental development. As this takes place, new insights into the problem may lead to a more refined knowledge base and even an alternative knowledge representation may become more attractive [Refs. 35, 7]. It is therefore difficult to ascertain whether one of these tools will be worth the investment in time and money once the prototype becomes better defined unless obvious advantages are apparent in the early stages. Other tools are so constrained in their options that an application has to be written based on the tool's limitations and not on the problem at hand.

For the prototype of ETUDES-INITIAL, the best balance between flexibility and control was found in using OPS5© [Ref. 12], [Ref. 6]. It is a production rule system which is part of the OPS family of AI languages originally developed at Carnegie-Mellon and used in one of the first major expert systems, R1[Ref. 24], which configured VAX super-minicomputers, R1's successor is actively used by DEC. OPS5 is rule-based, reasonably fast (it is a compiled language), and allows for communication with other software. However, it does not have the capability of true frame knowledge representation per se. In spite of this, some of the qualities of frames can be emulated by the basic knowledge entity found in OPS5 - a Working Memory Element. Its syntax consists of a class name followed by attribute names. Each attribute name is followed by the particular value bound to it. There can be as many attribute names as the memory of the host computer will allow. Usually, this number of attributes is more than adequate.

As an additional convenience, vector attributes can be defined which allow multiple values to be assigned to each attribute provided there is only one vector attribute per Working Memory Element. The structure of this knowledge entity is identical to that of frames even though the reasoning mechanism is not designed to be frame-based. Since OPS5 does have a certain degree of flexibility, the structure of the knowledge base and control strategy can be designed so as to permit a reasonable approximation to a frame-based inferencing engine even though it is perhaps more tedious than using a tool with this facility already built-in. The primary reason why this emulation is possible is that a given Working Memory Element can have several attributes associated with it. If each knowledge entity could have only one dimension (one attribute) associated with it (as in Prolog [Ref. 9]), this process would be very difficult or unfeasible for even a moderately sized system.

The basic architecture of the OPS5 language consists of a knowledge base made up of two components - working memory and production rules. An interesting feature of OPS5 is that the inferencing engine is part of the run-time system and thus allows complete interactive capability. The run-time system also includes other support routines for managing the working memory, debugging, and for evaluating system performance.

REASONING MECHANISMS - CONTROL STRATEGY

OPS5 has a built-in forward-chaining inference engine which can also be used to emulate a backward-chaining control strategy by the judicious use of condition elements. Essentially, the main control strategy used in ETUDES is forward-chaining. Backward-chaining is used as a demon-like device to change the flow of control when certain conditions are found in working memory. As will be discussed in the next Section, the knowledge base can become very complex if not structured properly due to the fact that there are many conflicting considerations which must be taken into account. To address these difficulties, the AI problem-solving strategy of problem reduction was implemented at the basic level. The main concept of this approach is to subdivide the problem into smaller independent sub-problems. Normally, these subproblems are solved by using a Boolean Logic Network where subproblems are interconnected with arcs so that the order in which they are solved depends on AND or OR nodes being true. For the present problem a simpler approach was implemented using the Means-End Analysis option in OPS5. This option puts special emphasis on the first condition element of each production rule. These condition elements contain goals corresponding to subproblems. By using this approach, the knowledge base is compartmentalized so that new productions can be added without affecting other parts of the knowledge base. This is important both in terms of maintenance (reduced probability of working memory-dependent bugs) and incremental development (modular compilation is available which reduces overall development time). From an AI viewpoint, Means-End analysis is a problem solving strategy whereby a problem is solved by applying operations on the current data so that the "distance" between the current state and the goal state is reduced. The OPS5 implementation is similar although the metric concept is not retained and the onus of monotonically approaching a goal state is on the knowledge engineer.

Insofar as ultimate control is concerned, the user is still in control. This is made possible by the fact that the user is interacting with the command interpreter which controls the execution of the entire system and also by the intrinsically user-friendly nature of the knowledge base. Others have found that an expert system receives much more use and trust if the user is always in control.

Problem Statement

The problem of choosing an appropriate timestep for a given simulation can be cast in the form of an optimization problem over a continuous space subject to multiple constraints. The constraints on the timestep arise from factors that include: computer resource utilization, accuracy as dictated by the fidelity of the dynamic model, user-imposed limits on error, and machine and method limitations. Thus, the problem is posed as

$$\text{Min } C = C(\text{CPU, memory requirements}) \tag{1}$$

subject to

$$|u^{cont} - u^{dis}| \leq \varepsilon_1 \tag{2}$$

$$|\bar{u}^{dis} - u^{dis}| \leq \varepsilon_2 \tag{3}$$

$$\Delta t \geq \Delta t^* \tag{4}$$

where

$u^{cont}, u^{dis}, \bar{u}^{dis}$ = continuous, ideal discretized, actual discretized solution

Δt^* = lower bound for timestep stipulated by numerical round-off error

C = computation cost

$| \; |$ = some norm or measure of distance in function space

The variables involved include: the type of integrator, explicit or implicit; the method implemented within each type; the fidelity of the finite element model as dictated by the spatial discretization of the continuum. Constraint (2) depends almost entirely on the refinement of the spatial discretization, whereas (3) depends on the timestep and the time integrator. We will focus on constraint (3).

The type of time integrator is fundamental to the determination of the time step. Explicit time integrators, of which the central difference method is the most common in finite element programs [Ref. 4], are conditionally stable in that the time step must be below a critical time step. Otherwise, the solution grows exponentially and is totally worthless. Implicit time integrators are unconditionally stable [Ref. 3], so the time step may be arbitrarily large without instability. The time step in implicit integrators is set on the basis of accuracy alone. However, explicit time integrators are computationally far more efficient per time step, because unlike implicit methods, no equations need to be solved.

As mentioned above, the exact solution of this problem may be more complex than the actual simulation. Thus, the role of heuristics in this expert system is to reduce the size of the search domain such that a near optimal solution can be found for the majority of simulations within the area of expertise found in the knowledge base. It is understood that for reasons of practicality, the domain of the knowledge base must be limited, but the architecture of the knowledge base is open so that additional scenarios may be added as they become necessary. This permits the expert system to have a graceful degradation when the user reaches the limits of the domain of the knowledge. New scenarios in effect extend the domain and provide behavior that is ductile yet consistent since consistency rules are applied.

The use of another AI technique, constraint management, enabled the reduction of the search space. Two types of constraints are used: upper bounds and lower bounds. The primary constraints are the upper bounds, which are based on error estimates or indicators and empirical knowledge. The error estimates used in this prototype are based on errors introduced by the time integrator due to frequency distortion and amplitude attenuation.

The use of error indicators is logical from an engineering viewpoint since accuracy is crucial. The lower bounds consist of heuristic knowledge accounting for machine and method dependent limits. These constraints are successively applied so that the possible range of timesteps is increasingly limited.

As is evidenced by the nature of the above constraints, the knowledge base can only contain apriori knowledge of the simulation. Whether the timestep and time integration method are chosen appropriately will be found in the accuracy of the subsequent structural response (for known simulations) and the cost of the calculation. However, in the general case, the structural response is not known at the beginning of a simulation. Neither of these knowledge types used exclusively can successfully solve the problem. Furthermore, since no two simulations are exactly alike, the experiential knowledge found in the knowledge base cannot possibly address all possible conditions. The fundamental characteristics of the problem must be abstracted so that experiential knowledge can be effectively combined with the present problem characteristics to arrive at a recommendation. At a higher level in the hierarchy of abstraction, the structure of the knowledge base should be such that all of the critical problem characteristics for an arbitrary simulation type can be rationally utilized by the experiential component.

The approach used to satisfy these objectives consists of two components: a knowledge structure - scenario; a mathematical model - discrete spectrum model. A scenario is a frame-based surface level representation of the critical characteristics which uniquely describe a simulation with regard to time integration. The general concept of a scenario as a means for representing simulation knowledge is discussed in [Ref. 38].

DOMAIN KNOWLEDGE

Error in Discretized Systems

In this work, only errors caused by the temporal discretization of the partial differential equations of motion of the structure will be considered. It is implicitly assumed that the error due to spatial discretization is small relative to the temporal discretization. The spatial discretization error then represents a lower bound on the total error i.e.,

$$\xi_{TOTAL} = \xi_{SD} + \xi_{TD} \tag{5}$$

$$\xi_{TD} = \xi_{TD}(\Delta t) \tag{6}$$

$$\lim_{\Delta t \to 0} \xi_{TOTAL} = \xi_{SD} \tag{7}$$

where ξ is a measure of error in either temporal or spatial discretization.

The error predicted for the integrator will not be the total error in solving the governing equations, because of the error in the spatial discretization.

Time integration error can be subdivided into method-independent error and method-dependent error. Method-independent error are those errors caused by implementation of numerical methods on a finite precision machine, roundoff, numerical truncation, etc. This type of error is assumed to be of much smaller than the method-dependent errors which will be covered in the next Section. This assumption is necessary to avoid masking the error of interest by external parameters not entering into the analysis .

Time Integration Errors

As discussed above, time integration error estimates will be used as one set of constraints in finding the optimal timestep and time integration method. The two types of errors considered , frequency distortion and amplitude attenuation, can be conveniently described and analyzed in the frequency domain.

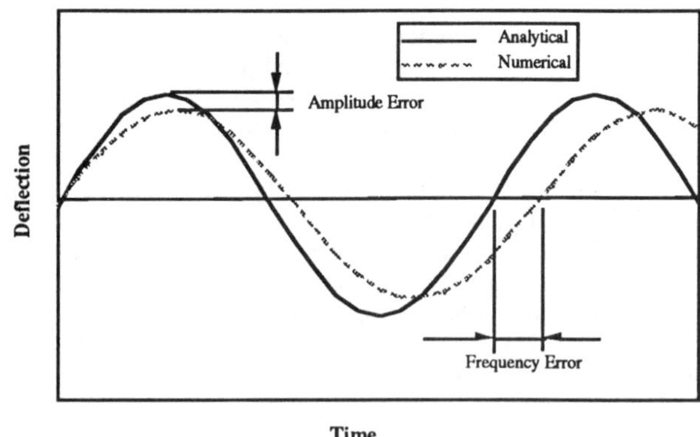

Fig. 3. Analytical vs. Numerical Behavior for System Response in the Temporal Domain

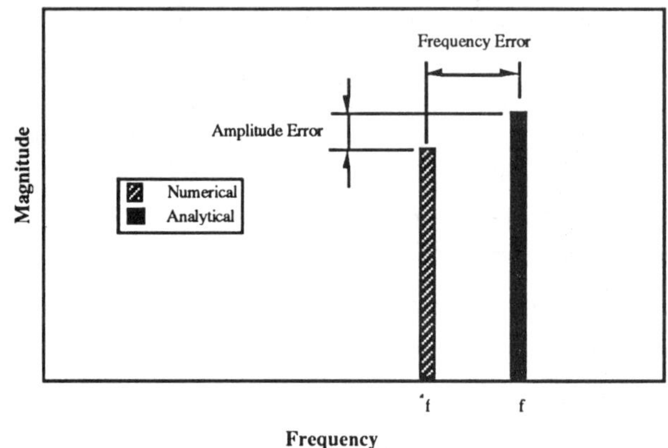

Fig. 4. Analytical vs. Numerical Behavior for System Response in the Frequency Domain

Fig. 3 is a comparison of the exact and numerical response of a simple 1 degree-of-freedom model in the time domain while Fig. 4 is a comparison of this response in the frequency domain. Frequency distortion affects the spectrum of the response by compressing or expanding the time axis in a nonlinear manner. Thus, in this example with both frequency distortion and amplitude attenuation, the error manifests itself in compression of the waveform and decrease of the amplitude in the temporal domain. Not all time integrators have amplitude attenuation but most have frequency distortion. It is evident from comparing Figs. 3 and 4 that the structural response and its attendant errors lends itself more readily to analysis when viewed in the frequency domain.

Error Measures

Error measures can be classified according to the nature of their sampling continuity in space and time (the number of sample points used in the error measure). These measures are:

* Continuous space, continuous time

  - This measure is used to determine the fidelity of a simulation in both space and time.

  - It is expensive to implement since all values and space and time are considered.

$$\xi = \int_t \int_V || y(\mathbf{x},t) - \hat{y}(\mathbf{x},t) || \, \mathbf{dV} \, dt \tag{8}$$

NOTE The integration in space is in all dimensions.

* Continuous space, discrete time

  - This measure addresses spatial variation and peak values in time. It is more appropriate for controlling error in spatial refinement i.e. adaptive mesh methods since it measures error in space. There is no guarantee that the temporal waveform is correct.

$$\xi = \int_V \| y(\mathbf{x},t^*) - \hat{y}(\mathbf{x},t^*) \| \, d\mathbf{V} \tag{9}$$

  where t^* represents a particular time of interest

* Discrete space, discrete time

  - This measure addresses peak or spectral values of displacement at a point of dominant response, i.e. for degree-of-freedom i. - There is no constraint on waveform accuracy since the coincidence of the numerical response with the exact at one discrete point does not guarantee fidelity of the overall response.

$$\xi = \| y_i(\mathbf{x}^*,t^*) - \hat{y}_i(\mathbf{x}^*,t^*) \| \tag{10}$$

* Discrete space, continuous time

  - This measure determines accuracy in the temporal domain for a degree-of-freedom, i, which is taken to be the point of spectral displacement. Spatial waveform accuracy is tacitly assumed since it does not enter explicitly into error control.

$$\xi = \int_t \| y_i(\mathbf{x}^*,t) - \hat{y}_i(\mathbf{x}^*,t) \| \, dt \tag{11}$$

Here $\| \bullet \|$ is an arbitrary vector norm which operates solely on a scalar quantity i.e. for the Euclidean norm, it corresponds to the distance between two points.

Definition of Error Measure

Of these choices, the last method: discrete space, continuous time, was adopted. It appears to be the best compromise between sufficient control and excessive cost in measuring and evaluating error relative to running the simulation. The precise form chosen is the continuous L_2 norm. The error, the cumulative squared error, is defined as follows,

$$\xi(t_e) = L_2^2[y_i] = \int_0^{t_e} (y_i - \hat{y}_i)^2 dt \tag{12}$$

where y_i = analytical response at the node of interest i
$\hat{y}_i$ = numerical response at the node of interest i
t_e = length of simulation

In order to normalize the error, the energy of the response is defined as follows,

$$E(t_e) = \int_0^{t_e} y_i^2 \, dt \tag{13}$$

The normalized cumulative squared error is then

$$\bar{\xi}(t_e) = \xi(t_e) / E(t_e) \tag{14}$$

This measure is most useful when the point of critical response is known or when one can anticipate the point of maximum response.

Model Assumptions

The assumption of sufficient spatial resolution allows the spatial accuracy to be addressed external to the present error analysis process. Notwithstanding this simplifying assumption, the spatial sampling point has to be chosen rationally to ensure that the entire model will have sufficient accuracy throughout its spatial domain. The most logical choice is to sample the degree-of-freedom where the maximum response is expected. In effect, controlling this error serves to bound the error for the rest of the spatial domain. An additional consideration is the choice of response variable which is used to measure the accuracy of the solution. For the present work, displacements were chosen; however, velocity and acceleration could also be used. In a sense by using a bound on the displacement, the error in the velocity, and acceleration is bounded.

SOLUTION STRATEGY

Motivation and Preliminaries

As will be discussed below, the problem of determining the effect of time integration error on a simulation involves knowing the characteristics of three system parameters: the structure, load, and load-structure interaction. The basic concept in estimation of time integration error is to approximate the spectra of these three characteristics through the use of a discrete band model, and then using a heuristic formula (with these spectra) to obtain an aggregate measure of the overall error. It is convenient to be able to approximate spectra as relative quantities since absolute quantities are not known with certainty.

Structure of Band Scenarios

The frequency content of a structural response is usually quite complex, especially in stiff systems having significant low and high frequency activity. To provide an approximate picture of the spectral content, a discrete spectral model is used in conjunction with knowledge in the knowledge base (scenario). In each discrete band model, the spectrum is divided into n non-overlapping discrete bands uniquely described by bandwidth and mid-band frequencies

$$\omega_1^m < \omega_2^m < \ldots < \omega_{n-1}^m < \omega_n^m \tag{15}$$

where

ω_n^m = mid-band frequency of n-th band

These bandwidths and mid-band frequencies are chosen so that a each band has approximately homogeneous characteristics while being significantly dissimilar in these characteristics from neighboring bands. These characteristics include amplitude, relative importance to overall response, relative importance to overall error and other considerations that may be simulation dependent, i.e. earthquake response may have characteristic bands of interest (Table 1)

Band	Frequency Range, Hz.	Mid-Band Frequency, Hz.
1	0 - 10	5
2	10 - 20	15
3	20 - 30	25

Table 1. Typical Seismic Discrete Band Model

In running the expert system, the user has a choice of either selecting a pre-defined scenario (which has a corresponding pre-defined discrete band model) or to request aid in defining a new one which is appropriate for the current problem. This approach avoids the needless repetition involved in redefining scenarios which are commonly used while at the same time providing the option for defining new scenarios, thus, increasing the capabilities of the expert system. Through the use of consistency rules, new well-defined scenarios can be constructed

Once a scenario is defined, it is tailored to the particular simulation through user queries. Since this query system is knowledge-based, it retrieves all input data which are relevant to defining the particular characteristics of the problem, while also providing a reasonable amount of explanation about why it is asking for that information. To promote user-friendliness, more experienced users are permitted to avoid the augmented interrogation which is intended for more inexperienced users.

Spectral Measures

The information obtained through these queries is used in conjunction with a heuristic knowledge structure to fine-tune the scenario to the particular simulation. The structure of the heuristic knowledge analogous to the form of the SRSS (square-root sum of squares) formula used in earthquake engineering

$$S^2_{spectral} = \sum_{i=1}^{n} s_i^2 \qquad (16)$$

where $S_{spectral}$ = spectral measure of a quantity , s
s_i = spectral component measure of a quantity, s

It is important to note that the knowledge utilized in this knowledge base is not the exact formula per se, it is the structure of that formula. This distinction is important because it highlights the fact that knowledge can exist as structure in addition to the usual knowledge types which do not possess structure. Eqn. 16, as it pertains to modal analysis, states that the maximum dynamic response of a linear multi-degree of freedom structure can be estimated by the square root of the sum of the squares of the individual modal component amplitudes. This formula is widely used in structural engineering because of its simplicity and because it provides a good upper bound approximation to the exact result for the majority of cases. Equation 16 can be derived by considering the phase angles of the modal components to be random. One reason for its success lies in the fact that it weighs the largest response to a greater degree (in a relative sense) than the lesser components because of the squaring function on each component. While each component represents a maximum for the entire time history, they occur at different times. Since the modal component which contributes most to the response of the structure is of major concern to the structural engineer, a reasonably accurate measure of this total response is required. If one were to consider the alternative of summing the absolute value of the modal components (the L1 norm as opposed to the L2 norm) to arrive at a total response, one would obtain a very conservative bound. Thus, Eqn. 16 provides a more moderate amount of conservatism. The use of the above heuristic structure is a viable way of casting some aspects of the problem. The quantities involved in the knowledge base are not modal response values, so the heuristic structure is an abstraction based on Eqn. 16 which shares the same properties, rather than a formal application of Eqn. 16.

A spectral measure of the error is obtained by using a weighted sum of the squares of band error components:

$$S_e^2 = \sum_{i=1}^{n_B} \alpha_i \, E_i^2 \qquad (17)$$

where α_i = weight for band i
E_i = error measure for band i
S_e = spectral error measure
n_B = number of bands

The band errors are obtained from the knowledge base: these correspond to acceptable levels of error within each band for that scenario. For a new scenario, a user is guided in setting these levels through the use of experiential and empirical knowledge. The weights are normalized measures of the magnitude and importance of structural response within each band. In a very approximate sense, they correspond to weighted Fourier coefficients of the response if the response were expressed in a Fourier space with a very small number of components corresponding to the mid-band frequencies; the weights reflect the importance of each frequency band to an engineering evaluation. In essence, this would be similar to reducing the number of degrees-of-freedom in the frequency domain. Likewise, the same heuristic formula is used to determine a dominant frequency to be used later in the selection of a timestep.

$$S_d^2 = \sum_{i=1}^{n_B} \alpha_i \, f_i^2 \qquad (18)$$

where f_i = mid-band frequency for band, i
S_d = spectral dominant frequency

Frequency Domain Analysis of Response

For a linear system, the Fourier transform of the response is equal to the product of the Fourier transforms of the transfer function and the loading function.

$$R(\omega) = H(\omega)F(\omega) \qquad (19)$$

where $R(\omega), H(\omega), F(\omega)$ = Fourier transform of structure response, transfer, loading function, respectively and the frequency domain is subdivided into n_B bands. For the purposes of simplicity the response is assumed to be periodic i.e. we neglect damping ([Ref. 33] discusses the damped case). An approximation of the response spectrum is obtained as described below from user input and the knowledge base. The transfer function is idealized as a limited bandpass filter. Its structure depends on heuristic knowledge from the knowledge base. To estimate the transfer function which is needed to set the limits of the discrete band model, assumptions have to be made and information provided by the user regarding the support available for each loading band. For example, in order for a frequency to have a nonzero (or numerically significant) response magnitude, there must be a load at that frequency or a mode at that frequency that provides sufficient support for that load (enough to be excited by the load). For an arbitrary dynamic response, the frequency domain representation can be obtained by modal superposition. The entire frequency spectrum can be obtained by summing these modal contributions over all significant modes and load frequencies. In this expert system, however, the main goal is to estimate the distribution of activity within each band so the weighting coefficients, α_i can be determined for use in Eqns. 17, 18. This will be addressed in the next section.

Estimation of Spectra

From Eqn. 19 and the assumption of the structure as a limited bandpass filter (the support of the transfer function is approximately the passband of the filter), the transfer function can be viewed as the attenuation distribution that multiplies the loading distribution and gives the response spectrum distribution. The transfer function incorporates information about the way the structural stiffness is distributed and implicitly how the structure interacts with the load, which we will call the load-structure interaction. It is therefore advantageous to analyze the distribution using this *three part* approach so that knowledge about each part can be obtained independently. Each part corresponds to an attribute of the scenario. The three attribute scenario consists of

1. the load spectrum, which is defined by a weight factor α_{pi} for each band.
2. the structural stiffness spectrum, which is defined by a weight factor α_{si} for each band
3. the load-structure interaction spectrum, which is defined by a weight factor α_{ri} for each band

The weight factors above will be defined subsequently. The importance of each response band is included implicitly in the level of acceptable error for each band, E_i

The heuristic formula for the weights, which emanates from a structural dynamic analysis of a simple system (zero initial conditions) (c.f [Ref. 30, 31])

$$\alpha_i = \alpha_{ip}^2 \alpha_{is}^2 \alpha_{ir} \tag{20}$$

where $\alpha_{ip}, \alpha_{is}, \alpha_{ir}$ = load distribution, structural stiffness, load-structure interaction weight factors for band i

They are given by

$$\alpha_{ip} = \frac{P_i}{\sum_{j=1}^{n_B} P_i}$$

$$\alpha_{is} = \frac{\frac{a_{im}^2}{K_i}}{\sum_{j=1}^{n_B} \frac{a_{jm}^2}{K_i}} \tag{21a,b,c}$$

$$\alpha_{ir} = \frac{\psi_i}{\sum_{j=1}^{n_B} \psi_i}$$

where

P_i, K_i = modal load and stiffness for mode i

ψ_i = Energy due to load-structure interaction during one period of response and is given by

$$\psi_i = \sum_{l=1}^{n_L} \frac{1 + r_{li}^2}{\left[1 - r_{li}^2\right]^2} \tag{22}$$

ψ_i is the weight factor which accounts for the interaction between the structure and the loading. For a given band the weight coefficient is computed at the mid-band frequency. Note that the weights need to be normalized. If the expert system cannot find a scenario that is sufficiently similar to the user's simulation, the above weights must be approximated. The transfer function is estimated by asking the user for the first three or four system frequencies in the scenario (the number depends on the scenario; it is assumed that the user knows at least this many frequencies for the response is to be meaningful, otherwise the user can not check the results). The load spectrum is estimated likewise by the first three or four dominant load amplitudes and frequencies. The weights for each band are determined using Eqns. 21a,b,c. If this information is not available then a very crude approximation has to be made: constant attenuation throughout the spectrum and with the response weights, $\alpha_i = \alpha_{ip}$

Determination of Timestep

The timestep is based primarily on error measures or error indicators. In these error measures, the errors depend on the frequency content of the structural response, or more specifically, the sampling frequency which is defined as

$$\Omega = \omega \Delta t. \tag{23}$$

For example, the error measure for frequency distortion for trapezoidal implicit time integration is

$$E(\Omega) = \frac{1}{\Omega} \tan^{-1}\left[\frac{2\Omega}{2 - \frac{1}{2}\Omega^2}\right] \tag{24}$$

The error measure, E, is a function of the sampling frequency and so it is also a function of the frequency of vibration found in the structural response and the timestep

$$E = E(\Omega) = E(\omega, \Delta t) \tag{25}$$

Given an acceptable level of error, a sampling frequency can be computed by solving the transcendental equation, Eqn. 24. By using this result in conjunction with Eqn. 23, the timestep can be computed. As discussed above, a value for the timestep can be obtained from knowing the acceptable error level (sampling frequency) and a frequency of vibration. For a system consisting of more than one frequency, heuristic formulae must be used to obtain an equivalent spectral error level and dominant frequency. The process consists of the following:

Determination of spectral error measure: Eqn. 17 is used to determine a heuristic overall error measure. The weights are determined by the expert system as are the number of bands and mid-band frequencies.

Determination of dominant frequency: Eqn. 18 is used to determine a heuristic dominant sampling frequency. The weights are the same as were used above. Other methods can be used to determine the dominant frequency but they must be consistent with the computation of the spectral error.

Determination of timestep: An equation of the type of Eqn. 24 is used, with the spectral error, $E_{spectral}$ to determine a weighted spectral sampling frequency. Eqn. 23 is then used to determine the timestep. The result provides an estimate of the timestep for the simulation for this error measure based on apriori heuristic knowledge.

The above process for timestep selection is to be repeated for each error measure considered. The remaining upper bound constraints include sampling considerations (Nyquist sampling), method dependent concerns (stability for explicit methods), and other heuristic limits set by experience. Constraints such as sampling are included because although related to the other error measures, they should be dealt with independently to insure a sufficient amount of redundancy. Redundancy is a desirable quality in knowledge bases since it aids in maintaining consistency. From the viewpoint of knowledge engineering, redundancy is a beneficial by-product of analyzing a problem from different perspectives. In addition to redundancy, other consistency constraints are used in conjunction with constraints representing machine and method-dependent limits. All of the above constraints are upper bounds and thus, the governing value of the timestep for these constraints will be the minimum obtained from these constraints.

The lower bound constraints will enforce minimum limits on the timestep based on machine precision (machine epsilon) and method-dependent concerns (ill-conditioning of matrices for implicit methods). The governing timestep limit for the lower bound constraints will be the maximum of the lower bound constraints. This lower limit will affect the final value of the timestep only if it is higher than the upper bound timestep, in which case the timestep will be equal to the lower limit. The latter situation is not a desirable one since it means that the simulation cannot be undertaken with the prescribed level of accuracy using the available machine and integration method resources. Either another method or another machine has to be used or possibly both. Alternatively, accuracy constraints can be relaxed but this requires knowledge of error sensitivity for each scenario which is not always known. Therefore, this condition is to be taken as a limitation caused by approaching the limits of the knowledge base. In order to simplify the development of this prototype other parameters have been constrained to be fixed or not significant for the simulations, i.e. element types, number of spatial integration points, etc... .

Method Selection

Once the choice of timestep is made, the choice between the explicit and implicit integration is strictly algorithmic in the sense that the correct choice is made by simply making computation estimates. The choice of method depends on which is more economical, taking into account operation counts and number of timesteps necessary for a simulation. Thus, this expert system, in addition to serving as a means to set the time step, is also a means to making the choice between explicit and implicit integrators. The choices of time integration methods have been limited to the explicit central difference method and the implicit trapezoidal rule, respectively. An explicit integrator will be selected, if the following statement is true,

$$\Delta t_e > \frac{CPU_e \, \Delta t_i}{CPU_i} \tag{26}$$

where

CPU_e, CPU_i = CPU time per timestep for explicit and implicit integration, respectively

$\Delta t_e, \Delta t_i$ = timestep for explicit and implicit time integration, respectively

The cpu estimates are overall measures based on operation counts and memory manipulation timings for the various time integration methods. Other absolute limits and considerations may also affect the explicit vs. implicit decision. In particular, the size of core restricts the size of stiffness matrix which can be solved without memory swapping using the implicit approach. Out-of-core techniques noticeably deteriorate performance in most CPUs. Considerations such as the difficulties in convergence for an implicit method are beyond the scope of the present prototype. However, since the user is in control this latter point poses no real limitation as the user can override the choice of method while still using the corresponding recommended timestep.

EXPERT SYSTEM IMPLEMENTATION ISSUES

The tradeoff between optimizing the temporal integrator operator and minimizing overall system resources must be carefully addressed since the critical quantity, the evaluation of error, must be measured in some manner with an accuracy commensurate to the degree of refinement of the model. It is not practical to use the same error evaluation procedure for all models and/or simulations. For example, preliminary models with coarse meshes may not require a high degree of refinement in the time domain if only low frequency detail is required. As the degree of refinement (referred to as resolution from this point on) of the model increases, the degree of resolution of the error evaluation must also increase. To this end several levels of error estimation are discussed which provide varying degrees of error resolution. The criteria used to judge the degree of resolution necessary for a model is indexed to the amount and quality of the information available about the dynamic characteristics of the problem. In most cases, the analyst does not have a complete body of knowledge regarding the structural dynamic behavior of a simulation. Yet for all practical simulations it is expected that the user has knowledge of the first three or four frequencies of natural vibration and a rough approximation of the way the response is distributed in the frequency spectrum of the model. The existence of this information in the analyst's knowledge base is warranted by the fact that the relative merits of the model's fidelity will depend on the numerical results being accurate in these lower frequencies and that the relative distribution of the error resembles that which is expected for that particular simulation type.

In the most extreme cases, no knowledge of the simulation exists and thus a very rudimentary error measure must be used. This situation should be avoided whenever possible since there is no method by which one can verify the fidelity of the results. At the other extreme, for very complex structures (and/or loadings) requiring high frequency resolution for significant high frequency activity, very stringent error tolerances must be enforced and thus the type of error estimator must be of a suitable type.

Form of Error Estimators

By using the form of the error caused by frequency distortion error it is feasible to estimate the error using approximations for the parameters that affect the error. The main objective of the error estimator should be to obtain a reasonably accurate measure of the error. Preferably it should be an upper-bound on the error. Given the fact that apriori knowledge does not contain information about the exact solution, i.e. the numerical solution would be redundant if one did have the exact solution, it is advantageous to make use of what knowledge is available that can applicable in determining the amount of error in the simulation. The characteristics of the knowledge that is of interest here are factors that affect the response and thus the error. The three main structural factors that affect the response are the load, stiffness, and load-structure interaction distributions in the frequency domain. These are known to some extent as relative quantities by the analyst and not necessarily as absolutes. The expert system uses these three factor distributions as a means for estimating the relative amount of energy in each of the various bands of the discrete band model of the structure. The form of this energy band participation factor introduced previously is given by Eqn. 20.

Note that the weights are normalized, so that the relative energy measure to be valid. That is, each $\bar{\alpha}_i$ represents a measure of the amount of energy in band i as a proportion of the sum of the bands.

$$\bar{\alpha}_i = \frac{E_i}{\sum_{j=1}^{n_B} E_i} \tag{27}$$

Ideally one would want to use this knowledge of the relative weighting of the response to arrive at a relative measure of how each band affects the total error. Consider the total error for a simulation as measured by the normalized cumulative squared error for a simple problem with two bands of response

$$\bar{\xi}_{TOT}(t_e) = \frac{\xi_{TOT}(t_e)}{E_{TOT}(t_e)} = \frac{\int_0^{t_e} (y_1 + y_2 - \hat{y}_1 - \hat{y}_2)^2 dt}{\int_0^{t_e} (y_1 + y_2)^2 \, dt} \tag{28}$$

Let us introduce, $e_i = y_i - \hat{y}_i$, and Minkowski's Integral Inequality for p=2,and dividing each side of the equation by the sum of the band energies, taken as the total energy.

$$\frac{\left[\int_a^b (e_1 + e_2)^2 \, dt\right]^{\frac{1}{2}}}{\left[\int_a^b (y_1 + y_2)^2 \, dt\right]^{\frac{1}{2}}} \leq \frac{\left[\int_a^b (e_1)^2 \, dt\right]^{\frac{1}{2}} + \left[\int_a^b (e_2)^2 \, dt\right]^{\frac{1}{2}}}{\left[\int_a^b (y_1 + y_2)^2 \, dt\right]^{\frac{1}{2}}} \tag{29}$$

Note that the left hand side of the Eqn. 29 is actually the Total normalized cumulative squared error, $\xi_{TOT}(t_e)$, while each term on the right hand side is

$$\frac{\left[\int_a^b (e_1)^2\, dt\right]^{\frac{1}{2}}}{\left[\int_a^b (y_1 + y_2)^2\, dt\right]^{\frac{1}{2}}} = \left[\alpha_1 \bar{\xi}_1\right]^{\frac{1}{2}} \tag{30}$$

where $\bar{\xi}_1$ is the proportion of the energy in a band that is in error, $\alpha_1\ \bar{\xi}_1$ is then the proportion of the total energy which is in error. This relation can then be used to obtain an upper bound approximation for the total normalized cumulative squared error. For this two band system the approximation is

$$\left[\bar{\xi}_{TOT}\right]^{\frac{1}{2}} \leq \left[\bar{\alpha}_1 \bar{\xi}_1\right]^{\frac{1}{2}} + \left[\bar{\alpha}_2 \bar{\xi}_2\right]^{\frac{1}{2}} \tag{31}$$

Given the fact that we have this relation, it can be used to estimate the total error by estimating the parameters on the right hand side. In the expert system ETUDES, this is done by a series of heuristics which require an extensive interrogatory of the user. Outside the framework of an expert system framework it is also possible to arrive at these quantities in a procedural fashion such that these quantities are reasonable. For example, the ability of an analyst to scrutinize the results of a numerical simulation is intrinsically dependent on knowledge of the dynamic characteristics of the problem. In reference to the loading, its distribution in frequency space usually is known to some degree because it is an input parameter. This knowledge is also needed in order to adjust the refinement of the mesh. Some knowledge of the structural dynamic properties should be known in advance in order to anticipate one expects the structural modes to be excited by the loading, and to what degree each range of frequencies is to have activity. Again these quantities need not be known in absolute terms only in relative ones since those are sufficient for use in this error estimation process. It is possible to consider each component distribution that is not known as being uniform in its effect on the overall frequency spectrum, although this will limit the amount of accuracy in the estimate.

Calibration of Error Estimators

If the error estimate is to have any validity, the amount of error in the error estimate must be of a much smaller magnitude than the error estimate. The mechanics of how this is accomplished will be covered in a subsequent section, but basically, the confidence with which one can predict error depends on the amount and quality of the information about the dynamic characteristics of the problem as provided by the user. The basic idea of error estimation is to develop a form of the error and determine the dependencies of this error on the system parameters. Using those dependencies a measure of the system parameters can be used to estimate the influence of each on the error.

Issues in Setting Error Tolerances

In Eqn. 31 an error tolerance for each band or indirectly for the entire spectrum is required. The resulting considerations in setting these tolerances are discussed with regard to how to rationally set these tolerances. The dominant variables that influence the error are used to establish the context of this study.

Should a High Frequency Dominant simulation have the same error tolerance as a Low Frequency Dominant simulation?
The most straightforward answer would be, yes if the mesh has sufficiently high resolution in the HFD case to actually have worthwhile (accurate) information in the upper frequency spectrum, the error tolerance should be at least as stringent as the LFD case if not more. However, a typical simulation does not have very good HF resolution so it would be counterproductive to constrain the error (and correspondingly the timestep) to resolve information which is not very accurate. This does not infer that all hope should be lost and that the high end of the spectrum is not important, it merely means that a simulation should not be excessively penalized for being HF dominant in the general case.

Should a long simulation (large number of timesteps) have the same error tolerance as a short one?
Ideally, it should. Particularly if the information of interest lies in the later part of the simulation as is often the case. For error which is non-random (or more precisely non-self-correcting) and cumulative, it is even more applicable to constrain the timestep to a smaller value for a longer simulation.

Two simulations are run with approximately the same CPU per timestep but one simulation has information which is of greater interest than the other. How do you account for this in setting the error tolerances?
First of all, the entire frequency spectrum for the more important simulation should be constrained to tighter tolerances than the less important simulation. This will be referred to as an importance factor approach from this point on and is dealt with as an inter-scenario problem. Secondly, within a given scenario there should be a means for setting importance factors for different bands within the scenario (intra-scenario importance distribution).

What happens when a mixed set of characteristics and importance factors applies to a particular simulation?
The existence of conflicting competing alternatives renders a pure discrete-state rule-based-only algorithm intractable or even impossible. An example of this is the following:

i. the LF has lower error tolerances than the MF or HF regions (or bands) because there is more valuable information and is therefore much more important;

ii. the loading activity is dominant in the MF, yet because of the distribution of load frequencies the HF still has some excitation;

iii. the structure has a typical stiffness distribution such that the lower frequencies are less attenuated.

To address this, the use of a mathematical model is very advantageous.

Error Tolerance Setting Methodology

From the previous Section's discussion it is clear that the problem of setting the timestep is governed by various conflicting factors which can be dealt with only by a framework which allows all of the various components to influence the result in a continuous manner. It is evident that each band of the spectrum must be dealt with independently yet the aggregate result of these bands gives the final result. To that end a mathematical model is adopted which makes use of the error bound form Eqn. 31 in its more general form for n_B bands. Each band has a normalized cumulative squared error measure associated with it, $\bar{\xi}_i$. This is a relative measure which is nondimensional and thus does not depend on the scale of the problem as dictated by the loading or structure. To set the timestep for a simulation given a particular integration method and scenario, Eqn. 31 can be used to determine an upper bound on the error given the participation of each band, as measured by the energy band participation factors, α_i. These procedures can either be single iteration estimates based on pareto-optimal parameters which cover a broad range of similar simulations (instances of a scenario type) or completely iterative procedures which optimize the timestep for the particular simulation not just the simulation type.

Fig. 5 depicts a flowchart for a single iteration estimate of the timestep that is considered appropriate for the problem given a limited set of knowledge about the problem. Note that this approach is a first approximation since it is based on heuristics and a pareto-optimal approach which may not exactly be appropriate for the current simulation. The conditions for a simulation having similar characteristics are derived from the factors which affect the error as given by Eqn. 20. These parameters are; the number and location of bands (mid-band frequencies), the loading, the load-structure interaction and structural stiffness (transfer function), and the total length of the simulation. It may be observed that if the set of parameters in the current simulation is sufficiently distant from those used to set up the reference scenario, the use of it to extrapolate for new instances may be somewhat inconsistent. Note that the use of energy participation factors merely allows for the interpolation of new combinations which represent adjustments to fine-tune the reference scenario to the current model characteristics. In the spirit of maintaining a robust model, which degrades ductily as the limits of its domain are approached, some extrapolations are possible yet it is not advisable to use a model for a simulation very far removed from that used to calibrate it.

Error Estimation Flowchart for Single Iteration Procedure

1. Obtain energy participation factors and scenario description .
   Check whether the current simulation is sufficiently close to an available scenario given the dynamic behavior being modeled, the number of bands, the frequency ranges of those bands, the error tolerances imposed on those bands, the simulation lengths common to those simulations. If the current scenario is not sufficiently close to a scenario in the knowledge base then a new scenario will have to be defined.
2. Use information obtained in step 1. in Eqn. 31 to compute an aggregate upper bound error estimate given the set of ncse's in the scenario. Repeat for every candidate integration method.
3. Choose method with lowest total CPU.

Fig. 5. Flowchart for single iteration procedure

For cases where the type of simulation is sufficiently critical in its accuracy or sufficiently large in its expense of CPU time an iterative procedure is more appropriate. The details of this are given in Fig. 6.

Error Estimation Flowchart for Iterative Procedure

1. Obtain energy participation factors and scenario description. If the current scenario is not sufficiently close to the available knowledge base of scenarios then a new scenario will have to be defined.
2. Use information obtained in step 1. in Eqn. 31 to compute an aggregate upper bound error estimate given a trial value of the timestep.
3. If this aggregate error is larger than the tolerance reduce the timestep and return to 2. Otherwise, use the indicated timestep for this integration method. Repeat for every candidate integration method.
4. Choose method with lowest total CPU.

Fig. 6. Flowchart for iterative procedure

Scenario Definition

If there is sufficient need (difference in dynamic characteristics) a new scenario will need to be defined. The process of defining a scenario is primarily heuristic and entails determining how many bands of distinct activity there are in a simulation, the location of these bands as dictated by the transfer function, typical loading, and the relative importance of each of these bands. The choice of the number of bands is sometimes dictated by the loading alone, i.e. significant modal excitation expected in only a few discrete ranges and transfer function assumed uniform. In others, the transfer function is sensitive to certain ranges and thus needs to be taken into account.

One very important consideration which must be addressed when setting up a scenario is the extent to which the scenario discrete band model is representative of the actual model. It is crucial that the actual model be characterized with sufficient fidelity to allow for meaningful error estimation. The two basic types of approaches which are possible in setting up a discrete band model are:

a. Reduced Band Spectrum - the approximate response spectrum is characterized by system and load frequencies which are representative of the response likely to occur within each band. These frequencies populate the major ranges of activity of the spectrum and they interlace with the actual modes of the structure. The "reduced" part of the name comes from reduced degrees of freedom in the frequency domain.
b. Truncated Band Spectrum - the approximate response spectrum is characterized as a subset of the actual model response spectrum. Essentially, frequencies in the upper end of the spectrum are truncated at the point where they do not participate significantly in the response.

In both approaches the energy of the system approaches an upper limit as the model is refined and more bands are added, yet because of the decreasing marginal returns nature of structural dynamic systems, additional contributions become increasingly less significant. Furthermore, the increase of spatial resolution may affect the distribution of the energy much more than the absolute amount of error, i.e. as degrees of freedom are added support is extended for loading (subsequent activity for increasingly higher frequencies) while the lower frequency spectrum remains equally populated but with frequencies closer to the exact solution. In finite element analysis, additional degrees-of-freedom are usually added usually not for the additional high frequency support but more so for their attendant increase in accuracy that they lend to the lower frequencies.

The fact that structural systems are weighted towards the lower frequencies in their energy distributions ensures that higher frequencies become less significant, making it possible for partial or reduced spectrums to be used and thus simplify the problem. Indeed even the full mathematical models are approximate subsets of their continuous real world counterparts. The principal effect of truncating or reducing the frequency spectrum is to leave a portion of the energy and error unaccounted for. The goal is to ensure that these unaccounted quantities are small relative to what is accounted for. In order to analyze this condition let us introduce the actual error in the model as

$$\xi_a = \xi_p + \Delta\xi \qquad (32)$$

where

ξ_p = Predicted error
$\Delta\xi$ = Error in approximation of error and the actual energy in the model as

$$E_a = E_p + \Delta E \qquad (33)$$

where

E_p = Predicted energy
ΔE = Error in approximation of energy

The condition which needs to satisfied for an approximate spectrum to be valid is

$$\frac{\xi_p}{E_p} \leq \frac{\Delta\xi}{\Delta E} \qquad (34)$$

PERFORMANCE EVALUATION

ETUDES-INITIAL was tested using simple structural models under the action of various loadings to evaluate its performance. These simple benchmark problems served as reference points for the expert system and as a testbed for tuning the knowledge base. Once the expert system mastered these benchmark problems, more general problems were tried. Some were permutations of the benchmark problems, while others were completely different than anything that it had seen before. The latter were used to determine the limits of the knowledge base. In the evaluation of this system, various problems were used to examine the behavior of the system when confronted with different scenarios. As part of this evaluation, the method presented in this paper was compared with another method which does not require any knowledge of the frequency content of the structural response. This latter method, which will be referred to as the N-th Frequency Error limit [Ref. 22], is a first order approximation (Level I) method in the expert system for use in simulations where virtually nothing is known about the frequency characteristics of the structure or the load. The N-th Frequency Error Limit method sets the timestep by limiting the amount of frequency based error in the highest frequency considered of importance i.e,

$$\Delta t = \frac{S_f^*}{2\pi f_N} \tag{35}$$

S_f^* is obtained from solving a transcendental equation such as Eqn. 24, for an acceptable level of error at the highest frequency, f_N. The resulting timestep is insensitive to the distribution of activity in the response spectrum and furthermore, will be conservative for many simulations. For example, the timestep can be very small if the highest frequency of interest is very large, which is wasteful if this frequency doesn't participate significantly in the response. Furthermore, because of the monotonic relationship between error and sampling frequency, S_f^* needs to be small for the error to be sufficiently bounded for non-zero high-frequency modes.

The test problem is three degree-of-freedom spring-mass model subjected to a sinousoidal load at the free end as shown in Fig. 7. A discrete system is used to ensure that time integration error is not masked by spatial discretization error.

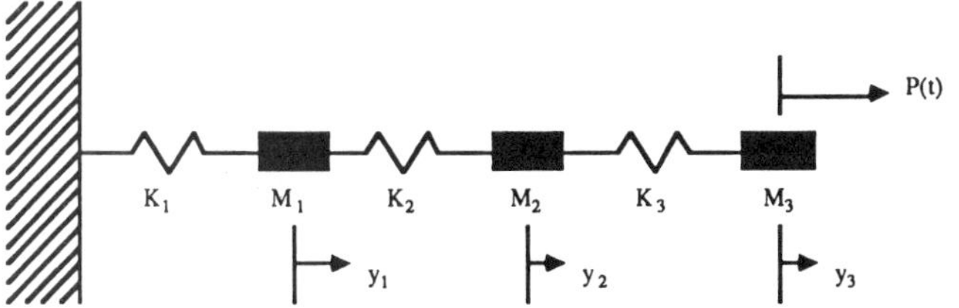

Fig. 7. 3 Degree-of-freedom Model

The measure of time integration error used in the evaluation is the continuous L_2 norm in time for the displacement of the node of interest (Eqn. 11). For this 3 degree-of-freedom model the node of interest is the node of maximum displacement, node 3. By choosing this node, the infinity norm is in effect being applied to the spatial domain. $\bar{\xi}(t_e)$ represents the portion of response energy which is in error after a time integration of t_e seconds. Since $\bar{\xi}(t_e)$ is a function of time integration error and response energy it can effectively measure the effect of timestep choice in relation to an absolute quantity; thus, different simulations can then be compared

For the purposes of comparison, the 3 degree-of-freedom model is considered with 3 different loading cases - sine functions having frequencies of 40 Hz, 20 Hz, and 15 Hz. The natural frequencies of the 3 degree-of-freedom system are 9.17 Hz, 30 Hz, and 54.4 Hz. The N-th Frequency Error limit method was used to obtain a timestep, given an acceptable level of frequency distortion of 0.36% at the highest structural frequency. Since the time integration method was the trapezoidal rule, Eqn. 24 was used to solve for the corresponding sampling frequency at the highest modal frequency, $S_f = .208418$. Eqn. 23 was then used to obtain the timestep of 6.098E-4 sec. The normalized error results of these three load cases are shown in Fig. 8.

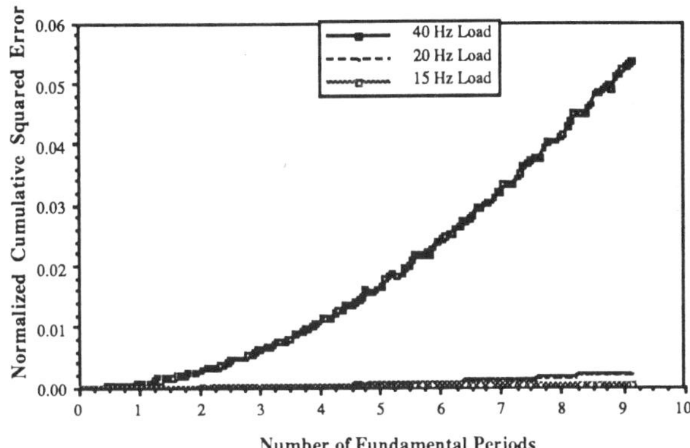

Fig. 8. Results of Time Integration Error, Δt = 6.098E-4 sec for the 3 degree-of-freedom System

Fig. 8 demonstrates the large range of normalized errors which is obtained with the same timestep but different loadings. It is not possible to effectively show the values obtained for the three cases on one plot so the results are summarized in Table 2.

Loading Case	Load Frequency, Hz	Normalized Cumulative Squared Error at 1.0 Second
1	40	0.0536
2	20	0.00227
3	15	0.00315

Table 2. Results of 3 degree-of-freedom model simulation using Δt = 6.098E-4 sec.

It is evident that the first case differs from the third by two orders of magnitude, yet the sampling frequencies were exactly the same since the same timestep was used (S_{f1} = .035132, S_{f2} = .114936, S_{f3} = .208418). Furthermore, a comparison of numerical vs. analytical time history responses at node 3 for the three load cases (Figs. 9, 10, and 11) show significantly different behavior and error levels. It is clear then from these results that the total error in the response depends on the load spectrum. This behavior is analyzed in [Ref. 32].

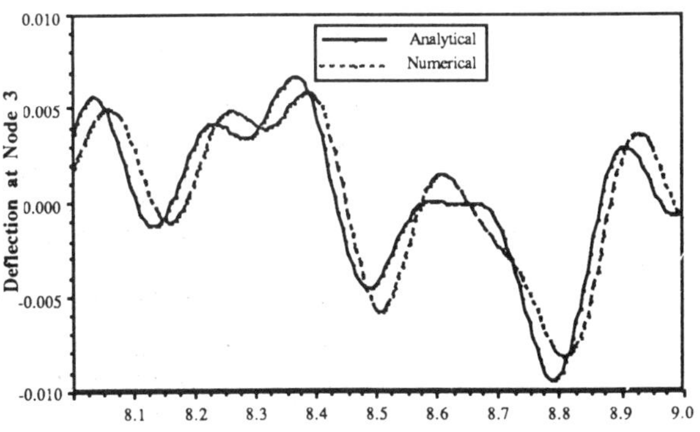

Fig. 9. Comparison of exact vs. numerically integrated results for loading case 1 for 3 degree-of-freedom system

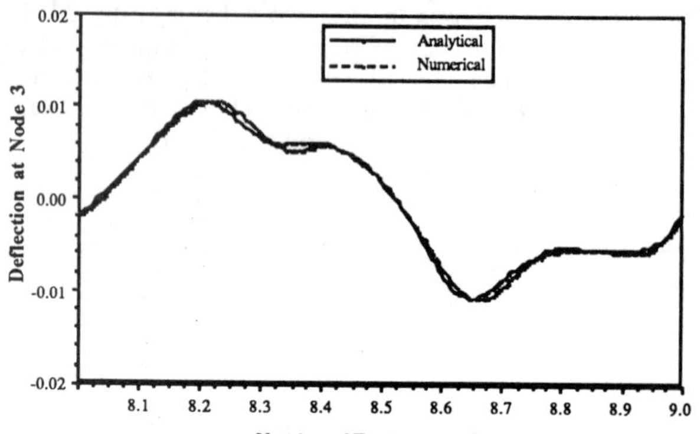

Fig. 10. Comparison of exact vs. numerically integrated results for loading case 2 for 3 degree-of-freedom system

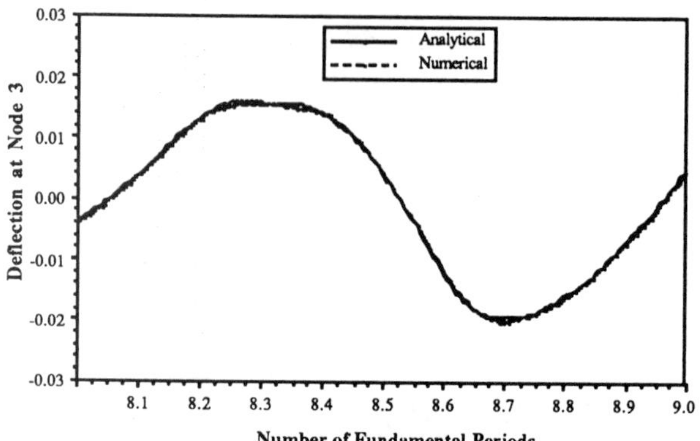

Fig. 11. Comparison of exact vs. numerically integrated results for loading case 3 for 3 degree-of-freedom system

The timesteps for the three load cases were then obtained by the expert system. The acceptable levels of frequency distortion were set at 0.15%, 0.05%, and 0.06% for the three bands of the model. It should be noted that this model fits naturally into the discrete band model structure with one mode per band. The weights were evaluated using the energy distribution algorithm described earlier and these values were then used in conjunction with Eqns. 17,18, and 24 to obtain the spectral dominant frequency, f_s, and the spectral sampling frequency, S_{fs}. With f_s and S_{fs} the timestep can be obtained by substituting f_s for f_N and S_{fs} for S_f^* in Eqn. 23. The results for these new timesteps are given in Table 3 along with the Normalized Cumulative Squared Error, $\bar{\xi}$, obtained at the end of the simulations (1.0 sec.). Upon comparison with the above set of data for a timestep of 6.094E-4 sec. the expert system results show case 1 has a lower level of error (6.094E-4 sec is too large) and cases 2 and 3 have a higher level of error (6.094E-4 is too small). Yet in an overall sense, all three cases have approximately the same level of error, the differences in the absolute level of each error is accounted for by the spectral energy distribution in each case. It should be noted that for a different set of acceptable error levels, these results would be different. The appropriate set of error levels depends on the particular model and the amount and importance of the activity occurring in a given band. Hence, the normalized cumulative squared error will depend on the energy distribution in a model. For example in Table 3, $\bar{\xi}$ for loading case 3 is smaller than the other two cases because there is more energy in band 1 than the other two bands; band 1 contains more important information and thus for a low frequency dominant simulation the error levels will be reduced accordingly. The error in a simulation depends on the distribution of energy in the response spectrum which in turn depends partially on the distribution of energy in the load spectrum The same effect would be apparent if instead of varying the load spectrum, the structural stiffness or load-structure interaction distribution were modified. This fact is the basis for the structure of the weight coefficients used in this expert system (load, structural stiffness, and load-structure interaction).

Loading Case	$\bar{\alpha}_1$	$\bar{\alpha}_2$	$\bar{\alpha}_3$	f_s	S_{fs}	Δt	$\bar{\xi}(1.0)$
1	.339	.450	.210	32.48	.101064	4.95	.0243
2	.812	.172	.016	16.44	.124477	12.05	.0215
3	.965	.030	.003	11.09	.131254	18.84	.010

Table 3. Weighting coefficients and timestep data (10^{-4}secs) from expert system for the 3 degree-of-freedom holzer Model

SUMMARY AND CONCLUSIONS

One goal in the design of the knowledge base was to allow for a graceful degradation in the reliability of the recommendations of the expert system. This ductile behavior is preferable to a brittle one which essentially involves having a point in the knowledge domain where there is an abrupt change in the reliability of the expert system. Brittle behavior is unacceptable because it is usually unforeseeable and can lead to completely erroneous results with no warning. In pragmatic terms, it is better to give no recommendation rather than a bad recommendation. If an expert system has been exposed to a reasonable number of uniquely different scenarios there is usually sufficient opportunity to find the limits of the knowledge base and thus modify the knowledge base accordingly to either handle those new scenarios or warn the user that no recommendation can be made. Notwithstanding, the simplifying constraints imposed on the prototype the development process has been sufficiently challenging.

This expert system, ETUDES, is primarily applicable to linear systems, although for slightly nonlinear response, where a single triangularization of the stiffness matrix is optimal, it would also be useful. The prototype, Fig. 2, although limited in scope, has provided some insights into the sources of error and the factors involved. The ability to make use of as much information as is available to the user in conjunction with heuristic knowledge stored in the knowledge base allows the user to exercise control on the magnitude of time integration error in the simulation. Since standard models are often used for a particular type of simulation, the scenario concept allows time integration related characteristics to be stored for future use. As the number of scenarios increases the capability of the expert system will also increase. Since the user has a means of selecting timesteps less conservatively, larger timesteps can be used and therefore permit more computationally intensive simulations to be run. This approach provides a method by which the level of error in a simulation can be quantified and thus controlled.

From the AI viewpoint several accomplishments have been made. On a strictly pragmatic level, the numerical functionality of FORTRAN has been combined with the symbolic capabilities of OPS5. Non-procedural and procedural processes have been meshed to allow a procedural component - the Finite Element Program which relies on a deterministic flow path to be controlled by a non-procedural component - the expert system which allows for different goals to be addressed according to the data that drives it. In the context of knowledge types, the concept of knowledge as structure allowed for heuristic data to be combined with non-heuristic data to achieve a result that is a combination of the two. The modularity of the knowledge base allows for the system to continuously adapt to new models and knowledge as it becomes available.

ACKNOWLEDGEMENTS

The support of NASA-Langley Research Center under grant no. NAG-1-650 to Northwestern University and the support of the

Air Force Office of Scientific Research under grant no. F49620-88-C-0011 to Northwestern University are gratefully acknowledged. One of the authors (MRR) would like to acknowledge the support of the National Science Foundation through their fellowship program.

REFERENCES

1. Apte, C. V., "A Framework for Expert Control of Interactive Software Systems," Ph.D. Diss., Dept. of Comp. Sci., Rutgers, 1984.
2. Balzer, R. L., Erman, P., London, P., and Williams, C., "HEARSAY-III: A Domain-Independent Framework for Expert Systems," Proc. 1st Ann. Natl. Conf. on AI , Stanford Univ., 1980, pp. 108-110.
3. Belytschko, T., Schoeberle, D. F., "On the Unconditional Stability of an Implicit Algorithm for Nonlinear Structural Dynamics," *J. of Appl. Mech.,* ASME,**17**., 1975, pp. 865-869.
4. Belytschko, T., R. Mullen, and N. Holmes, "Explicit Integration, Stability, Solution Properties and Cost, in": *Fin. Elem. Analy. of Trans. Nonlinear Struct. Behavior,* ASME, 1975, pp. 1-21.
5. Bennett, J., Creary, L., Englemore, R., and Melosh, R., SACON: A Knowledge-Based Consultant for Structural Analysis, Tech. Rep. STAN-CS-78-699, Stanford Univ., Sept., 1978.
6. Brownston, L., Farell, R., Kant, E., and Martin, N., *Programming Expert Systems in OPS5*, Addison-Wesley, 1985.
7. Buchanan, B. G., Shortliffe, E. H., *Rule-Based Expert Systems*, Addison-Wesley, 1984.
8. Cagan, J., Genberg, V., "PLASHTRAN, An Expert Consultant on Two-Dimensional Finite Element Modeling Techniques, " *Eng'g. with Comp*., 1987.
9. Clocksin, W. F., Mellish, C. S., *Programming in Prolog*, Springer-Verlag,Berlin, Heidelberg, NY, 1981.
10. Duda, R. O., Gaschnig J. G., and Hart, P. E., "Model Design in the Prospector Consultant System for Mineral Exploration," in D. Michie (ed.), Expert Systems in the Microelectronic Age, Edinburgh: Edinburgh Univ. Press, 1979, pp. 153-167.
11. Duda, R. O., Reboh, R., "The PROSPECTOR Experience," *Artificial Intelligence Applications for Business*, Ablex Pub. Corp., 1984.
12. Forgy, C. L., *The OPS5 User's Manual,* Carnegie-Mellon Univ., Dept. of Comp. Sci., Tech. Rep.1981.
13. Forgy, C. L., McDermott, J., "OPS, A Domain-Independent Production System Language, " *Proc. Fifth Internat'l. Joint Conf. on Artificial Intelligence,*MIT, 1977, pp. 933-939.
14. Fain, J., Hayes-Roth F., Rosenschein, S. J., Sowizral, H., and Waterman, D., *The ROSIE Language Reference Manual*, The Rand Corp., N-1647-ARPA, Dec., 1981.
15. Fain, J., Hayes-Roth F., Sowizral S. J., and Waterman, D., *Programming in ROSIE: An Introduction by Means of Examples*, The Rand Corp., N-1646-ARPA, Feb., 1982.
16. Greiner, R., Lenat, D. B., "A Representation Language Language, " *Proc. First Nat'l. Conf. on Artificial Intelligence*, Stanford Univ., 190, PP. 165-169.
17. Hart, P., "Directions for AI in the Eighties," SIGART News. 79:1-16, 1982.
18. Hawla, D. L., Neishlos, H., "Knowledge Acquisition for Effificient Use of Finite Element Software, *Eng'g. with Comp*, 1990.
19. Hayes-Roth, B., "*A Blackboard Model of Control*", HPP Rep. No.HPP 83-38, Rev. Aug., 1984.
20. Hayes-Roth, F., Gorlin, D., Rosenschein, S. J., Sowizral, H., and Waterman, D. A., *Rationale and Motivation for ROSIE,* The Rand Corp., N-1648-ARPA, Nov. 1981.
21. Hayes-Roth, F., Waterman, D. A., and Lenat, D. B., (eds.), *Building Expert Systems*, Addison-Wesley Pub. Co., Inc., Reading, MA, 1983.
22. Hughes, T. J. R., *The Finite Element Method: Linear Static and Dynamic Finite Element Analysis,* Prentice-Hall, Inc. 1987.
23. Huntington, D., EXSYS: Expert System Development Package, Albuquerque, New Mexico, 1985.
24. McDermott, J., "R1: An Expert in the Computer Systems Domain, " *Proc. of the 1st Ann. Conf. on AI*, 1980, pp. 269-271.
25. Michie, D., et al., "RuleMaster, A Second Generation Knowledge Engineering Facility, " Radian Tech. Rep., MI-R-623, Dec., 1984.
26. Mitall, S., Dym, C. L., "Knowledge Acquisition from Multiple Experts," *AI Magazine* **6**, No. 2, Summer 1985, pp. 32-36.
27. Newell, A., Simon, H. A., "GPS, A Program that Simulates Human Thought." *Computers and Thought*, E. A. Feigenbaum and J. A. Feldman (eds), NY, McGraw-Hill, 1963.
28. Nii, H. P., Aiello, N., "AGE (Attempt to Generalize): A Knowledge-Based Program for Building Knowledge-Based Programs," *Proc. Sixth Internat'l. Joint Conf. on AI*, Tokyo 1979, pp. 645-655.
29. Nilsson, N. J., *Principles of Artificial Intelligence*, Tioga Pub. Co., 1980
30. Ramirez, Martin R., Belytschko, Ted, "An Expert System for Setting Timesteps for Dynamic Finite Element Programs," **5**, No. 3&4, pp. 205-219,1989, *Eng'g with Comp*.
31. Ramirez, M.R., "An Expert System for Setting Timesteps in Dynamic Finite Element Programs," Diss. sub. for the part. fufillment of the req. of the Ph.D. in the prog. of Theor. and Appl. Mech., Northwestern Univ., Evanston, IL., 1990.
32. Ramirez, M.R., Belytschko, T., "Analysis of Time Integration Error in Forced Response", in preparation.
33. Ramirez, M.R., Belytschko, T., "Analysis of Time Integration Error for Damped Systems", in preparation.
34. Rich, E., *Artificial Intelligence*, McGraw-Hill, Inc., 1983.
35. Scown, S., "The AI Experience: An Introduction, " DEC, 1985.
36. Stefik, M., "Planning with Constraints (MOLGEN 1)," *AI:* 16:111-140, 1981
37. Stefik, M., "Planning and Meta-Planning (MOLGEN 2)," *AI:* 16:141-169, 1981
38. Tangen, K., Wretling, U., "Intelligent Front Ends to Numerical Simulation Programs", in *Research and Development in Expert Systems III,* M. A. Bramer ed., British Computer Society Workshop Series, Proc. of Expert Systems '86,Cambridge Univ. Press, Cambridge 1987.
39. van Melle, W., *System Aids in Constructing Consultation Programs, Univ. of Michigan Res.Press, 1981.*
40. Waterman, D. A., Hayes-Roth, F., "An Investigation of Tools for Building Expert Systems", Rep. R-2818-NSF, The Rand Corp., Santa Monica, CA., 1982.
41. Weiss, S., Kulikowski, C., Apte, C., Uschold, M., Patchett, J., Brigham, R., and Spitzer, B., "Building Expert Systems for Controlling Complex Programs," In *Proc. of the 2nd Ann. Nat'l. Conf. on AI,* pp. 322-326, Carnegie-Mellon Univ. and the Univ. of Pitts., 1982.
42. Weiss, S. M., Kulikowski, C. A., "EXPERT: A System for Developing Consultation Models, " Proc. Sixth Int'l. Joint Conf. on AI, Tokyo, 1979, pp. 942-947.
43. Winston, P. H., *Artificial Intelligence*, Addison-Wesley Pub. Co., NY 1984.

QSTRUC: AN APPROACH FOR QUALITATIVE STRUCTURAL ANALYSIS

R. Fruchter†, K.H. Law‡ and Y. Iwasaki⋆
†Center for Integrated Facility Engineering, Department of Civil Engineering, Stanford University,
‡Department of Civil Engineering, Stanford University,
⋆Knowledge System Laboratory, Department of Computer Science, Stanford University,
Stanford, United States of America

Understanding the behavior of a structure and its components in the preliminary design phase is important since decisions made at this time can have a significant influence on the quality of the final design. It can also potentially reduce the number of alternative solutions and avoid costly design revisions. However there are few tools available for modeling of a structure and performing qualitative analysis and interpretation of a conceptual design.

We describe a framework **QStruc** for *qualitative structural analysis* which combines first principles in structural engineering and experiential knowledge of structural behavior. The purpose of QStruc is to generate refined qualitative models and to infer the qualitative response of a given preliminary structure. The results are expressed by qualitative deflected shape, moments, and reactions. The proposed qualitative analysis strategy is a "greedy" depth-first algorithm that tries to expand the response of behavior as much as possible from known parameter values. This strategy makes use of: (1) causal ordering mechanism, (2) qualitative calculus, and (3) Quantity Lattice [Ref. 7] to reason about partial ordering among physical parameters.

The paper discusses the following aspects of QStruc: (1) *representation* of structures, fundamental principles, and experiential knowledge, (2) *model generation*, and (3) *qualitative analysis*. We provide a simple example to illustrate the performance of the implemented prototype QStruc. The paper is concluded with a discussion on preliminary findings of this research and future work.

1 BACKGROUND AND MOTIVATION

Determining the behavior of a structure and its components in the early design stages is important since decisions made at this time strongly influence the quality of the final design. The traditional approach to the analysis of a structure has been mostly based on numerical methods. To apply a numerical method, a structural designer must supply numerical values to the loads and dimensions of the structure and its components; the method then determines the numerical results of reactions, moments, deflections, etc. In the preliminary design stage the analyzed structure is given as a rough, qualitative sketch and there is not sufficient data to warrant the use of numerical analysis tools. Experienced engineers, however, are able to reason about structural performance during the preliminary design stage (such as approximating bending moments, deflected shapes) based on physical principles and experiential knowledge. This qualitative approximation is then used for the selection of structural components. This is an important task which can potentially reduce the number of alternative solutions and redesign efforts. One difficulty in formalizing a qualitative analysis strategy is that there is no a priori known sequence of steps in the qualitative approach. The most effective solution may emerge from different points for different problems. Thus, a flexible qualitative approach is needed to adjust the solution strategy as more information is obtained.

Although there has been some success in developing strategies for using qualitative analysis in the recent years [Refs. 2, 3, 4, 8], there is a need to develop an approach that can model the flexible reasoning strategies that a human designer uses in the initial design stage. We propose a framework for qualitative structural analysis (**QStruc**), based on fundamental principles of structural behavior and experiential knowledge. Furthermore, a flexible "greedy" reasoning strategy is developed to determine the response behavior from the known parameter values efficiently. The motivation for the present work is the desire to provide a useful tool for the complex task of structural modeling, analysis and interpretation at an early stage of the design such that it may be integrated into a larger CAE framework.

In the last decade, the topic of qualitative reasoning about physical systems has received much attention in artificial intelligence research [Refs. 1, 9]. Qualitative reasoning attempts to draw useful conclusions about the behavior of a physical system using mostly qualitative knowledge about its structures and the governing physical principles. The mathematical tool often used in qualitative reasoning is qualitative calculus, in which functional relations among the variables are expressed in terms of qualitative variables and operators. Qualitative variables are variables whose values are one of a small number of predefined intervals, for example positive (+), zero (0), or negative(-). Qualitative operators such as addition, subtraction, and multiplication can be defined to act on such qualitative values. Appendix A shows examples of such operators over qualitative values (positive, zero, negative), which we will denote as *Q+*, *Q0*, and *Q-*. Qualitative calculus enables a program to draw interesting conclusions from imprecise information about quantities and their relations. Because of the impreciseness of the information used, qualitative reasoning can give rise to ambiguous conclusion, which is a serious problem that must be solved by use of additional information.

Conventional knowledge-based systems (KBS), which rely on task-specific heuristic knowledge, tend to fail ungracefully when confronted with a problem even slightly outside their narrow domain of expertise. One reason for such "brittleness" is their lack of fundamental knowledge of the domain to fall back on when the heuristic knowledge fails. Lack of fundamental knowledge also results in poor quality of explanations that KBS systems can provide. *Model-based reasoning* attempts to overcome these problems of conventional KBS's by providing the system an explicit domain model, consisting of fundamental principles of the domain and an ability to reason from such knowledge. In the present work, we use explicit models and first principles, as well as heuristic knowledge, to analyze the behavior of a structure. Some of the heuristic knowledge used has been demonstrated to be useful in guiding structural analysis [Ref.3] and is used to focus on relevant problems in a given model.

There are three main stages in model-based reasoning about possible behavior of a structure - *construction of an appropriate model* for reasoning about the particular type of behavior in question, *analysis* of the model, and *interpretation* of the result. Many powerful CAE tools exist for analyzing a model numerically when the model has enough quantitative details to make such an analysis possible. In contrast, few tools are available to help with model construction and interpretation. Likewise, there are no analysis tools that can be used when there is insufficient quantitative information. In this work we have tried to provide a framework to help with all three stages - model construction, analysis, and interpretation - at the initial design phase when only incomplete and qualitative information about the design is available.

The organization of the paper is as follows: Section 2 presents an overview of the approach. Section 3 describes the domain representation. Section 4 describes our method for generating qualitative models. This is followed by a discussion of our qualitative analysis strategy in Section 5, where a "greedy" algorithm is proposed and the techniques that we employ for qualitative reasoning are described. Finally, in Section 6, we summarize the preliminary findings of this research and discuss future work.

2 SYSTEM OVERVIEW

The goal of the QStruc system is to:

(1) generate models necessary for deriving qualitative deflected shape tendencies, reactions, and moments, and
(2) perform qualitative analysis, given an incomplete description of a structure.

The input to the system is a schematic description of the structure, as a designer would first sketch it, in terms of structural objects (e.g., topology, supports, information about loads) and their relations (e.g., connectivity). The output is represented by the qualitative response of the given structure, in terms of a deflected shape, moments and reactions. The top-level conceptual structure of the **QStruc** system is shown in Fig. 1. The following sections discuss the various modules of the system. QStruc has been implemented using an object-oriented knowledge engineering tool, KEE[1].

3 DOMAIN REPRESENTATION

3.1 Representation of structures

Structural models are represented by instances of the following classes:

- *Components:* This class has two subclasses - *beam* and *column*. Components can be combined to create more complex structures or decomposed into subcomponents to highlight a relevant region from the behavior perspective [Ref. 3].
- *Terminals:* This class contains the information about the end sections of a component, like topological (e.g., *location_x*) and behavioral (e.g., moment, shear force, displacement) information.
- *Connections:* This class provides the means to relate two terminals of adjacent objects. It also represents a basic unit that facilitates the transfer of information (e.g., forces, deflections) from one object to another. It has two subclasses, *continuity-connection* and *interior-hinge-connection.*
- *Nodes:* This class is used to connect any number of terminals of any adjacent type of connection to that node. It serves as a means for information transfer or redistribution among the connected objects.

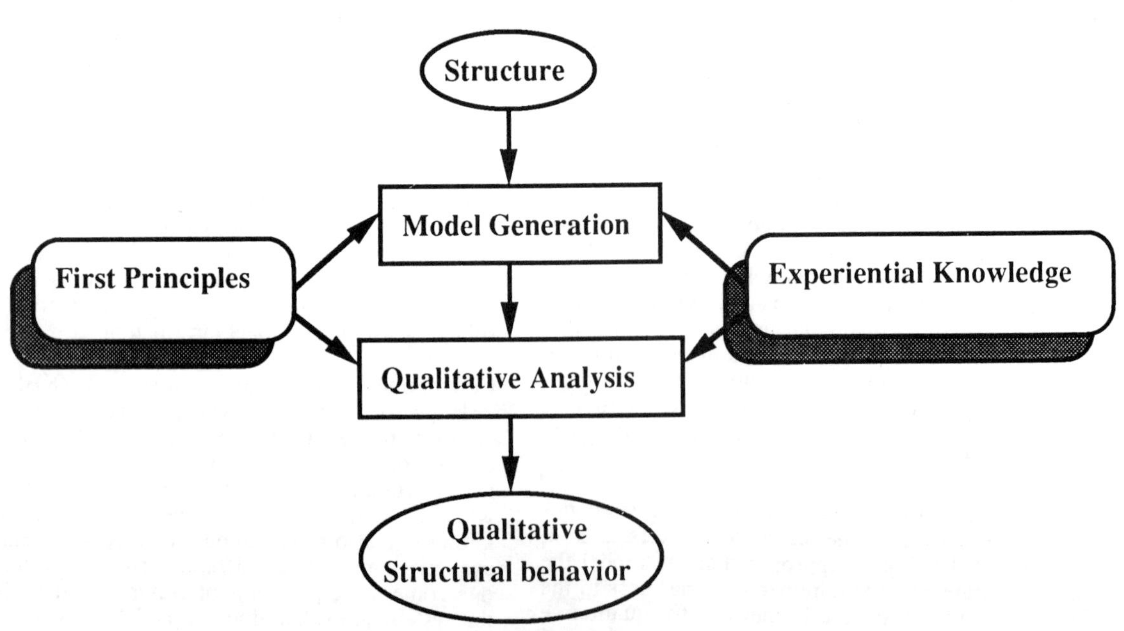

Fig. 1 System Overview of QStruc

[1] KEE is a trademark of Intellicorp Inc.

- *Supports:* This class has a set of subclasses including *fixed_support*, *pinned_support*, *free_end*, and *roller_support*. They contain information about the boundary conditions and possible reactions that can develop.
- *Loads:* This class has as subclasses load types, such as *concentrated_load* and *uniformly_distributed_load.*

These structural objects contain slots for the following types of information:

- *Geometrical information* includes the location for the end sections of a component, the application point of a load, and the location of the point with respect to which the qualitative equation of moment equilibrium is derived.
- *Contextual information* describes the loading situation.
- *Behavioral information* includes existing boundary conditions (e.g., rotation, displacement) in the case of a connection or support object, possible existence of reactions or internal forces, values derived from the qualitative analysis, information about processes (e.g., bending, compression), laws of information processing and transfer (e.g., behavior equations such as laws of equilibrium), and a list of active forces.

It may be worth noting that for this representation scheme, if numerical information is available for all the objects, traditional numerical analysis tools can be used to derive the structural response.

An example of the representation of a continuous beam is shown in Fig. 2. In the example, *B1* is an instance of *beam*. *Term.A*, *Term.B*, etc. are instances of *terminals*. *Fixed.support.A* is an instance of the subclass *fixed_support*, etc. *P* is an instance of *concentrated_load.*

Each parameter (e.g., moment, force, rotation) associated with an object is represented by a slot whose value is an instance of the class *qualitative_values*. *Qualitative_values* has instances, *Q-*, *Q0*, and *Q+*, representing a negative value, a zero, and a positive value respectively. These three qualitative values are used uniformly as values for all variable types, but are interpreted differently depending on the parameter type. For example, the value *Q+* of a force variable means the direction of the force is upward, while it means clockwise rotation for a moment variable. Also in the class of *qualitative_values* are two qualitative identifiers, *C+* and *C-*, indicating respectively the existence and non-existence of a constraint.

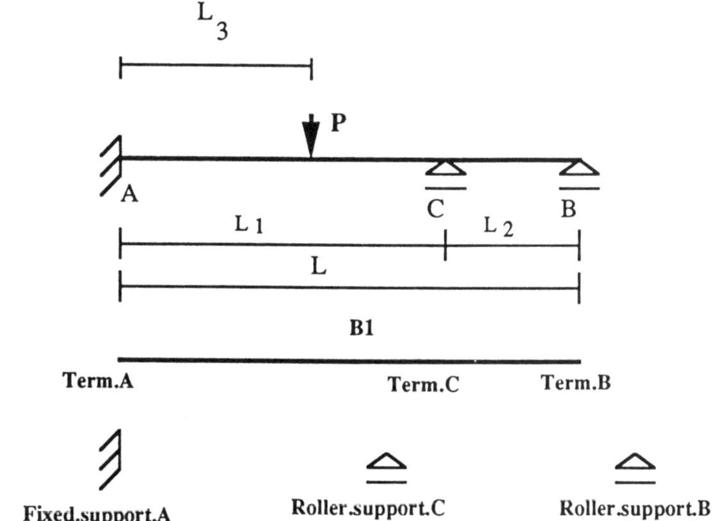

B1
Location_i: ***TermA***
Location_j: ***Term.B***
Load: P
Possible_bending_process: unknown
.
.
Possible_rotation_at_i: unknown
.
.
Moment_equilibrium_equation: unknown
.
.

Term.B
Internal_force_N: unknown
Internal_force_S: unknown
Internal_force_M: unknown
Location_x: ***10***
Location_y: 0
Point_location: unknown
Q_x_displacement: unknown
Q_y_displacement: unknown
Q_rotation: unknown
.
.

Fixed.support.A
Location: ***Term.A***
Supports: ***B1***
N_reaction: ***C+***
S_reaction: ***C+***
M_reaction: ***C+***
Q_N_reaction: unknown
.
.
Allows_x_displacement: ***C-***
Allows_y_displacement: ***C-***
Allows_rotation: ***C-***
Q_x_displacement: unknown
.
.
.

P
Load_direction: ***Q-***
Load_magnitude_qlattice: unknown
Supports: ***B1***
Load_location_x: ***3***
Load_location_y: 0
Point_location: unknown
Force_moment_direction: unknown
.
.
.

Fig. 2 Example of the Representation of a Continuous Beam

3.2 Representation of domain knowledge

QStruc uses fundamental domain knowledge as well as heuristic knowledge. Knowledge is represented explicitly as object classes:

- *Equations and relations:* This class has subclasses, *equations* and *relations*. *Equations* represent the various physical laws (e.g., *Moment-Equil-Eq*), and *relations* represent experiential knowledge (e.g., *Load-Direction-Displacement-Direction relation*). Instances of these subclasses contain pointers to the structural component to which they belong and to the physical quantities (parameters) involved.
- *Physical.Quantities*: This class has subclasses representing the different possible parameters (e.g., *Vertical-Force, Displacement*). Instances of these subclasses have slots such as: *causal-dependents, causal-dependent-on, causal-exogenous*. They also have pointers to the structural objects to which they belong and to equation(s) and/or relation(s) that they are involved in.

A general method that represents a specific first principle is placed at the level of structural objects. For example, when the moment equilibrium slot of a component object receives a message from a behavior process (e.g., *bending*) this will cause the *Moment-Equil-Eq* to be generated for the specific component together with the corresponding instances of the involved physical quantities.

Other physical principles, such as the definition of a moment generated by a force with respect to a location, are represented as "If-Then" rules, which can be activated in a forward or backward chaining manner. Heuristic knowledge about model generation is represented as methods for transforming and updating the qualitative information of the structure. Knowledge about experiential relations among parameters is encoded as rules.

4 MODEL GENERATION

The objective of the **model generation** module is to transform the input structure, which we will call *schematic model*, into several models, each of which is used to analyze a particular behavioral aspect of the structure. Since our goal was to emulate the reasoning strategy of a human designer, we identified which relevant types of models are used in reasoning about structural behavior, and what types of knowledge are used in generating such models. Fig. 3 illustrates the models that are prior to qualitative analysis and the sequence of transformation.

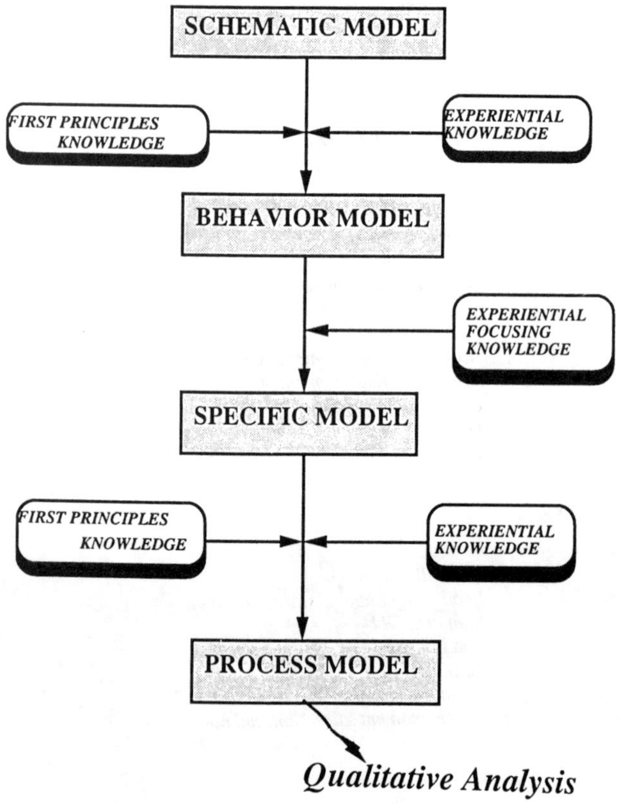

In the following, we describe each transformation step using the beam example of Fig. 2 and illustrate graphically the results generated by the system implemented in KEE. The figures were manually generated from the symbolic results of the system.

4.1 Behavior model

The first transformation derives a *behavior model* from the input *schematic model*. A behavior model describes the identified existing interactions (e.g., forces) among structural objects. To generate a behavior model, each structural object is represented by its free body diagram. With the knowledge of how information is transferred by the support and connection objects, the system infers the existence of possible reactions and boundary conditions. An interaction map of the whole structure is generated by combining the free body diagrams of components showing the information flow (load transfer) among them. From this interaction map, the behavior pattern of the whole structure is identified.

The explicit representation of interactions among the structural objects provides a way to model the structural behavior that is often implicitly used in the reasoning process of a designer. This representation enables the user to inspect the inferred interactions of the various structural objects. If the user finds a discrepancy between the interaction map and the interactions he intended, he will be able to rectify it at this initial stage by changing the schematic model.

Fig. 4a illustrates the interaction map of the beam example. Based on the component and support types (and also connection types if any), the system identifies the behavior pattern of the structure as a *continuous_beam*.

4.2 Specific model

Next, the *behavior model* is transformed into a *specific model* by identifying the regions of interest with respect to the behavior of the structure. The purpose of this transformation is to enable subsequent qualitative analysis at these regions. The system uses experiential knowledge to focus attention on such relevant regions. The model description is updated by adding new objects as specified by such rules.

The system first considers the implication of the given loading context, and searches for a focusing rule to refine the interaction map in the region where the load is applied. In our example, this step results in the introduction of a *continuity-connection JC1* at this point as shown in Fig. 4b. Then, *B1.comp1* and *B1.comp2* are added as substructures of beam *B1*. Next, the system searches for focusing rules relevant to the behavior pattern of the behavior model. This will result in further decomposing of *B1.comp2* into *B1.comp2.comp1* and *B1.comp2.comp2* and adding the *continuity-connection JC2* to the interaction map at the location of *Roller.support.C*. The structural representation is also updated in terms of new terminal instances corresponding to the new beam substructure instances (e.g., *Term.B1.comp1*, *Term.B1.comp2*, etc.). The final result of the *specific model* generation is illustrated in Fig. 4c.

Fig. 3 Model Generation Sequence

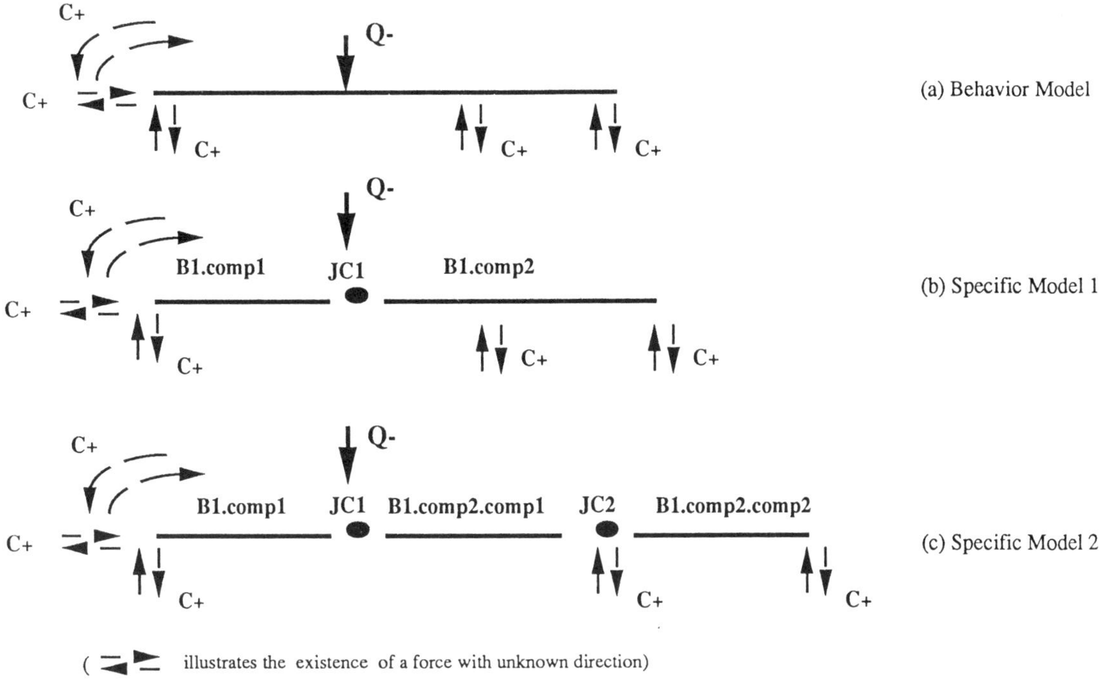

Fig. 4 Model Generation Example

4.3 Process model

A *process model* is defined by the underlying domain model represented by the set of applicable qualitative equations and relations, together with the set of involved physical parameters. The purpose of transforming the *specific model* into a *process model* is to identify active processes of the overall structure and the components. A process is defined as the possible behavior (e.g., bending, compression) of a structure in the context of the loading and boundary conditions. For a given specific model the system checks the preconditions of possible processes associated with each component to identify the active processes. QStruc represents processes as objects. Each process has references to applicable equations and relations.

In our example, *bending_process* is identified as an active process in the beam component. As a result, the applicable experiential knowledge and first principles are instantiated for each of the subcomponents of the specific model. The process model generated in terms of *equations, relations*, and *physical quantities* is partially shown in Fig. 5.

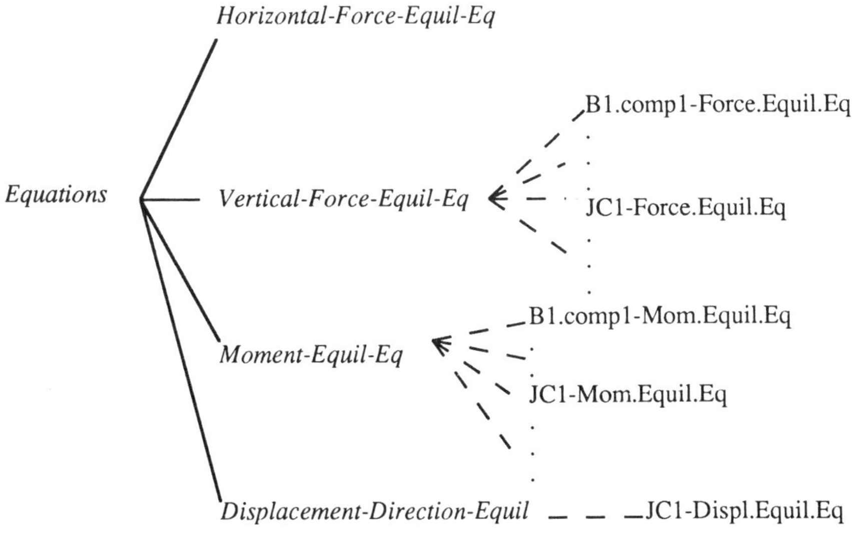

Fig. 5 Process model instantiation - equations, relations, physical quantities

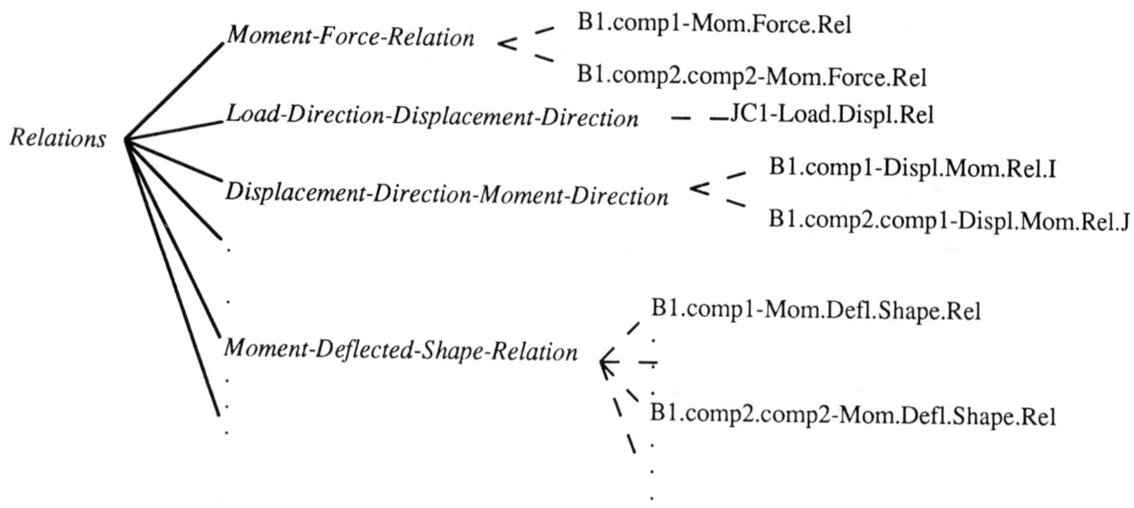

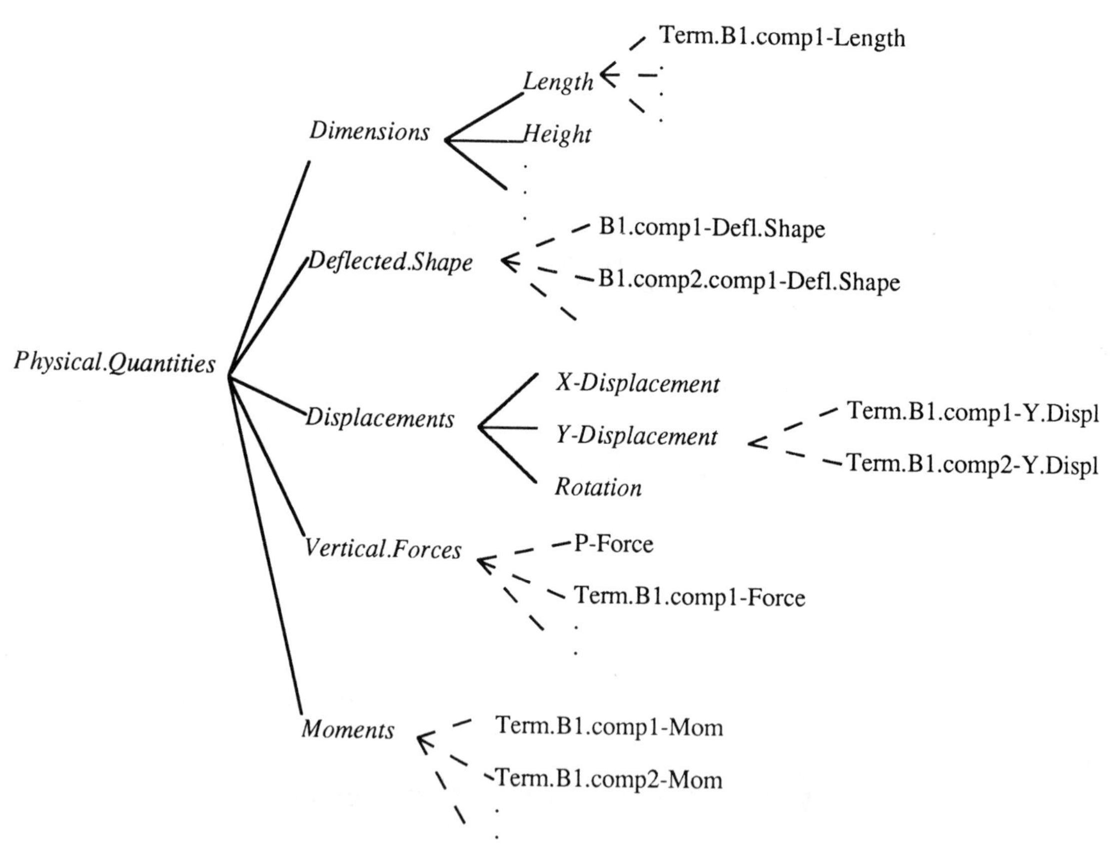

Fig. 5 Process model instantiation - equations, relations, physical quantities (continued)

5 QUALITATIVE ANALYSIS

5.1 Conceptual approach

A significant difference between qualitative analysis strategies employed by human engineers and quantitative analysis methods is that there is no fixed sequence of steps in the qualitative approach. In the numerical approach the sequence of solution steps is often predefined. In contrast, when analyzing a preliminary design, it is not obvious from what point the solution will emerge most effectively and in what direction the solution process will proceed. It may be best to start with the deflected shape in one case, but with the bending moment in another case. Also, depending on the amount and type of information available, it may be more effective to apply first principles in some situations and experiential knowledge in others. Since we do not know in advance what types of information will be available, we need a flexible solution strategy that can infer as much as possible from whatever types and amount of information.

To achieve such a flexible analysis strategy we employed the following mechanisms:

(1) a causal reasoning mechanism [Ref.6] that is used to identify a solution path from an arbitrary set of relations and equations that compose the process model;
(2) an agenda mechanism that consists of a sequence of tasks for qualitative analysis, where each task represents one reasoning step expressed as a rule, a method or a qualitative evaluation of an equation;
(3) qualitative calculus [Ref.5] that is used to derive the structural response;
(4) Quantity Lattice [Ref.7], a mechanism that is used to keep track of the information about the ordering relations among physical quantities. It is used to reduce some of the ambiguities in the qualitative reasoning.

5.2 Representing behavior of structures by causal dependency networks

A causal model is used to describe the behavior of a structure in terms of physical quantities and causal interactions among them. The physical parameters (e.g., moments, forces, displacements) can be represented as nodes in a graph. The causal interactions (e.g., moment equilibrium equation, load displacement relation) can be represented as arcs in such a graph (Fig. 6). Interaction arcs are directional and denote strict causal precedence (e.g., "*P-Force* ——► *Term.B1.Comp2-Y.Displ*" means *Term.B1.Comp2-Y.Displ* depends on *P-Force*). Analyzing the behavior of a given process model involves identifying the initial conditions of input (in our case the parameters with known values, called exogenous parameters), inferring the *causal dependency network* among the physical parameters, and propagating the information across the causal interaction arcs. A causal dependency network for the example problem under consideration is shown in Fig. 6.

The explicit representation of the behavior of a structure by the causal dependencies among the physical parameters is a useful tool in assisting the user in focusing on the influences and interactions among components, processes and/or specific parameters. This capability could be useful in bounding the qualitative evaluation on a focused object.

5.3 Deriving the qualitative behavior by qualitative evaluation

Having generated the causal dependency network, the system transforms the information into a sequence of qualitative evaluation tasks to determine the structural response. For this purpose we further defined:

- a *cluster* as the subgraph of the causally connected physical parameters belonging to the same structural component,
- a *cluster root* as the physical parameter that is the root node of a cluster and is shared with another cluster,
- a *priority of a cluster* as the number of new parameter values that can be derived in a cluster,
- a *task* as any causal link (equation or relation) between two nodes (parameters) that has to be executed in the qualitative evaluation step to derive the qualitative value of the dependent parameter. The naming convention for the tasks was chosen such that each name will be self explanatory. The first part of the name denotes what structural component is considered and the second part of the name describes the task.
- an *agenda of tasks* as a list of tasks generated from the cluster hierarchy ordered in ascending order of priority.

Fig. 6 shows examples of these concepts.

After the agenda of tasks has been generated, the system proceeds to perform qualitative evaluation to determine the deflected shape tendency, qualitative reactions, internal forces and moments. Since all such attributes are represented as slots of objects in the specific and process models, performing qualitative evaluation means filling as many of these slots as possible with qualitative values. The system infers unknown parameter values from the known ones by means of equations, relations, and methods.

The way qualitative equations are used to infer unknown values can be illustrated as follows: Suppose we have in the agenda a task *B1.compl-Force.Equil.Eq* that will trigger the force equilibrium equation of *B1.compl:*

$$Term.A\text{-}Force + Term.B1.compl\text{-}Force = 0$$

The substitution of the parameters with their current qualitative values gives:

$$(Q+) + (C+) = 0$$

Recall the value C+ indicates that the parameter exists but the value is unknown. The unknown physical quantity *Term.B1.compl-Force* can be evaluated by using the definition of qualitative negation operator (see Appendix A), which gives:

$$Term.B1.compl\text{-}Force = Q-$$

In general, if we know the values of all but one of the variables in an equation, we can attempt to determine the value of the unknown variable from the known values based on the definitions of qualitative arithmetic operators. However, we cannot always uniquely determine the qualitative value of the remaining variable because of the ambiguity of qualitative calculus. In our system, we use a program called Quantity Lattice [Ref.7], which keeps track of information about the ordering of the relations among the quantities and uses the information to resolve ambiguity. By consulting Quantity Lattice, our system can reduce some of the ambiguities in qualitative reasoning. An example of using Quantity Lattice for resolving ambiguity is given in Appendix B.

5.4 Solution algorithm

We employ a "greedy" algorithm for inferring a causal solution path and executing the tasks in the agenda. The tasks are executed on a last-in-first-out basis. Processing a task (such as firing a rule, calling a method, or qualitatively evaluating an equation) produces a new parameter value. In this algorithm the system first picks a task that is most promising based on existing information about involved parameters (exogenous parameters and/or cluster root), applicable knowledge and causal ordering among parameters. This task is executed to produce as many additional conclusions as possible that may result from the exogenous parameters and/or cluster root. The proposed *"greedy" depth-first algorithm* can be described as follows:

(1) Identify all the parameters of the models such that their values are already known.
(2) For each such parameter, identify the possible analysis tasks (i.e. equations and relations that are associated with the parameter and that can be used to derive the value of other parameters). Perform a causal ordering analysis for the generated process model and represent the dependency among parameters explicitly in a *causal dependency network*.
(3) Cluster the derived causal dependency network and identify the cluster roots together with the priority of each cluster.
(4) Generate the cluster hierarchy in ascending order of the priorities.
(5) Push all the tasks onto an agenda in ascending order, such that the tasks of the cluster with the highest priority should be at the top.
(6) Derive the qualitative behavior of the given structure by executing these tasks in a last-in-first-out manner until the agenda is empty.

5.5 Example

In this section, we describe the solution strategy with the example

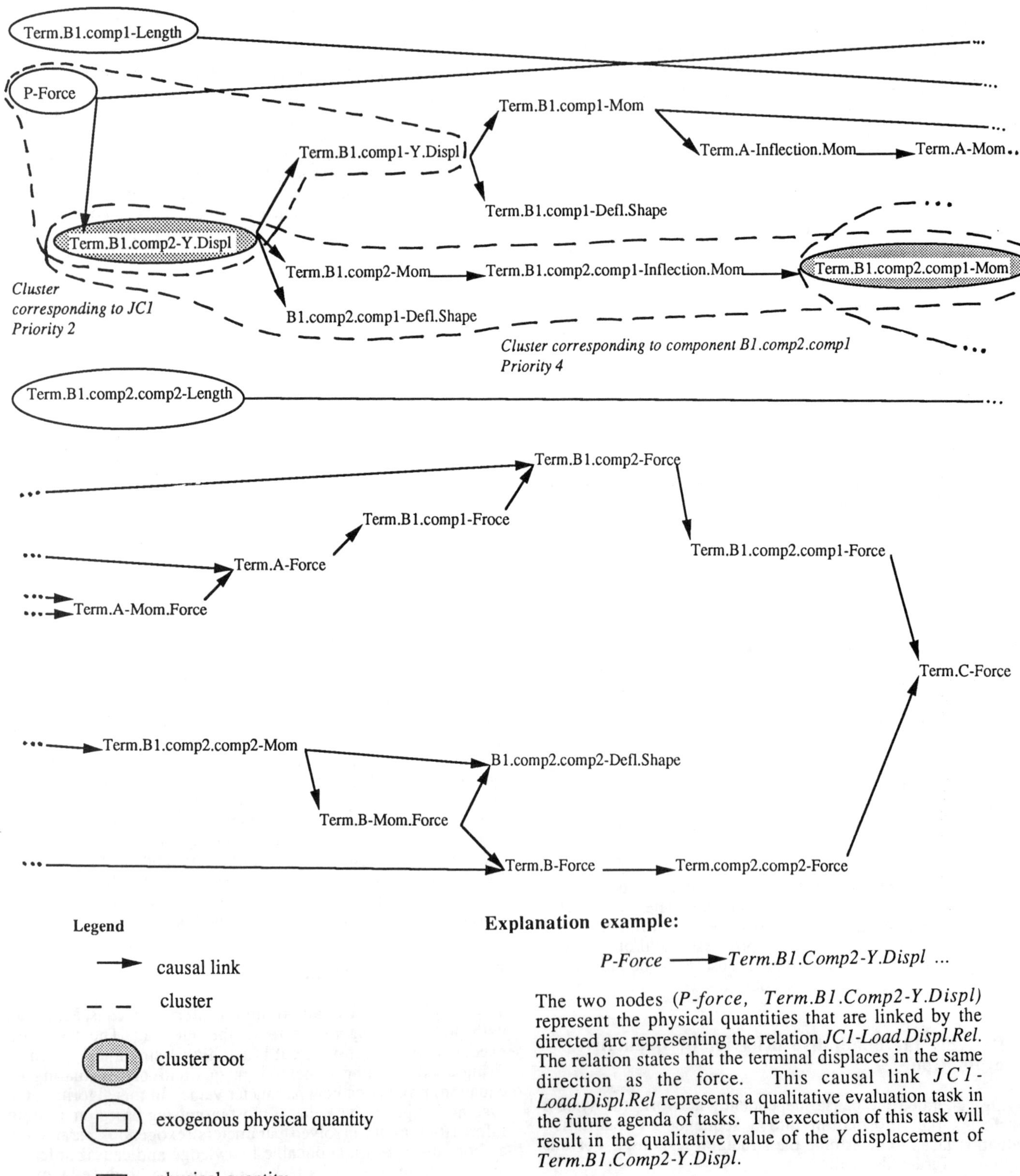

Fig. 6 Causal Dependency Network

shown in Fig. 3. Fig. 6 illustrates part of the causal dependency network of the process model derived in the model generation step. The nodes represent the physical parameters that describe the behavior of the structure. The directed arcs represent the equations or relations among the connected nodes. The left most physical quantities represent the identified exogenous parameters.

After a solution path is identified by the causal reasoning mechanism, the system generates the agenda of tasks from the hierarchy of clusters that are ordered in ascending order of priority. Each task is carried out by applying qualitative calculus and/or Quantity Lattice. Consider the tasks of the qualitative evaluation steps in the agenda illustrated in Fig. 7. Processing part of these tasks is explained and the qualitative evaluation results are illustrated schematically in Table 1, where the dashed arrows represent unknown parameters derived at that step.

JC1-Load.Displ.Rel
JC1-Displ.Equil.Eq
B1.comp1-Displ.Mom.Rel.I
B1.comp1-Displ.Defl.Shape.Rel
B1.comp1-Inflection.Rel.I
B1.comp1-Inflection.Rel.J
B1.comp1-Mom.Equil.Eq
B1.comp1-Mom.Force.Rel
B1.comp1-Force.Equil.Eq
B1.comp2.comp1-Displ.Mom.Rel.J
B1.comp2.comp1-Displ.Defl.Shape.Rel
B1.comp2.comp1-Inflection.Rel.I
B1.comp2.comp1-Inflection.Rel.J
JC2-Mom.Equil.Eq
B1.comp2.comp2-Mom.Equil.Eq
B1.comp2.comp2-Mom.Defl.Shape.Rel
B1.comp2.comp2-Mom.Force.Rel
B1.comp2.comp2-Force.Equil.Eq
.
.
.

Fig. 7 Agenda of Tasks

Table 1 Examples from the Qualitative Evaluation Results

Task	Applied Knowledge and Result	Qualitative evaluation
JC1-Load.Displ.Rel	Represents the relation between load direction and displacement direction at a *continuity-connection*. Derives the displacement direction *(Q-)* at *Term.B1.comp2* of *JC1*.	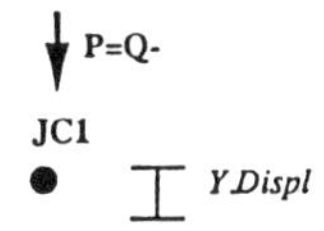 P=Q- JC1 *Y.Displ*
JC1-Displ.Equil.Eq	Represents the displacement equilibrium principle of the two adjacent terminals to a *continuity-connection*. Derives the displacement direction at *Term.B1.comp1* of *JC1*.	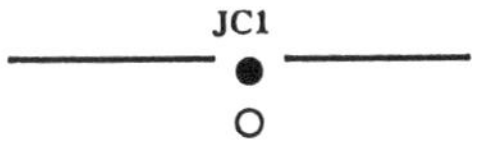 JC1
B1.comp1-Displ.Mom.Rel.I	Represents the relation between displacement direction at a *continuity-connection* and moment direction. Derives the moment direction at *Term.B1.comp1* of *B1.comp1*.	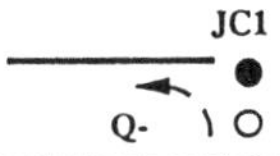 JC1 Q-
B1.comp1-Displ.Defl.Shape.Rel	Represents the relation between deflected shape, support condition and load context. Derives the deflected shape of *B1.comp1*.	
B1.comp1-Inflection.Rel.I	Represents the relation between inflection point and moment. Derives the moment direction of the inflection point near *Term.A* of *B1.comp1*.	 Q0 Q-
B1.comp1-Inflection.Rel.J	Represents the relation between inflection point and moment. Derives the moment direction at *Term.A* of *B1.comp1*.	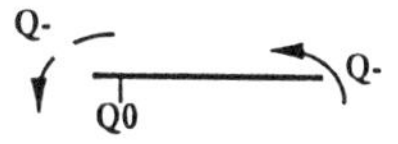 Q- Q0 Q-
B1.comp1-Mom.Equil.Eq	Represents the moment equilibrium principle on a component. Derives the moment *Term.A-Mom.Force* of the vertical force at *Term.A* of *B1.comp1*.	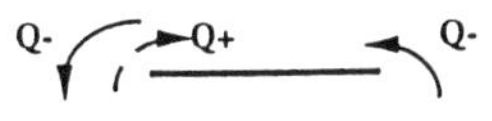 Q- Q+ Q-
B1.comp1-Mom.Force.Rel	Represents the definition of a moment given by a force with respect to a point. Derives the direction of the force at *Term.A* of *B1.comp1*.	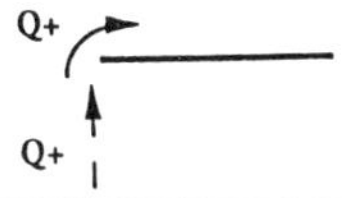 Q+ Q+
B1.comp1-Force.Equil.Eq	Represents the force equilibrium principle on a component. Derives the force at *Term.B1.comp1* of *B1.comp1*.	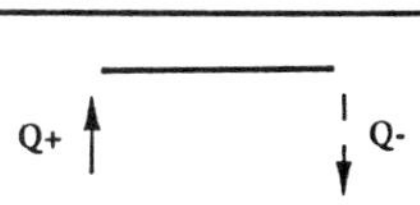 Q+ Q-
. . .	. . .	. . .

6 CONCLUSIONS

In this paper we have presented a framework QStruc for qualitative analysis of structures that could be a useful tool for preliminary design of structures. This framework is based on first principles as well as experiential knowledge. We have described the representation scheme for structures and the domain knowledge. We have discussed a series of model transformations, each of which is useful for analyzing a specific behavior aspect of a structure. The system enables the user to explicitly focus on specific structural behaviors by means of an interaction map, on regions relevant for the behavior of interest, and on the active processes for the given scenario.

We proposed a "greedy" algorithm for qualitative analysis for:

- identification of a solution path based on the causal ordering among parameters,
- expansion of the response behavior as much as possible from known parameters, and
- focusing at different levels of abstraction (e.g., the dependency of specific parameter, processes and physical laws, components).

From our knowledge, such an explicit representation of the causal dependency among physical parameters has not been used extensively in structural engineering analysis. This representation scheme seems to work effectively and is useful to explain the analysis of moments, reactions, and deflected shape tendencies. Furthermore, when defining the solution strategy, we resolve some ambiguities by employing Quantity Lattice together with additional information about the physical parameters.

In this preliminary work, we have demonstrated the potential use of a qualitative approach in structural modeling and analysis. We believe that this approach and the proposed framework, when fully validated, will bc a useful tool for supporting preliminary analysis and design of structures.

Future directions for improving the present framework include:

- extending the QStruc prototype so that it will be able to analyze more complex structures;
- integrating QStruc in a larger CAE framework to enhance the conceptual design process - alternative configuration generation, interpretation of design alternatives, and preliminary sizing of components;
- developing a friendly and expressive user interface that will enable the engineer to dynamically use the modeling, analysis, and interpretation capabilities of QStruc in the conceptual design stage;
- extending the current solution strategy with new reasoning methodologies that would enable the transition from a qualitative model to a quantitative model.

ACKNOWLEDGEMENT

This research was partially sponsored by the Center for Integrated Facilities Engineering and was conducted at the Knowledge System Laboratory, Stanford University .

REFERENCES

1. *Qualitative Reasoning about Physical Systems*, ed. Bobrow, D.G., MIT Press, MA, 1984.
2. Cagan, J. and Agogino, A.M., "Innovative Design of Mechanical Structures from First Principles." *AI EDAM*, Vol.1, No.3, pp169-186, 1987.
3. Fruchter, R. and Gluck, J., "Application of AI Techniques to the Analysis of Structures." *Computers & Structures,* Vol.30, No.3, pp747-753, 1988.
4. Ganguly, J., Kausel, E. and Sriram, D., "A Knowledge Based System for Deriving Qualitative Seismic Behavior." *Computer Utilization in Structural Engineering, Proc. of the ASCE Structures Congress,* ed. Nelson J.K., pp418-427, 1989.
5. Iwasaki, Y., "Qualitative Reasoning." *The Handbook of Artificial Intelligence,* Vol. IV, ed. Barr A., Cohen P.R. and Feigenbaum E.A., Addison-Wesley, pp323-414, 1989.
6. Iwasaki, Y. and Simon, H.A., " Causality in Device Behavior." *Artificial Intelligence,* Vol.29, pp3-32, 1986.
7. Simmons, R., "Commonsense Arithmetic Reasoning." *Proc. of AAAI86*, Morgan Kaufmann, San Mateo, CA, pp118-124, 1986
8. Slater, J.H. "Qualitative Physics and the Prediction of Structural Behavior." *Proc. of ASCE Symposium Expert Systems in Civil Engineering,* ed. Kostem C.N. and Maher M.L., pp239-248, 1986.
9. *Readings in Qualitative Reasoning about Physical Systems*, ed. Weld, D.S. and deKleer, J., Morgan Kaufman, San Mateo, CA, 1990.

Appendix A Examples of qualitative operators

The following tables define qualitative operators on qualitative values. For a complete discussion of qualitative calculus see [Ref.5].

Table A-1 Qualitative Addition

[y]

[x] \ +	Q-	Q0	Q+
Q-	Q-	Q-	?
Q0	Q-	Q0	Q+
Q+	?	Q+	Q+

Table A-2 Qualitative Negation

[x]	[y]= -[x]
Q-	Q+
Q0	Q0
Q+	Q-

(? represents an unknown qualitative result;
[x] and [y] stand for two qualitative parameters)

Appendix B Use of Quantity Lattice in reducing qualitative calculus ambiguity

Consider a force equilibrium equation consisting of three forces, an unknown quantity $F1$ and two known quantities $F2$ and $F3$.

$$F1 + F2 + F3 = 0$$

$$(C+) + (Q-) + (Q+) = 0$$

Relying only on qualitative calculus would require the consideration of all three cases $Q-$, $Q0$, $Q+$ for the unknown $F1$, leading to ambiguous solutions. This can be resolved if there is a means to reason about the relative magnitudes of the two other forces based on first principles. For example, knowing that the relationship among the lever arms $L2$ and $L3$ (of the forces $F2$ and $F3$) is $L2 < L3$, and that the moment equilibrium equation with respect to the *application point* of $F1$ is:

$$F2 L2 + F3 L3 = 0$$

Using Quantity Lattice, the relationship between the two forces $F2 > F3$ can be derived. This information can be further used to decide that the resulting qualitative value of $F1$ is $Q+$.

THREE-HINGE ARCH AND ASTEROID TRUSSES PROGRAMMED IN LOGIC

V.K. Koumousis
Institute of Structural Analysis and Aseismic Research,
National Technical University of Athens,
Athens, Greece

Classical methods of Structural Analysis play an important role in understanding the behavior of structures because during the process of solution the conditions of equilibrium and compatibility are used explicitly. Computationally these methods are problem dependent and thus are not suitable for algorithmic programming. Logic programming can be used to implement Classical methods of Structural Analysis. This can formulate the problem on a modern base from which design considerations can be tackled by using Artificial Intelligence methods and Expert Systems. It is anticipated that classical methods as more physical and conceptually clear will provide the required information in a more flexible way that will suit better the problem at hand, avoiding the complete solution or reanalysis of the structure. In this paper three-hinge arch and asteroid trusses are analyzed using logic programming. These topologies are also important for the analysis of statically indeterminate trusses.

INTRODUCTION

The direct stiffness method of Structural Analysis formulates the complete solution of a structure in terms of its nodal displacements and the corresponding nodal loads. In the context of this method the distinction between Statically Determinate and Indeterminate structures is insignificant. Based on geometrically admissible displacement fields the method enforces the equilibrium at the nodes of a structure to establish the required stiffness relations. These relations determine the solution in terms of the nodal displacements.

The direct stiffness method is esoteric in terms of pointing out equilibrium and/or compatibility. It consists of interconnected flexible members of predefined exact geometrical displacement fields, where the equilibrium at the nodes has been enforced through the formation of the total stiffness matrix of the structure. Consequently neither the equilibrium of the nodes nor the compatibility of the displacements are directly exhibited to the engineer during the process of solution.

The compactness of the method may be interpreted as its major advantage in performing the analysis but also as a disadvantage concerning the ability to interfere in the process of solution. Equilibrium or compatibility checks can be performed only after the completion of the solution.

By focusing on the modern integrated systems that will assist the engineer in all aspects of design, analysis tools that perform a transparent analysis will be preferable. This will permit the interaction at any level of the analysis process, allowing also for partial changes of the structural system.

Most of the expert systems for the design of truss structures focus on the design part and use for the analysis the direct stiffness method (Ref.1 and 2). In more recent papers (Ref.5 and 9) some attempts to reformulate the direct stiffness method in a way suitable for the partial synthesis of trusses were presented.

In the present work an attempt has been made to deal with problems of Classical structural analysis. In a previous paper (Ref.7) simple and compound statically determinate trusses where programmed in Prolog. In what follows three - hinge arch and asteroid trusses are solved. These topologies are important for a subsequent treatment of statically indeterminate trusses following the force method of analysis.

THE THREE HINGE ARCH OR THE TWO DISK ARCH

The three hinge arch is a statically determinate configuration consisting of two statically determinate bodies or disks that are linked by one hinge between them, while each disk is supported by another hinge with the assumed as fixed disk. Representing every hinge with two truss members the general form of a three hinge arch is presented in Fig.1

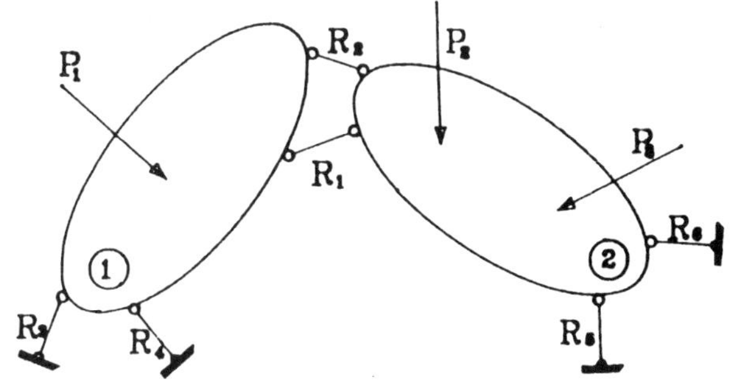

Fig. 1. A Three-Hinge Arch

A disk is defined as an assemblage of members and nodes having a base triangle and members and nodes connected to it by adding a new node and two members at a time such that the two new members together with an already member of the disk form a triangle.

What is important to notice in comparison with a Gerber type structure is that the unknown reactions are four in both cases but they are grouped in a way that for the Gerber structure there is always a solvable disk with three unknown reactions where in the case of the three hinge arch both disks are supported by four members. Thus none of the disks can be solved independently and coupling is present through the two common members. This means that the forces are not distributed independently and directly from the one disk to the other but an interaction is present between the disks as to assure the equilibrium.

To determine the magnitude of these interaction forces besides the standard methods that solve the coupled system of equations one can use the following method which is also suitable for the general case of asteroids treated in the next paragraph. The method is recursive i.e it invokes itself and as such is easily programmed in Prolog.

We know that a solvable disk may have maximum three supporting members. In order to make both disks solvable and reveal the interaction we may think the two disks supported as in Fig.2,

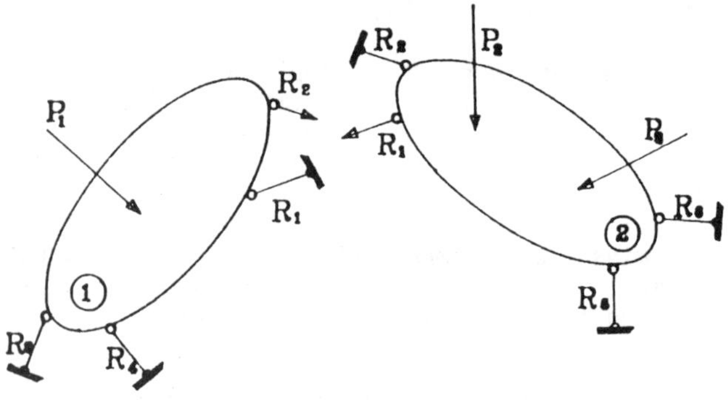

Fig. 2 Isolated disks

where R_1 is a reaction for disk 1 but an external force at disk 2, and R_2 is the reaction for disk 2 and an external force at disk 1. Thus, this concept poses the problem as follows : find a force R_2 that if imposed to disk 1 results a reaction R_1 which in turn imposed into disk 2 results the opposite force R_2 as a reaction of disk 2. In other words this is a fixed point problem for these forces because we input a force to an operator, which here is the equilibrium through the two trusses and actually we require the output to result the same force. Thus we can write down the following cyclic relations

$$\begin{aligned} R_1 &= R_{10} + R_{12}R_{20} + R_{12}R_{21}R_1 \\ R_2 &= R_{20} + R_{21}R_{10} + R_{21}R_{12}R_2 \end{aligned} \qquad (1)$$

where R_{10} is the reaction at member 1 due to the external loads of disk 1, R_{20} is the reaction at member 2 due to the external loads of disk 2, R_{12} is the reaction at member 1 due to a unit force $R_2=1$ and R_{21} is the reaction at member 2 due to a unit force $R_1=1$. In other words R_{12} and R_{21} express the rate of change of the magnitude of the reaction at member 1 or 2 respectively, due to every additional unit load in the direction of members 2 or 1 respectively. These coefficients represent the percentage of the external load which is transmitted through the specific support when a force is acting at a specified place. In words relation (1) says that R_1 is composed from the reaction R_{10} of disk 1 due to the external loads at this disk, from the reaction at member 1 due to R_{20} which is transmitted to disk 1 and finally from the reaction at member 1 due to R_2 that comes through disk 1 and R_1.

Following this line of reasoning in order to calculate the axial forces of the members that connect the two disks we need to solve each disk twice, once with the external loads and another time with a unit load in the direction of the other member. In this way we circumvent the coupling of the equations of equilibrium.

Having established the axial forces in the members that connect the two disks it is easy to determine the other two reactions of the two disks by invoking standard predicates that perform this task. This is accomplished by adhering to the following steps. The axial forces R_1 and R_2 are treated as external loads at disk 2 and 1 respectively and the reactions are determined by considering the moment equilibrium about the points of intersection of the lines of the remaining reactions. After that every disk can be solved as a simple truss i.e by searching every time for a node with two members, solving this node, inserting the resulting axial forces as external loads at the opposite nodes and deleting the node and the members until there is no member left in the database for that particular disk.

MULTI DISK ARCHES - ASTEROIDS

Extending the basic concept of the three hinge arch we may consider more complicated topologies by grouping more than two statically determinate disks in a loop while in the same time every disk is supported at the fixed disk with exactly two members. These topologies are called multi disk arches or asteroids and are characterized by their order which is the number of the participating disks. In Fig.3 an asteroid of order 3 is presented.

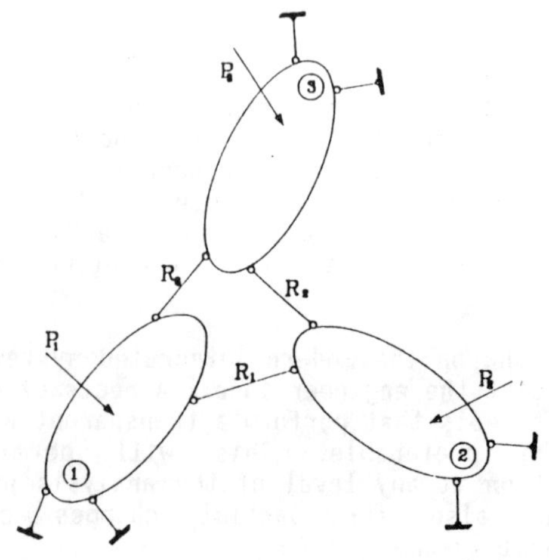

Fig. 3 Asteroid Trusses

As in the case of a three hinge arch we have to identify first the participating disks that form the interconnection ring. Every disk is supported by four members. The total number of unknown axial forces of an asteroid of Nth order is : two external supports for every disk of the asteroid and N members forming the interconnection ring that is 2*N + N = 3*N unknowns. By considering the equilibrium of the N disks three equations can be written for the unknown reactions. Thus the 3*N unknowns are determined from a system of 3*N algebraic equations of equilibrium confirming also that the problem is statically determinate. In order to decouple these equations every disk is considered having only three from its four supports leaving the remaining as external force. By doing this in a cyclic way anticlockwise considering the same member first as member and then as external force in the subsequent disk it turns out that every of the N members of the interconnecting ring is a third support of a distinct disk Fig 4.

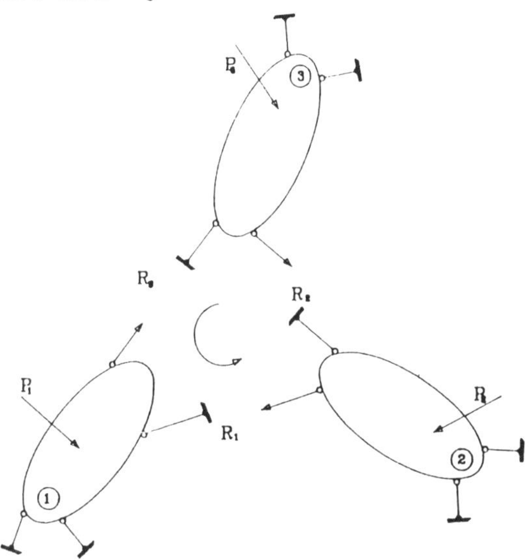

Fig. 4 Isolated disks

With this setup focusing on every disk and its reaction that participates in the ring we can write the following cyclic relation,

$$\begin{aligned} R_1 &= R_{10} + R_{13}\,R_3 \\ R_2 &= R_{20} + R_{21}\,R_1 \\ R_3 &= R_{30} + R_{32}\,R_2 \end{aligned} \tag{2}$$

and from these directly the axial forces of the members of the ring as:

$$\begin{aligned} R_1 &= R_{10} + R_{13}\left[R_{30} + R_{32}\left[R_{20} + R_{21}R_1 \right]\right] \\ R_2 &= R_{20} + R_{21}\left[R_{10} + R_{13}\left[R_{30} + R_{32}R_2 \right]\right] \\ R_3 &= R_{30} + R_{32}\left[R_{20} + R_{21}\left[R_{10} + R_{13}R_3 \right]\right] \end{aligned} \tag{3}$$

where again R_{i0} represents the reaction at member i due to the external loads of the disk that contains member i as a support member and R_{ij} represents the reaction at the support member i due to the unit tensile force in the direction of the next member j in the ring.

The above relations can be used in a recursive scheme to compute the axial forces in the interconnection ring. After establishing these forces every disk is solved as a regular disk for the other two reactions and then the procedure follows the path of determining the axial forces of all the members of all disks following the standard node method.

IMPLEMENTATION IN PROLOG

To establish the solution for this problem one has to accomplish the following:

1. Identify the topology.
2. Establish the interconnection forces.
3. Solve the decoupled disks as simple trusses.

First the program searches the database to establish the different disks. This is performed by applying the definition of a statically determinate disk considered in this paper which consists of two logical actions. The first is to establish a basic triangle and then to add new nodes and members to this triangle in a way that for every new node two new members exist which with a third member of the already formed disk form a triangle. The relevant predicates are :

```
find_triangle(DN):-
     member(0,M1,N1,N2),
     member(0,M2,N2,N3),M2<>M1,
     member(0,M3,N3,N1),M3<>M1,M3<>M2.

add_triangle(DN):-
     member(DN,_,N1,N2),
     member(0,M1,N,N1),member(0,M2,N,N2).
```

The remaining members are members that support the disks with each other or to the ground which is considered as fixed disk. A common node between two disks is treated as two supporting members. Every supporting member is considered in relation to both disks that supports. In this way a data base is formed that contains all the supports of all the identified disks of the problem.

```
support_members:-
     member(0,M,J,K),
     node(N1,J,_,_),N1<>0,
     node(N2,K,_,_),N2<>0,
     not(sup(N1,M)),not(sup(N2,M)),
     assertz(sup(N1,M),data),
     assertz(sup(N2,M),data),
     retract(memb(0,M,J,K),data),
     support_members.
support_members.
```

Having the database in this form the identification of the disks with three supports can be performed in the following way :

```
find_disk_with_three_supports(DN,M1,M2,M3):-
     sup(DN,M1),
     sup(DN,M2),M2<>M1,
     sup(DN,M3),M3<>M2,M3<>M1,
     not(fourth_sup(DN,M1,M2,M3)).

fourth_sup(DN,M1,M2,M3):-
     sup(DN,M4),M4<>M1,M4<>M2,M4<>M3.
```

The identification of a three hinge arch can be performed as follows :

```
find_three_hinge_arch(N1,N2,M1,M2,M3,M4,M5,M6):-
      sup(N1,M1),sup(N2,M1),N1<>N2,
      sup(N1,M2),sup(N2,M2),M2<>M1,
      not(third_link(N1,N2,M1,M2)),
      other_two_sup(N1,M1,M2,M3,M4),
      other_two_sup(N2,M1,M2,M5,M6).

third_link(N1,N2,M1,M2):-
      sup(N1,M3),sup(N2,M3),M3<>M2,M3<>M1.

other_two_sup(N1,M1,M2,M3,M4):-
      sup(N1,M3),M3<>M1,M3<>M2,
      sup(N1,M4),M4<>M3,M4<>M2,M4<>M1,
      not(fifth_sup(N1,M1,M2,M3,M4)).

fifth_sup(N1,M1,M2,M3,M4):-
      sup(N1,M5),M5<>M1,M5<>M2,M5<>M3,M5<>M4.
```

The interconnection forces in a three hinge arch or an asteroid truss of order N are established by applying the recursive relations (1) or (3).

Having established the interconnection forces the remaining member forces in each disk are determined by treating each disk as a simple truss.

SIMPLE TRUSSES

In order to analyze simple plane Statically determinate trusses we proceed by performing certain cuts and enforcing the requirements for equilibrium of the resulting parts. These parts must satisfy the rules stated subsequently using the notation: <--> denotes equivalence, ---> means implies, <---means if, $\wedge$ denotes conjunction and $\vee$ denotes disjunction of clauses. Thus the clause that the solution of a Simple Statically Determinate Truss (SSDT) for a specific loading is established if its reactions and the axial forces in all its members can be established can be stated as:

solution of *SSDT* <-- *solution - reactions*
$\wedge$ *solution-all members.*

This clause represents the goal of the whole process.

The structure has three reactions which do not pass from the same point. Thus, it is true that the reactions can be established and moreover if this is done the part of solution that concerns the evaluation of the reactions is completed.

rule 1 :
solution - reactions <----- three reactions determined

In order to establish the axial forces in all members of the truss nodes with two unsolved members are isolated recursively. Thus, the solution for the axial forces in all members is true if the solution of the remaining members is concluded.

rule 2 :
solution - all members <------ solution - remaining members.

If there are remaining unsolved members in the structure then there exists at least a node with two or one unsolved members.

rule 3 :
solution - remaining members <----
exists node with two members
$\vee$ *exists node with one member,*
$\wedge$ *solution - remaining members.*

The clause that there exists a node with two members implies that the axial forces at these two members can be determined and then this node can be eliminated from the list of unsolved nodes. Moreover the two members can be eliminated from the list of unsolved members. The remaining structure subjected to the computed axial forces and the given external forces is in equilibrium. This fact is used to solve for the forces in the remaining members.

rule 4 :
exists node with two members ---->
node equilibrium
$\wedge$ *node is eliminated*
$\wedge$ *two members are eliminated*

The clause that there exists a node with one unsolved member implies that the axial force in this member can be determined and that this node can be eliminated from the list of unsolved nodes and that the member can be eliminated from the list of unsolved members.

The number of members of a simple truss is always an odd number M = 2N - 3. Thus by eliminating two members when solving a node ultimately we have a node with one (unsolved) member.

rule 5 :
exists node with one member ---->
node equilibrium
$\wedge$ *node is eliminated*
$\wedge$ *member is eliminated*

Following the afore stated five rules the *solution of a simple statically determinate truss* can be proved to be true. Notice that the nodal loads do not affect the solution strategy.

It is important to mention that the logical interrelations of the above rules point to specific actions and in that sense they represent a set of procedural logical interrelations that as a whole point out to the solution strategy.

EVALUATION OF REACTIONS

The first step in the solution of a statically determinate simple plane truss is to determine its reactions. We distinguish two possible support cases a) three roller supports whose directions of rolling are not parallel b) a hinge and a roller support. A hinge support is equivalent to two non parallel roller supports with their axes intersecting at the hinge. Moreover a roller support is equivalent to a reaction member in the direction normal to the plane of rolling. Thus all the supports can be treated as additional members (reaction members). This approach simplifies programming considerably. A small logic program can be written to convert the various types of supports to reaction members and the axial forces in these reaction members to reactions.

The equilibrium of a properly supported plane structure can be insured in a number of ways. In this paper we set equal to zero the moments of all the forces acting on the structure about three characteristic points of intersection of two reactions. This leads to the following relations :

$$\sum M_i + R_i * D_i = 0 \quad \text{or} \quad R_i = -\frac{\sum M_i}{D_i} \qquad (4)$$

$$i=1,2,3$$

where $\sum M_i$ $(i = 1,2,3)$ represents the sum of moments of all the given external forces about point O_i $(i = 1,2,3)$.

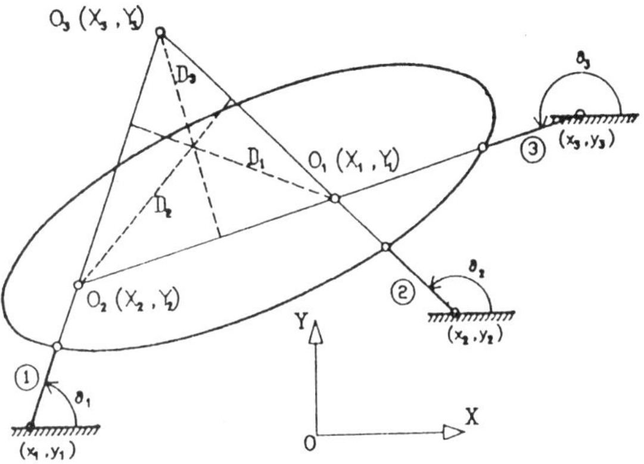

Fig.5. Characteristic Points of Reactions

The coordinates X_k and Y_k of the characteristic points (see Fig.2) can be derived easily from analytical geometry as the intersection of two lines:

These relations are not valid in case of two parallel reactions $[(\theta_i - \theta_j) = 0, \pi]$. In this case the corresponding equation of equilibrium (4) is replaced by one expressing the vanishing of the sum of the given external forces along the non parallel direction.

The above calculated reactions are considered as loads at the supported nodes. The truss subjected to the given external forces and the computed reactions is in equilibrium. In the sequel every disk is solved following the node method described in the five rules mentioned above.

COMPOUND TRUSSES

There are cases where statically determinate trusses are composed from different disks which are interconnected as disks having three supports or three hinge arches or asteroid trusses. In these cases a general strategy for the solution must be determined. In this paper first the disks supported by three supports are identified and solved first, and the three hinge trusses or asteroids are processed subsequently.

Example 1

As an example the three hinge arch of Fig.6 is considered.

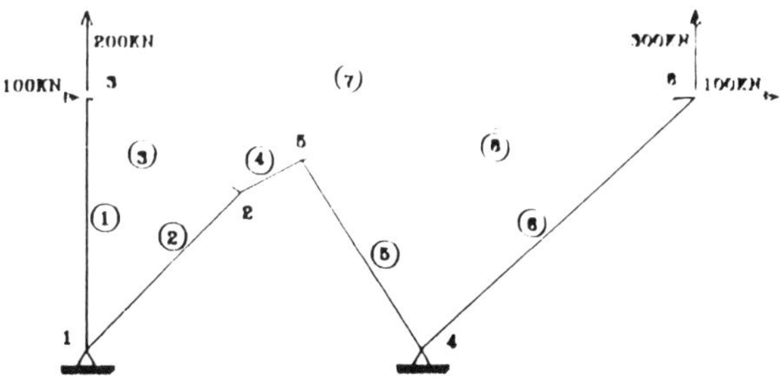

Fig. 6. A Three hinge Arch

Following the steps described before the program first identifies the two disks. This instance is illustrated in Fig.7.

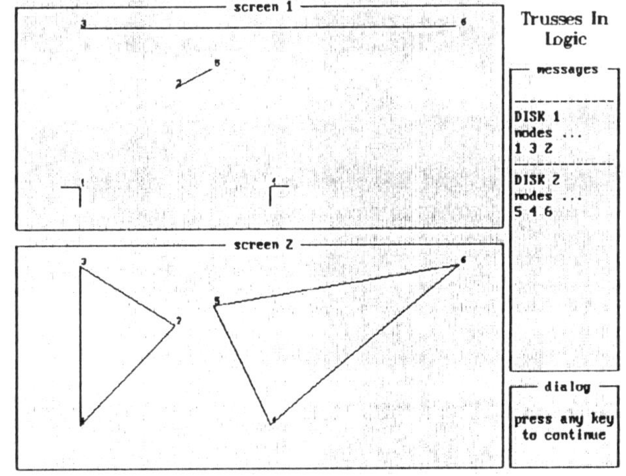

Fig. 7 Identification of disks.

Then the interaction forces are calculated by resolving the coupling and each of the two disks is solved as a simple truss.

The axial forces in the members of the three hinge arch are:

```
axial_force(1,252.5),     axial_force(2,-272.24)
axial_force(3,-102.04),   axial_force(4,-313.05)
axial_force(5,100.62),    axial_force(6,560.27)
axial_force(7,-12.5),     axial_force(8,-328.82)
```

Example 2

As a second example a more complicated case is considered which is illustrated in Fig.8.

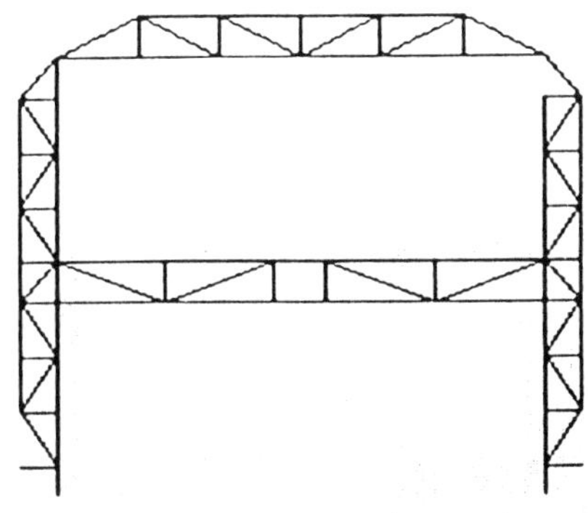

Fig. 8 A compound Truss

The identification procedure determines the three disks presented in Fig. 9.

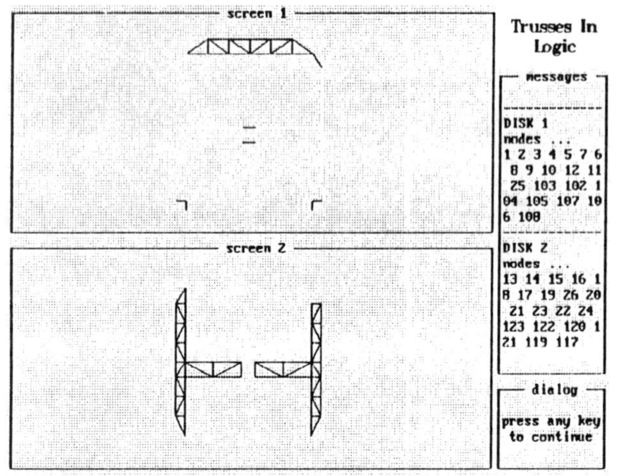

Fig. 9 Identification of disks.

The solution proceeds first by solving disk 3 which has three supports. By removing the already solved part, it performs a new identification and decides to solve disks 1 and 2 as a three hinge arch. This completes the solution of the compound truss.

CONCLUSIONS

In this work a logic program is presented for the solution of three hinge arches and asteroid trusses. The method is recursive and resolves the coupling of the forces by performing two separate solutions for each disk of the truss. The method is based on the identification of the statically determinate disks that form the truss. The specific way that the disks are connected is identified by the logic program and the solution is decomposed to the solution of simple trusses. The method is transparent in the sense that reveals the interconnection and succeeds in decomposing the whole structure.

REFERENCES

1. Adeli, H. and Balasubramanyam, K. V. "A Knowledge-based System for Design of Bridge Trusses" Journal of Computing in Civil Engineering, 2, No. 1. 1988.

2. Adeli, H. and Al-Rijleh, M. M. "An Expert System for Design of Roof Trusses", Microcomputers in Civil Engineering, 2, No. 3. 1987.

3. Armenakas, A. E., "Classical Structural Analysis - A Modern Approach" McGraw Hill, 1988.

4. Bratko, I., "PROLOG Programming for Artificial Intelligence" 2nd Edition, Addison-Wesley Pub. Co. 1990.

5. Chan, W., T. and Paulson B., C., Jr.,"Exploratory Design Using Constraints", Artificial Intelligence for Engineering Design, Analysis and Manufacturing 1(1), 1987.

6. Clocksin, W.F. and Mellish, C.S., Programming in Prolog, 3rd Edition, Springer-Verlag, New York, 1981.

7. Koumousis V.K, Georgiou P.G, "Statically Determinate Trusses Programmed in Logic", (Submitted for publication).

8. Kowalski, R., Logic for Problem Solving, North-Holland, 1979.

9. Lakmazaheri, S. and Rasdorf, W., J.,"The Analysis and Partial Synthesis of Truss Structures via Theorem Proving", Engineering with Computers 6(1), 1990.

10. Ross, P.,"Advanced Prolog : Techniques and examples" Addison-Wesley Pub. Co. 1989.

11. Sterling, L. and Shapiro, E., "The Art of Prolog: Advanced Programming Techniques", MIT Press, 1986.

12. Turbo Prolog Manual , Ver. 2.0 , Borland Int.

2 FINITE ELEMENT MODELLING AND ANALYSIS

AN EXPERT SYSTEM FOR FINITE ELEMENT MODELING AND ANALYSIS

R.K. Rajan†, R.V. Nambiar‡ and K.L. Lawrence‡
†General Dynamics, Fort Worth, Texas,
‡Mechanical Engineering Department, University of Texas at Arlington, Arlington, United States of America

Finite element analysis is an integral part of the structural analysis process in the design and development of parts and assemblies. Various levels of integration exist between design systems and finite element modeling systems. The feasibility of a knowledge based system to tightly couple product design systems and finite element analysis systems was investigated. A framework incorporating feature based design systems, heuristics based planning modules, procedural pre-processing/solving/adaptive refinement schemes was developed. The pertinent modules of the framework were organized into a prototype and implemented in the VAX environment utilizing the EUCLID geometric modeling system, CLIPS expert system development tools, PATRAN 2.3 finite element pre-processing system, a finite element solver and a unique adaptive refinement system developed specially for the framework. The scope of the research work was limited to static stress analysis. The results of the research indicated that design systems can be successfully integrated with finite element analysis systems in a knowledge based framework. The framework was also found to be suitable for finite element based structural optimization and self-learning processes.

1. INTRODUCTION

1.1 The Finite Element Method

The Finite Element Method (FEM) is a general numerical procedure for solving a system of differential equations. Even though FEM had its roots in structural and solid mechanics, the technique has been adapted into various disciplines such as fluid mechanics and dynamics, thermal sciences, control systems analysis and electromagnetism.

In applying the finite element method, some continuous quantity such as displacement, temperature or pressure over a complex region is expressed as piecewise continuous functions over a finite number of its sub-regions. These sub-regions are usually referred to as elements. The piecewise continuous functions over each of the sub-regions are interpolated from values of the function at a finite number of points over the sub-regions. These points are referred to as nodes and the matrix of nodes is collectively referred to as the finite element mesh. The simple sub-regions are assembled together systematically to obtain the approximate behavior of the desired quantity over the entire complex region. This procedure results in a number of discrete equations which contain the values of the desired quantity at the nodes and the solution of these discrete equations yields those values. This whole process is usually referred to as Finite Element Analysis (FEA).

1.2 Geometric Modeling Systems in Structural Design and Analysis

Product design may be thought of as a collection of information related to a part or assembly that resides in the mind of the designer and also in the culture of an enterprise. This information is structured differently depending on the individual and the enterprise. Product definition data is an abstraction that is used to formulate, communicate and apply such information related to a part (Ref. 10). Computer based systems that enable the capture of this data may be referred to as product definition data systems (PDDS).

Geometric Modeling Systems (GMS) have replaced the traditional engineering drawing as the method of capturing and conveying product definition data to the structural analysis function. Since FEM is an integral part of structural analysis, tight integration of GMS and FEM is a highly desired functionality in the structural design/structural analysis arena.

Product Definition Data is usually put into a structured format for communicating the information to interested parties. Various standards such as MIL STD 1000, IGES and PDES define this format. These formats contain only a subset of the total product definition data. Existing finite element analysis process is based on this subset of product definition data. There are several disadvantages in this scenario:

1. The FEM process is blind to useful and important sets of information of the total product definition data.

2. A FEM process that works on one representation is incapable of operating on another representation.

3. It is difficult to directly interpret the results of the analysis against aspects of the design for applications such as design optimization and design evaluation.

There are various levels of product definition data. They may be broadly categorized into two dimensional drawings, three dimensional solid models and the emerging feature based models. Features are sets of elements of product definition data used in creating, evaluating and producing a design (Ref. 11, Ref. 5). They may be related to the functionality of a part (shear webs, stiffeners, flanges), shape related (depression, protrusion, passage) or non-shape (tolerances). Features are interpretations of product definition from different perspectives and provide a consistent way of capturing the design intent and allow a method of capturing interpretations of product data (Ref. 7). A standardized set of features and their underlying meanings hold the promise of direct integration of PDDS and structural analysis.

1.3 Research Objectives

The objective of this research was to develop the framework for an environment in which product definition data systems and finite element based structural analysis are tightly coupled. This research focuses on the methods of integrating FEM based structural analysis and PDDS, utilizing feature based product definition data in a knowledge based framework. Throughout this paper, a typical wing spar is used to illustrate the concepts and methods. This spar is typically a machined structural aluminium part used in the construction of aircraft wings and other control surfaces. The spar was chosen for its complexity though the approach of this work is meant to be general.

2. THE FRAMEWORK

The structural analysis function encompasses several tasks such as loads analysis, static stress analysis, dynamic analysis, buckling analysis, fatigue/fracture analysis and thermal analysis. The finite element models that are generated depend on the type of analysis, experience of the analyst and the tools that are available. The approach presented here focuses on static stress analysis only, even though the concepts may be adapted for other types of analysis also. The framework incorporates a feature based design environment and a finite element modeling and analysis system. The concepts of the framework have been implemented in the VAX environment at General Dynamics/Fort Worth division. The top level software components of the system and the information flow are shown in figure 1. The concepts were rapidly prototyped with commercially available packages such as EUCLID (Ref. 6), CLIPS (Ref. 2), PATRAN (Ref. 9) and additional software written in 'C' and PASCAL languages. To validate the concepts, the static stress analysis task for a shear web with a complex cutout was prototyped.

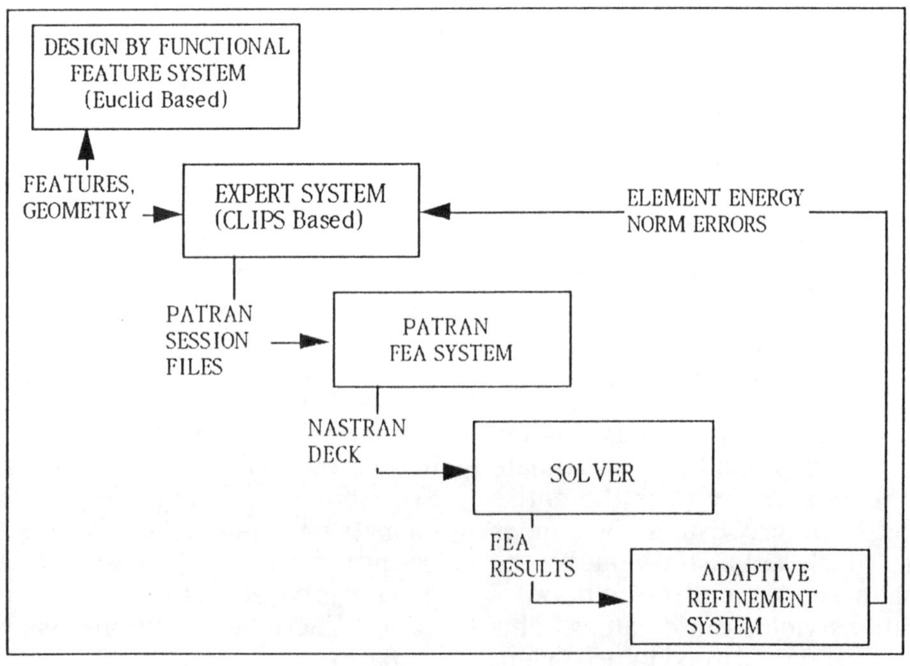

Figure 1. The Prototype Architecture

2.1 The Functional Feature based Design Environment

A prototype version of a functional feature based design system (Ref. 7) was obtained to simulate a product definition data system that incorporates functional feature definitions and interpretations. The functional feature based design system uses EUCLID for specific geometric operations and manipulations. The formal access to the product definition data in the prototype design system is still under development. Therefore, the required data was transferred to the finite element modeling system thru an ASCII neutral file. The neutral file is only an interim solution and in the future, feature based product definition data shall be obtained thru formal access methods.

2.2 The Finite Element Modeling and Analysis System

The finite element modeling and analysis system is composed of four main blocks: Planning, Pre-processing, Solving and Evaluation/Adaptive Refinement.

2.2.1 Planning the Finite Element Model

Planning the finite element model is one of the most crucial and time consuming steps in building the analysis model. The physical model is translated into a mathematical model during this step. This is accomplished by expert analysts with considerable knowledge about the finite element process itself and also utilizing prior experience with similar problems. For static stress analysis, this experience consists of knowledge regarding the variation of stresses and strains in models with functionally similar features and associated parameters such as boundary conditions and forcing functions. Utilizing this approach in the framework, the functional features of the product design are transformed into geometrically defined finite element regions and the stresses in those regions are characterized in terms of their relative magnitudes and variations. Once the regions are formulated and the stresses are characterized, appropriate elements are selected and mesh densities are stipulated. The individual regions are condensed into a whole FEM model by a graph theory based approach to propagate nodal and element constraints. Thus, a mathematical model is distilled from the feature based description of a product. Therefore, this block of the framework has five modules that perform the above mentioned activities:

2.2.1.1 Mapping Functional Features into Finite Element Regions
2.2.1.2 Stress Characterization of Regions
2.2.1.3 Selecting Elements for Regions
2.2.1.4 Specifying Mesh Density
2.2.1.5 Assembling Individual Regions into a Whole Model

2.2.1.1 Mapping Functional Features into Finite Element Regions

The finite element method is the process of expressing a continuous quantity such as displacement, temperature or pressure over a complex domain into piecewise contiguous functions over a finite number of its sub-regions. Each of these sub-regions is modeled by a specific type of finite element. Most finite element modeling systems are able to manage collections of these sub-regions as logical groups. These groups will be referred to as regions. Mapping the definition of a problem into these regions is the most difficult step in planning the finite element model.

Expert designers and engineers are able to manage a quantity of interest such as stresses or strains over a complex domain by characterizing them over sub-domains of the part definition. For this process, they utilize knowledge gained from prior experience with functionally similar designs and an understanding of the appropriate fundamental theories. A person who has managed to abstract such knowledge based on the functional features of a part is usually called an expert. In the framework, this knowledge is effectively used in mapping the part definition into finite element regions.

The task of mapping a definition involves obtaining data regarding the functional features and their relationship to each other and translating them into regions based on heuristic knowledge. A specific feature is mapped into one or several regions based on the complexity of its intended function, the loading and boundary conditions and its relationship to other features. The regions are formulated to depict the gradients of the strain energy potentials (Ref. 12). The regions themselves may be simple or complex. One such region is formulated as shown in figure 2.

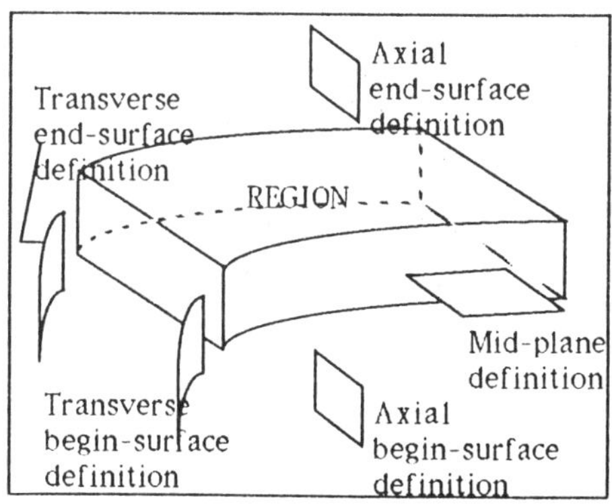

Figure 2. Definition of a simple region

This type of region is referred to as a parametric plate region and is characterized by a volume of constant thickness defined by a definition of the middle plane and definitions of surfaces bounding the axial and transverse directions. The planes and surfaces are defined geometrically and combined with a numerical value for the thickness, defines the region completely.

An axis system and co-ordinate directions is defined on the region by convention. An abscissa direction is defined isoparametrically from the axial begin-surface to the axial end-surface. The ordinate is defined from the transverse-begin-surface to the transverse end-surface. A positive and negative Z-direction is defined from the middle plane of the region. These conventions are needed for defining finite elements over the regions and later for refining the region with more elements.

2.2.1.2 Stress Characterization of Regions

A quantity of interest such as stress is now defined as a varying quantity over this region. The stress vector components can be described as variations with the directions defined by the co-ordinate directions at a specific point in the region. The objective of the finite element analysis is to obtain the nominal values of the stress vector at each point. The variation of the stresses is classified as linear, quadratic, cubic or more complex definitions.

The utility of such a region is realized when modeling a part such as a wing spar. The spar is an important structural member used in the construction of wings and other control surfaces. The objective of a spar is to resist shear loads and bending moments experienced by the wing during various maneuvers of an aircraft. Functionally the important features of a spar are its shear web and flanges. In an ideal spar, the shear web resists the shear loads, and the flanges in unison, resist the bending moment.

Table 1. Stress Characterization of a Region from a Shear Web

Stress Components	Axial (x) Distribution	Transverse(y) Distribution	Thru (z) Distribution
σ_x	linear	linear	constant
σ_y	nil	nil	nil
σ_z	nil	nil	nil
τ_{xy}	constant	quadratic	constant
τ_{xz}	nil	nil	nil
τ_{yz}	nil	nil	nil

The shear web, in its standard configuration, distributes the shear load along the web, developing shear stresses that are quadratic in nature. The flanges resist the bending moment by developing axial stresses across its cross-section. Thus, for a spar, the shear web may be mapped into one isoparametric plate region. The flanges map into similar regions also. The variation of the stresses are formulated and captured based on heuristic knowledge regarding the functional features and their relationships to neighbouring features. The relative magnitude of the stresses, if known, are also formulated. The stress characterization of a shear web region is portrayed in table 1.

The software module to perform these tasks were built using the EUCLID geometric modeling system and the CLIPS rule based programming language. EUCLID was used for the geometric operations needed to define the regions and CLIPS was used to implement the heuristics based inferencing schemes for formulating the regions and characterizing stress variations.

The geometry of the region in conjunction with the type of stresses present and the variation of these stresses, is used to automatically select the appropriate elements and the proper mesh density. The information regarding the relative magnitude of the stresses is used to further optimize the selection of elements.

2.2.1.3 Selecting Elements for Regions

Selecting finite elements and specifying a mesh density is the second most important step in formulating a physical problem into a finite element problem. In this step, the objectives are to determine the type of elements to be used, the complexity of the elements, the grading of the mesh, aspect ratio of elements and other related parameters affecting the performance of an element.

The important criteria used in the selection of elements is the stress characterization of a region and the geometric characteristics of the region. The stress characterization includes definition of the stress variations, symmetric nature of the problem and possible stress singularities that may exist in the region. The geometric parameters of the region definition define the shape of the region. The objective of this step is to build the optimum model to give reasonable results.

The optimality of the model is determined by the time it takes to obtain the solution and the accuracy of the results. This factor is directly influenced by the formulation of the element and the computer resources available for the solution. Benchmark tests are usually published by the vendor that provide insight into the characteristics of each element even though the actual formulations are usually proprietary.

The accuracy of the results is dependent on the context of the analysis. Static stress analysis during conceptual design does not require high degrees of accuracy. Fatigue and fracture analysis, on the other hand, require accurate stress results to predict the life expectancy of a part or assembly.

Deciding the optimality of an element is facilitated by utilizing declarative knowledge. For example, the flanges of a wing spar, which experience axial stresses only, may be formulated with solid elements, shell elements or rod elements. All of these elements are capable of simulating axial stresses. The simplicity of the rod element makes it an ideal element to compute the axial stresses on the flange for preliminary sizing of this feature. The capability and complexity of each element is declared in a repository and an inferencing mechanism selects all elements that are appropriate for a region based on their capability to model the prevailing stresses, and geometrical compatibility with the region. The selected elements are also ranked based on their degrees of freedom.

In the prototype, a library of elements have been compiled to describe the stresses and their variations over each element. Since most elements are based on an assumed displacement field, characterizing stresses is an indirect process. It was learned from experience that knowledge gained from experience with the elements is disseminated in terms of their stresses rather than the displacement fields. The capabilities of each element to model stresses are described in a nine by nine matrix.

2.2.1.4 Specifying Mesh Density

Once an element is selected, an appropriate mesh density needs to be specified to create a finite element model of the region. The density may be calculated using simple procedures based on a desired error criteria. If the stress variation is known to be quadratic and an element such as a NASTRAN three noded triangular element is selected, then the error in the calculation may be formulated based on piecewise approximations of the quadratic stress with constant stresses.

The viability of an element is also determined by its geometrical characteristics such as aspect ratio and curvature. Such factors are computed before finalizing this step of the modeling process. The type of elements and the initial mesh density selection are used to calculate figures of merit for the finite element formulation of the region. The specification of the mesh density is revised to improve the aspect ratio of the elements. Some figures of merit such as the curvature may dictate the selection of higher order elements to properly formulate the region. an error statistic (er) (Ref. 3) was used to approximate the relative quality of the mesh from the following equation:

$$er = p1\, p2\, \{1/N^{1/n}\}^{q-r}$$

where:

p1-> largest element aspect ratio
p2-> ratio of the characteristic length of the largest to the smallest element
N-> number of elements in the mesh
n-> stress characteristic dimensionality of the region
r-> 1 for stress analysis
q-> 1 plus the degree of the displacement field of the element

For preliminary static stress analysis, an error of 10% is usually acceptable and for this criteria, an er value of 1 is preferred. The current values of the aspect ratio and characteristic lengths are obtained by processing the region/element/density parameters. The evaluation of the mesh quality and optimization of the various parameters are accomplished with a procedural iterative scheme. Once the elements are selected and the proper mesh densities have been specified for each region, all the regions need to be assembled to create the final model. This step imposes additional constraints during the assembly of the entire model.

2.2.1.5 Assembling Individual Regions into a Whole Model

A part is usually composed of a multitude of functional features which are related to each other in complex relationships. After each feature has been processed into its respective finite element regions, they need to be assembled to generate the finite element model for the complete part or assembly. When the regions associated with each feature are assembled together, the interfaces between regions have to satisfy many constraints, mainly: nodal and element compatibility.

When a functional feature is processed for formulating regions, selecting elements and specifying mesh density, the adjacent feature is generally not taken into account. Each feature is processed to obtain the most efficient finite element model representation for that particular feature. When the regions representing each feature are finally assembled, interface constraints dictated by adjacent regions may require adjusting the parameters of the finite element representation. This process will be referred to as propagating nodal and element constraints. Nodal compatibility is achieved by arranging and equivalencing proper nodes along region interfaces. Elemental compatibility requires elements with compatible fields to exist at the region interfaces. These requirements ensure accurate results from the finite element analysis.

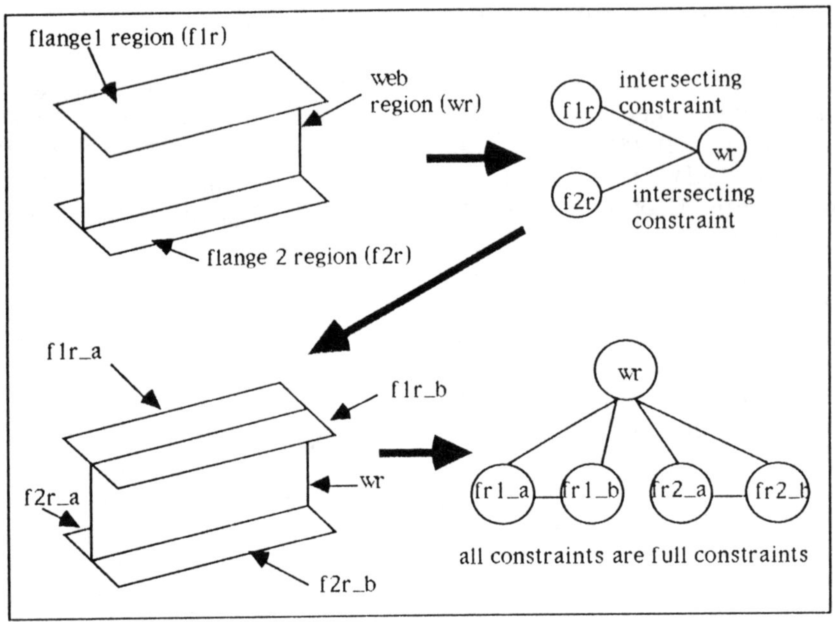

Figure 3. The Process of Generating a Graph and Propagating Constraints

To propagate nodal and element constraints, the geometric relationships between functional features are utilized to form a graph. In this graph, the nodes represent the regions and the edges represent the geometric relationships. These geometric relationships are enumerated into a discrete list. The graph formulation for a wing spar is graphically shown in figure 3. Ideally, a graph of just simple constraints is desired due to the ease of propagating constraints across boundaries. The presence of complex constraints need to be resolved into simpler constraints. Heuristics based on the capabilities of the pre-processing system are utilized at this step to generate regions to facilitate more simple constraints. Certain pre-processing packages like PATRAN can accommodate more complex constraints.

Many techniques exist in the literature to utilize such a graph to propagate constraints (Ref. 4). The simplest approach is a heuristics based method. Based on heuristics, the enumerated list of geometric relationships are organized into a hierarchy. Based on this hierarchy, the graph is reformatted into a tree (figure 3). Once this is accomplished, the nodal and element constraints are propagated from the child nodes to the parent nodes till the whole tree is traversed.

In the framework, nodal constraints are propagated simply by specifying the higher densities for both regions. Element constraints are governed by the complexity of the field (displacement, velocity etc.) at the region interfaces. The order of the admissible functions specified by the elements have to be compatible within each region and also at the regional boundaries. The compatibility is determined from a compatibility matrix. In the case of the spar, if shell elements were selected for the shear web (in the case of a thick web), and rods were selected for the flanges, then these elements are not displacement compatible. Therefore, shell elements will have to be dictated for the flanges to preserve displacement compatibility between elements. This step is required to preserve the quality of the results, even though finite elements models without such compatibility can still be analyzed.

In the prototype, this operation was accomplished with the help of the EUCLID geometric modeling system. The open architecture of EUCLID allows its capabilities to be exploited in an integrated environment. The FORTRAN based EUCLID functions were called from C based CLIPS by utilizing the VAX common language environment (figure 1). The geometric constructs were interrogated against the constructs of each related feature to determine the type of constraint. The graph is defined in CLIPS by associated lists of related regions.

The graph depicting the arrangement of the regions is not an ideal method for propagating nodal and elemental constraints. The presence of loops in the graph leads to the possibility of infinite loops in the propagation process. The methods described in the literature may be organized into three groups: network flow, linear programming, heuristics based. The heuristics based method, though less sophisticated, was chosen for rapid prototyping the system.

For the class of problems attempted in the prototype, a simple strategy of subdividing regions was incorporated as rules in CLIPS. Region constraints of intersecting and partial types were transformed into full constraints by subdividing the pertinent regions into two or more at the intersecting vertex or edge. This method, though simple, performed adequately in the prototype system. The other methods were not explored fully due to time and resource constraints.

Once the tree is formulated, the nodal and elemental constraints are propagated by starting at the leaves of the tree and proceeding towards the root. Nodal compatibility was preserved by propagating the higher density to the regions at the next higher level. Elemental constraints were propagated by specifying compatible elements for regions at the next higher level. The compatibility matrix and the propagation rules were implemented as facts and rules in CLIPS. The sub-division was performed with the help of EUCLID and the resulting geometry was tagged and stored in EUCLID.

Once the nodal and element constraints are propagated, the mathematical model is complete. Next, the bulk data deck is generated to submit for analysis.

2.2.2 Pre-processing

Pre-processing is the realization of the actual finite element model. The concepts of the mathematical model is translated into a suitable format such as the NASTRAN bulk data deck with the help of pre-processing tools such as PATRAN or GIFTS. NASTRAN data formats have become a defacto standard in the sense that most finite element analysis systems provide a translator from this format into their indigenous format. The pre-processing activity is unique to each pre-processing system. These systems facilitate groups of finite elements to be generated with a single command by providing specialized utilities. The finite element model may be pre-processed in a number of different ways to generate the identical data deck.

The pre-processing activity is a system specific activity. In this particular block of the framework, a specific system is chosen for pre-processing based on the complexity of the mathematical model. System specific commands are issued in the form of a script. The framework utilizes knowledge about the pre-processing tool to carry out this operation in an efficient manner. The output of this block is the bulk data deck that can be input to a finite element solver.

In the prototype, the translation step into an analysis model was accomplished with PATRAN (Ref. 9). PATRAN is a complete pre-processing, analysis and post-processing system. PATRAN also provides indirect access to its internal utilities for managing the analysis model. The access to PATRAN utilities was accomplished by generating session files which are processed by the PATRAN interpreter. The session files mimic the keyboard input typically provided by a user in generating a finite element model. Since PATRAN was selected for the prototype, this step required defining the methods for transforming certain regions into PATRAN entities. This requirement was facilitated by rules that were coded in CLIPS. There may also be incompatibilities between the geometric entities utilized in the region constraints and the geometric elements supported in PATRAN. Therefore, some complex geometries such as bezier curves and surfaces had to be broken down into simple constructs such as parametric cubic curves and surfaces. The transformation was implemented as rules in CLIPS.

The models that were generated thru the session files were graphically displayed in PATRAN for visual verification. The graphical display of the shear web model is shown in figure 4. PATRAN also has a utility to generate a NASTRAN bulk data deck that is invoked thru the session files. This format was chosen for the solver (FRAME3) obtained from Dr. Kent L. Lawrence of U. T. Arlington.

2.2.3 Solving

In structural analysis, the solution of the finite element model is a transparent activity. The finite element model in the form of a bulk data deck is submitted to a solver such as NASTRAN or PATRAN/FEA to obtain the desired static stresses. The framework is able to utilize knowledge about available solvers and interrogate the aspects of the mathematical model to select the most efficient solver. This capability was not modeled in the prototype.

2.2.4 Evaluation/Adaptive Refinement

The finite element method of stress analysis is only an approximation to the exact solution of a physical problem. Therefore, the difference between the finite element solution and the exact solution is of paramount importance in structural analysis. Since the exact solution is not known, the difference between the approximate solution and the exact solution cannot be derived. Various techniques exist in the literature (Ref. 1) to approximate the magnitude of this difference on a global or elemental level. Such an approximation can be used to assign a confidence factor to the results and to refine the mesh for a desired confidence factor. This capability is implemented in the framework to preserve the robustness of the results thru a closed loop approach. The framework utilizes a simple error estimator based on a smoothed stress solution (Ref. 13). This technique has been implemented in closed form for a selected class of elements (Ref. 8). This error estimator is converted to a refinement factor and applied in a unique algorithm to refine the mesh for a desired accuracy in the results.

for the design and development of parts and assemblies. The automation of this step is hindered by the lack of integration between product definition data systems and finite element analysis systems. The functional feature description of a part can be

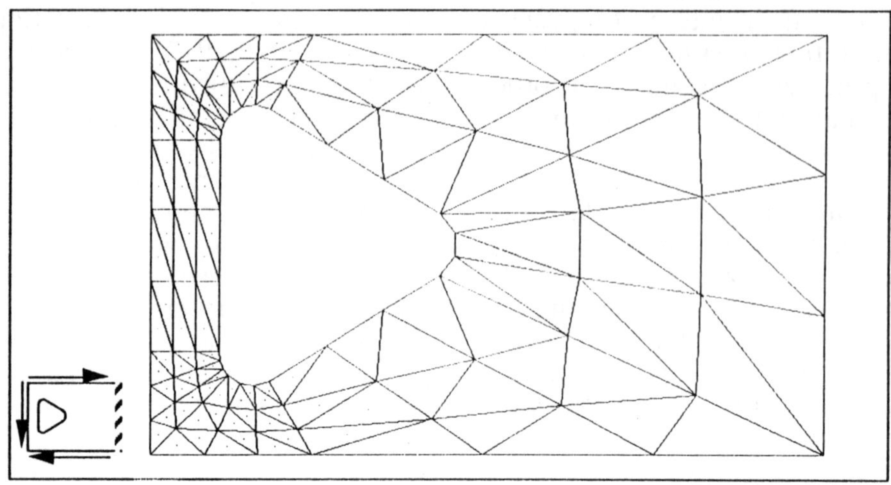

Figure 4. Initial Mesh for the Shear Web with Complex Cutout

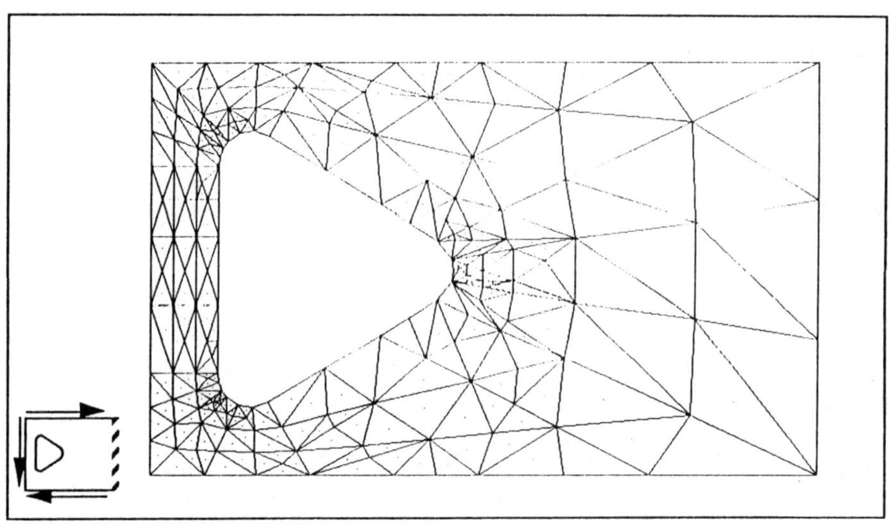

Figure 5. Final Mesh for the Shear Web with Complex Cutout

The finite element model for the complex shear web was analyzed for the prevailing stresses and the error in the energy norms were computed. The refinement factors for a 10% error in the energy norm were then derived. The mesh was refined based on the refinement factor densities and the final mesh is shown in figure 5. This final model is ready for analysis for the desired stresses and displacements.

3. SUMMARY AND FUTURE DIRECTIONS

3.1 Conclusions

The generation of finite element models is one of the most difficult steps in utilizing finite element analysis utilized effectively to provide the required integration.

Traditional methods lack direct correlation between the geometric aspects of a part design and its corresponding finite element model. This lack of correlation makes it difficult to utilize the results of the finite element analysis to improve aspects of the design. The proposed framework provides a strong correlation between the functional aspects of a design and its finite element model.

Existing finite element modeling processes are driven from the geometric representation of a part. The representation is also a subset of the needed design information for the FEM process. As the methods of representation evolve (drawing, wireframe, solids etc.), the FEM process for a particular representation becomes obsolete for the advanced methods of representation. In the proposed method, the FEM process is driven from the product design information and avoids the dependence on the representation. Generating an accurate finite element model determines the success of downstream steps such as analysis and mesh refinement. The proposed framework utilizes the design intent embedded in the functional feature descriptions to provide an accurate and efficient model. The mesh refinement schemes developed in this prototype close the loop and provide a means to optimize the mesh for a desired criteria in the results. The improved accuracy of the initial model will also reduce unnecessary mesh refinement iterations.

Even though, the functional feature based system can only handle the enumeration of features in its library, the permutations and combinations of these features can provide an infinite number of part classifications. Continuing research in this arena will extend the classifications even more.

3.2 Future Research Directions

The effects of part symmetry, boundary conditions and external loads have not been addressed in this dissertation. Part symmetry may be classified into material symmetry and shape symmetry. Part symmetry may be utilized with symmetric boundary conditions and external loading symmetry to reduce the size of the finite element model.

The framework that is proposed may be extended to other aspects of the finite element analysis such as design optimization. Mesh refinement may also be enhanced by reducing redundant elements and moving the nodes towards higher error densities (r refinement). The efficiency of the model may be increased an order of magnitude by an intelligent combination of these techniques. The direct correlation between the design and the finite element model that is captured in the proposed framework will enable more intelligent mesh refinement and design optimization schemes. Finite element based design optimization based on the attributes of the functional features can be considered true optimization compared to conventional finite element model optimization.

REFERENCES

1. Babuska I., Zienkiewicz C., Gago J., Oliveira E. R.; "Accuracy Estimates and Adaptive Refinement in Finite Element Computations"; Wiley, 1986.

2. CLIPS Version 4.2; Artifitial Intelligence Section, Lyndon B. Johnson Space Center, 1989.

3. Cook R. D.; "Concepts and Applications of Finite Element Analysis"; Third Edition, wiley, 1989.

4. Dechter R., Pearl J., "Tree Clustering for Constraint Networks"; Artificial Intelligence, pp 353-365, Volume 38, 1989.

5. Drake S., Sela S., "A Foundation for Features"; Mechanical Engineering, pp 66-73, January 1989.

6. EUCLID Version 2.1; Matra Datavision; 91961 Les Ulis Cadex, France, 1989.

7. Magleby S.; "Design by Functional Feature for Aircraft Structure"; Ph. D Thesis, University of Wisconsin - Madison; 1988.

8. Nambiar R. V.; "Closed Form Expressions for Hierarchic Triangular and Tetrahedral Elements"; Doctoral Dissertation, Univ. of Texas at Arlington, 1989.

9. PATRAN Version 2.3; PDA Engineering, Los Angeles, California 92626, 1989.

10. "Product Definition Exchange Specification working group documents", National Bureau of Standards, United States; 1988.

11. Shah J. J.; Sreevalsan P., Rogers M. T., Billo R., Mathew A.; "Current Status of Features Technology"; Report, Dept. of Mechanical and Aerospace Engineering, Arizona State University; August 1988.

12. Turke D. J., McNeice G. M.; "Guidelines for Selecting Finite Element Grids based on an Optimization Study"; Computers and Structures, Vol. 4, 1974, pp 499-519.

13. Zienkiewicz O. C., Zhu J. Z.; "Effective Adaptive h-p Procedures for Practical Engineering Analysis"; Int. J. Num. Meth. Eng. 24, pp 337-357, 1989.

A KNOWLEDGE BASED SUPPORT SYSTEM FOR FINITE ELEMENT COMPUTER PACKAGES

H. Urakami and T. Sugiyama
Chubu Electric Power Co Inc., Nagoya, Japan

In this paper, we propose a knowledge based support system for finite element computer packages. This system is divided largely into two parts. One is the analysis guide system for selecting an apporopriate program by means of expert system. The other is the support system for making geometric model, and for setting up material propeties. As to explain the developed system, we give an application, which is the compaction settlement analysis of embankment.

INTRODUCTION

For rational design and examination of safty on civil engineering structures, numerical analysis has been used, such as finite element analysis. Numerical analysis is applied, not only to the design of many important structures, but to that of more everything.

We developed or purchased a number of computer programs, finite element method and boundary element method, and these programs have been effectively used on structural design. Such programs have been used by experts in various branch offices of civil engineering. But, only a few problems have existed, when these programs are used by inexperienced persons. One of the problems is that the user can not choose an appropriate program or an analytical method, without the theoretical background of program, from a number of general and special analysis programes available. Another is that the user takes a long time to understand the method to use each program, as the determination of material property and geometric model.

Several studies have been conducted on the development of support system in structural analysis to solve these problems. The support systems for the generation of finite element modeling and mesh were made by Miyoshi,T. [1], and Holt,R.H. [2], Eisenberger,H. [3]. Also a few researchers have studied applications of expert system for consulting of finite element analysis, for example, studies were made by Benete,J.S. [4] and Hukuda,S. [5], Yagawa,G. [6]. Although little research have been undertaken on the support system for finite element analysis on civil engineering structures.

Therefore, the knowledge based support system for finite element program has been devoloped for efficient design of civil engineering structures by present authors. Most of civil engineering structures of electric power companies are soil and concrete structures. In this paper, as the objects of developmment, the soil structures are selected. We give the applications, which are the stress-deformation analysis of embankment and excavation, and seepage-flow analysis.

PROCESS OF NUMERICAL ANALYSIS IN ACTUAL DESIGN

The procedure of computer aided engineering is broken down, in order to clarify the problems of finite element analysis and other numerical analysis in structural design(Fig.1)

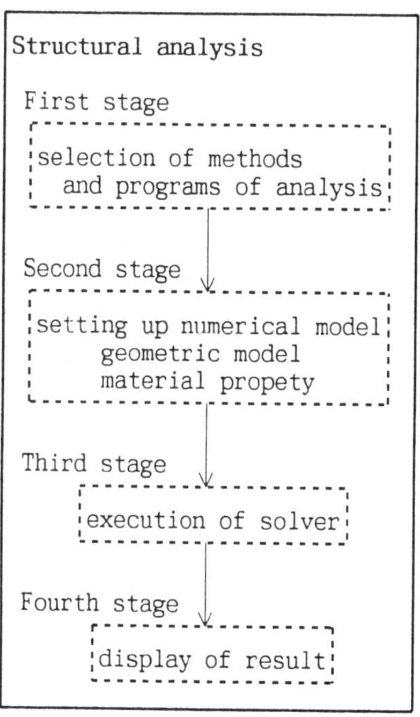

Fig. 1 Process of structural analysis

First stage, a designer makes a list of the items, which should be investigated to design the structure. He selects the pertinent analysis method for each item, and determines the general or special analysis programs suited to the analysis methods selected. The second stage involves the determination of material property data, preparation of the analytical model of the structure, preparation of the geometry data and determination of the boundary and loading conditions. The data are formulated in accordance with the code manual for the analysis program. In the third stage, the commands are fed into the computer using the formulated data and the numerical analysis was performed. The fourth stage, the results are put out on color graphic display or printer, and the designer checks on the results. In all these stages, the designer is required to make high-level decisions, based on expert knowledge and experience, and it takes a long time to reach a stage where one is able to make use of analysis programs with ease. For these stages in numerical analysis, we develop support system through a use of expert system.

Development of user environments that include such expert systems for numerical analysis will be indispensable in order to overcome the problems faced by non-expert users and promote the use of numerical analysis in design.

OUTLINE OF THE PRESENT SYSTEM

This system is being constructed using on an IBM 4381 general-purpose computer at the Electric Power Research & Development Centre of the Chubu Electric Power Co.. Chubu Electric Power Co. has developed and purchased a number of numerical analysis programs on general-purpose computers to date. The application programs to develop this system were formulated through an efficient use of the expert system construction tool (Knowlege-Tool) and automatic mesh preparation and charting program (CAEDS).

On this system, the knowledge base has been created from the knowledge and know-how of experienced engineers and previous analyses in the Research & Development Centre. The knowledge rules consist of production rules and inferences by means of the forward reasoning function. The rules are divided into blocks according to their functions to facilitate verification and maintenance of the knowledge in the system.

The developed support system can be divided largely into two parts. One is the guide system to select the analysis methods and programs for designing structures and investigation items. The other is the input and output support system for the FEM program of ground analysis. These are illustrated in Figure 2.

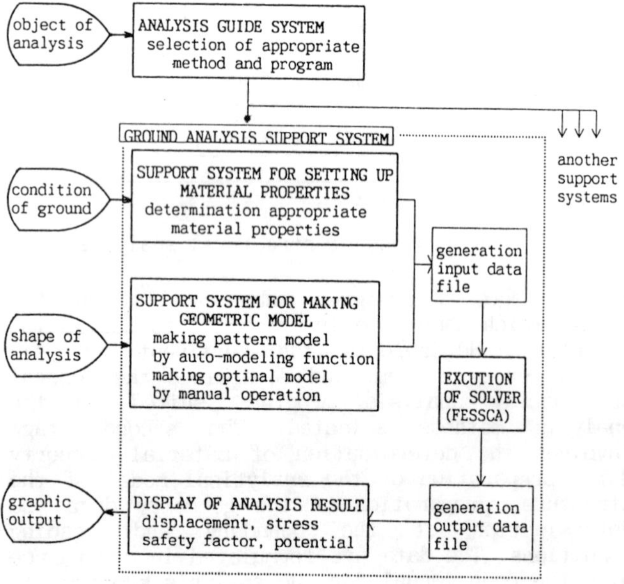

Fig. 2 Structure of the system

ANALYSIS GUIDE SYSTEM

There are 12 analysis programs at Chubu Electric Power Co..

No.1 Program for Ground Displacement Analysis Program Which is Possible to Consider Seepage Flow (FEM) : FESSCA
No.2 Program for Slope Analysis of Stability (FEM)
No.3 Ground-Structure Dynamic Analysis Program(FEM)
No.4 Boundary Element 3D Potential Program
No.5 Boundary Element 3D Thermo-Elasticity Program
No.6 Automatic Dynamic Incremental Non-Linear Analysis (FEM)
No.7 One Dimensional Earthquake Response Analysis
No.8 Two Dimentional Elasticity Finite Element Method Program
No.9 Dynamic Analysis of Non-Homogenous Ground Movements by the Boundary Integral Equation Finite Element Hybrid Method
No.10 A Structural Analysis Program for Earthworks Considering a Transformation of Structure System (FEM)
No.11 A Excuvate Analysis Program for Non-Linear Calculation of Relaxation Factor (FEM)
No.12 Stability Analysis of Soil-Structure Interaction

Each program has its own characteristics. The expert selected and used the program which is best suited for the particular problem. The analysis guide system on the basis of expert knowledge and experience makes decisions on the analysis method and program which is appropriate for the problem.

The knowledge base has a tree structure made up of analysis types, structures analysed, analysis methods and programs. The tree structure for the analysis of foundation ground is given in Figure 3 as an example.

On this system, the user can select the appropriate analysis method and program through dialogue.

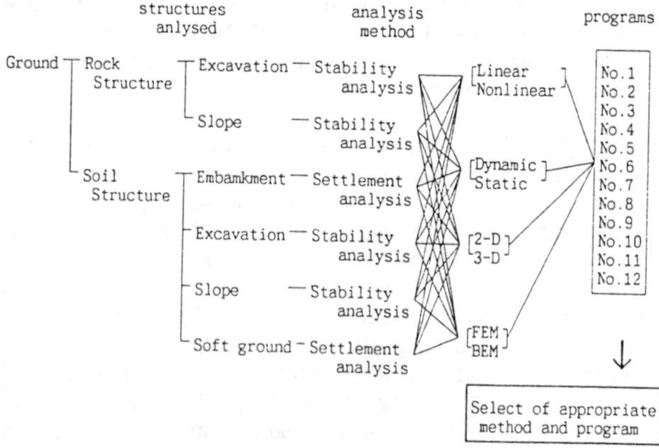

Fig. 3 Tree structure for the analysis of ground

SUPPORT SYSTEM FOR GROUND ANALYSIS PROGRAM

This prototype of support system has been created with the aim of preparation of the input data and display of the analysis results. The analysis program which the system intended to support is normally called FESSCA , which is the analysis program for grand deformation taking account of seepage flow. This analysis program has been developed in the Chubu Electric Power Co.. The principal types of analysis for which FESSCA can be used are listed below.

1. Compaction settlement analysis for embankment work taking account of construction procedure
2. Stability analysis of slopes subjected to cutting and caverns
3. Stability analysis of excavation surfaces, for example, in cofferdam construction
4. Seepage analysis for investigation on uplift underneath dams

Support system for setting up material properties

This system aims at reduction of the work for determining material properties, reduction of the work for data cording, structuralisation of expert knowledge and storage of know-how.We made the knowledge base for determination of material properties. The knowledge base consiste of items of knowledge structuralised as shown in figure 4, on the basis of opinions given by experts on methods of determining material properties. Therefore, it is easy to verification and maintenance of the knowledge.

The system is adopted database of material properties by arranging the following items in tables.

1. Casagrande soil classification
2. Relationship between elasticity coefficient and N-value of soil
3. Uniform classification of soil and sandy gravel materials and relationship between various properties
4. Approximate values for permeability coefficients
5. Critical flow velocities corresponding to particle sizes (Justin's equation)

This system makes use of interactive processing. The user inputs the conditions of analysis, the conditions of construction, and the conditions of ground, following the input instructions that appear on the display of the computer. Examples of the analysis conditions are the object of the analysis, the analysis type, and the analysis grade. Examples of the construction conditions are the details of the construction work and loading steps for embankment and excavation. And examples of the ground conditions are the soil properties obtained from the boring data and the groundwater levels. The system makes inferences on the material properties from the conditions inputted.

In Figure 5 examples is given for the process of inference, which are modulus of deformation and coefficient of permeability. The inference results are then displayed and the data can be corrected directly, if need be.

Support system for making geometric model

We considered that the finite element models of soil structures follows several patterns, when the analyzed cases were tidied up. Examples of cases were the finite element models under embankment and excavation. This is illustrated in three patterns in Figure 6.

Using the patterned model in the computer and auto-modeling function in the CAEDS, this system makes possible automatic preparation of the numerical models. This system takes in knowledge concerning procedures for using the CAEDS and that concerning preparation of the finite element models, such as determination of analysis regions and coordinates. The system is also capable of taking in the material properties inferred. With this system, the user can prepare the basic configuration of the analysis model by inputting the numerical data while referring to the predetermined default values that appear on the screen.

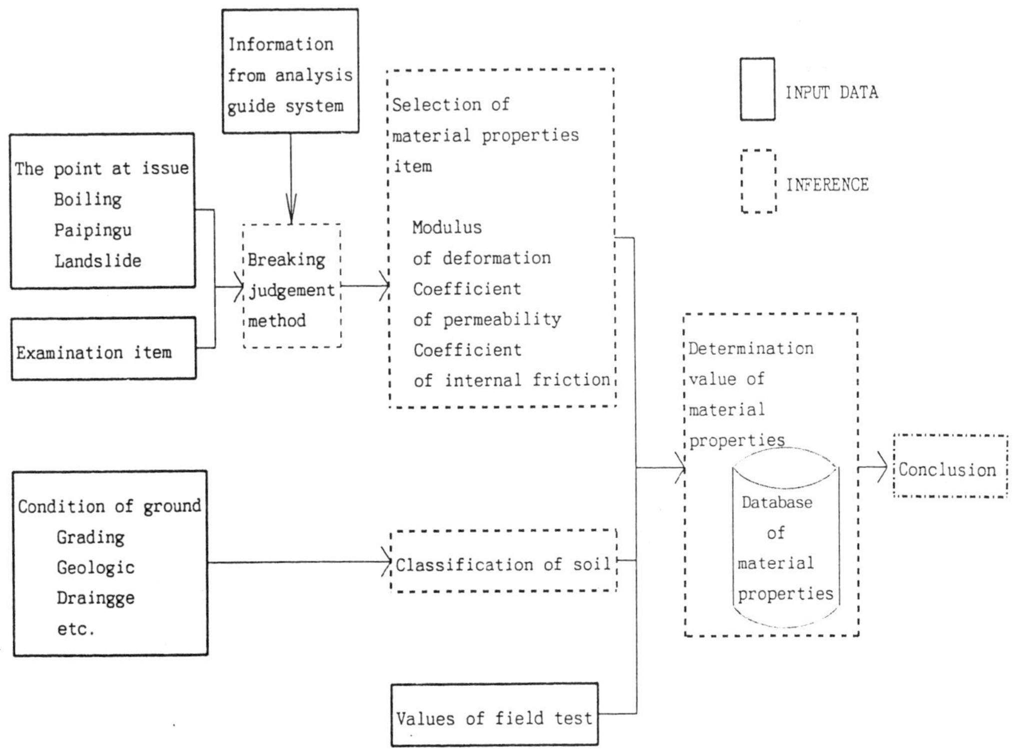

Fig. 4 Structure of knowledge

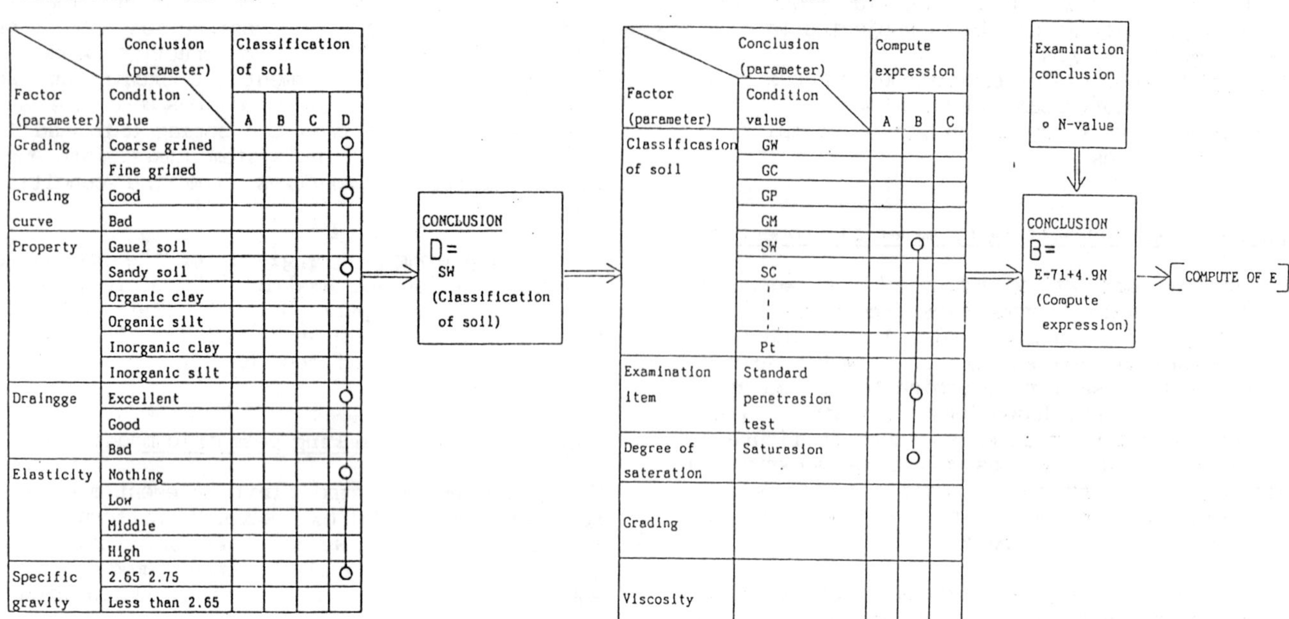

(Modulus of deformation)

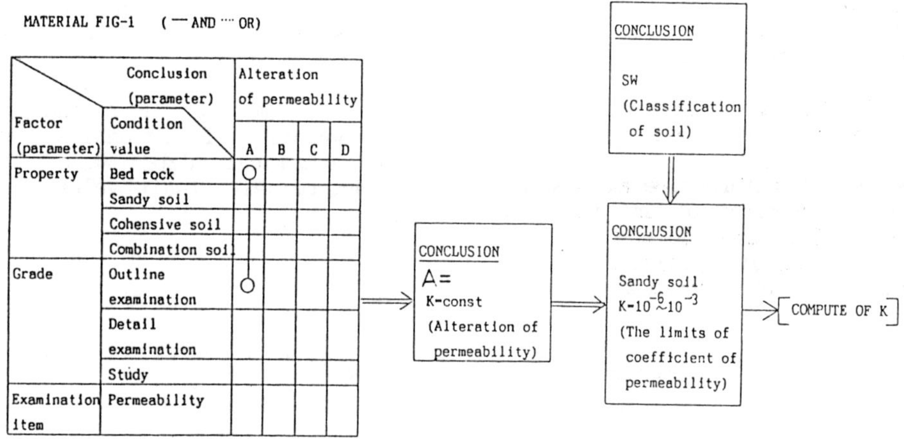

(coefficient of permeability)

Fig. 5 Determination method of material properties

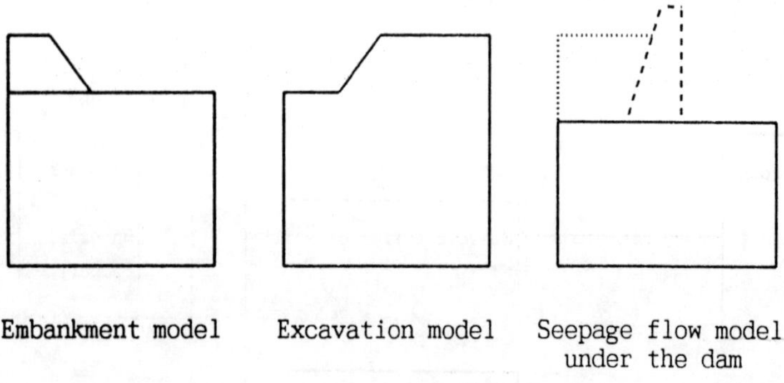

Fig. 6 Patterns of geometric model

Execution of analysis

The input data for finite element analysis, prepared using the support systems, are rewritten in formats peculiar to the analysis program package, and the input data file is created. The input data file is then transmitted to the main processor and computation automatically begins. The output data is converted automatically into the format for the graphic program, and the analysis result data file is created. The data file created is saved as an analysis case file for accumulation of know-how.

Graphic output of analysed result

In order to have the analysed results reflected in the design, there must be a function for having the results displayed accurately and speedily in a form that can easily be understood. This system makes a graphic output of analysis results on items such as displacement, stress, water pressure and flow velocity.

EXAMPLE USE OF THE SUPPORT SYSTEM

An operation for the analysis on embankment work is given below, as an example of the use of this system. The solver is selected using the analysis guide system(Fig. 7).

FESSCA is selected as the analysis program and we now move on to the support system for setting up material properties. The analysis, construction and ground conditions are inputted into the support system for setting up material properties, which makes inferences and determines the material properties(Fig. 8).

On the support system for making a finite element model, the region subjected to the analysis of settlement due to the embankment work and the configuration of the embankment are inputted and the analysis model and mesh data prepared by the auto-modelling function(Fig. 9).

Confirmation is then made on the connection between the physical property and configuration data, after which the data are converted into the FESSCA format and analysis is implemented. As results of the settlement analysis, drawings of the stress distribution and displacement are displayed on the display of the computer(Fig. 10).

CONSIDERRATIONS

The support system developed allows a user who is not an expert in analysis to carry out embankment settlement, excavation stability and seepage flow analyses in short periods of time (1 to 2 hours). The problems and points of note observed in developing the system are discussed below.

```
**** Analysis Guide System ****
------ Structure analysed -----

  ==>

        1 - Slope
        2 - Embankment
        3 - Excavation
        4 - Soft ground
        5 - the others
```

(input object)

```
**** Analysis Guide System ****
------ Apporopriate solver ----

    Consided next programs

        1 - FESSCA
        2 - FEM1
        3 - SUPPER-FLUSH
        4 - SLOPAN
```

(result of inference)

Fig. 7 Sample of analysis guide system

```
- Stratum No. 1 -- Parameter of Inference -
   Parameter : Property of Ground (B)

        bentonite              -> 0
        organic clay           -> 0
        organic silt           -> 0
        inorganic clay         -> 0
        inorganic silt         -> 0
        ice clay               -> 0
        ice silt               -> 0
        gravelly soil          -> 0
        normal soil            -> 0
        sandy clay             -> 1
        clay                   -> 0
        silt                   -> 0
        clayey soil            -> 0
        combination soil       -> 0
```

(input condition)

```
Stratum No. 1 -- Inferential results
Linear analysis(unit : MKS)
  Young's modulus          => 1.1798 E+06
  Poisson's ratio          => 3.0000 E-01
  Unit weight              => 2.6499 E+03
  Initial void ratio       => 3.7000 E-01
  Yield cohesion           => 4.0000 E+04
  Breaking angle of        => 3.8000 E+01
   internal friction
  Coefficient              => 3.1599 E-04
   of permeability
  Fixed number A           => 2.0000 E-01
  Fixed number B           => 1.0000 E-01
```

(result of inference)

Fig. 8 Sample of support system for setting up material properties

Planning of the system

Selection of the program which the system is intended to support and the range of its application are the most basic points in construction of the expert systems. The knowledge structure required in numerical analysis for design work is extremely complex and there have only been few support systems developed so far. In view of these limitations, a decision was made to develop a prototype, selecting just one program, which is the one most frequently used at Chubu Electric Power Co.. As for the range of its application, typical analysis problems were selected out of the cases of analysis implemented on the program.

Determination of the program to be supported and the range of application of the system requires preliminary studies and much time is required for investigations. In determining the program to be supported, one needs to conduct studies on the length of the program source, input and output formats and adaptability of the program, as well as rearrange the code manual. As regards the application range, one needs to select problems for which there are ample past cases and ensure that experts who have conducted analysis on the problem are available for questioning.

Development of the system

In developing the system, one needs always to bear in mind the possibility of expanding the system. "Prototyping" applied in the development of the system discussed above is a procedure typically used in the development of expert systems. The system which begins its life with a limited application range is constantly evaluated and adjusted, so as to expand its application range until it can be applied to practical use. This "prototyping" has been found effective in development of systems for which the structure of the problems involved are unclear.

Collection and arrangement of knowledge

Collection and arrangement of knowledge and know-how were the most laborious parts of the work involved in developing the system. As mentioned above, FEM and similar numerical analyses depend heavily on the experience of experts. For the purpose of developing the system described above, the thought processes of the experts were examined through interviews using typical questions. This is considered an effective method for analysis work that involves strong experiential elements. Further time was

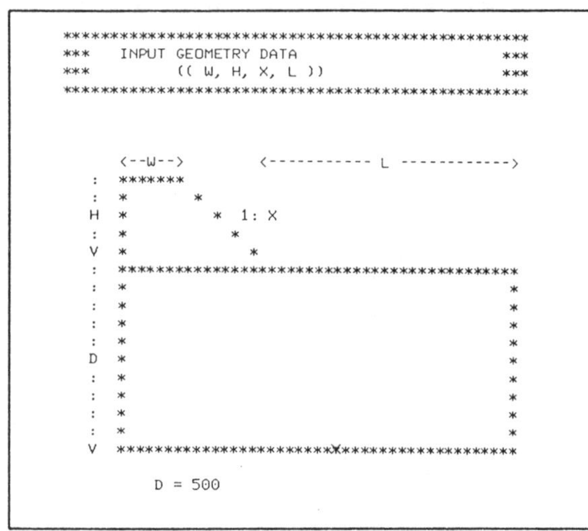

(input geometry data)

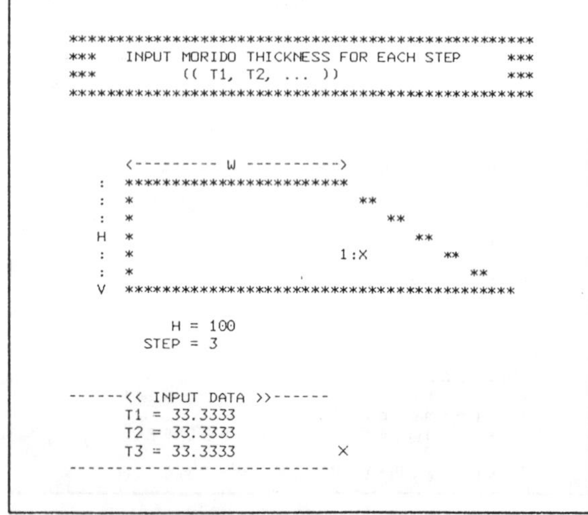

(input embankment thickness for each step)

Fig. 9 Sample of support system for making geometric model

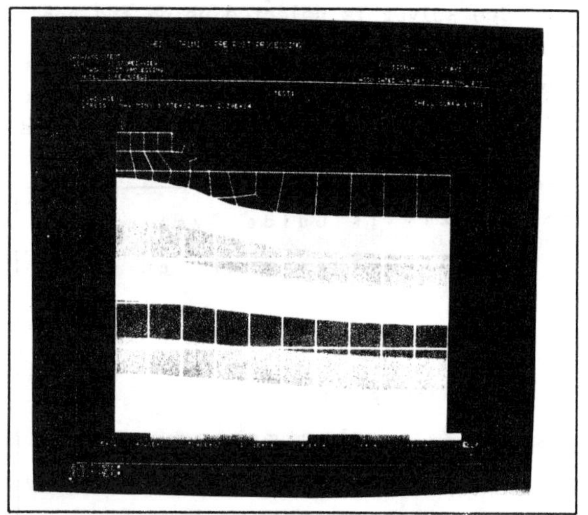

(result of stress)

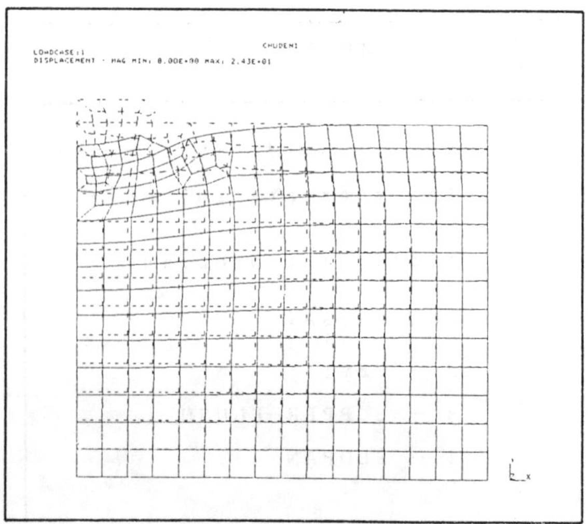

(result of displacement)

Fig. 10 Sample of display of analysis result

required in collection and arrangement of knowledge and know-how, because of the lack of understanding over technical terms between the experts and the system engineers. An important task of the system engineers is to have an understanding of the work in question.

Circumstances of development

Since most of the numerical analysis programs owned by Chubu Electric Power Co. operate on general-purpose computers, the support system too was developed on a general-purpose computer. Also,in consideration of that the system will be large system in the future, the system are developed on a general-purpose computer, which is full disc capacities. With regard to the software, linkage of the expert system to the picture processing programs and conventional control-driven programs is also an important point.

CONCLUSION

This paper presents the support system for structural analysis. The system is used patterns of geometric model, auto-modeling function, and determination material properties through a use of the expert system. The system helps the user to efficiently execute finite element programs. As the system described above sees wider application, we hope to improve the system to facilitate its use and to expand the range of analysis, as well as to develope new programs. While the rapid improvements being made in the performance of hardware is expected to cause numerical analysis to be used increasingly in various fields of research and design and eventually to replace model tests, the gap that exists between the tools used (numerical analysis programs) and human capacity will continue to necessitate the development of expert systems for interface between the two.

ACKNOWLEDGMENT

The authers wish to express heartful thanks to Mr.Toyoji Higashigawa, Chubu Electric Power Co.,who gave us the chance to research this subject, and Mr. Shirou Kato, Toyohashi University of Technology, who gave us many precious suggestions to describe the paper. The authers also thank Mr. Suzuki, Chuden Construction Co., who give us great knowledge about analysis programs, and also thank IBM Japan Ltd. and NS&I Ltd..

REFERENCES

1. Miyoshi, T. and Yoshida, Y., Syudy of finite element mesh generation method through a use of expert system, Proc.67th J.S.M.E., 149-150, 1989.

2. Holt, R.H. and Narayana, U.V.L., Adding intelligence to finite element modeling, Expert System Gov. Symp., 326-337, 1986.

3. Eisenberger, M., Finite element translator and critic for design system, Development in structual engineering, 1348-1359, 1990.

4. Bennett, J.S., Creary, L., Engelmore, R. and Melosh, P., SACON : A knowledge-base consultant for structural analysis, Proc. 6th I.J.C.A.I., 1979.

5. Hukuda, S., Nakagiri, S. and Ohkubo, M., Development of finite element analysis advisor expert systems, J.S.M.E., Vol.56, No.521, 168-172, 1990.

6. Yagawa, G. and Yoshimura, S., Some applications of knowlege engineering to computational mechanics, Jour.J.S.M.E., Vol.92, No.847, 1989.

AN EXPERT SYSTEM FOR THE EFFICIENT NUMBERING OF FINITE ELEMENT NODES

D.Givoli†, L.Manevitz‡ and M. Margi‡

†Department of Aerospace Engineering, Technion, Haifa, Israel,

‡Department of Mathematics and Computer Science, University of Haifa, Haifa, Israel

In this paper we describe the design and implementation of a preliminary (MARK I) version of an expert system, employed in numbering the nodes of a finite element mesh in such a way as to minimize the average bandwidth of the stiffness matrix. The numbering procedure is heuristic and is designed to mimic the methods of a human expert numberer. We demonstrate the performance of the expert system by means of a two-dimensional example.

INTRODUCTION

The finite element method (FEM) has become during the last three decades a major and very widely used tool for the numerical solution of partial differential equations. This is especially (but not exclusively) true in structural engineering and solid continuum mechanics. The finite element procedure has been proved most effective when applied to linear boundary value problems in spatial domains which are complicated geometrically. (See e.g. Hughes [1].)

It is well appreciated among finite element users in industrial and scientific communities that the successful application of this technique requires a fair amount of experience and expertise in order to make the computation feasible and the results accurate at the same time. This is true with regard to any of the large commercial finite element packages available currently, although various codes may have a different amount of flexibility. The user expertise is needed because applying the FEM to a problem requires making choices, such as the choice of mesh and the type of elements to be used, for which there are no known effective algorithms that can replace a human user in all cases.

In this paper the methodology of expert systems is applied to a process in this category: *the optimal numbering of nodes*. Over the last twenty years, expert system methodology has emerged as a way to automate non-algorithmic human expertise (see e.g. [2] and [3]). Many such systems have been implemented in various fields of application, as described e.g. in [4–9].

The system developed here is what we choose to call MARK I, following [2]. Namely, it is a preliminary design and implementation meant to work only on a selected subset of problems in order to establish the feasibility of the idea and to provide a laboratory for testing the heuristics extracted from the user. Hence, the results obtained are preliminary, albeit promising for relatively simple meshes chosen by us for testing. In addition, a major example of a mesh which has been used in real applications [10] was tested and the results, presented in this paper, compared favorably with those of a human expert.

PROBLEM DESCRIPTION

When applied to linear boundary value problems, the finite element discretization of the governing partial differential equations leads finally to a linear system of algebraic equations (see Hughes [1]),

$$\boldsymbol{K}\boldsymbol{d} = \boldsymbol{F} \ . \tag{1}$$

Here $\boldsymbol{K}$ is the so-called *stiffness matrix*, $\boldsymbol{F}$ is the load vector, and $\boldsymbol{d}$ is the vector of unknowns. The dimension of the system (1) in typical industrial applications is of order 100 or 1000 or even 10000. The computational effort involved in solving such a large system is reduced considerably owing to an important property which the matrix $\boldsymbol{K}$ possesses: it is *sparse*, i.e. most entries K_{ij} are zero. If the original partial differential equation being solved by the FEM is self-adjoint then $\boldsymbol{K}$ is also *symmetric*. In fact, even for problems which are not self-adjoint (where $\boldsymbol{K}$ is non-symmetric) the sparseness structure of the matrix is symmetric, that is the locations of the nonzero entries form a symmetric pattern with respect to the diagonal (see Irons and Ahmad [11]).

The degree of sparseness of $\boldsymbol{K}$ depends on the specific finite element discretization of the spatial domain, or more explicitly on the *finite element mesh*. The spatial domain is divided into *elements*, which are connected to each other by *nodes*. In two dimensions, the elements may be quadrilateral or triangular, and the nodes are their vertices. Points on the edges and in the interior of the elements may also be defined as nodes. Depending on the physical problem under consideration a number of degrees of freedom may be defined at each node, but for the sake of simplicity (and without any loss of generality) we shall assume a single degree of freedom per node.

Two nodes are said to *interact* if they both belong to a common element in the mesh. If there is a total number of N nodes in the mesh and they are numbered from 1 to N, then $K_{ij} \neq 0$ only if nodes i and j interact. Thus the structure (or skyline; see below) of the stiffness matrix is determined by the node interaction. Since the mesh is typically imposed on a physical domain and thus can be

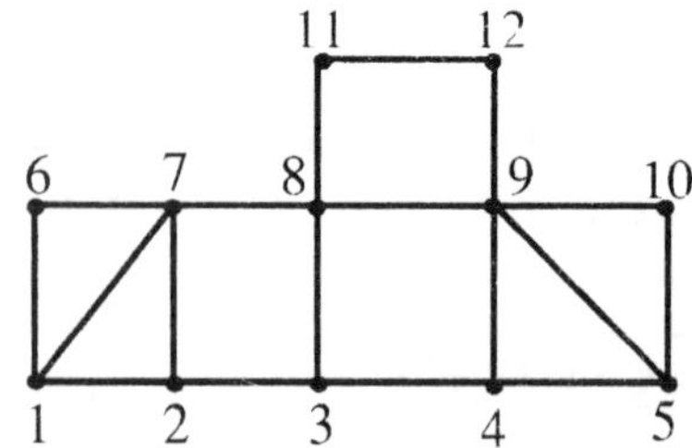

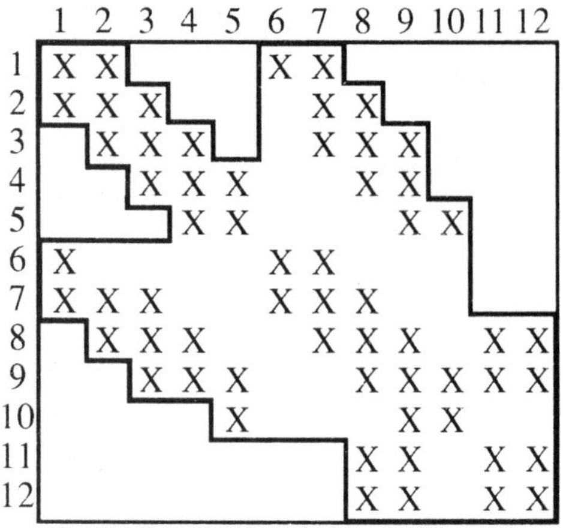

Fig. 1(a)

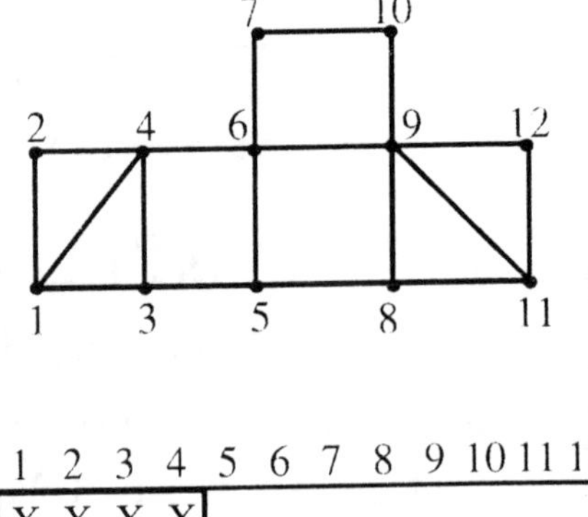

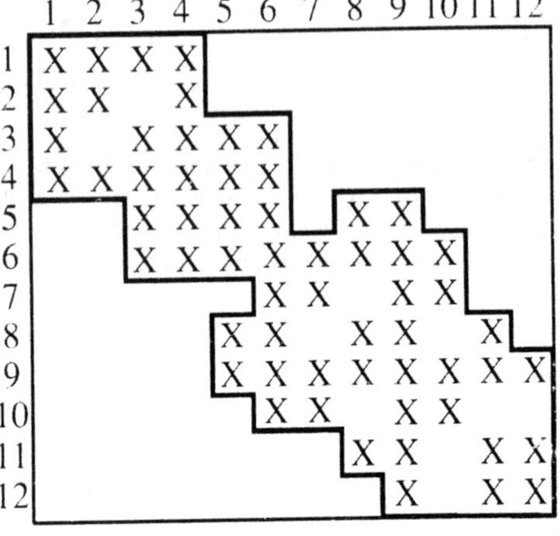

Fig. 1(b)

thought of as a planar or spatial graph, most K_{ij} are zero, i.e. $\boldsymbol{K}$ is sparse.

The discussion above implies that the order in which the nodes are numbered is crucial, since this numbering determines how sparse $\boldsymbol{K}$ is. The present paper attacks the problem of finding a good numbering system. Figure 1 shows an example of a finite element mesh numbered in two different ways and the structure of the two stiffness matrices $\boldsymbol{K}$ corresponding to these numbering systems. Each X denotes a nonzero entry. Also illustrated in the figure are the upper and lower *skylines* of $\boldsymbol{K}$. The upper skyline is the border indicating the highest nonzero entry in each column, whereas the lower skyline indicates the leftmost nonzero entry in each row.

The linear solver used in many finite element programs to solve the system (1) employs the *skyline method* of storage, in which only the portion of $\boldsymbol{K}$ between the skylines, called the *band*, is stored and operated on during the solution process. If $\boldsymbol{K}$ is symmetric, only the portion between the diagonal and the upper skyline need be stored. One should note, however, that any zero entries *within* the band are stored since they may become nonzero during the solution process (the "fill-in" effect; see Johnson [12]).

The description above is equally relevant to skyline (band) solvers based on direct Gauss elimination and to skyline solvers which use some iterative procedure (see [12]). In both cases the mesh nodes should be numbered such as to minimize the band area. This would ensure a minimal amount of storage requirement and floating point operations. Another way to state the same goal is that the nodes should be numbered so that in average, the maximal difference of node numbers in each element would be minimal.

An alternative to the skyline scheme is the *frontal scheme*, in which the stiffness matrix $\boldsymbol{K}$ is assembled simultaneously with the solution process. In this method it is the numbering order of the elements, not the nodes, which is important. However, a procedure for element numbering which has proved effective in practice is to first number the nodes in an optimal manner, as in the skyline method, and then to number the elements in ascending sequence according to their lowest node number; see e.g. Sloan and Randolph [13]. Thus, the necessity for an optimal nodal numbering system is common to the skyline (band) and frontal schemes. For a discussion and comparison of these two methods see Irons and Ahmad [11] and George and Liu [14].

In order to state the optimization problem mathematically we introduce some notation. Let the dimension of the matrix $\boldsymbol{K}$ be $N \times N$. Let b_i be defined as

$$b_i = i + 1 - \max_{K_{ji} \neq 0} j \ . \tag{2}$$

In other words, b_i is the height of column i starting from the diagonal and up to the skyline. Now, let the *average half bandwidth* AHB and the *root-mean-square bandwidth* RMSB be defined as

$$\text{AHB} = \frac{1}{N} \sum_{i=1}^{N} b_i \ , \tag{3}$$

$$\text{RMSB} = \left(\frac{1}{N} \sum_{i=1}^{N} b_i^2 \right)^{1/2} \quad . \tag{4}$$

The column heights b_i for the two numbering systems of the model shown in Figure 1, as well as the corresponding AHB and RMSB values, are given in Table 1. It is apparent that the bandwidth depends on the nodal numbering system. This dependence becomes strong for large meshes.

It can be shown that a complete factorization of $\boldsymbol{K}$ using the skyline or frontal schemes requires $O(\sum_{i=1}^{N} b_i^2)$ floating point operations, namely the *computing time* needed is determined by RMSB. From the definition of b_i it is clear that the *storage* requirement is $\sum_{i=1}^{N} b_i$ for symmetric matrices, and hence it is determined by AHB. The optimization problem under consideration can thus be stated as follows: Find a numbering system for the nodes (from 1 to N) such that AHB (or RMSB) will be *minimal*.

PREVIOUS METHODS AND THE EXPERT SYSTEM APPROACH

For a mesh containing a large number of nodes and elements it is of course impractical to find an exact solution for the optimization problem by performing a full enumeration. Various algorithms have been proposed and been incorporated in commercial finite element codes for obtaining a nearly-optimal solution. These include the methods described by Cuthill and McKee [15], King [16], Collins [17], Akin and Pardue [18], Gibbs [19], Razzaque [20], Pina [21], Sloan and Randolph [13], Fenves and Law [22] and Sloan [23]. The implementation, testing and comparisons of these algorithms were reported e.g. by Everstine [24], Lewis [25], Armstrong [26], Sloan [27] and Duff et al. [28].

In practice, the user always provides the finite element program with an *initial list* of node numbers and coordinates and with the *element connectivity*, i.e. the node numbers associated with each element. These geometric and topological characteristics of the mesh are either provided directly by the user or sometimes are the product of an automatic mesh generation program which is run as a preprocessor. The node ordering algorithm receives as input the element connectivity and generates as output a new list of node numbers, which hopefully has a low AHB and RMSB associated with it. Some of the reordering algorithms work in an iterative fashion to gradually improve the initial list of nodes. In this case the end result depends on (although it is not necessarily very sensitive to) the initial numbering system. Other algorithms, based on graph theory, regard the initial node numbers merely as symbols, and generate a final list which is independent on the initial one.

Here an *expert system* approach is used, based on artificial intelligence principles, for obtaining a good nodal numbering system. This is in contrast to the algorithmic approach adopted in all the previous works mentioned above. Although we are still not in a position to say that the expert system approach is consistently superior in performance to the algorithmic approach, there seem to be a number of reasons why the former has a promising potential of being very effective:

1. None of the algorithms mentioned previously will give exact optimal results in general. In fact, there are theoretical reasons to believe that no such algorithm exists. Hence the finite element programs rely on algorithms which give what is hoped to be "reasonably good" results rather than optimal ones. Since no optimal solution can be hoped for, trying to mimic heuristic expertise may be useful. Typically, in situation such as this it is appropriate to apply the technology of expert systems.

2. For large meshes with simple mesh topologies, e.g. large rectangular grids, the way to number the nodes in an optimal way is usually very obvious to a human user, while in the best case the reordering algorithm would generate the same result after a significant amount of computational work. The reason is that a human user *sees* the mesh. This suggests the use of AI tools, possibly with pattern recognition capabilities.

3. An experienced finite element analyzer who compares the AHB results obtained by manual node numbering and by the use of an automatic procedure, even for complicated mesh topologies, would come to the conclusion that the algorithm rarely produces results which are significantly better than the manual ones. Thus, the main role of current automatic procedures is to *replace* the user in the tedious task of numbering the nodes in a good manner, not so much as to *outperform* him/her. Another role is to improve the node numbering system produced by an automatic mesh generator, which in general may be quite poor. It is reasonable to expect that an expert system would do a better job in replacing a human user than a code based on an algorithmic approach.

Case	$i =$	1	2	3	4	5	6	7	8	9	10	11	12	AHB	RMSB
(a)	$b_i =$	1	2	2	2	2	6	7	7	7	6	4	5	4.25	4.80
(b)	$b_i =$	1	2	3	4	3	4	2	4	5	5	4	4	3.42	3.62

Table 1

4. Sometimes there is in an advantage in having a node numbering scheme which is simple and systematic, since this may give the user a fast and convenient access to reading the output and modifying the input related to the physical problem being analyzed by the FEM. The node numbering system generated by a reordering algorithm is typically not simple, and is characterized by some amount of winding. Therefore, some finite element packages translate the node numbers back to their original counterparts at the finite element output stage. When an expert system is used there is no need for such translation, since the system provides a numbering scheme which is simple and systematic.

Following these indications, a prototype of an expert system is being developed which is designed to follow the numbering heuristics of a human expert. This paper reports on the description and results of the first stage in the construction of the system. For this first MARK I version we limited ourselves to relatively simple situations, as explained in detail in the next section.

DESCRIPTION OF THE EXPERT SYSTEM CHARACTERISTICS

One of the hardest steps in producing an expert system is to construct the heuristics according to which the expert system will perform, and which will mimic the considerations of a human expert. To this end, we note the following functions typically performed when numbering the nodes of a finite element mesh manually.

0. Preparation of numbering strategies for "simple blocks" with different geometries and topologies. Varying parameters of freedom are associated with each strategy. These strategies are achieved by experience, by trial and error, and sometimes by full analysis.

1. Subdivision of the mesh into a disjoint union of simple blocks.

2. Choosing the order in which these blocks are picked for numbering.

3. For each block, choosing a numbering strategy, depending on the topology and the geometry of the block. That is, choosing the best match to a paradigm simple block.

4. For each block, determining the free parameters associated with the strategy chosen for that block, such as the node from which to start the numbering.

We now comment on these steps.

Step 0

This step is not "performed" for each new mesh separately, but as a continuous mental process in which knowledge is being accumulated as the user experiences with an increasing number of examples. By "simple blocks" we mean geometrically and topologically simple meshes, e.g. in two dimensions pseudo-rectangular meshes, pseudo-annular meshes and pseudo-disc meshes. See Figure 2 for a standard numbering method of the first two of these types of meshes.

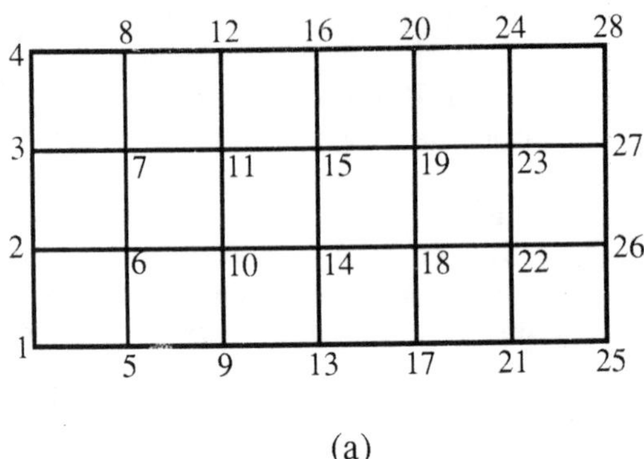

(a)

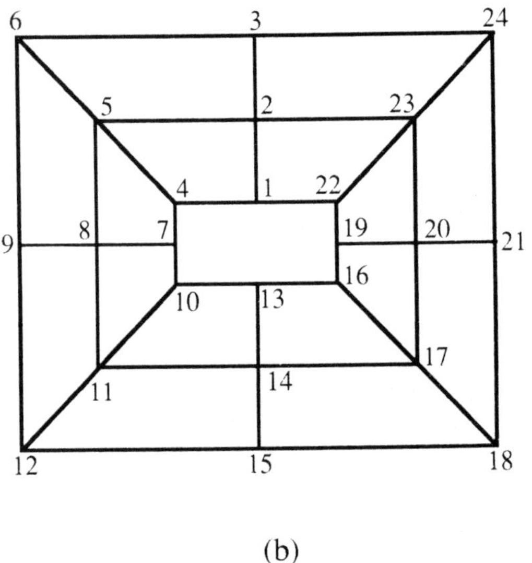

(b)

Fig. 2

Probably the most obvious example is that of a rectangle with regular net. In this case one numbers the nodes lexicographically by coordinates, with the coordinate in the shorter direction being the major coordinate, as shown in Figure 2. This leaves the corner where the numbering starts as a free parameter, to be determined in step 4. However, when the simple block under consideration is not precisely aligned as a rectangle, one usually still wishes to number the nodes across the shorter dimension. This strategy is demonstrated in the example of Figure 1(b) and Table 1. In this more general case, the expert system has to decide in which direction to proceed. The heuristics for such decision are described below, in the remarks to step 3.

Step 1

In this step, the mesh is subdivided into simple blocks, for each of which an appropriate strategy can be chosen from the strategy collection. It is interesting and crucial to note that while the mesh is a purely topological object as far as node numbering is concerned, the user decides on the division based not only on topological criteria, but also on *geometric* criteria. This is an obvious logical fallacy and it is easy to devise artificial mesh examples where the results are extremely bad. Nonetheless for real meshes the results seem acceptable.

The reason why human users can be so successful in numbering meshes although basing their decisions on geometric considerations is that they intuitively rely on the fact that *actual meshes are designed with implicit geometric constraints.* For example, a good practice in mesh design is to use elements which have aspect-ratios close to unity, namely quadrilaterals and triangles which are nearly equilateral. Another good practice is to pass from a crude region (large elements) to a refined region (small elements) in a gradual manner. (See e.g. Irons and Ahmad [11].)

Our expert system tries to mimic a human user; therefore it will make geometric considerations. This is in contrast to the algorithmic approach, where only the topological properties of the mesh play a part in the node reordering procedure. Notwithstanding, topological considerations are still important for the human user. For example, consider a pseudo-disc block and a pseudo-annular block. Interestingly, the numbering methods for these two blocks are usually chosen differently. Since the methods must be in some sense insensitive to small perturbations, this is a case where the topological information seems very important. Even in this case, however, the geometric relative size of the hole plays a part in the strategy of the pseudo-annular block.

The division into simple blocks (step 1) can be performed in principle either by algorithmic methods of computational geometry or by an artificial intelligence-based pattern recognition scheme. This step was not incorporated into the MARK I version; the present work concentrated on steps 2–4. Thus, MARK I receives the division into blocks as an input. One should note that this setup is not so artificial as it might seem, since subdivision into blocks is usually needed already in the mesh generation stage, which precedes the node numbering stage.

Step 2

In this step the order of the simple blocks is determined. To optimize the numbering a user tries to keep the node numbers as continuous as possible. However when passing from block to block this is often impossible. Thus the goal is to keep the "jumps" across block interfaces as small as possible. In the discussion that follows, a *component* is a topological component, i.e. a maximal connected sub-body.

In the MARK I system the procedure for block ordering is as follows:

- Choose the first block to be numbered as the largest simple block, based on the number of nodes.
- Remove this simple block from the mesh. (This may cause the remaining mesh to have several disconnected components.)
- Place all resulting separate components in increasing order in a stack.
- Until the stack is empty do the following:
  - Remove the smallest component from the stack.
  - From the chosen component, choose the next simple block to be numbered in the following way:
    - Consider all simple blocks in the component with an interface with the previously numbered block (there are always such).
    - Choose the one with the largest interface.
  - Remove the chosen block from the component. (This may divide the remaining blocks into several connected components.)
  - Place the new components on the stack in increasing order.

Figure 3 demonstrates this process by showing a sequence of "snapshots" for a typical example. Roughly speaking, the heuristic is to try to number the blocks continuously with small "glitches" along the way. This is achieved primarily by preferring smaller components and secondarily by considering the size of the interfaces between neighboring components.

Step 3

In this step the numbering strategy is chosen for each block. Essentially it has two parts: (1) solution of a pattern recognition problem, i.e. given a simple block find which of the paradigm simple blocks it is closest to; (2) after identification, choosing a numbering method for the block from amongst the possible methods mentioned in step 0. In general, the choice of the strategy may be context-determined, e.g. may depend on the dimensions of the block, the interfaces it has with its neighboring blocks, etc. In the MARK I version, all the blocks are assumed to be pseudo-rectangles and there is only one numbering strategy for numbering rectangles, so this step is to a large extent vacuous.

Notwithstanding, it remains for the system to perform the following analysis: (a) to decide on the orientation of the block, regarded as having a rectangular shape; (b) to be able to ignore small perturbations. The first task is done in the following fashion:

- Construct the *convex hull* of the block. This is the smallest convex polygon enclosing the block.
- Find by regression (i.e. best curve-fit by least squares) the slope of the line that fits the vertices of the convex hull.

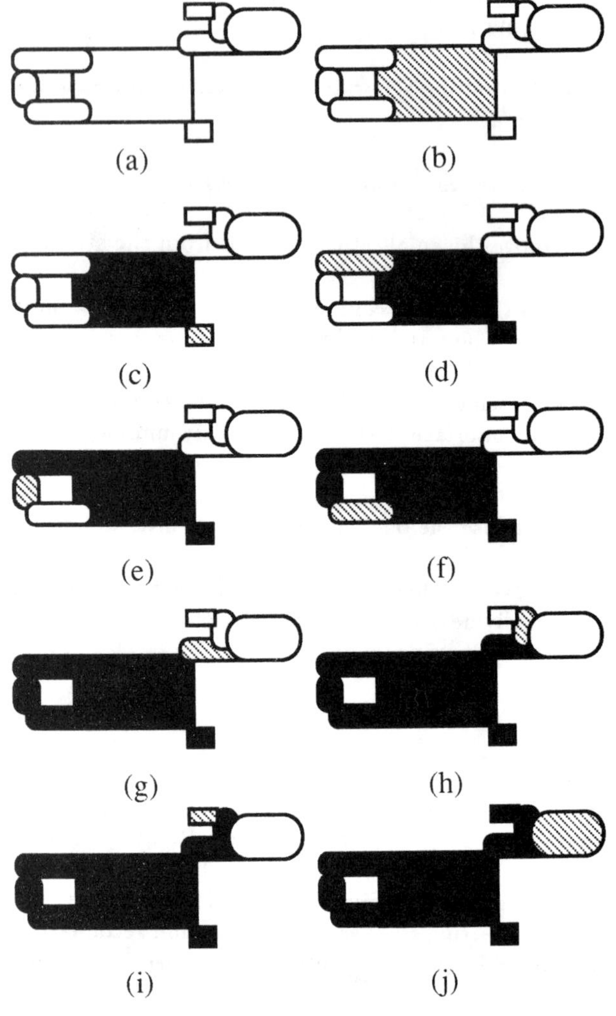

Fig. 3

- Rotate the block rigidly so that this line ends parallel to the x axis.
- Compute the height and width by taking the maximal differences in the x and y directions.

Figures 4(a)–(d) demonstrate this process. The numbering of the block in Figure 4(d) is finally performed in the "natural" order for a rectangle, i.e. across the smaller dimension.

We now comment on the second task mentioned above, i.e. that of detecting small perturbations in order to ignore them. Consider for example the simple block shown in Figure 5(a) after rotation. The order of numbering desired is ABCDEF. However, lexicographic ordering (by $y-x$ coordinates) does not work because of the small perturbations of nodes B and E from the vertical lines AC and DF. The human user ignores such perturbations; here we account for them by the following procedure.

Essentially one has to define a "grain" size, and to allow deviations smaller than this "grain". To this end, the convex hull of the block is considered again. For each element containing a node on the convex hull one calculates the maximal coordinate difference in the larger dimension. One then takes the minimum over all such elements. This minimum is the "grain" size, denoted δ. When numbering lexicographically one ignores changes larger than δ.

For example, if the x-dimension of the pseudo-rectangle is larger than the y-dimension, then one calculates

$$\delta = \min_{e} \max_{p,q} |x_p - x_q| \,. \tag{5}$$

Here e is an element in the convex hull, p, q are any two consecutive nodes in e ("consecutive" means that one follows the other in a cyclical listing of the nodes of e), and x_p, x_q are the respective x coordinates. Then, when numbering the nodes, a node with coordinates (a,b) is numbered before a node with coordinates (c,d) if either of the two following conditions is satisfied:

(i) $a+\delta < c$;

(ii) $|c-a| \leq \delta$ and $b < d$.

Similar expressions can be written for the case where the y-dimension of the pseudo-rectangle is larger than the x-dimension.

Consider the example shown in Figure 5(b) after rotation. Here the x-dimension is larger than the y-dimension, and δ is calculated as follows. In element I one chooses the maximum between the x-coordinate differences of the segments AB, BC, CF, FE, ED and DA. According to the figure, CF is chosen. In element II it is DG which is chosen. Now δ is the smaller of the two, in this case CF. Then when numbering, since the displacement of A to the right of B is less then δ, A is numbered before B and C, and similarly G comes before

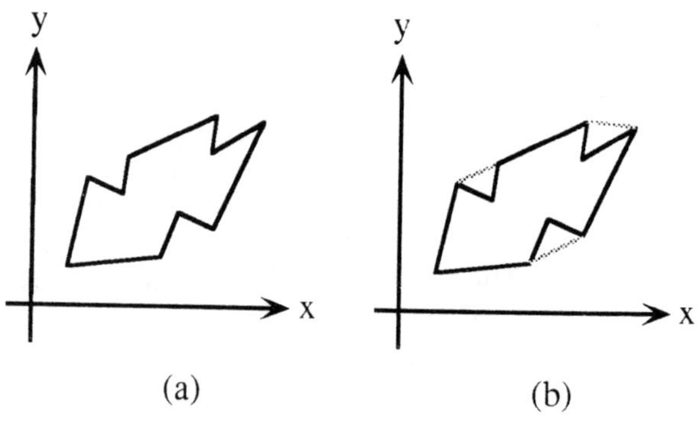

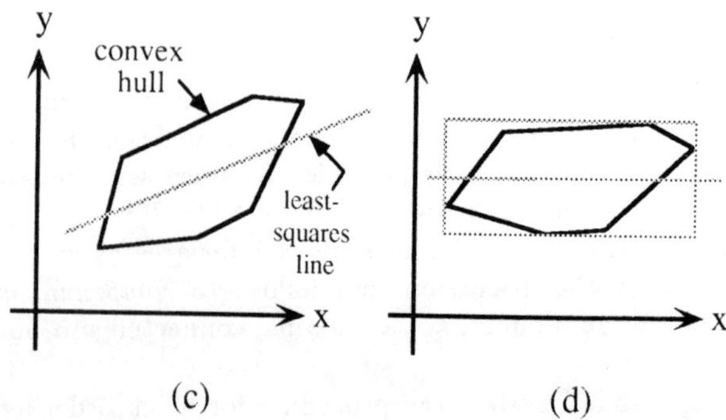

Fig. 4

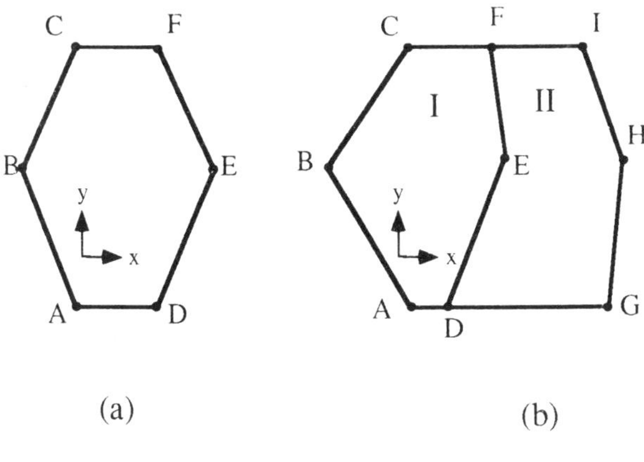

Fig. 5

H and I. Thus, the numbering sequence is ABCDEFGHI, as seems to be desired.

Finally, we note that in MARK I this step is separate from step 1 where the mesh is subdivided into blocks. Actually, it may be better to combine step 3 with step 1. In other words, deciding on the type of a simple block may be part of the reasoning process in creating the block.

Step 4

In this step, the system determines the free parameters associated with the strategy chosen for each block. For the rectangular strategy which is handled in MARK I, the only free parameter is the node from which to start the numbering, and there are four possibilities, corresponding to the four corners of the pseudo-rectangle. We call this starting point n^*, and it is selected in the following way:

- From amongst all the simple blocks which have already been numbered and which have an interface with the current block, choose the largest one (based on the number of nodes) and call it I.
- Consider n^* of block I (this starting point has already been found).
- Choose n^* of the current block as the corner which is geometrically closest to n^* of block I.

Note that the decision on n^* of the current block is based on both *topological* and *geometric* considerations (see comments to step 1). Note also that in the present version the appropriate choice of the starting corner *for the very first block* is user determined, i.e. MARK I does not make this decision.

EXAMPLE AND RESULTS

Figure 6 shows the mesh used as a major testing example. This mesh is similar to the one which appears in [10] in another context. There are 359 nodal points. The mesh was divided (manually; see comments to step 1 in previous section) into 6 simple blocks named A, B, C, D, E and F, as shown in the figure. The thick lines in Figure 6 indicate *disconnecting borders*. It is given that elements and nodes on the two sides of such border do not interact across the border. The line denoted a-b in the figure was a regular line connecting nodes in [10], but here we consider two cases: that in which a-b is a disconnecting border (case (i)), and that in which it is just a regular line (case (ii)).

Table 2 summarizes the results. The AHBs corresponding to three numbering systems are compared: that obtained by a human expert, that obtained by the MARK I expert system, and that obtained by numbering the nodes in a random order. The order of numbering the blocks in the first two numbering systems is also indicated. The results are presented for both case (i) and case (ii).

The results show that although the human expert performs better than the expert system, the differences between the two AHBs are reasonably small. As expected, the AHBs are larger for case (ii), in which the topology of the mesh is that of a pseudo-annulus. In this case the difference between the AHBs obtained by the human expert and by the expert system is larger, since the MARK I system does not handle

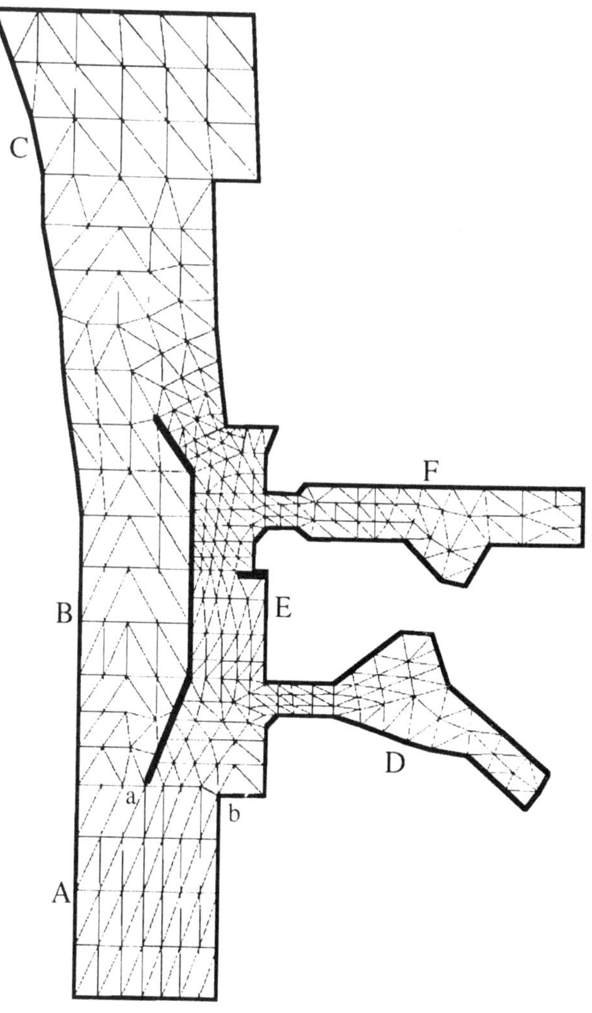

Fig. 6

Case	(i)	(ii)
Human user	AHB=15.9 CEDFBA	AHB=18.2 CBAEDF
MARK I system	AHB=19.7 EDFCBA	AHB=26.1 EDFCBA
Random numbering	AHB=210.2	AHB=216.1

Table 2

pseudo-annular topologies properly. The results obtained by random numbering are given for reference. The expert system numbers the blocks in a different order than that of the human expert. This shows that the heuristics for block ordering must be improved, as already noted in the comments to step 2 in the previous section.

Other examples have been tested as well and the results compared favorably with those of a human expert.

CONCLUSIONS AND REMARKS

The results presented in the previous section and the results obtained for other testing examples demonstrate the feasibility of the approach of constructing an expert system for optimal and systematic finite element node numbering. They also indicate that the method has the potential of being very useful when employed in the preprocessing stage of a commercial finite element code.

Of course, one does not expect a MARK I version to be fully competitive in "real" applications. In this light the results of the tests performed in the present work are quite encouraging. However, many more improvements are needed. We list some here that will be analyzed in the MARK II version:

- Adding more strategies to the strategy collection (step 0).
- Subdivision into simple blocks (step 1). This problem can be attacked by computational geometry or by pattern recognition methods. In both cases it is best treated simultaneously with mesh generation.
- Improving the heuristics of block ordering (step 2). Currently, the choice made emphasizes the size of the components attached to the simple block over the size of the interface between the current and the next block. However, it is clear that some balance should be made between the size of the component and the size of the interface. Heuristics on judging this balance have to be extracted and implemented, thereby complicating the method of block ordering.
- Combining the process of deciding on the type of a simple block with the process of creating the block (i.e. combining steps 3 and 1).
- Improve the heuristics for choosing the free parameters associated with a specific strategy (step 4). E.g., adding the sizes of the interfaces to the considerations, numbering the first node automatically, etc.
- Making the system capable of handling general meshes. These include three-dimensional meshes and meshes containing nodes not only at the vertices of the elements.
- Additional tools are needed for ease in working with the expert system: fully-graphic display, means for fast mesh definition, etc.
- Intensive testing with examples of various types and levels of difficulty. Also, comparing the results of the expert system to those of commonly-used algorithms.

Finally we mention two additional projects which are related to the present one and which can be handled by expert system techniques. First, computerized learning techniques can be applied for acquiring knowledge in strategy design (step 0). Thus, the expert system will accumulate experience and will gradually improve its own performance. Second, one may attempt to design an expert system for the automatic generation of finite element meshes. This is also a field where the algorithmic approach currently dominates, and where a human expert outperforms common algorithms in many cases.

ACKNOWLEDGMENTS

This work was partly supported by the Joint Haifa University-Technion Research Fund. The third author has received the 1991 Computer Science Akavia Award for graduate students, for his contribution in the present work.

REFERENCES:

1. T.J.R. Hughes, *The Finite Element Method*, Prentice-Hall, Englewood Cliffs, New Jersey, 1987.
2. F. Hayes-Roth, D.A. Waterman and D. Lenut (eds.), *Building Expert Systems*, Addison-Wesley, 1983.
3. D.A. Waterman, *A Guide to Expert Systems*, Addison-Wesley, 1986.
4. R. Duda, J. Gashnig and P. Hart, "Model Design in the PROSPECTOR Consultant System," in D. Michie (ed.), *Expert Systems in the Micro-electronic Age*, Edinborough University Press, 1979.
5. J. McDermott, "R1: An Expert in the Computer Science Domain," *AAAI*, 1, pp. 269–271, 1982.

6. B.G. Buchanan and E.A. Feigenbaum, "DENDRAL and meta-DENDRAL," *Artificial Intelligence*, 11, pp. 5–24, 1978.

7. E. Shortliffe, "Consultation Systems for Physicians," in *Proc. Canadian Soc. for Comput. Studies of Intell. (CSCSI)*, University of Victoria, Victoria, British Columbia, 1980.

8. T.C. Walker and R.K. Miller, *Expert Systems*, SEAI Technology, Madison, Georgia, 1988.

9. E. Turban and P. Watkins (eds.), *Applied Expert Systems*, North-Holland, Amsterdam, 1988.

10. O.C. Zienkiewicz, *The Finite Element Method*, 3rd ed., McGraw-Hill, New York, 1977.

11. B. Irons and S. Ahmad, *Techniques of Finite Elements*, Ellis Horwood and John Wiley, Chichester, England, 1980.

12. C. Johnson, *Numerical Solutions of Partial Differential Equations by the Finite Element Method*, Cambridge University Press, Cambridge, England, 1987.

13. Sloan and Randolph, "Automatic Element Reordering for Finite Element Analysis with Frontal Schemes," *Int. J. Numer. Meth. Eng.*, 19,pp. 1153–1181, 1983.

14. A. George and J.W.H. Liu, *Computer Solution of Large Sparse Positive Definite Systems*, Prentice-Hall, Englewood Cliffs, New Jersey, 1981.

15. E. Cuthill and J. McKee, "Reducing the Bandwidth of Sparse Symmetric Matrices," *Proc. ACM National Conf.*, Assoc. for Computing Machinery, New York, 1969.

16. I.P. King, "An AUtomatic Reordering Scheme for Simultaneous Equations Derived from Network Systems," *Int. J. Numer. Meth. Eng.*, 2, pp. 523–533, 1970.

17. R.J. Collins, "Bandwidth Reduction by Automatic Renumbering," *Int. J. Numer. Meth. Eng.*, 6, pp. 345–356, 1973.

18. J.E. Akin and R.M. Pardue, "Element Resequencing for Frontal Solutions," in J.R. Whiteman (ed.), *Mathematics of Finite Elements and Applications*, Academic Press, New York, 1975.

19. N.E. Gibbs, "A Hybrid Profile Reduction Algorithm," *ACM Trans. Math. Software*, 2, pp. 378–387, 1976.

20. A. Razzaque, "Automatic Reduction of Frontwidth for Finite Element Analysis," *Int. J. Numer. Meth. Eng.*, 15, pp. 1315–1324, 1980.

21. H.L. Pina, "An Algorithm for Frontwidth Reduction," *Int. J. Numer. Meth. Eng.*, 17, pp. 1539–1546, 1981.

22. S.J. Fenves and K.H. Law, "A Two-Step Approach to Finite Element Ordering," *Int. J. Numer. Meth. Eng.*, 19, pp. 891–911, 1983.

23. S.W. Sloan, "An Algorithm for Profile and Wavefront Reduction of Sparse Matrices," *Int. J. Numer. Meth. Eng.*, 23, pp. 239–251, 1986.

24. G.C. Everstine, "A Comparison of Three Resequencing Algorithms for the Reduction of Matrix Profile and Wavefront," *Int. J. Numer. Meth. Eng.*, 14, pp. 837–853, 1979.

25. J.G. Lewis, "Implementation of the Gibbs-Poole-Stockmeyer and Gibbs-King Algorithms," *ACM Trans. Math. Software*, 8, pp. 180–189, 1982.

26. B.A. Armstrong, "Near-Minimal Matrix Profiles and Wavefronts for Testing Nodal Resequencing Algorithms," *Int. J. Numer. Meth. Eng.*, 21, pp. 1785–1790, 1985.

27. S.W. Sloan, "A FORTRAN Program for Profile and Wavefront Reduction," *Int. J. Numer. Meth. Eng.*, 28, pp. 2651–2679, 1989.

28. L.S. Duff, J.K. Reid and J.A. Scott, "The Use of Profile Reduction Algorithms with a Frontal Code," *Int. J. Numer. Meth. Eng.*, 28, pp. 2555–2568, 1989.

3 CONCEPTUAL DESIGN IN STRUCTURAL ENGINEERING

CONCEPTUAL DESIGN: PUSHING BACK THE BOUNDARIES WITH KNOWLEDGE BASED SYSTEMS

J.C. Miles and C.J. Moore
School of Engineering,
University of Wales Cardiff, Wales

The use of Knowledge Based Systems in Civil Engineering design is examined, with especial reference to bridge design. The effectiveness of the systems which have been built to date is assessed in the light of past successes such as finite element analysis and reasons are put forward as to the slow acceptance rate of Knowledge Based Systems by industry. Assuming these reasons to be correct, the current use of heuristics in Knowledge Based Systems is analysed and 5 broad groupings of heuristics are identified. These groupings are then assessed in terms of their being replaced by either more detailed heuristics or by more fundamental procedures usually called deep knowledge. It is suggested that by so identifying and replacing current heuristics that the utility of Knowledge Based Systems will be enhanced to the point where they are valuable design aids rather than as at present being simulations of current expert thought processes. The idea of pushing back the boundaries arises because the above substitution will probably increase the amount of analysis in design at the expense of pure conceptual design.

INTRODUCTION

Since its introduction the digital computer has found increasing use for a wide variety of applications. Mostly these applications involve using the attribute of the computer to undertake a lot of repetitive tasks at high speed. This ability is the greatest strength of such machines and by the utilisation of this aptitude man has been able to relieve himself of many boring tasks and also undertake operations which were previously impossible.

Artificial Intelligence has existed for about the same length of time as the digital computer. AI has been put to a number of uses but increasingly it has been used to try and simulate the behaviour of "experts" in so called expert systems. Generally expert systems, more accurately called Knowledge Based Systems (KBS) are acknowledged to be one of the most successful areas of AI and yet the number of such systems in every day use is relatively small compared to the total number which have reached the prototype stage. Is this reluctance to use KBS a natural time delay in the acceptance of a new technology or is there a deeper seated problem? Without a lot of research it is impossible to give a definitive answer but this paper looks at some of the possible answers in relation to Civil Engineering design and suggests possible ways forward. It is not intended to imply that the thinking in this paper is necessarily the correct or the only way forward. The ideas are presented to promote discussion.

The work that has lead to the thoughts presented here has largely involved the development of a number of KBS for aspects of Civil Engineering design, and in particular the work of Soh(1990) and Moore(1991) and as yet unpublished work on the design of waste disposal facilities. The examples given below are based on the work of Moore(1991) on the Conceptual Design of Bridges and the main thrust of the arguments is in terms of Conceptual Design (also known as Preliminary Design) as applied to Civil Engineering. To fully develop the arguments it is necessary to initially look at how design is currently undertaken in Civil Engineering, how these processes have been developed and what the implications are for the developers of KBS. This then will be developed into a discussion on knowledge and the types of heuristic used and from an analysis of the latter a possible way of enhancing the performance of Conceptual Design KBS is suggested.

STATUS QUO

Civil Engineering Design can be roughly split into two components, these being Conceptual Design and Analytical Design. The relationship between the two is shown in Fig 1 (Moore and Miles, 1991b) where it can be seen that Conceptual Design contains few calculations and is mainly based on heuristics whereas analytical design is mainly calculations based on well defined algorithms. Calculations in large numbers are boring and repetitive and it is no coincidence that to date the main area of application of computing techniques has been to analytical rather than conceptual design.

The work that has lead to the thinking behind this paper has however largely concentrated on conceptual design and in particular the use of AI techniques for the development of KBS for conceptual design. This is a relatively new area for the application of computer systems.

Conceptual Design is widely agreed to be that area where the type and overall form of the artefact is decided and it is also recognised that there is then an area of overlap with analytical design where preliminary member sizing is undertaken. Conceptual Design in Civil Engineering is largely undertaken by a single person or a small group of people (Moore,1991) who using their past experience, fairly rapidly arrive at a suitable solution. Experience is a vital aspect of this process, conceptual design is not described in text books largely because the processes used are not based on rigourous theory but instead on heuristics. Miles and Moore(1989) found that for bridge design each expert designer tended to have developed his/her own heuristics for the conceptual design process even though the solutions produced were very similar to those of other experts using different heuristics.

In undertaking a conceptual design many factors are considered such as structural behaviour, geotechnics, economics and aesthetics. Current practices have developed to suit the capabilities of human beings and so all these factors tend to be dealt with by heuristics, some of which it will be shown below are short cuts.

In any development process be it the creation of paintings or KBS, initial attempts tend to mimic the existing work in the area. Early efforts to create KBS for Civil Engineering design are no exception and from the pioneering work of Maher(1984) through to more recent contributions such as that of Kumar and Topping (1988) and Moore(1991), the systems have been created in such a form that they largely attempt to mimic the behaviour of design experts. Even where such work has used texts and design Codes such as notably in the work of Maher(1984), these are quite properly written in such a form that they tend to favour human capabilities rather than those of computers and the resultant systems are therefore based on human behaviour.

It can therefore be argued that at present the current technology of KBS for Civil Engineering design has created systems which have no visible gain over the performance of human experts. Indeed such systems provide a rather inadequate simulation of human performance, and one which rapidly degrades once it reaches the limits of its Knowledge Base. Such systems do have potential uses in, for example training junior engineers in aspects of design but they cannot exhibit advantages over the performance of expert human designers. In an industry such as Civil Engineering which has low profit margins and is therefore wary of changes which do not have an obvious economic benefit, it is probable that the acceptance rate of such systems will be relatively slow.

HEURISTICS : DEEP AND SHALLOW KNOWLEDGE

The use of Heuristics is central to the way that human beings perform their everyday tasks. In AI a fairly narrow definition of heuristics is used in that they are said to be experience based rules of thumb which constitute expertise, the factors that enable an expert to go beyond factual "textbook" knowledge and exhibit "expert" behaviour.

It is generally accepted that current KBS technology based on these heuristics uses so called shallow knowledge, that is the underlying fundamental laws of behaviour, known as deep knowledge, have been supplanted by heuristics. Furthermore it has been stated that if KBS are to progress beyond their current capabilities then they will have to be capable of reasoning from first principles (i.e deep knowledge) (Grimson and Patil, 1986). It is argued that systems which contain deep knowledge will be able to exhibit real machine learning and will be able to tackle any problem within their domain as opposed to the current shallow knowledge systems which can only cope with those situations which are explicitly defined in their knowledge bases.

Implicit in the above arguments for deep knowledge systems is the premise that underlying all heuristics are fundamental laws of behaviour and that somehow these can be described and utilised in a useful manner within a KBS.

Classification of Heuristics

Moore and Miles (1991) have undertaken an analysis of the types of heuristic used in the Conceptual Design of Bridges. Although it is difficult to put heuristics into neat groups, 5 different types have been identified. It is not pretended that the classification is unique and many heuristics belong in more than one group but if the types are examined then it becomes possible to pick out certain groups where the short cuts taken by human experts could possibly be replaced by more rigourous procedures which might in some cases be classed as deep knowledge. On the other hand there are some heuristics which it would be very difficult to replace by fundamental laws.

The classification is as follows

Type 1 : Short Cut Heuristics

When one thinks of heuristics, what usually comes to mind is a process which is a "short cut" to a lengthy process, that is some way of reducing the amount of work required down to an easily manageable package. For example in bridge design when choosing the form of the deck, there is a wide variety of structural forms to choose from. Bridge design experts have heuristics which they have developed over a period of years which enable them to assess which is the most economical deck option for a given span range. Originally the experts had gone through the laborious process of sizing the various options and then costing them but with practice and the experience of examining tenders from contractors the expert becomes able to state in simple rules what type of structure is economic for a given situation. The use of such heuristics by bridge designers enables them to make substantial reductions in design costs because they are able to use their past experience to make rapid judgements rather than ploughing through the mass of work that otherwise would be needed. The savings in terms of work input per design runs into hundreds of man hours. However the experts themselves would be the first to admit that in some areas the procedures they are forced to use to save time, are not always entirely accurate or up to date.

Type 2 : The Original Concept is not Known by the Expert

These are heuristics which have been passed on to the expert by others who are knowledgeable in a related area and consequently the expert is not always completely aware of the underlying theories or assumptions behind the heuristics. In bridge design the main area to which this sort of heuristic applies is the foundations where the heuristics have been developed by Geotechnical experts.

Type 3 : No Formal Method Exists

There are some situations where no known method or rule exists for providing a solution and experts have therefore been forced to develop heuristics based purely on experience to help them make a judgement. There are a surprising number of such judgements in bridge design. For example when a bridge is over 20m wide it is usual to use an abutment rather than a bank seat because the abutment enhances the tunnel effect created by such a wide bridge and this is preferable to the large covered void created if a bank seat is used. Other heuristics of this type are based on construction difficulties. For example when the clearance height of a bridge is more than 10m then a bank seat would be used because it is difficult to construct tall abutments. Other areas where such heuristics are used are when purely factual data are involved, as for instance in the transport of prefabricated or precast beams which becomes difficult for lengths of over 27.5 metres due to legal restrictions on road movements. Also there are topics like for instance allowing for services such as water and gas mains and how these are best dealt with both in terms of the bridge deck and the foundations.

Type 4: Heuristics Based on Background Knowledge

This class of heuristic is based on fundamental known theoretical knowledge. Experts become so familiar with many of the tasks they routinely carry out that they implicitly 'know' what the answer is without recourse to the underlying theory or calculations. Unlike type 1 heuristics, the intermediate steps have not been replaced because of the length of time needed to undertake the work but because the expert is so familiar with the work it is unnecessary to follow all the steps involved. This sort of heuristic is particularly applicable to areas where small structural calculations are involved. A typical example is that for the sort of loadings imposed on UK road bridges the span : depth ratio of the bridge deck is normally about 20 : 1. Such a heuristic enables the designer to rapidly assess the depth of any deck and determine whether or not special measures will be required to maintain clearances.

Type 5 : Heuristics from Experimental Evidence

Sometimes where the existing theory is inadequate, experimental work is used to provide guidance and rules to follow. No examples of this type of heuristic were identified during the work on bridge design but historically there are well documented examples of such techniques being used as for example in the work of Robert Stephenson on the Menai Bridge and I.K. Brunel on the Saltash Bridge. It is suspected that with the improvement of our understanding of structural behaviour that the need for such heuristics has diminished but the group has been included for completeness. It is also possible that such heuristics are implicitly contained in design codes where such things as the load factors or factors of safety may well be based on experimental evidence.

HEURISTICS WHICH COULD BE REPLACED/IMPROVED

In current conceptual design, all the processes used have been developed to suit the capabilities of the human brain with its limited ability to undertake repetitive operations and small capacity short term or process memory. Expert Systems which set out to replicate these processes are disadvantaging themselves by not identifying the areas where the heuristics have been developed to avoid operations which human beings find difficult but which a serial computer could readily undertake.

The above grouping of heuristics into 5 types enables broad areas to be identified where possibly the heuristics could be replaced or "improved". Some types can be immediately declared as being inviolate so far as this replacement process is concerned. These types are numbers 3 and 5. It is possible that with further research work to determine the underlying "deep" knowledge behind these heuristics that they could be replaced but for the time the discussion here will keep to what is possible with current knowledge.

Taking the remaining 3 types in turn starting with No. 1 :-

Type 1 is ripe for replacement, and by providing suitable computer based routines for such an area even the expert would benefit from this. At the moment many of the heuristics used to assess the economics of various choices of structural form and the decision to use steel or concrete are based on either out of date or limited data. For example, ideally every time a design is undertaken the economics should be examined in detail but Civil Engineering designers obtain their work by fee competition. In such circumstances design costs have to be cut to a minimum. In an advanced KBS of the sort being proposed here, facilities could be included where the system could undertake the preliminary design of a number of options, including approximate member sizing and costing. Such work would possibly involve the derivation of new heuristics suited to the computer rather than the human designer but the ability to rapidly assess the economics of a variety of options would give the designer greatly enhanced abilities and should lead to better designs.

The benefits to be gained from altering or replacing the type 2 heuristics are less clear cut but nevertheless could be appreciable. These heuristics are to enable the expert to cope with another area of expertise without having detailed knowledge of that area and so the possibility of replacement would depend on the type of knowledge being dealt with.

For the expert, the replacement of type 4 heuristics by the underlying process would bring no benefit. If a KBS is being used by a non-expert, then it would be useful for them to appreciate the underlying knowledge, but they also need to learn the heuristic as this saves time. So for this group there are no great gains to be made but an appreciation of the underlying theory would help beginners.

So the main gain is the replacement of type 1 short cut heuristics with either new purpose built heuristics or methods based on more fundamental precepts. It is interesting to note that for bridge design many of these replacement methods will be algorithmic in nature and so ironically the development of a KBS has highlighted areas where the use of algorithmic techniques could be extended. To an engineer this should cause no difficulty. Engineers are people who have a reputation for pragmatism, solving problems by whatever techniques are most appropriate. In terms of Fig.1 what is being proposed can be conceived as a movement or pushing back of the boundary between the "mainly heuristics" and "mainly algorithmic" portion so that some of the conceptual design would consist of algorithmic procedures undertaken on a computer. Possibly this could be interpreted as Conceptual Design being replaced by analysis.

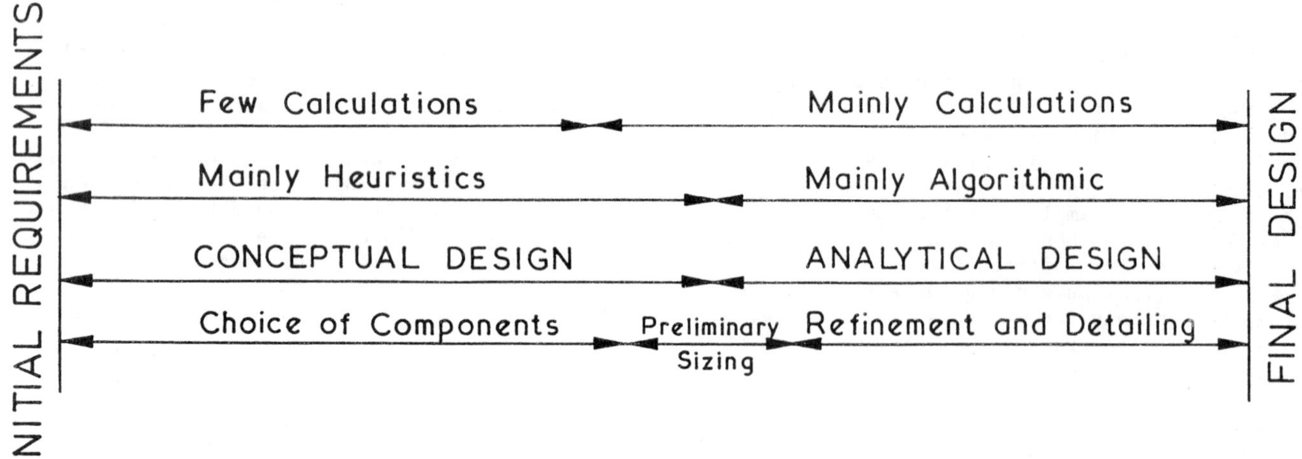

FIGURE 1

EXAMPLE

As yet the thoughts contained herein are hypotheses rather than facts and so to give fully validated examples is not possible but feasible areas are very easy to identify, again as before relying on Conceptual Bridge design as a suitable domain.

The area of the choice of deck type for a bridge has been mentioned above as a possible area and this will be concentrated upon in this section.

When choosing a suitable type of deck for a given span it is necessary to consider a number of factors such as the allowable on-site construction time, the restrictions on the use of temporary formwork or other supports and access to the site. These are mainly non-economic factors that cannot be classified as type 1 heuristics but having gone through a series of constraints to remove various structural forms from the reckoning there is usually a choice between several different types of material and structure.

First to consider the choice of material (although it has to be recognised that in practice such a choice cannot be separated from other considerations). The common choices for bridge decks are either steel, concrete or steel/concrete composite. The cost of each type of material can vary according to a number of factors. For example if the house building sector in Britain is buoyant then the cost of cement rises and hence concrete costs rise. The price of steel also rises due to the increased demand for reinforcing bars but the two prices change at different rates. Hence the balance between the choice of materials is subject to relatively short term price variation and thus fixed heuristics based on long term experience and the evidence of past tenders can be misleading. The experts used for the development of the Conceptual Bridge Design KBS recognised this problem but said that given the current constraints of time and money they were unable to regularly update their heuristics.

At the same time as considering the choice of material type, the bridge designer has to consider the structural form of the bridge. For example should it be prestressed, precast concrete beams, steel plate girders with concrete deck, insitu concrete with a voided deck or a box girder in either steel or concrete. Many factors have to be considered in making the choice but having determined the feasible options at present the bridge designer uses heuristics to determine the most economic alternative. These heuristics tend to be fairly simple although they are based on a lot of experience. Time and money constraints dictate that the expert designer does not spend a lot of time considering alternatives but rapidly converges to a single solution which is then analysed and detailed.

Given current technology it is perfectly possible to formulate a system for bridge design which would reduce the number of deck options to a manageable number of feasible alternatives and then determine preliminary member sizes and an approximate costing. Indeed the former process could more easily be undertaken by the expert but the member sizing and costing section of such a system would be useful to a wide range of bridge engineers. The amount of computing time to undertake such an operation would depend on the techniques used. For example it might be possible to derive new heuristics for the costings which would enable the amount of calculation to be kept to an easily manageable level. It is likely that a satisfactory assessment of the costs of the various forms could be made without the need for a very detailed design so the process rather than using "deep" knowledge would be "deeper" than current methods.

CONCLUSIONS

Based on the hypothesis that the current slow acceptance rate of KBS by the construction industry is because, at present, KBS offer no great economic or other readily visible gain over current practice, a possible way of enhancing the attractiveness of such systems has been proposed. The construction industry is noted for being conservative, possibly because of the low profit margins, and without a readily tangible gain in the adoption of KBS, it is probable that most of the industry will continue to use current technology.

The types of heuristic used in a typical design KBS have been classified into broad groups. It is recognised that the classification is not unique and that many heuristics will belong to more than one group but nevertheless the grouping is useful in that it enables the identification of heuristics which have been developed to avoid the limitations of human capabilities.

It is suggested that such heuristics should be examined in detail and where it is thought that computer based techniques would be more powerful and useful then suitable replacements could be formulated. The development of such replacements would give KBS a tangible improvement over current practice and hence increase the acceptability of such systems.

The ideas suggested are simple and are capable of being implemented with current technology. It is probable that the sort of KBS proposed will involve the development of 'new' heuristics to avoid the need for excessive execution times, although experience with finite elements has shown that if the gains are sufficiently great then users are prepared to wait for results. It is suggested that also the introduction of such systems may lead to the replacement of some areas of heuristics by algorithmic procedures.

Finally in their arguments for the incorporation of deep knowledge, Grimson and Patil(1986) stated that the introduction of such knowledge would enhance the development of effective machine learning. It is not thought that what is suggested in this paper would create such an advance.

ACKNOWLEDGEMENTS

The authors would like to acknowledge the contribution made by their former colleague, now Professor Howard Wright of Strathclyde University to the thoughts contained in this paper.

REFERENCES

Grimson, W.E.L. and Patil, R.S. (eds), 1986. AI in the 1980's and Beyond : An MIT Survey. MIT Press, MIT, USA.

Kumar, B. and Topping, B.H.V., 1988. INDEX : An INDustrial Building Design EXpert, Civil Engineering Systems, 5(2), 65-76.

Maher, M.L., 1984. A Knowledge Based Expert System for the Preliminary Design of High Rise Buildings, PhD Thesis, Dept of Civil Eng, Carnegie Mellon Univ. Pittsburgh, USA.

Miles, J.C. and Moore, C.J., 1989. An Expert System for the Conceptual Design of Bridges: in Topping B.H.V (ed), AI Techniques and Applications for Civil and Structural Engineers, CIVIL-COMP89, City University, London, 171-176.

Moore C.J., 1991. An Expert System for the Conceptual Design of Bridges, PhD thesis, School of Engineering, Univ of Cardiff, UK, 330pp.

Moore, C.J. and Miles, J.C., 1991a. In Depth Analysis of Heuristics in KBS Development, Under Review.

Moore, C.J. and Miles, J.C., 1991b. Conceptual Design and Design with especial reference to Road Bridges, Under Review.

Soh, C.K., 1990. An Approach to Automate the Design of Offshore Structures, PhD thesis, School of Engineering, Univ of Cardiff, UK, 330pp.

ARCHITECTURE OF A DESIGN SYNTHESISER FOR KBES

C.S. Krishnamoorthy, C. Srinivasa Rao and S. Rajeev
Department of Civil Engineering,
Indian Institute of Technology, Madras, India

Preliminary design involves the synthesis of potential structural configurations satisfying certain basic constraints. Many of the conceptual aspects of structural design are embodied in this phase. Creativity on the part of the designer is vital in obtaining desirable solutions. The problem of creativity is compounded by the fact that at this stage of the design, every parameter of design may correspond to a fairly large set of options. Structural systems in general are complex and the variations possible in synthesising structural configurations can be quite tedious. Deficient reasoning at this stage results in iterations at later stages of the design process since the basic form of the solution is identified in this phase. In this paper, a design synthesiser is presented, which provides a systematic and structured approach to preliminary design.

INTRODUCTION

The use of computers across the spectrum of activities constituting structural design process has not been uniform. This is due to the varying degrees of formalization of the problem statements involved. While extensive use is made of CAD in analysis and drafting, the design phase has lagged behind. At the preliminary design stage when the most important decisions are made, little use is made of CAD. Knowledge-based expert systems(KBES) have been the object of attention of engineers in recent years, in view of their ability to address the various difficulties arising out of applying traditional computational techniques to design problems. The use of generic components offers a powerful method of building real life expert systems (Ref. 1). This paper describes the implementation of a knowledge-based system for design synthesis. The role of the module in building KBES prototypes is discussed.

SYNTHESIS

Synthesis is the process of selecting components to form a system that meets the design objectives while satisfying all the constraints governing the selection. The components may themselves be complex entities which need to be synthesised first. Therefore, the artifact is arrived at progressively by building the smaller components first and then using these to build bigger components and so on, until the entire solution is achieved. To enable the selection process, the domain knowledge needs to be represented heirarchically. The representation concerns the characteristics describing each component and the relationship between the various components. Such a representation is termed the decomposition of the knowledge of the domain. The approach is based on the systematic depth-first search of a space, which includes all possible solutions, using heuristic knowledge in the form of constraints. Essentially, this involves resolving the design goal into a set of subgoals. Each subgoal in turn may consist of further subgoals and so on, until eventually terminating in making a functional or physical selection. The system provides a framework to define a design space to form alternate feasible solutions. The design space is defined by an expert in the specific domain. All possible solutions are synthesised by the synthesiser. A solution violating any constraint is discarded as soon as the violation is apparent. The next solution is then attempted and so on, until no further attempts can be made.

The architecture of the Design Synthesiser is shown in Fig. 1. The system is structured to be generic.

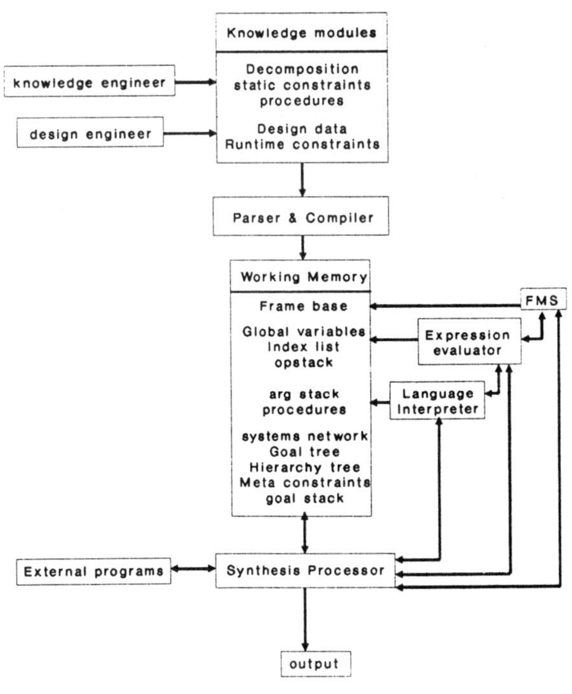

Fig. 1. Architecture of the Design Synthesiser

The algorithm is separated from the knowledge modules. The system is capable of functioning in two different ways. It is a stand alone shell for processing design knowledge to generate feasible solutions. Alternatively, it can be used as a tool or generic component of knowledge-based expert systems. For instance, an inference engine can call upon the system from within a production rule to generate alternate solutions for a specific design goal.

Consider, for example, an expert system for designing a steel industrial structure. The expert system can arrive at a structural scheme while inferencing. It is not possible for the inference engine to consider all the alternatives possible in choosing a structural scheme for the structure. This is due to the nature of the chaining process itself. Also, if the design activity extends to the component design stage, then every member in the structural scheme needs to be created as an object first and arranged in a suitable fashion. In a structure with several hundred individual members, the amount of work required to create the objects and arranging them can be quite tedious. It is obvious that the rule base must necessarily contain a large number of rules to just accomplish this task alone. If, however, a synthesiser is available as a generic module, the inference engine can pass the basic information required to the synthesiser and obtain all or selective configurations in return. The knowledge concerning the various structural schemes, which can be quite large in size, can now be kept separate from the essential knowledge controlling the inference process. Also, it may not be convienient to express knowledge about structural schemes in the representation framework of the calling knowledge base. The inference engine can control the way the synthesiser functions, and also pass on to it necessary constraints to eliminate unwanted solutions. The inference engine can activate the synthesiser repeatedly by modifying the data and constraints passed to the synthesiser to obtain varying solutions. This ability is essential in simulating the iterative nature of this phase of design process. The synthesised configurations are in the form of a hierarchy of objects. The various solutions can then be evaluated by either another generic module or a knowledge base and the best one selected.

COMPONENTS OF THE SYNTHESISER

Knowledge modules

Whenever the synthesiser is activated, the names of the files containing the knowledge are specified. The first file, called the decomposition file, contains the domain knowledge. The second file called the preconditions file, contains problem specific knowledge. The domain knowledge consists of descriptions of all the objects that can possibly constitute a part of the solution to the problem. Problem specific knowledge contains design data and constraints applicable to the particular application run.

The structure of a decomposition file appears as:

```
title
declarations
system descriptions
static constraints (if any)
procedures
```

The structure of a preconditions file appears as:

```
global variables
runtime constraints
```

Each object in the decomposition is termed a system. Systems describe objects and relations with other objects. Systems consist of attributes. An attribute is a characteristic of the object that can be assigned a value. The set of all attributes describes the object completely. There are five categories of attributes. These are described below.

The ONE_OF attribute definition has the form:

```
<attrname> ONE_OF <symbol1><,symbol2> ...
```

The symbols are literal values that can be assigned to the attribute alternatively.

If the attribute is a numeric quantity derived from the design data, attributes of other systems, and mathematical functions, it is of the type EXPRESSION. The definition has the form:

```
<attrname> EXPRESSION <expression>
```

The attribute will be assigned the value returned by the expression at synthesis time.

An attribute can be assigned the value returned by a procedure. The definition has the form:

```
<attrname> PROCEDURE <proc-name><(arg1,arg2,...)>
```

An attribute can similarly be assigned the value returned by an external program. The definition is similar to that of procedure attributes. The keyword PROGRAM should be used instead of PROCEDURE.

A SUBSYSTEM attribute indicates that the system attached to the attribute would have to be synthesised completely when the attribute is taken up. The form of the attribute definition is:

```
<attrname> SUBSYSTEM <subsystem-name>
```

Sometimes, an attribute may be associated with a set of alternative systems. In such a case, the attribute may be declared as a ONE_OF attribute. The list of symbols is replaced by the list of subsystem names.

A system description may also contain heuristics that help modify the knowledge depending on the context. This type of knowledge is stated as planning rules embodied within the system description. Each rule has an IF part containing a conjunction of logical expressions and a THEN part containing an action. When the IF part evaluates to TRUE, the action in the THEN part is performed. The action can be either a reordering of the attributes within the system or creating multiple copies of a SUBSYSTEM attribute. It is also possible to specify the planning rules unconditionally, i.e. without any conditions in the antecedent part. There can be any number of planning rules within a system. The synthesiser applies the rules sequentially as long as conflicts do not occur. Fig. 2 shows a sample system description for a steel industrial structure.

The first system defined in the decomposition defines the whole artifact and is the one synthesised ultimately. It encompasses the entire design. The

```
SYSTEM bldg_frame

length          EXPRESSION   length
width           EXPRESSION   width
height          EXPRESSION   height                                                /* eave height */
roofing_matl    ONE_OF       ACC,GI                                        /* Acc_sheets,GI_sheets */
siding_matl     ONE_OF       ACC,GI                                       /* material used for sides */
end_framing     ONE_OF       brick_wall,gable_frame
type            ONE_OF       fink,saw_tooth                                    /* literal strings */
rise            PROCEDURE    rise(width,bldg_frame.roofing_matl)              /* rise of the truss */
nobays          PROCEDURE    nobays(length,width,side_opening)                      /* no of bays */
bay_spacing     PROCEDURE    bspacing(length,bldg_frame.nobays)                  /* bay spacing */
nscols          PROCEDURE    nscols(length,bldg_frame.nobays,bldg_frame.end_framing)/* side columns */
ngcols          PROCEDURE    ngcols(width,bldg_frame.type,bldg_frame.end_framing) /* gable columns */
ntruss          PROCEDURE    ntruss(bldg_frame.nobays,bldg_frame.end_framing)    /* no of trusses */
s_column        SUBSYSTEM    s_column                                            /* side column */
g_column        SUBSYSTEM    g_column                                           /* gable column */
truss           ONE_OF       Fink,Saw_tooth                                 /* truss subsystems */
roofing         SUBSYSTEM    roofing                                       /* roofing subsystem */
siding          SUBSYSTEM    siding                                         /* siding subsystem */
gable           SUBSYSTEM    gable                                            /* end girts etc. */
bracing         SUBSYSTEM    bracing                                 /* entire bracing system */

IF ( bldg_frame.nscols > 1 ) THEN FOR sc_count = 1 , bldg_frame.nscols REPEAT s_column
IF ( bldg_frame.ngcols > 1 ) THEN FOR gc_count = 1 , bldg_frame.ngcols REPEAT g_column
IF ( bldg_frame.ntruss > 1 AND bldg_frame.type == fink ) THEN
                                 FOR tr_count = 1 , bldg_frame.ntruss REPEAT Fink
IF ( bldg_frame.ntruss > 1 AND bldg_frame.type == saw_tooth ) THEN
                                 FOR tr_count = 1 , bldg_frame.ntruss REPEAT Saw_tooth
FOR  si_count = 1 , 2 REPEAT siding
FOR  ga_count = 1 , 2 REPEAT gable

END_SYSTEM
```

Fig. 2. Sample system description for a steel industrial building

physical ordering of the other systems in the decomposition is not significant. In addition to systems, the decomposition file may contain knowledge that describes incompatible combination of components in a solution or specific elimination criteria that are always true irrespective of the problem statement. This knowledge is expressed as constraints. Since these constraints do not change with problem statement, they are termed static constraints.

Consider for example, the constraint:

```
IF( bldg_frame.roofing_matl == ACC_sheets)
THEN roofing.sheathing == none
```

The constraint says that if the roofing material chosen is ACC_sheets, then no sheathing should be provided. Typically, the attribute 'sheathing' of the system 'roofing' would be a ONE_OF attribute having values like none, wood, etc. In any solution having roofing_matl as ACC_sheets, any value other than 'none' assigned to the attribute 'sheathing' would indicate a non feasible solution. Constraints are the governing factors in synthesis process. The synthesiser adopts a constraint directed, depth-first search of the solution space in each solution run. For efficient performance of the synthesiser, proper formulation of constraints is very important. To enable flexible representation of various types of constraints, three forms of constraints are provided. The various forms of constraints available are:

attribute constraints
meta constraints
constraint functions

Constraints are also classified on the basis of their location. Constraints given in the decomposition file are called static constraints and constraints in the preconditions file are called runtime constraints. Each individual constraint, static or runtime, can be expressed in any of the above three forms. Attribute constraints have the form:

<system.attribute> <operator> <expression>

Such a constriant is attached to the attribute by the synthesiser. As soon as an attribute is synthesised, all attribute constraints attached to the attribute are checked for. If any one constraint is violated, the solution being synthesised is aborted and the next one is taken up. An expression can consist of any algebraic quantity expressed in terms of global variables, attributes of already synthesised systems, and other data.

Meta constraints have the form:

IF (set of conditions) THEN set of conditions

The set of conditions in the THEN part are imposed on the solution only if the set of conditions in the IF part evaluate to TRUE. All meta constraints are imposed at the end of each solution run. Therefore, meta constraints are more expensive computationally.

Constraint functions have the form:

function_name()

A constraint function is the most powerful form of

constraint available. It can express constraints of any complexity. A constraint function returns either TRUE or FALSE to the synthesiser. All constraint functions are checked for at the end of each solution run. Constraint functions are procedures coded using the inbuilt programming language provided by the implementation. The syntax of the language is similar to that of C language. The body of a constraint function is included along with the other procedures. Only the name of the function is specified in the constraints.

Since the synthesiser can function in two different ways, the context available to it can come in two ways. If the synthesiser is used as a stand alone environment, then, the context is supplied by the user. If it is called from an expert system, the context is supplied by the expert system. Context is of two forms. The first is the data necessary to process the knowledge. Data provides the initial state of the problem and is constant through out the synthesis run. It is referenced in expressions and passed onto external programs. This data is referred to as preconditions. As synthesis progresses, new data is generated. The data generated during synthesis is added to the context. This data, however, does not remain static and may be qualitatively and quantitatively different in each solution run.

The second form of context is the knowledge that is specific to the particular application of the domain knowledge and is similar to the constraints in the decomposition. This part of the knowledge is not static and applies only for the specific application and hence is referred to as runtime constraints. An additional feature available to runtime constraints is the ability to deactivate systems. When a system is deactivated, it ceases to be a subsystem of the parent system and is converted to a literal. All subsystems of the deactivated system and their subsystems etc. are automatically deactivated. Deactivation is very useful because the decomposition file is expected to remain unchanged even if the problem statement or the solution requirements change. But in some cases, the fine granularity of the decomposition causes the synthesiser to generate solutions differing only trivially whereas the objective may be to study the variations in certain other parameters. Deactivation helps in moulding the available knowledge to suit specific applications without actually changing the knowledge itself.

Compiler

The compiler performs several functions. Firstly, it verifies that the decomposition and the context are syntactically and semantically correct. The compiler issues relevant messages when it finds errors in the input files. Since the knowledge is represented as a network, infinite loops are possible. These and other logical errors are checked for. Such errors are fatal and the synthesis process is not initiated. The context may also contain errors which are pointed out and if the errors are not fatal, synthesis continues. Since the context is variable and is only valid for one application, no output file is generated by the compiler. The results are directly stored in memory.

Synthesis processor

The synthesis processor is an algorithm that synthesises the solution by first designing the smallest parts and then combining them to build bigger parts until the entire solution is achieved. It generates all possible solutions of the problem as stated by the context. It starts at the first goal, breaks the goal into subgoals, and then addresses the first subgoal, repeating the process on each subgoal. It is a depth-first search because it moves deeper into the hierarchy before completing a level. It is a constraint-directed search because it prunes a branch of the search space as soon as that branch violates a constraint.

Language interpreter

The system provides a language for writing simple procedures. Procedures resemble C functions generally. The language has facilities for variable declarations, conditional branching, compound statements, looping with 'BREAK' and 'CONTINUE' features. IF and WHILE blocks can be nested to any level. The compiler compiles the code of all procedures along with the rest of the decomposition and stores the code in interpretable form. It is possible for procedures to access the frames created by the synthesiser, loop over object instances, recognise current instances etc. The importance of the interpreter lies in the fact that certain constraints can only be expressed as constraint functions that need to be interpreted at synthesis time.

Expression evaluator

The syntax of the decomposition knowledge is such that attributes need not be declared as to their data types. The synthesiser assigns suitable data type to an attribute after synthesising the attribute value and finding its data type. At compile time, therefore, it is not possible to distinguish between variables, attributes, literal strings etc. occuring in expressions. All expressions encountered during compilation are broken up into interpretable code for later evaluation at synthesis time. The code is written to a binary file and the offset to the start of code for each expression is stored with the variable or data structure associated with the expression. Before synthesis starts, the entire code is read into memory. Expressions can be of any complexity involving arithmatic operators, unary minus, power function, trigonometric functions etc. Parentheses can be used to override precedence. Expressions are broken up into individual actions that need to be performed to evaluate them. Each action is assigned an opcode. If an action requires arguments, then the arguments follow the opcode. The opcodes and the arguments dictate what operands are to be loaded onto the opstack and what actions are performed on them. Both the synthesiser and the language interpreter use the same expression evaluator. Due to differences in the syntax of expressions that can be used in the decomposition and the programming language, identity of the module calling upon the expression evaluator is supplied to the evaluator. The behaviour of the evaluator depends on this. Errors in the decomposition such as unknown variables, illegal operands, use of attribute values in expressions before they have been synthesised etc. cause errors at evaluation time. At such times, the evaluator abandons the expression and returns a special code to the calling module.

Goal tree

The goal tree depicts the final hierarchial form of the knowledge. The root node represents the design artifact and the nodes below it are the immediate subgoals. The goal tree does not constitute the search space because of planning rules controlling creation of multiple instances of subsystems in a system. Since the rules might involve conditions with values of attributes of systems already synthesised, different rules may fire in different solution runs. Hence, the form of the search space is different in each solution attempt. It is, therefore, not known at the start of the synthesis as to how many solution attempts will be made. The goal tree identifies the form of the search space. It remains static during the entire synthesis process.

Hierarchy tree

The hierarchy tree represents the form of the search space for a single solution run. It retains all the information contained in the goal tree but it does not remain static during the synthesis run. In each solution attempt the hierarchy tree is built anew gradually until the solution pertaining to the solution attempt is achieved or a constraint is violated or an error has occured causing the attempt to be abandoned.

Frame management system

During synthesis, systems are stored in memory as frames. The synthesiser stores the solution in memory as a frame base with the help of a Frame Management System. It dumps the solution in both text form and machine readable form to disk files. If the synthesiser was activated from within an expert system, the Frame Management System has to regenerate the actual frame base in memory when control returns to the expert system using these disk files. The user can look up the file containing the solutions in text form whether or not the synthesiser was invoked as a stand alone shell or by an expert system.

External programs

The current directory must contain executable modules of all the external programs referred to in the decomposition. External programs are activated by the synthesiser through a system call. The synthesiser passes the required parameters to the programs through a disk file. Programs return values to synthesiser through the same disk file. Obviousely, excessive use of external programs affects the performance of the synthesiser but in many cases their use is unavoidable. It is important that external programs should be well behaved for otherwise, the entire environment may be affected.

IMPLEMENTATION DETAILS

Data structures

The implementation is done using C language. The representation of the knowledge and data in memory and their manipulation during synthesis are discussed below. The compiler scans the input files and stores the information in suitable data structures. Initially the systems are arranged sequentially in a linked list. At the end of the compilation a network is created to incorporate the interdependencies between various systems. The data structure of a system is shown in Fig. 3.

Each system has a unique name. The name of the system identifies the corresponding system data

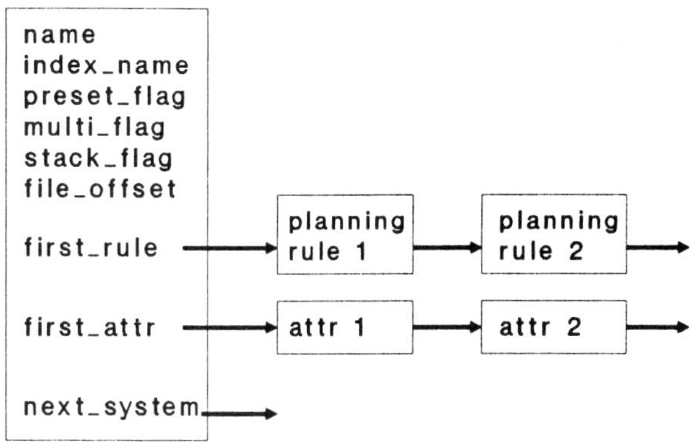

Fig. 3. System data structure

structure. Planning rules within a parent system may cause a system to be repeated at synthesis time. Every **repetition** planning rule must have a unique index of repetition. Indices appear often in expressions. The name of the index corresponding to a system is stored as **index_name** with the system. All planning rules stated within a system are stored in a linked list and attached to the system. The data structure of a planning rule is shown in Fig. 4a.

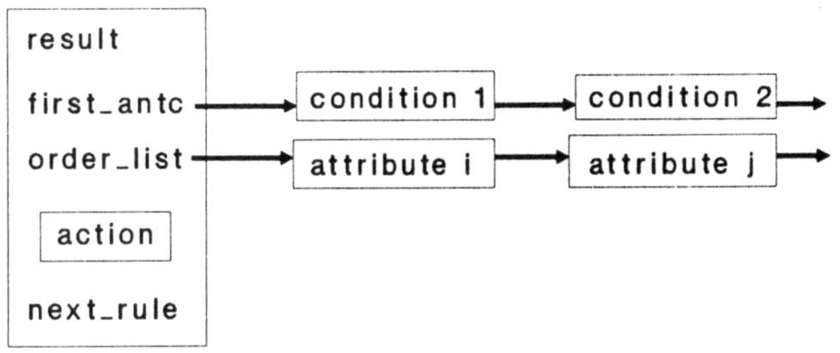

Fig. 4a. Data structure for a planning rule

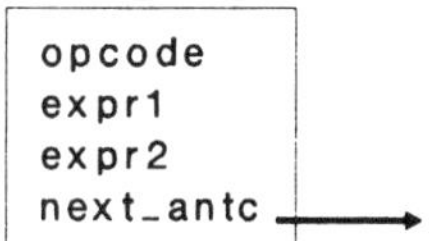

Fig. 4b. Planning rule condition

All conditions in the antecedent part of the planning rule are stored in a linked list. The data structure of a condition is shown in Fig. 4b. Each condition has a relational operator(opcode) and offsets to the start of code of two expressions(expr1, expr2). If the planning rule is of RESHEDULE type, the rule contains the order in which the attributes are to be reordered. The names of the attributes are stored in a linked list and attached to the planning rule. If the planning rule is of REPEAT type, the details of the action part are stored as a data structure **action** within the planning rule. This data structure is shown in Fig. 4c.

A system can contain any number of attributes. All attributes in a system are stored as a linked list and stored with the system. The data structure of an attribute is shown in Fig. 5a. When the compiler scans the constraints, (static and runtime) it categorizes them as attribute constraints, meta constraints and constraint functions. All attribute constraints are attached to the attributes they pertain to in a linked list. As soon as an attribute is synthesised, all the constraints attached to it are checked for. If any constraint is violated, the solution run is aborted. The data structure of an attribute constraint is shown in Fig. 5b.

```
index_name
syst_name
start
end
```

Fig. 4c Planning rule action

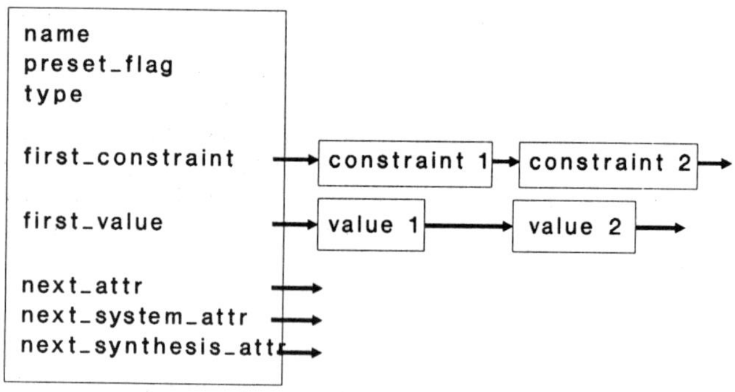

Fig. 5a. System attribute

An attribute can have one or many values depending on its type. All values are stored in a linked list and attached to the attribute. The data structure of an attribute-value is shown in Fig. 6a. Depending on the type of the attribute, the value of an attribute-value is stored in a union called value within the attribute-value data structure. The data structure of the union is shown in Fig. 6b. Depending on whether the type of the attribute is SUBSYSTEM, PROGRAM, PROCEDURE, EXPRESSION or ONE_OF, the union contains a pointer to the corresponding data structure of the system, program, procedure, offset to the code of the expression or a string. The data structure of a program is shown in Fig. 6c.

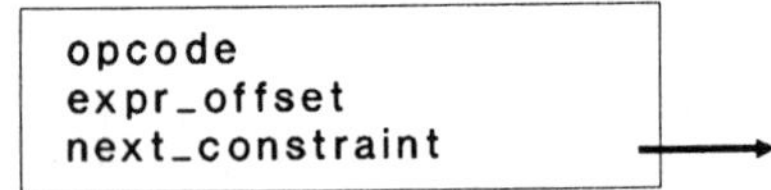

Fig. 5b. Attribute constraint

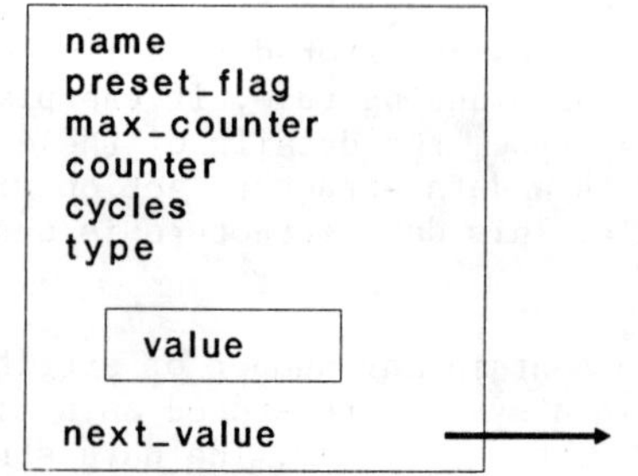

Fig. 6a. Attribute value

```
union
{
   struct syst *System;
   program_type *Program;
   procedure_type *Procedure;
   long expr_offset;
   char String[110];
}
value;
```

Fig. 6b. Common storage for value

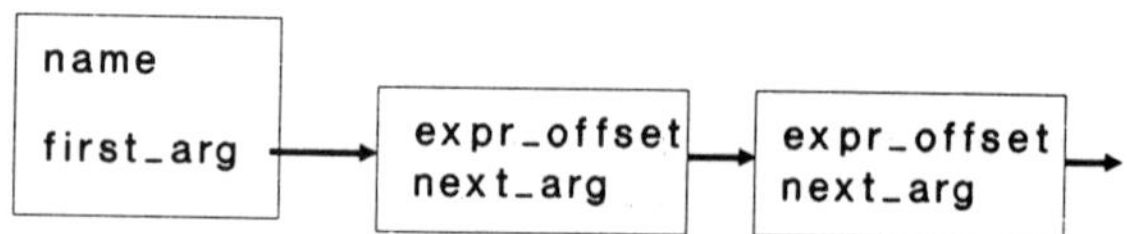

Fig. 6c. Data structure for external program

Meta constraints are stored separately as a linked list. Since meta constraints may involve attributes of different systems in the expressions, it is not possible to associate a meta constraint with any one attribute. All meta constraints are imposed at the end of each completed solution run. The data structure of a meta constraint is shown in Fig. 7.

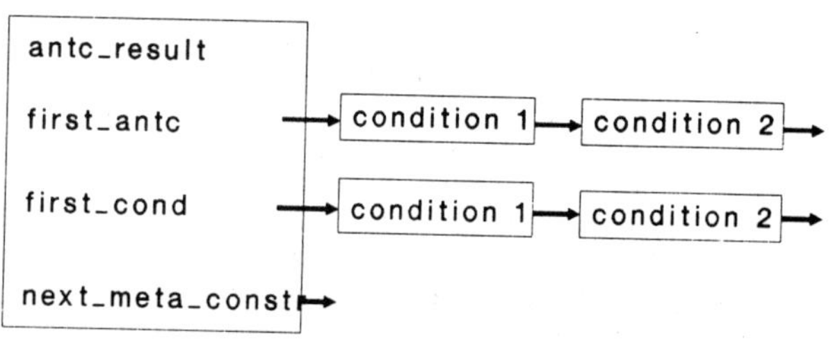

Fig. 7. Meta constraint

All indices of REPEAT planning rules are stored as a linked list. The data structure of an index is shown in Fig. 8. Index values can be used in expressions and are often passed to external programs. Since there may be several planning rules concerning the repetition of a particular system, there may be several indices corresponding to the system but every index corresponds to exactly one system.

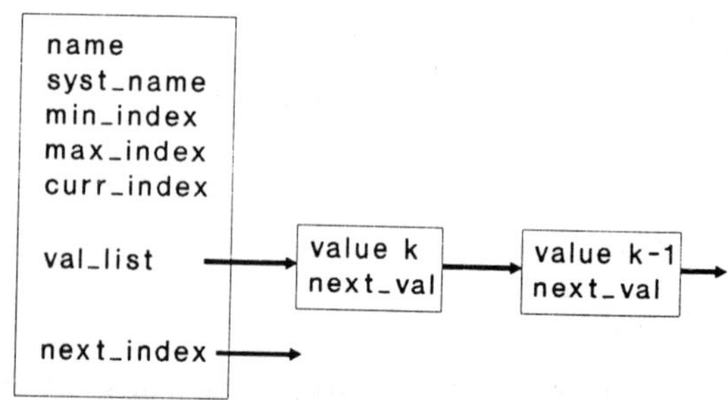

Fig. 8. Index of a planning rule

Options

Various options can be used while invoking the synthesiser. Options affect the way the synthesiser behaves. The -c option causes the synthesiser to check the input files for syntax and semantic correctness. No solutions are generated. The -s option causes the synthesiser to synthesise all solutions after the compilation is over. The solutions are written to the output file in text form only. The -a option causes the synthesiser to present each feasible solution to the user in text form on screen. The user can terminate the synthesis any time he finds a good solution. The -d option causes the synthesiser to write each solution to disk file in text form, present the solution to the user, and asks him whether he wants to dump the frame base in machine readable form to a disk file. These options are necessary since the way the synthesiser is used to build the prototype may vary.

Synthesis processor

The term synthesis processor applies to that part of the implementation of the Design Synthesiser that is specifically concerned with the assemblage of the solutions. After the compiler creates the network of systems and other data structures, a few tasks are taken up before the synthesis processor is invoked. First, the entire code generated for expressions is read into memory. This creates a correspondence between the file offsets stored in the data structures in place of expressions and the actual code. Then, all RESCHEDULE planning rules are enforced. It may be noted that while REPEAT planning rules are invoked in each solution run, the RESCHEDULE planning rules will be enforced only once before the synthesis process begins. From the network of systems, a tree of systems is created called the goal tree. The goal tree reflects the exact nature of the knowledge in the decomposition. This tree forms the basis from which the search space is created for each solution attempt. The synthesis processor tries every solution possible. Some of the solution attempts will be abandoned due to violation of constraints etc. To see that patterns of solutions arc not repeated, two regulating values are assigned to all attribute-values in the goal tree. To understand the role of these values, consider the simple decomposition shown below.

```
SYSTEM A
A1    ONE_OF       a1,a2
A2    ONE_OF       a3,a4
A3    SUBSYSTEM    B
END_SYSTEM

SYSTEM B
B1    ONE_OF       b1,b2
B2    ONE_OF       b3,b4
END_SYSTEM
```

There are 16 solutions to the problem. If we list them systematically, we note that each value occurs a certain number of times and in several cycles. For instance, the value a3 occurs 4 times consecutively at any place and in two cycles in the set of solutions. So, if each value is assigned a counter that specifies how many times it occurs in successive solutions and a cycles that specifies how many such times it repeats the process, we have a means of generating the solutions. However, the actual situation is not so trivial. Planning rules in the decomposition fire depending on the values of attributes already synthesised. This causes the search space or the hierarchy tree to be indeterminate and counters and cycles cannot be assigned. Therefore, cycles and counter values are assigned to values in a system ignoring the presence of all subsystems. They then reflect only the variations with respect to the particular system and not the entire tree. The synthesis processor is invoked in a loop as many times as the number of solutions possible for the topmost system of the goal tree.

```
SOLUTION NO: 1
----------------------
bldg_frame.length=16.000000
bldg_frame.width=8.000000
bldg_frame.height=5.000000
bldg_frame.roofing_matl=ACC
bldg_frame.siding_matl=ACC
bldg_frame.end_framing=gable_frame
bldg_frame.type=fink
bldg_frame.rise=1.333333
bldg_frame.nobays=4
bldg_frame.bay_spacing=4.000000
bldg_frame.nscols=10
bldg_frame.ngcols=0
bldg_frame.ntruss=5
s_column[1].bnx=0.000000
s_column[1].bny=0.000000
s_column[1].bnz=0.000000
s_column[1].fnx=0.000000
s_column[1].fny=0.000000
s_column[1].fnz=5.000000
s_column[2].bnx=0.000000
s_column[2].bny=4.000000
s_column[2].bnz=0.000000
s_column[2].fnx=0.000000
s_column[2].fny=4.000000
s_column[2].fnz=5.000000
s_column[3].bnx=0.000000
s_column[3].bny=8.000000
s_column[3].bnz=0.000000
s_column[3].fnx=0.000000
s_column[3].fny=8.000000
s_column[3].fnz=5.000000
s_column[4].bnx=0.000000
s_column[4].bny=12.000000
s_column[4].bnz=0.000000
s_column[4].fnx=0.000000
s_column[4].fny=12.000000
s_column[4].fnz=5.000000
s_column[5].bnx=0.000000
s_column[5].bny=16.000000
s_column[5].bnz=0.000000
s_column[5].fnx=0.000000
s_column[5].fny=16.000000
s_column[5].fnz=5.000000
s_column[6].bnx=8.000000
s_column[6].bny=0.000000
s_column[6].bnz=0.000000
s_column[6].fnx=8.000000
s_column[6].fny=0.000000
s_column[6].fnz=5.000000
s_column[7].bnx=8.000000
s_column[7].bny=4.000000
s_column[7].bnz=0.000000
```

Fig. 9. Sample text output from synthesiser

APPLICATION

The use of design synthesiser is illustrated with an application to the synthesis of feasible structural configurations for a steel industrial building. The input data includes length, width and eave height of the building, plinth level, skirting wall height etc. The solution desired should include the location of all memebers in the skeleton, including truss members, columns, purlins, side and gable girts and bracing members. The text form of a part of a solution is shown in Fig. 9. The machine readable form of the solution dumped to a disk file was used to generate the graphic images(Fig. 10).

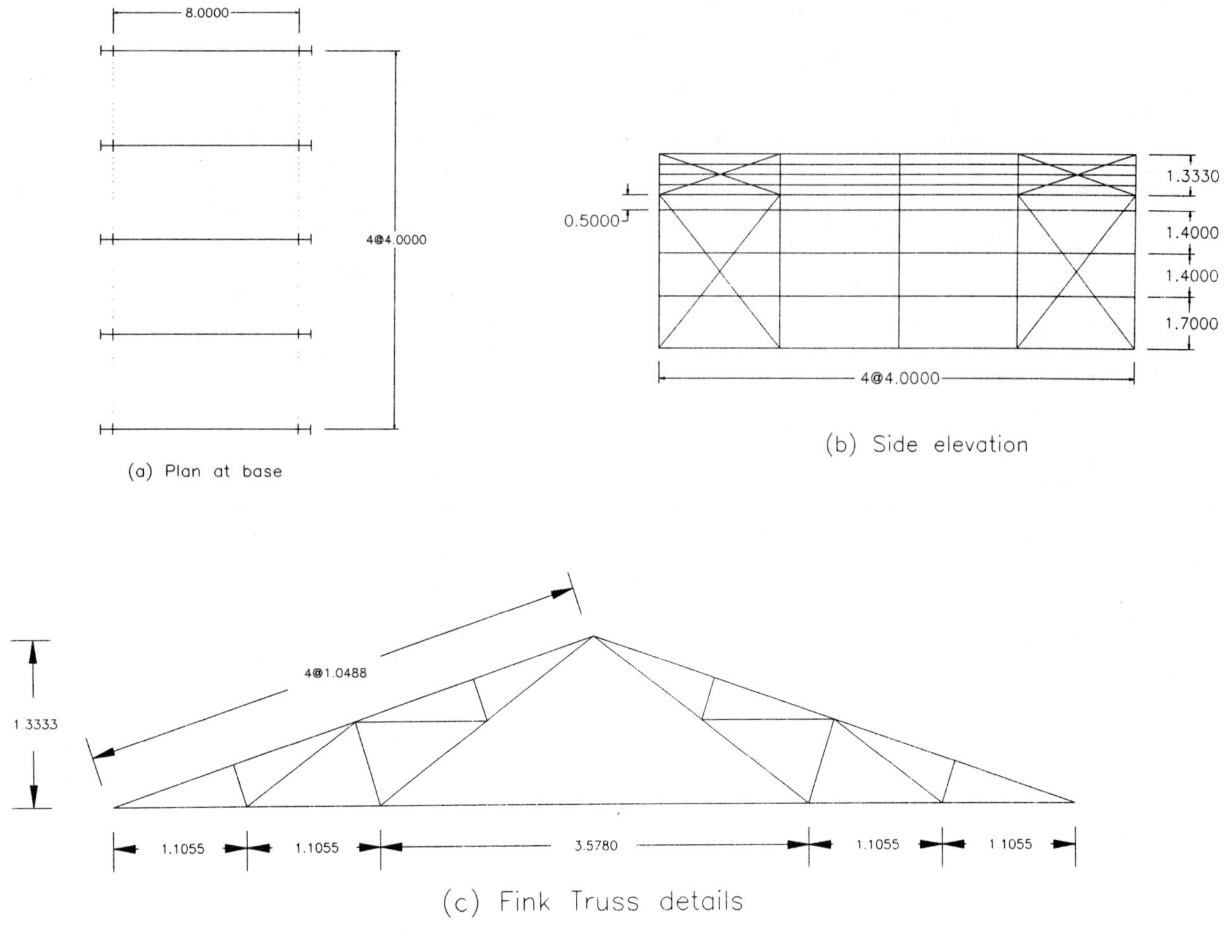

Fig. 10. Details of the synthesised structural scheme

CONCLUSIONS

The use of design synthesiser provides a systematic and structured approach to the conceptual phases of design process. Generic tools such as Design Synthesiser provide a powerful way of building real life expert systems. The implementation discussed in this paper meets all the necessary requirements of a generic tool that facilitates building expert systems. The use of generic components facilitates experimentation with different solution strategies by selecting suitable combinations of the components for diverse design subgoals.

ACKNOWLEDGEMENT

The work reported in this paper is based on the Indo-US research project on "KBES for Civil Engineering Design" under NSF Grant nos. INT-8819520 and INT-891386 currently in progress at Indian Institute of Technology, Madras in collaboration with Carnegie Mellon University, Pittsburg, USA. The authors are grateful to Prof. S.J.Fenves for his advice and suggestions in the development of the software for this project.

REFERENCES

1. Rajeev, S., Fenves, S.J., Krishnamoorthy, C.S., Architecture and Generic Components of Knowledge-Based Expert Systems for Civil Engineering Design, Research Report No. R-90-186, Dept. of Civil Engineering, Carnegie Mellon University, USA, and Dept. of Civil Engineering, Indian Institute of Technology, Madras, India.

2. EDESYN: Users's manual, Dept. of Civil Engineering, Carnegie Mellon University, USA, January, 1990.

3. Srinivasa Rao, C., Krishnamoorthy, C.S., Rajeev, S., Engineering Design Synthesiser: Implementation Manual, Research Report No. CE-91-06, Dept. of Civil Engineering, Indian Institue of Technology, Madras, India, January, 1991.

4. Srinivasa Rao, C., Krishnamoorthy, C.S., Rajeev, S., Engineering Design Synthesiser: User's Manual, Research Report No. CE-91-07, Dept. of Civil Engineering, Indian Institute of Technology, Madras, India, January, 1991.

SUPPORTING CONCEPTUAL DECISIONS IN STRUCTURAL DESIGN

A. Borkowski†, N. Fleischmann‡ and K.U. Bletzinger⋆
†Institute of Fundamental Technological Research, Warsaw, Poland,
‡Zublin AG, Stuttgart, Germany,
⋆Institut fur Baustatik, Universitat Stuttgart, Germany

The paper deals with conceptual decisions taken during the preliminary design of structures. Aiming at the computer support of such decisions one comes up inevitably to the application of AI-techniques. The following issues are discussed: intelligent access to the previous design experience stored in the database, automatic generation and comparison of plausible alternatives, acquisition of new knowledge through algorithmic structural optimization. The difficult task of supporting innovative solutions plays the main role in the considerations. Several knowledge-based computer programs illustrate the implementation of the proposed ideas.

1. INTRODUCTION

A process of structural design includes three main phases: the initial design, the verification and the detailed design. It is commonly accepted that the quality of the final solution depends substantially upon conceptual decisions taken at the first phase. The circumstance that such decisions are based upon the designer's experience and the rules of thumb dictated by the engineering practice stimulates the efforts to apply AI-techniques for creating software tools that would support the initial design. Thus numerous expert systems for preliminary design of structures and machine parts are reported in the literature (Refs. 9,10).

Most of them address the so-called routine design. It amounts to the selection of the best alternative from a finite set of à priori known possible solutions. Though conceptually modest, this approach still offers advantages when compared to the systems that do not provide an automatic generation of the plausible alternatives. Theoretically every conceptual decision should be taken after a thorough evaluation of alternatives. In practice, however, the time period available for preparing the preliminary project is limited. If no computer support is available, few alternatives are taken into account and a possible advantageous solution may fall out of consideration. It seems, therefore, that knowledge based programs generating and evaluating all plausible solutions of a given design problem should prove to be useful in practice.

Another obviously reasonable tool for a preliminary design is a database system allowing the designer to take advantage of the experience stored in the form of previously accepted solutions. Each established design office holds the descriptions of many projects erected in the past. Casting such records into a properly designed database enables the effect of expanding experience to come into appearance. Such a database should include an intelligent front-end that allows the user to make queries based upon incomplete and possible vague request.

Despite their practical merits, both already mentioned tools have a common principal drawback: they do not provide innovative solutions. Since creative design is the most important part of an engineering activity, efforts to provide computer support for it are made at almost every research center involved in the development of experts systems for structural design. Though several techniques provided by the machine learning seem to be applicable in the area, few practically important results were achieved up to day (Refs. 1, 2, 11, 17). The reason for that might be the gap between the shallow and deep knowledge observed in many proposed learning systems. In our opinion, shallow knowledge methods like the case-based learning or learning via analogy should be enhanced by the structural optimization.

It is the great advantage of structural design that at a certain level of abstraction the best shape of the structure or the optimum values of certain parameters can be found algorithmically. The simplifications necessary on the way to such an abstract model may prevent the direct practical application of the optimum solution. It should be further transformed and adapted according to the heuristic rules. Hence, a natural scheme for expert systems supporting the innovative design seems to be the mixed declarative-procedural knowledge representation with the elements of deep knowledge implemented algorithmically. The above mentioned problems of knowledge-based support for structural design were the subject of joint effort taken at the Laboratory of Adaptive Systems belonging to the Institute of Fundamental Technological Research, Warsaw, Poland an the Institute of Structural Analysis, University of Stuttgart, Stuttgart, Germany. Preliminary results obtained in the period 1990-91 are reported in the present paper.

In the next section we consider an intelligent interface to the database that contains the design experience in a certain area. Exploiting such an experience can be seen as a simplest form of machine learning - the rote learning. Section 3 deals with a prototype generator of alternative designs for the load carrying structure of a single bay hall. This project, closely related to the professional activity of the second author, aims at the improvement of the design proposals prepared at the phase of competing for the contract.

In section 4 the crucial problem of supporting innovative elements in design is discussed. The problem itself remains far from being solved, but at least general requirements imposed on the knowledge-based systems able to provide such a support are formulated and the directions of further research are pointed out.

The subject of section 5 is the relation between the conventional algorithmic structural optimization and the knowledge based approach. We argue that structural optimization should be regarded as an important and relatively cheap source of new knowledge. Hence, an integrated structural design environment should include both algorithmic and heuristic optimization tools.

2. ROTE LEARNING APPROACH

Usually, after a new design task has been specified, one tries to find out whether a similar problem has been already solved in the past. Storing the information about all preceding designs in the database falls into the category of rote learning: the knowledge base grows continuously and the probability of finding a similar solution increases in time. The advantages of properly organized archivation of the projects are especially significant in large engineering offices, where less experienced members of the staff can learn from the solutions proposed by leading designers.

Storing knowledge in project databases seems to be useful in addition to storing knowledge in form of rules. Rules must be acquired by a special knowledge engineer who tries to extract knowledge from the experienced planner by asking questions. The disadvantages of this kind of knowledge acquisition are:

- lack of experienced planners who have time to answer questions,
- difficulty of formulating questions,
- impossibility to reproduce design knowledge without being faced with the actual design context,
- ambiguity of design knowledge between several design experts,
- difficulty to maintain rule-bases containing more than several hundred rules,
- high costs for the work of the knowledge engineer.

Obviously, the project database should be implemented in one of the commercially available environments. They provide comfortable editing of the data records and give the user access to efficient data retrieval techniques. However, relational query languages like SQL are not flexible enough to allow much freedom in the data request. Such a freedom is needed, when one is going to search for similar designs, since the information about the design goal is typically vague and incomplete. Therefore, it seems to be reasonable to place an intelligent interpreter between the user and the relational database. Such an interpreter should be able to supplement the information provided by the user by an additional information derived from the general knowledge about the design domain.

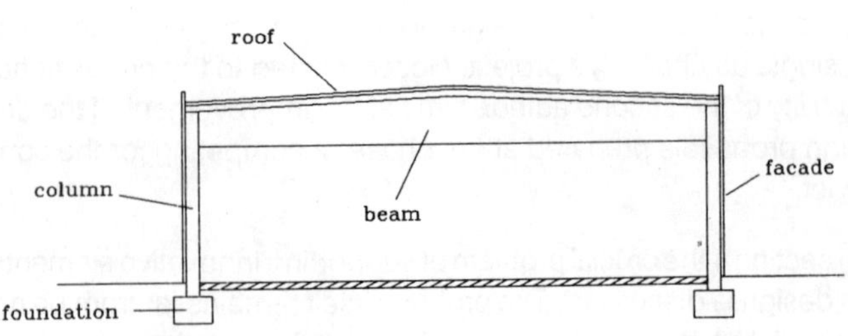

Fig. 2.1. *Cross section of a single-bay rectangular hall*

In order to check the feasibility of this approach an experimental database FIND with an intelligent front-end has been implemented using PROLOG on a MS-DOS computer. The knowledge domain of FIND covers the design of single bay halls with rectangular ground shape. A symmetric plane cross-section of the hall is considered. The load carrying structure of the hall can be divided into the foundation, the column and the beam (Fig. 2.1). The database consists of about 50 industrial halls with rectangular floor plan. The project data worth to be stored in the database include:

- general data,
- geometric data,
- costs,
- materials,
- form, type and description of the structural components,
- data about manufacturing and erection on site.

In the first step, these data are acquired by the structural engineer via a special user interface and stored in a relational database. The utilities offered by commercial database systems, like indexing and setting filters, are useful for maintaining the database. In the second step, however, in order to implement an intelligent database access, a frame-oriented representation was used instead of the relational representation, because the attributes of the projects are very complex. The frames form a taxonomy which is shown in Fig. 2.2.

Each frame contains a set of slots representing the attributes. The slots consist of several facets:

- parent frame,
- actual value of attribute,
- type of attribute (symbolic or numeric),
- unit (for numeric attributes),
- upper bound (for (numeric attributes),
- lower bound (for numeric attributes),
- set of legal values (for symbolic attributes),
- default value (for all attribute types).

The program FIND works in two phases. In the first phase the user is asked to enter the values of an arbitrary set of attributes. Some attributes, like span and load are obligatory. The given set of attributes is checked using a rule-based knowledge base. In the second phase, the requested values of the attributes are compared against the values stored in the database in order to find most similar solutions. A weighted difference between the requested value a_{i0} and the database value a_{ij} is defined as

$$\alpha_{ij} = w_i \; |a_{i0} - a_{ij}| \quad ; \quad \begin{array}{l} i = 1,2, \; \dots \; , N_{rq} \\ j = 1,2, \; \dots \; , N_{db} \end{array} \qquad (2.1)$$

where N_{rq} is the number of requested attributes and N_{db} is the number of solutions stored in the database. Numerical values of attributes are kept non-dimensional. The differences between symbolic values belong to the range [0, 1] and are taken from a distance table. For example the difference between material *prestressed_concrete* and material *reinforced_concrete* is assumed to be 0.2, whereas the difference between material *prestressed_concrete* and material *steel* is taken as 0.4.

In order to differentiate the significance of attributes, the weight factors w_i [0, 1] are introduced. The highest factors $w_i = 1.0$ are assigned to the span, load and material. Most of the attributes however are considered less important and their weight factors w_i lie between 1.0 and 0.4.

The distance between the request and the j-th entry of the database is calculated from the formula

$$D_j = \sqrt{\sum_{i=1}^{N_{rq}} a_{ij}^2} \qquad (2.2)$$

After D_j were found for all entries, ten "nearest" solutions are sorted out. Thus the result of the procedure **FIND** is a list of ten examples of halls, nearest to the requested one in the sense of similarity measure (2.1), (2.2). Fig. 2.3 shows the screen after a sample run. The user has set only two values, *span* = 20 m and *load* = 10 kN/m. A database entry with the smallest D_j is displayed as the result of the search. Subsequent matches can be browsed upon request.

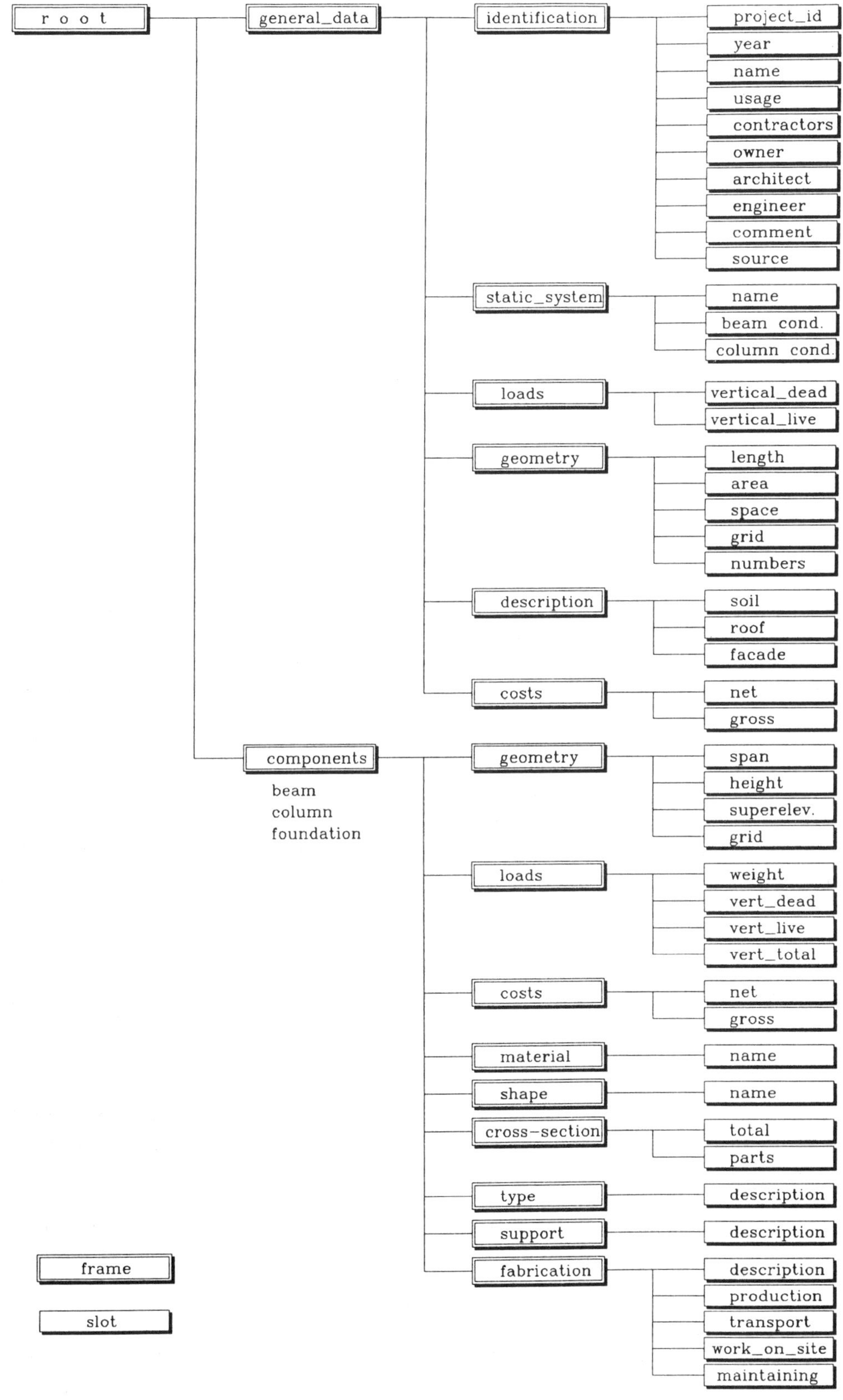

Fig. 2.2. *Taxonomy for rectangular halls*

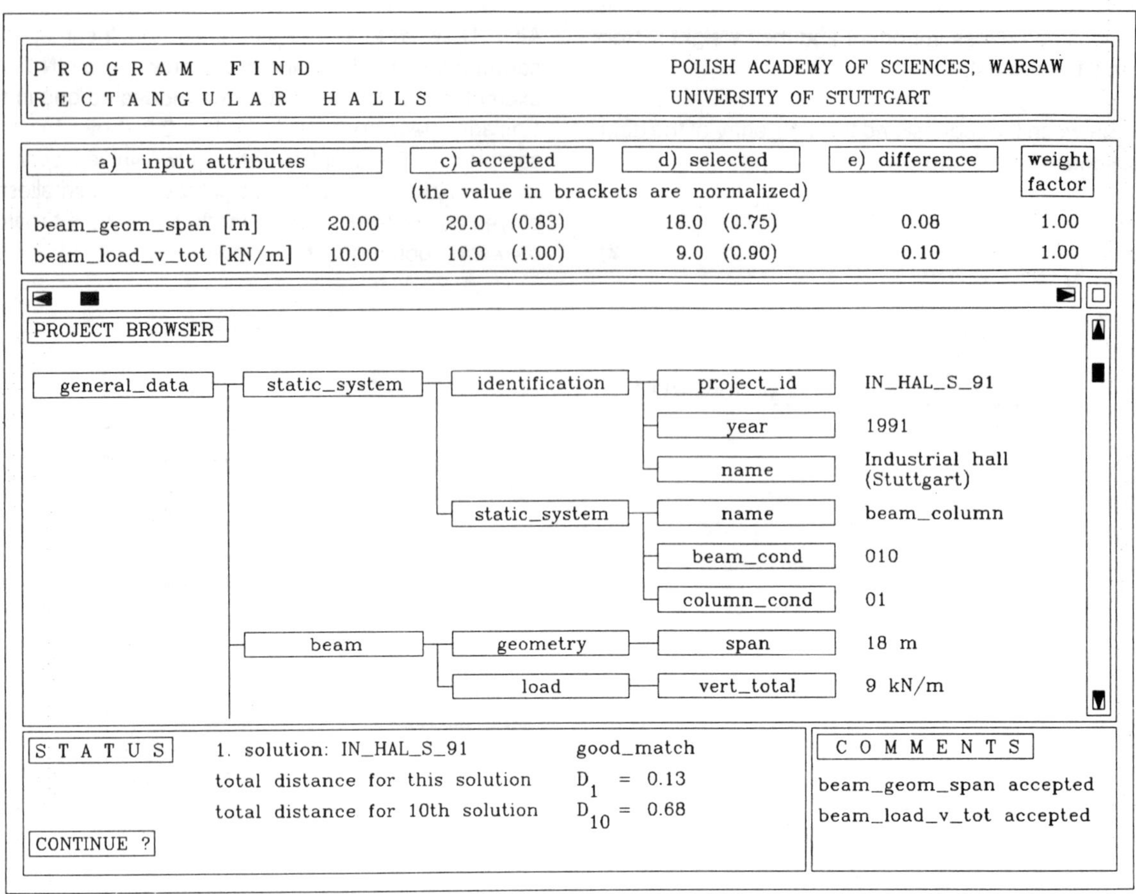

Fig. 2.3. *Sample screen output of program FIND*

The project databases can also be used to derive new knowledge from the raw experience stored in the database. This seems to be a promising field of study in the future and falls within the subject of machine learning - the rapidly developing branch of Artificial Intelligence. Current efforts to apply such a procedure in the context of structural design proceed in various directions. One of them is the case-based learning (Refs. 1, 2). It has been shown that after a certain training a program is able to predict rational cross-sections of a given structure, without carrying out analysis of optimization. This kind of qualitative reasoning is very useful in the stage of preliminary design, when conceptual decisions are to be taken.

3. GENERATE AND COMPARE APPROACH

Usually, the architect creates the initial design for a building, choosing general shape and dimensions of the object. Then, the task remains to the structural engineer to convert the architect's idea into a technical solution that satisfies the requirements of strength, rigidity, manufacturing etc. However, the structural engineer should take part in the design process, too. The need for the structural engineer's design activities exist in the early phase of a project, when he acts as a consultant engineer, as well as in the later phase, when he works for contractors tendering for an order.

It is very important for contractors during tendering to find special proposals. These proposals can differ from the architect's specification, supposed they offer other advantages. A special proposal should be competitive in comparison to rival proposals but should not underestimate the real cost of the object. In order to prepare a good proposal in short time, one needs a quick access to a large number of alternative solutions. Contrary to the detailed design performed after winning the competition, the preparation of the proposal does not require an exact analysis and dimensioning of the structure. It is much more important to know, which alternatives are available for the specific characteristics of the structure and which combination of those alternatives leads to the most economic solution.

It is worth pointing out that practical design goals for finding variant designs and special proposals almost always fall out of the scope of Structural Mechanics. The main goals of practical optimization are:

- reducing manufacturing costs,
- reducing manufacturing and construction time,
- reducing maintenance costs,
- improving quality and durability.

Additional aspects to be considered in optimizing structures are the design codes that are to be unified in European countries.

As a step towards computer supported conceptual design the program TREE (Fig. 3.1) has been developed in PROLOG. It is a prototype for a design guidance tool allowing the user to generate alternative solutions of a structural design problem and to evaluate them with respect to the measure of merit.

The design process is modelled as a tree according to the MSD-scheme (Ref. 14). The user is guided during a top-down design process in three steps:

- *Decomposition:* Decompose the structure in components and find all possible solutions for the components of the structure within a given search space according to a given set of input attributes. The search space is given through a object taxonomy, similar to that one of FIND.The process is divided into a *develop* process and a *select* process, both being carried out in two levels, the structural level and the conceptual component level.
- *Aggregation:* Assemble the components found in the previous step to overall alternatives using a knowledge base.
- *Evaluation:* Evaluate the overall alternatives using the knowledge base and evaluation tables. Produce a list of final results consisting of a set of ordered overall alternatives.

Fig. 3.2 shows the design tree for rectangular halls in detail. We see that at a *develop* node (AND-node) the program generates the *select* nodes, i.e. the items, which have to be designed next. For example, at the structural level of the design of a hall the pro-

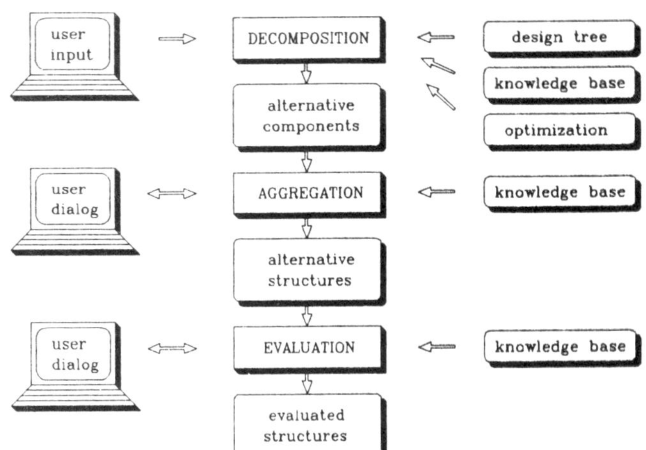

Fig. 3.1. *Flow chart of program TREE*

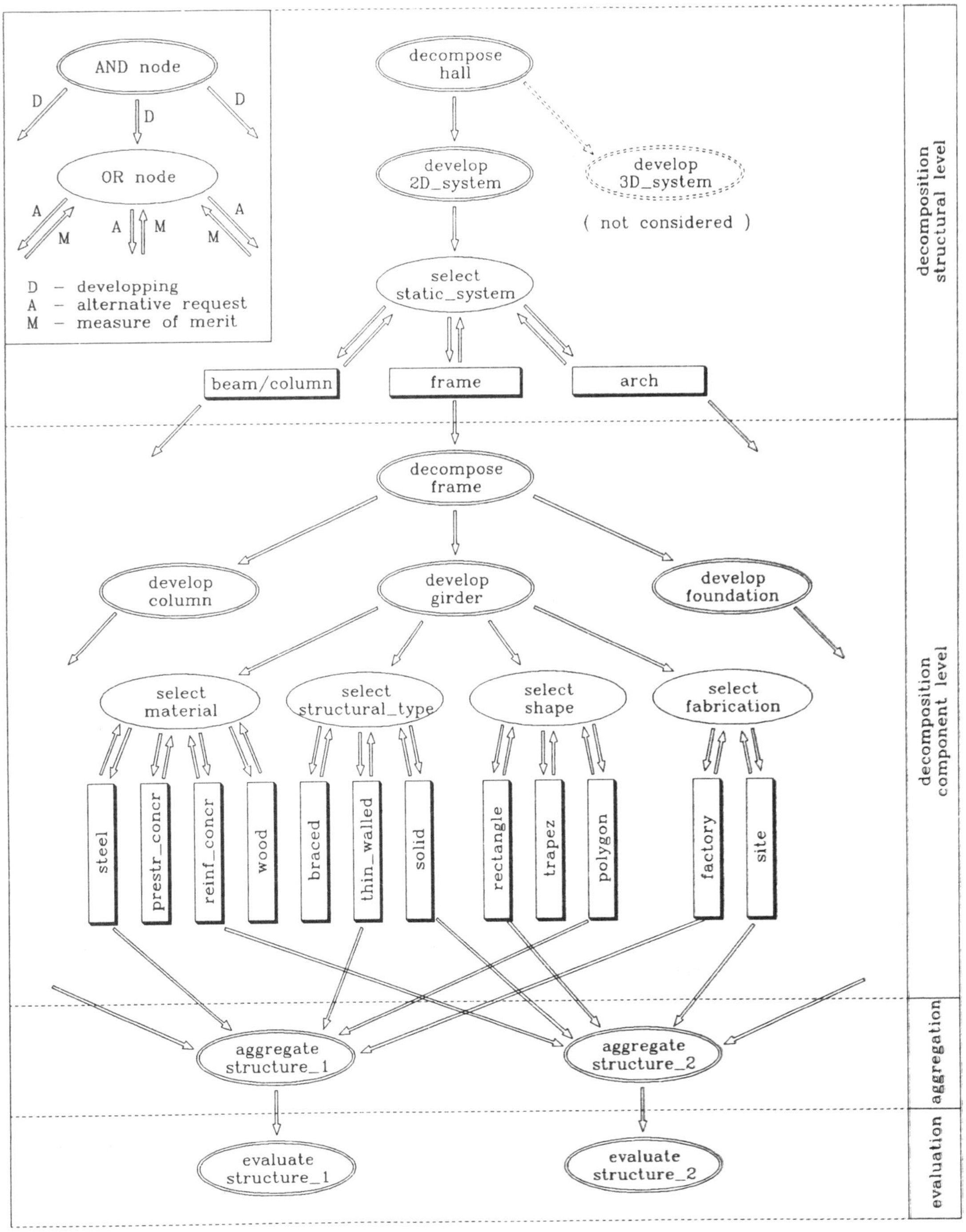

Fig. 3.2. *Design tree for halls of rectangular plan*

gram defines the static system to be found. At the conceptual component level, the program proposes material, shape, type, and manufacturing technique of the components columns, beam, foundations. If several decisions are to be made at the same level, the user is allowed to choose the order of selected nodes. This feature, absent in the original MSD-model, seems to be rather important.

At each *select* node (OR-node) TREE presents possible alternatives and asks the user to accept them or to prune some of them. Pruning is important to keep the number of stored alternatives as small as possible. The user's decision is supported by the computer using a rule-oriented knowledge base. Previous decisions obviously influence the alternatives that have to be considered at a certain *select* node. If, for example, *steel* and *wood* were pruned as beam material, then *thin-walled* and *braced* should not be considered as structural type. At the structural level we define only one *select* node

```
static_system      -   beam_column
                   -   frame
                   -   arch,
```

whereas we have four *select* nodes at the component level

```
material           -   steel
                   -   reinforced_concrete
                   -   prestressed_concrete
                   -   wood
shape              -   rectangle
                   -   trapeze
                   -   triangle
structural_type    -   braced
                   -   thin-walled
                   -   solid
manufacturing      -   site
                   -   factory.
```

Of course, some nodes or alternatives are not relevant like for example *material wood* for the foundation or the node *shape* for the column and foundation. TREE removes such nodes or alternatives by itself.

As an example let the user start the dialog, giving some obligatory input such as span, allowed maximal height, and maximal loads. The program asks for the first decision to be made:

```
> choose static_system from        { frame
                                   | beam_column
                                   | arch }.
```

The user prunes alternative *arch*. TREE then decomposes the remaining two alternatives *frame* and *beam_column*. The user selects *frame* to be decomposed at first. Note that the order of the select menus is not arbitrary but results from rules in the knowledge base. As the result of decomposition, TREE shows the 3 components to be developed next:

```
> choose next component for frame from { column
                                       | girder
                                       | foundation }
```

Let us consider *girder* to be developed then TREE shows the decisions to be made:

```
> choose next select item          { material
        for frame girder from      | shape
                                   | structural_type
                                   | fabrication }
```

If the user chooses *material* as the select item the program presents the following alternatives, which will all be stored unless the user prunes some of them:

```
> select material                  { prestressed_concrete
        for frame girder from      | reinforced_concrete
                                   | steel
                                   | wood }.
```

Before presenting the possible alternatives, the program prunes non-plausible alternatives by itself, for example, if the span is very large, *wood* wouldn't be presented as girder material. Thus the decomposition process produces a set of alternative static systems and of possible components for a hall.

During the decomposition process it is very important for the user to be informed about the conclusions embedded in the computer program. Therefore, design tools like TREE should offer as much explanation utilities as possible. A very powerful feature of TREE is the possibility to support decisions by structural optimization. At several select nodes it is possible to consult algorithmic optimization programs (see section 5). For example at the node *girder/shape* the program can generate new shapes by using a separate optimization algorithm.

```
> select material                  { rectangle
        for frame girder from      | triangular
                                   | trapezoid
                                   | optimize }.
```

After the complete tree of alternatives has been generated and accepted by the user, plausible combinations of components, which can be regarded as leaves of the tree, are taken as alternative proposals for the overall structure. The aggregation knowledge base eliminates nonfeasible cases such as a concrete beam supported by wooden columns. The user is allowed to reject some proposals as well.

The last phase of the consultation requires less interaction with the user. TREE computes the total cost for each alternative structure and presents ten best, i.e. cheapest, solutions. Thus the user gets valuable information about the conceptual alternatives embodied in this particular design task.

4. ROUTINE AND INNOVATIVE DESIGN

A question that occupies almost everyone involved in the development of knowledge based design tools is, how to make their scope broader than the multiplication of standard solutions or the refinement of prototypes. There is no doubt that most currently available expert systems cover this kind of the design activity. If one calls it "routine design", the adjective "routine" should not be understood as depreciating. The gradual improvement of existing solutions prevails in the everyday practice and the knowledge based computer programs can substantially improve this process. Fortunately, such a slow progress is accelerated from time to time by the technological breakthroughs, engineering inventions, and scientific discoveries. The gains due to innovations are so significant that the possible impact of the knowledge based systems upon the creativity of their users must be properly evaluated.

Since the first generation, knowledge based systems use the static knowledge, they lead the user inevitably towards routine solutions. This property is by no means introduced by the new software technology: each book devoted to the design of specific object promotes certain recipes. They are based upon the judgement of what is good or what is correct. Such a conservative knowledge keeps the progress on the rails of evolutionary improvement until somebody succeeds in bringing up a new idea. If the proposal is genuinely original, then it falls out of the frame depicted by the current static knowledge. Hence, the idea is perceived at the beginning as being false, impossible to implement etc. After a certain period of struggle one of the alternative events happens: either the non-orthodox idea is rejected, or the knowledge base is modified.

Naturally, we would like our software tools to be able to support the qualitative jumps in the design process. The dynamic knowledge base implies learning. If that base is improved by the program itself, then we are speaking about the machine learning. Up to nowadays, there were many attempts to generate conceptually new ideas automatically. Alas, results are rather unsatisfactory. They were obtained exclusively in very narrow and easy to formalize fields, like a search for the new compound substances in chemistry.

Therefore, it seems reasonable to us to begin with a less ambitious goal. Leaving the control and initiative to the user, we would like to encourage him to leave the safe track of routine solutions. This can be implemented in various ways, and we do not aspire for the general solution of this task.

Usually, the design process is modelled as a computational search in the fixed space X of design variables (attributes). Such a search is driven by a given goal. The requirements imposed upon the expected solution determine an admissible domain D belonging to X. It is clear that for fixed X and D such a model can hardly produce a conceptually new solution. This is especially true for the combinatorial type of search, with the à priori given sets of admissible values for each attribute. In the next section we show that the optimization based upon continuous variables can eventually lead the designer to conceptually novel ideas (e.g. an unexpected shape of the structure). However, in general we strongly support the position taken by Smithers and Troxell in their recent paper (Ref. 18) that the innovative design should be modelled as an exploration of the design space rather than as a search.

It seems to us that software tools supporting design should enable the user at least to modify D and preferably also to change X. A simple heuristic for the former task is to relax temporarily a particular constraint and to inspect then the solution. Fig. 4.1 illustrates such a procedure. Looking for the minimum weight of the rigid plastic frame, one obtains the yield moment of the middle column equal to the lower bound imposed for technological reasons on elements in bending. When that constraint is removed, the optimizer suggests zero bending resistance for this column. Thus, one is encouraged to try the conceptually different static scheme depicted in Fig. 4.1 b.

In order to discuss the modification of the design space itself, let us consider the following simple representation of the design object:

$$\text{structure} \; (\; [x_1, \, x_2, \, \ldots, \, x_m] \; , \; [x_1, \, x_2, \, \ldots, \, x_n] \;) \qquad (4.1)$$

Here x_i are the attributes divided into two groups: fixed input data $x_1, x_2, \ldots, x_m$ and variable output qualities $x_1, x_2, \ldots, x_n$. Taking the value of a certain attribute as given à priori means that the dimension $m + n$ of the design space is diminished by one. Thus, if we assume:

$$\text{structure} \; (\; [\text{load}], \; [\text{material}, \; \text{shape}] \;) \qquad (4.2)$$

then the search is to be conducted on the plane load = const. instead of considering the entire 3-dimensional space (Fig. 4.2 a). Looking for a structure capable to carry the given load is the routine way of thinking. One of the possible sidesteps would be:

$$\text{structure} \; (\; [\text{material}, \; \text{shape}] \; , \; [\text{load}] \;) \qquad (4.3)$$

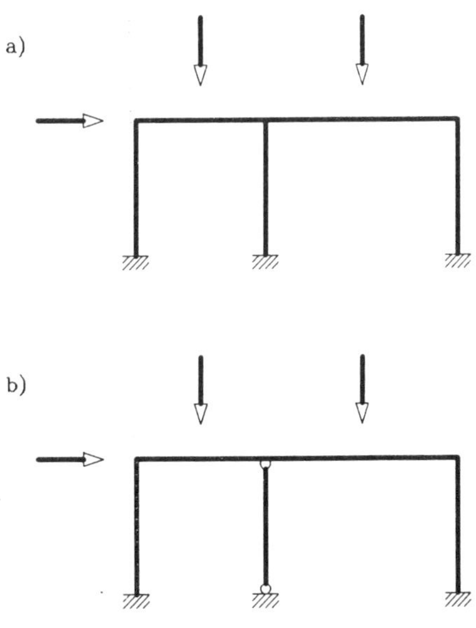

Fig. 4.1. *Modification of admissible domain in minimum weight design: a) initial static scheme; b) scheme suggested by optimization procedure*

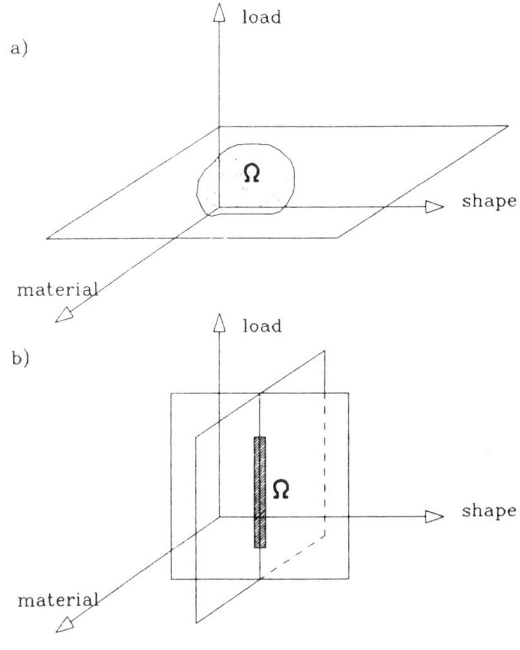

Fig. 4.2. *Modification of design space: a) given loading; b) given shape and material*

i.e. the search for the most favorable loading for the structure of given shape and material (Fig. 4.2 b). Such a problem makes sense when a complex system is decomposed into substructures and one wishes to optimize the stress flow between them.

Modern AI-languages, like PROLOG, give the user much freedom in declaring input and output data. It is therefore relatively easy to develop software tools that support the dynamic definition of the design space (Ref. 12). In the hands of motivated designers such programs may lead to interesting solutions.

5. OPTIMUM STRUCTURAL DESIGN - A SOURCE OF NEW KNOWLEDGE

At present, optimum structural design is understood as a synthesis of several individual disciplines (Refs. 4,8): (i) design modeling, e.g. by methods of computer aided geometric design; (ii) structural analysis, e.g. finite element method; (iii) behavior sensitivity analysis; (iv) mathematical optimization; and (v) interactive computer graphics. The methods have reached a remarkable standard and have been used as design tools to improve structural quality in many industrial applications especially in aircraft and automotive as well as in mechanical industries (Ref. 16). Their potentials for alternative design in civil engineering is not yet exploited to the last extent.

Structural optimization is divided into three levels of increasing complexity: sizing, i.e. design of cross sections, shape optimal design, and optimization of structural configuration or topology. Problems of structural sizing or shape optimization can be solved very satisfactory by methods of mathematical programming (Ref. 7). Applications to topological optimization are limited because of the highly discontinuous nature of those problems. Nevertheless, an interesting method based on material re-distribution is reported for the design of Michell-type structures (Ref. 3). Conceptional decisions of more severe consequences on the choice of structural configuration appear to be a matter of knowledge-based procedures. But even within the frame of a given principal class of structures, especially shape optimization can lead to so far unknown solutions and valuable new classifications of structures which are thought to be well known.

A typical problem of structural optimization is characterized by an objective $f(\mathbf{x})$ and constraints $\mathbf{g}(\mathbf{x})$ and $\mathbf{h}(\mathbf{x})$ which are nonlinear functions of the optimization variables $\mathbf{x}$. It can be stated as:

$$\begin{array}{lll} \text{minimize} & f(\mathbf{x}) & \\ \text{subject to} & h_j(\mathbf{x}) = 0 & ;\ j = 1,\ldots,m_e \\ & g_j(\mathbf{x}) \leq 0 & ;\ j = m_e+1,\ldots,m \\ & \mathbf{x}_L \leq \mathbf{x} \leq \mathbf{x}_U; & \mathbf{x} \in \mathbf{R}^n \end{array} \qquad (5.1)$$

where $\mathbf{x}_L$ and $\mathbf{x}_U$ are lower and upper bounds on the variables respectively. Most of sizing and shape optimization problems can be assumed to be continuous in gradients and curvature of the problem functions and are effectively solved by gradient oriented methods.

Modern program systems like CARAT (Computer Aided Research Analysis Tool, Ref. 6), which was used within in this study, allow us to formulate and to solve the above mentioned optimization problem as a standard procedure. Optimization variables $\mathbf{x}$ can be chosen from the set of pre-processor data which describe the structural layout. These can be coordinates and structural thicknesses which are interpolated by shape functions to yield FE data. Objective functions and constraints are evaluated by Finite Element Methods and include:

- structural weight and volume
- strain energy (to maximize stiffness)
- stress leveling (to yield a smooth stress distribution)
- rotational inertia
- frequencies
- stresses
- displacements
- geometrical conditions

Optimizers can be chosen from a library containing different methods among them two were used here: a sequential quadratic programming method (SQP, Ref. 15), NLPQL, and an own implementation of the method of moving asymptotes (MMA, Ref. 19). The SQP method appears to be superior with mildly constrained problems like shape optimization due to minimization of strain energy and stress leveling, while MMA gives very satisfactory results with highly constrained problems like weight minimization.

CARAT has been used successfully for the optimization of all kinds of structures especially free formed shells subject to one or more load cases (Refs. 5,13). Two examples show the capabilities of structural optimization as innovative part of a general approach to structural design as discussed in this contribution.

T-joint

A T-joint loaded by a vertical load and supported at the bottom is idealized as plane stress problem. The original structure (Fig. 5.1a) with sharp reentrant corners has high stress concentrations at the corners as well as in the upper fibres. Due to symmetry only one half of the structure has to be analysed. The contour of the structure is to be determined which exhibits a balanced v. Mises stress state with given amount of available material. For that instance a stress leveling function is taken as objective function which minimizes stress differences with respect to a given value. Reference value is the mean of von Mises stresses at the initial stage. First only the outer contour is adapted. Because the final design (Fig. 5.1b) still has a low stressed internal part the joint is provided with an internal hole. This could be initiated by consultation of a knowledge base which allows us to compare the optimization result with other already known solutions. The shape is optimized again so that the stress state is homogenized (Fig. 5.1c,d). Again, the optimization results are reported to the data base to be used in further runs. The structure was modelled by 6 and 5 design elements, respectively. The finite element analysis uses 2x2 integrated 8-noded plane stress elements. As optimizer NLPQL was used.

Note that it was not possible to obtain the finite shape in a continuous manner. Instead of that, two distinct runs of mathematical optimization were performed, linked together by a phase of conceptual reasoning. This kind of mixed procedural/knowledge-based approach to structural optimization seems to be most promising.

Shells of Revolution

The example demonstrates the principle capabilities of shape optimization in a qualitative way. The shape of nearly bending free shells of revolutions are to be designed. For this instant the objective is either weight or strain energy. Starting from a spherical shell of uniform thickness (radius 10.0 m) with an apex hole of radius 2.5m the optimal shape is obtained for four different load cases:

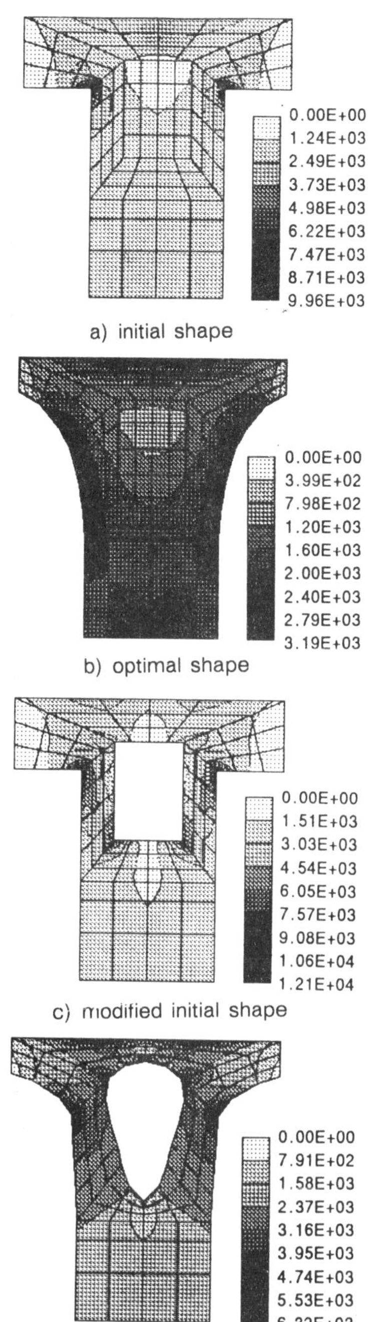

Fig. 5.1. *Shape optimization of T-joint (stress leveling of v. Mises effective stresses)*

concentrated ring load c = 255 kN/m on the upper free edge, dead load (γ = 78.5 kN/m^3), uniform vertical load of 0.75 kN/m^2 (snow) and a uniform horizontal load of 0.72 kN/m^2 with pressure on one side and suction on the other side roughly simulating wind. Either a fixed hinged or a horizontal roller support is assumed. The material properties of steel are chosen: E = 2.1 x 10^5 N/mm^2; υ = 0.3. Stress constraints with a von Mises stress limit of 160 N/mm^2 are considered.

The surface can be defined by one single Coons element described in cylindrical coordinates using a 6-node B-spline and a 2-node Lagrange edge in meridional and hoop direction, respectively (Fig. 5.2). Since the optimization for the horizontal load leads to a non-axisymmetric shell two Bézier elements are applied for one half of the structure. In both cases only 4 geometrical design variables are introduced. In the case of weight minimization the uniform thickness is taken as extra variable. Convergence was reached in 5 to 15 iterations using MMA.

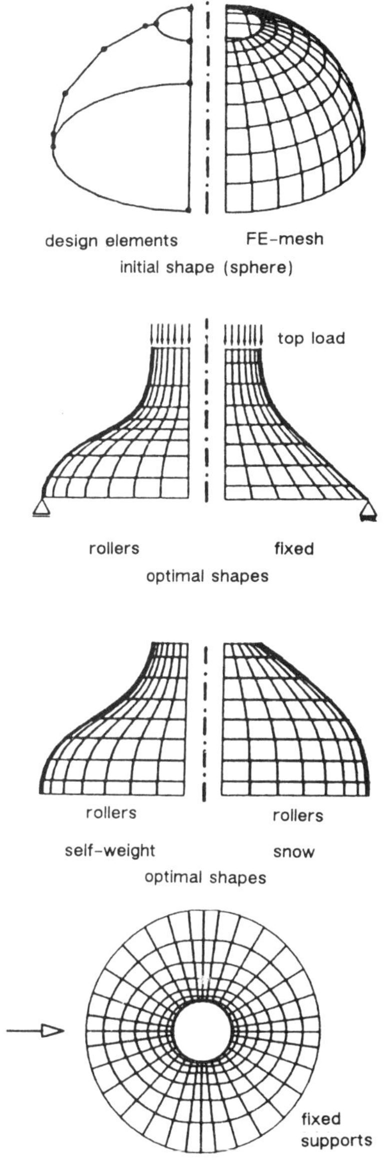

Fig. 5.2. *Shape optimization of shell of revolution*

Figure 5.2 shows the optimal shapes for different loads and boundary conditions. Except for the "wind load" the final shape allows the shell to carry the load by almost pure membrane action. This holds for weight (concentrated and snow load) as well as strain energy (dead load) minimization.

Starting with a rather trivial initial solution, we were able to obtain valuable portions of new knowledge from mathematical optimization. The change of the sign of curvature in the vertically loaded and simply supported shell, as well as the conical shape of the clamped shell, could be described as expected. However, the qualitative difference between the optimal shapes for the self-weight and the snow loading are far from being obvious. This knowledge has been completely automatically acquired.

6. CONCLUDING REMARKS

Expert systems based upon static knowledge are able to support the routine art of design. In order to promote innovative solutions one has to implement automatic knowledge acquisition mechanisms into such systems. Despite the progress achieved in machine learning, still many difficult problems must be solved before the innovative design will obtain a substantial help from the computer. Fortunately, a major interdisciplinary research effort in that direction is under way in many countries at present.

ACKNOWLEDGEMENTS

Applications mentioned in sections 2, 3 were partly based upon the real-life cases, kindly submitted by Ed. Züblin AG, Civil Engineering Contractors, Stuttgart. In section 5 we present the results obtained at the University of Stuttgart under the research project SFB 230 "Natural Structures - Light Weight Structures in Architecture and Nature", supported by the German Research Foundation (DFG). The support is gratefully acknowledged.

REFERENCES

1. Arciszewski, T. and Ziarko, W. "Structural optimization: a case-based approach". ASCE Journal of Computing in Civil Engineering, to appear.
2. Arciszewski, T. and Mustafa, M. "Inductive learning process: the user's perspective". In: "Machine learning" (edited by R. Forsyth), Chapman and Hall, New York, 1989, pp. 39-61.
3. Bendsoe, M.P. "Optimal shape design as a material distribution problem". Structural Optimization, 1 (1989) 193-202.
4. Bennett, J.A. and Botkin, M.E. (eds) "The optimum shape - automated structural design". Plenum Press, New York, 1986.
5. Bletzinger, K.-U. "Formoptimierung von Flächentragwerken". Ph.D. Dissertation, Institut für Baustatik, Universität Stuttgart 1990.
6. Bletzinger, K.-U., Kimmich, S. and Ramm, E. "Efficient modeling in shape optimal design". to appear in Computing Systems in Engineering, 1991.
7. Borkowski, A., Jendo, S. and Reitman, M.I. "Mathematical programming, Vol. 2". series "Structural optimization", ed. M. Save, Plenum Press, New York, 1990.
8. Braibant, V. and Fleury, C. "Shape optimal design and CAD oriented formulation". Engineering with Computers, 1 (1986) 193-204.
9. Coyne, R.D., Rosenman, M.A., Radford, A.D., Balachandran, M. and Gero, J.S. "Knowledge-based design systems". Addison-Wesley, Reading, 1990.
10. Garrett, J.H., Jr. "Knowledge-based expert systems: past, present and future". IABSE Surveys S-45/90, 21-40.
11. Hajela, P, "Genetic algorithms in automated structural synthesis". Proceedings, NATO ASI Conference on Optimization and Decision Support Systems in Civil Engineering, Edinburgh, 1989.
12. Jozwiak, S. "Improving structural programs using AI concepts". Engineering Optimization, 2, 12 (1987) 155-162.
13. Kimmich, S. "Strukturoptimierung und Sensibilitätsanalyse mit finiten Elementen". Ph.D.-Dissertation, Institut für Baustatik, Universität Stuttgart 1990.
14. Sause, R. and Powell, G.H. "A design process model for computer integrated structural engineering". Engineering with Computers, 6 (1990) 129-143.
15. Schittkowski, K. "The nonlinear programming method of Wilson, Han and Powell with an augmented Lagrangian type line search function". Numerische Mathematik, 38 (1981) 83-114.
16. Schmit, L.A. "Structural synthesis - its genesis and development". AIAA Journal, 19 (1981) 1249-1263.
17. Schmitt, G., Chen-Cheng Chen "Lernfähige Entwurfssimulatoren". In: KI-Forschung im Baubereich, ed. J. Gauchel, Ernst & Sohn, Berlin, 1990, 38-55.
18. Smithers, T. and Troxell, W. "Design is intelligent behaviour, but what's the formalizm?". AI EDAM, 2, 4 (1990) 89-98.
19. Svanberg, K. "The method of moving asymptotes - a new method for structural optimization". Int. J. Numerical Methods in Engineering, 24 (1987) 359-373.

THE DEVELOPMENT OF AN INTERFACE FOR AN EXPERT SYSTEM USED FOR CONCEPTUAL BRIDGE DESIGN

B.T. Philbey†, C. Miles† and J.C. Miles‡
†School of Psychology, ‡School of Engineering,
University of Wales Cardiff, Wales

The usability of any computer system is determined to a large degree by the reactions of users to that system. Both the interface to the system and the system itself determine these reactions. The difficulty for the interface designer stems from the necessity of integrating the physical features of the system with the psychological features of the human user. This paper considers both the psychological and the computer science philosophies of human-computer interface design. In addition, throughout the paper, the need for a user-centered design philosophy is stressed. An expert system for conceptual bridge design is described, which, although refined at the functional level, fails at the level of human-computer dialogue. We present arguments and data to show how a user-centred design philosophy can enhance the utility of such a system.

INTRODUCTION

"No one should design an interface without knowing the technology, without thinking about keystroke counts, or without planning for measurement and redesign" (Ref. 8).

For some years the School of Engineering at the University of Wales College of Cardiff (UWCC) has been developing an expert system for application in the domain of conceptual bridge design. The present paper outlines a joint Economic and Social Research Council/Science and Engineering Research Council funded project with the aim of developing a "user-friendly" interface for the system. The project is funded for three years and represents a joint venture between the Schools of Psychology and Engineering. We provide brief introductions to expert systems and user interfaces. Details of the various approaches to interface design found in both the psychology and the computing literature then follow. A description of the present system, presentation of initial empirical data, and details of how the design process will proceed are also provided.

Expert systems

Expert systems are, in essence, programs which comprise the knowledge which human experts use to solve problems. Within the expert system, such knowledge is expressed as rules. Expert systems are therefore also known as "knowledge-based systems" or "rule-based systems". A rule specifies an action to be taken whenever specific criteria are fulfilled e.g., If the central reservation of a carriageway is less than one metre wide, a central support may not be accommodated within that central reservation.

Human experts have a great deal of knowledge and it takes a great many rules for the performance of an expert system to approach that of human experts. Humans rarely perform perfectly so an expert system providing correct solutions 80% or so of the time is considered acceptable (Ref. 9). This level of performance reliability has been achieved by the Conceptual Bridge Design Expert System under discussion in this paper (a suitable design solution being provided for 80% of the scenarios).

The user interface

The means by which the system prompts the user to enter the information it requires, and the means of providing feedback to the user constitute the "user interface". The interface typically comprises a program which displays questions or information on the visual display unit (VDU) and accepts input from the user via keyboard or mouse. Involving the user in the development of the interface is essential if the finished system is to be an acceptable product. The correct balance between speed of interaction between user and system, and efficiency at performing the task for which the system is used, must be achieved during the course of the design process. The design of the system has to accommodate programming requirements, but at the same time it must match the expectations of the fallible human user. Consequently, interface design must involve consideration of both the system itself and the system user.

The difficulty of the design process is exaggerated by the number of variables and potential actions possible by both user and computer, and also by the limitations of the technology. The technological limitations include the nature and type of input/output mechanisms, the amount of available memory, and the computational power. Although information concerning the design process may be obtained from both the psychology and the computer literature, the emphases of the two tend to be very different. For instance, the psychological literature has much to say about the user's needs and assessment; computer literature details structured design methodologies and interface design principles. What is lacking in the literature is a methodology that facilitates knowledge integration from the two disciplines into a semblance of a structured and comprehensive design plan.

Psychologists study human behaviour and just as knowledge of the computer hardware and software must be incorporated into the design, so must knowledge of the "humanware" (the human and the human's behaviour). The numerous approaches outlined in both literatures leave any designer with an unenviable choice, but as interface design requires putting theory into practice, real value lies only in those theories which can readily be realised. Details of some of the more pertinent theories of how such knowledge may be incorporated into interface design now follow.

PSYCHOLOGICAL APPROACHES TO INTERFACE DESIGN

People and computers are complex entities and therefore it is unlikely that any single theory will provide sufficient information for the designer with respect to how a user will perform when using a particular interface. Interface design is therefore a difficult process. Additionally, any design will involve compromises. For example, providing a comprehensive help facility for the novice user must be balanced against the possible frustration caused to an expert user by the same facility. Norman (Ref. 10) has argued that although an ideal interface design may not be achievable, it is at least possible to avoid a "bad" interface design by adopting various design principles. These principles incorporate knowledge of the computer equipment, knowledge of human-computer interaction and knowledge of the problem domain.

The psychological approach to the task of interface design which places strong emphasis on the user is known as "user-centred design". Initially, the goals and needs of the user are determined, as are the tools required, the kinds of task to be performed, and the preferred methods for performing the task. If a user-centered approach is adopted, the interface is designed primarily to serve the user. As Norman writes, "The needs of the users should dominate the design of the interface, and the needs of the interface should dominate the design of the rest of the system." (p.61).

One of the main requirements of any interface is that it adequately fulfils its intended function. Hooper (Ref. 6) draws a number of illuminating parallels between the design of architectural and computational artifacts in order to aid consideration of such issues. She outlines topics common to both architectural and interface design, using examples of the former to illustrate issues in the latter. For instance, she points out that although designers of buildings in the U.S.A. did consider the psychological impact of buildings on their inhabitants, this approach did not provide designers with the type of guidelines required for building design. Thus, the emphasis of design moved back from psychological considerations to those of strict functionality. Hopefully, interface design will not suffer the same fate.

Although there are methods of specifying hardware and software requirements, just as architects are able to specify details such as square footage requirements, there is no systematic methodology for approaching interface design. However, it has been suggested that some order may be given to the interface design process after quantitative rules have been obtained which predict the relative success of interface features (Ref. 4). An empirical approach to interface design was proposed by Gould and Lewis (Ref. 5). Information about the user is obtained in order to arrive at an initial design. Methods of obtaining this information are detailed later in this paper. Empirical measurements of the success of the design are obtained and used for further re-design. This process becomes an iterative one and requires the development of ways of obtaining information about users and methods of detecting, measuring and diagnosing problems in design. Hooper (p.13) stresses that although an interface may be regarded as a facade, "the face that a computer presents to the world", and that "well-designed interfaces present orderly and attractive views to the user", a more detailed analysis of interfaces is required.

The visual display is a vital component of any interface and is the place where the user most often experiences the consequences of his/her actions. The nature of these actions need thorough investigation and understanding. Hooper illustrates this point with consideration of false facades. An interface will not be effective if it does not accurately represent the behaviour or functionality of the system. One may form an impression of the interior of a building after viewing it from the outside and then enter the building to ascertain the truth. Unfortunately, the computer user may only evaluate the system from the "outside"; i.e., via the interface. The importance of the many roles of the interface thus become abundantly clear.

The ideal user interface should achieve transparency; that is, the user should not be aware that the interface acts as some kind of input censor or filter. A successful interface design should allow the user to feel in total control of the system, reflecting Norman's view that the primary role of the interface is to serve the user. The goal of achieving a transparent interface is stressed by Norman and Draper (Ref. 11) who take the argument to its logical conclusion. They suggest the ideal scenario whereby the user exerts "direct engagement" with the computer, that is, the user has the feeling of directly manipulating the objects of the task domain. The user thus feels part of the scenario which the computer depicts and that the computer is in fact invisible. One may imagine a situation where the user receives computer generated images that give the sensation of being at a bridge site. The perceived dimensions and structures may be altered by the user interacting with the system by using voice input, a mouse, and so on.

Although an interface designer may have adequate information about the user, it is also necessary to have information concerning the intended task (Ref. 1). Bannon argues for a socially conscious design process that takes full account of the user's needs as well as highlighting the current misplaced design principle (that of designing a system to be "idiot-proof"). Such an approach implies that the system must be protected from any damage the careless user may inflict and is contrary to the principle of designing systems for humans (the principle of user centered system design). The authors of the present paper argue that an interface designer should adopt a user-centered approach, resulting in a refined interface which treats the user as having common sense whilst rendering it impossible for any harm to be done to the system from input to the interface. User-centred interface design and damage prevention are not and should not be regarded as mutually exclusive design goals.

The ultimate usability of a system in a given environment is largely influenced by the role the system plays in the workplace. Both levels of worker satisfaction and the overall use of new technology have been shown to be sensitive to the way in which work is organised (Ref. 3). Successful integration of new technology into the user's workplace and the subsequent satisfaction of the user may be predicted by the variety present in work (Ref. 2). Whilst performing work, the user has goals. Discussion of these goals follows.

At the same time as the user has goals to be achieved, the system is at a certain state. The difference between a desired goal and current state requires an intention on the part of the user which must lead to an action sequence. The action sequence specifies the physical acts to be performed upon the mechanisms of the system. The journey from intention to action requires consideration of the mapping between physical mechanisms and system state, and between system state and the resulting psychological interpretation. The mapping between system state and psychological interpretation is complex.

A theory to understand the user's actions and to distinguish among different stages of actions has been offered by Norman (Ref. 10). The user's goals are expressed in terms relevant to the user, that is, in psychological terms. The system's mechanisms and states are expressed in terms relative to it, that is, in physical terms. The resultant discrepancy between psychological and physical variables creates a major difficulty which must be solved in the design, use, and evaluation of the system. This discrepancy is illustrated by Norman's notions of the "gulf of execution" and the "gulf of evaluation". These gulfs refer to the discrepancy between the physical system and the user's goals, the methods by which the discrepancy is corrected, and are unidirectional. The gulf of execution runs from the psychology side to the system side and is bridged by the user's formation of intentions relevant to the system, the determination of a sequence of actions, and by the designer building the input facilities of the interface. The gulf of evaluation runs from the system side to the psychology side and is bridged by the user's perception of the system state and its interpretation, which is then compared to the original goals, and by the designer building the output facilities of the interface. Norman himself concedes that real activity does not always proceed in such ordered stages. Nevertheless, his theory of action illustrates the virtue in making the system more usable by the designer providing an interface that matches the user's needs and in a form that can be readily interpreted and manipulated.

Just as Bannon (Ref. 1) stresses the need for designers to understand the user's task, so Norman states the benefits to the user of understanding the task and how it relates to the overall system - "The correct conceptual model can transform confusing, difficult tasks into simple, straightforward ones." With complex tasks or non-expert users, the user must plan how to proceed from intentions to a sequence of actions. This planning process is assisted if the user has an accurate conceptual mental model of the system. Mental models evolve through interaction with the world and with the system. They are affected by the nature of interaction and by the user's prior knowledge and understanding of the system, but are unstable, fragile and incomplete (Ref. 7).

COMPUTER SCIENCE APPROACHES TO INTERFACE DESIGN

When designing the user interface, careful attention must be paid to existing computing design practice in order to ensure a professionally produced product. The following section summarises the main issues.

The interface is regarded as the means by which the user accesses the capabilities of a software system. The interface is designed first and then the operational/functional parts of the system program are designed to support the interface. This modular approach to system design makes maintenance and enhancement easier in the future and allows for re-usability (taking parts of existing software and using them in future designs). Factors which influence the quality of the interface include ease of learning to use the system, speed of using the system, frequency of user errors, user satisfaction and knowledge (of using the system) retention. The designer should therefore be aware of which properties of the interface are of greatest importance.

Different types of interface are appropriate for different types of user and so it is necessary to establish the requirements of the user prior to design. It is imperative to know the types of user for whom the system is intended. For instance, Shneiderman (Ref. 13) suggests three categories of user:- "novice", "knowledgeable intermittent user" and "frequent user", but whatever the categorisation method used, there must be flexibility present within it. It is not always possible to categorise a user precisely, nor does a user necessarily remain within a particular category. The design of the interface must allow for such cases. Indeed, the present system will be designed to provide extensive help and feedback facilities for the novice user, so that whilst performing the task of conceptual bridge design, the user learns the actual principles of design. The user who has acquired more experience of the system requires less help and feedback. The interface design will therefore provide less help and feedback to allow such a user to answer the questions, and therefore perform the design, at a quicker pace. The expert user will use the system primarily to accomplish a particular task, not to learn how to use the system.

Interface design principles

There are certain interface design principles which should be followed regardless of the type of interface or the experience of the user. Shneiderman suggests eight rules of design;

1)Strive for consistency in syntax, terminology, actions and layout.
2)Enable frequent users to take shortcuts.
3)Provide informative feedback after every user action.
4)Design for closure:- each set of actions should have a discernible beginning, middle and end.
5)Provide simple error handling and informative error messages.
6)Allow easy reversal of actions in order to reduce the time and effort required for undoing actions.
7)User-centered interaction should be supported so that the user feels in control of the system.
8)Reduce short-term memory load in order to minimise the mental workload of the user.

In addition to considering the above principles, the interface designer must also select the most appropriate method of interaction with the interface. Brief outlines of such methods follow.

Interface interaction methods

1. Menu selection interfaces are appropriate for novice and intermittent users because they inform the user of

all the choices available and require simple selection by mouse or keypress. A menu is simply a list of items, each of which represents an action that the system will take if a particular item is selected. A major disadvantage of menus is the number of items which they may prudently contain (up to about ten items is usually appropriate).

2. Form fill-in interfaces are more appropriate for inputting large amounts of data which may be grouped or otherwise organised. The design of the screen-displayed form is analogous to a paper form, having perhaps rows and columns for data to be entered under various headings. For example, a form fill-in interface would be most appropriate for an invoicing task, where the screen representation would be exactly the same as the layout of the paper invoice and information entered in the same relative positions on screen as on paper.

3. The command-based interface is perhaps the most appropriate type of interface for the frequent or expert user. Users enter commands with little or no prompting, with speed of operation the main advantage. However, the need to remember the structure and syntax of commands is the main disadvantage.

4. In contrast, direct-manipulation interfaces present the user with icons (pictorial representations of tasks). The icons can be moved or selected using a mouse or other pointing device. Such an interface is useful for novice and intermittent users because of the low memory load, providing the icon is close to the action it represents, either pictorially (a filing cabinet representing the file system) or semantically (a question-mark representing the help facility).

Any screen display should present information in a clear and well-structured way. Logically related items should be grouped together by the use of blank lines and spaces, text styles, colour, and so on. Although the system should be designed to minimise the occurrence of errors, errors will undoubtedly occur. An error message should inform the user of what the error is, why it occurred, and the appropriate corrective action. One time-efficient method of testing various interaction methods is rapid prototyping. This allows different interfaces to be experienced by users before final design but does not necessarily guarantee results useful for designers. Accurate evaluation of an interface can only take place with real users using the real system in the real working environment. The interface for the present system will be evaluated in just such a context. Following on from this discussion of the psychology and computer literature, are details of the present system.

THE CONCEPTUAL BRIDGE DESIGN EXPERT SYSTEM

Current work on the system at U.W.C.C. involves the use of established psychological procedures and techniques in order to improve both the utility and usability of the system. The system, written in LPA PROLOG, provides advice on the design of both steel and concrete bridges of up to four spans. The system has been developed using personal computers as these are widely available in design offices and so the software may be easily transferred from the development laboratory to the workplace of the users.

At present, a user enters information in response to a series of questions appearing on screen in six major sections: width being crossed, topography of site and bridge width, pier positions, ground conditions and bridge characteristics, end supports and wing walls, deck construction and materials. The various design issues that the user ought to consider and the system's recommendations are displayed at various stages of the dialogue, depending upon the user's responses. Comprehensive summaries of the system's design recommendations are displayed at the end of each design dialogue. The order of particular questions is determined by the user's responses to previous questions. In this way, the system "tailors" the user-system dialogue according to the information entered, with the advantage that each user-system session is orientated towards solving a specific and perhaps unique problem. As well as suggesting a final design, the system also contains help, explanation and glossary facilities. The system has been informally tested in field trials and experts agree that the engineering knowledge built into the system and the design solutions it provides are correct. However, one major criticism of the system concerns the nature of the interface. Present work concerns the development of a highly-refined user interface for the system. Initially, system usability and utility will be examined.

System usability and utility

If questions are phrased so their meaning is easily misinterpreted by the user, incorrect data may be input, and no matter how comprehensive the rule-base, incorrect data can only lead to incorrect solutions. We plan to employ a number of techniques in order to assess the usability and utility of the system. The techniques will include diary studies, interviews, questionnaires and keypress recording. Each of the subjective techniques has its weaknesses, but when the techniques are taken as a whole, they provide a powerful means of data collection. The methods to be used are detailed below.

Subjective assessment methods

Diary study - This can prove a surprisingly rich source of data. The idea is that the user is provided with a "blank" diary, and any difficulties or problems encountered whilst using the system are recorded separately each day over a period of perhaps four weeks. Entries can include the user's attitudes towards, and opinions of, the system, the nature of problems encountered and any corrective actions attempted. The diary may be of a free format form which does not restrict or pre-determine the nature of any entries made. Additionally, an extended form of diary, as developed by Reason (Ref. 12) may also be used. With the extended form, the user is required not only to provide the basic details of problems encountered and actions taken, but also to answer a series of standardised questions with respect to each problem. Users are therefore prompted to consider aspects of the problems which might otherwise be ignored. The obvious advantage of this technique is that it avoids the inaccuracies associated with retrospective reports of behaviour. The corpus of data collected from a number of users can be classified "post hoc" and each cause of difficulty remedied.

Interview - The purpose of the interview is for the interviewer to direct and manage questions concerning the user's attitudes and opinions of the system and to record the responses for later analysis. Interviews vary in the degree of structure present in the questions asked and this affects the precision of the response. It is our intention to employ semi-structured interviews. The advantages of this technique include the ability to tailor questions according to previous answers given and to allow clarification of questions and/or responses. However, the interview is a costly method of data collection in the sense that only one user may be interviewed at a time. Further difficulties, such as non-verbal cuing of responses and conflicting personalities, may arise due to the personal nature of the interaction between user and interviewee.

Questionnaire - This is a written collection or set of items (questions) to which the user responds. A questionnaire in an organisational setting is typically used to measure any combination of attitudes, opinions and demographic characteristics of the user. Among the important dimensions along which questionnaires vary are degree of structure, scales and scale anchors and method of administration. Advantages of using questionnaires in the project are that they are a relatively inexpensive method of data collection; they can be self-administered, thus obviating the need for the researcher to be present on site; the questionnaire presents a uniform stimulus to all users; the anonymity associated with questionnaires may lead users to give more honest and more comprehensive answers than they would perhaps be in an interview situation. One possible disadvantage is that clarification of responses is not as immediately possible as with interviews. Each of the subjective techniques has its weaknesses, but when the three techniques are taken as a whole, they provide a powerful means of data collection.

Objective assessment method

Keypress recording - Reprogramming of the present system will utilise this technique which facilitates objective measurement of user behaviour when interacting with the system. For each user the system is able to record full details of questions asked and answers given and their exact sequence, in addition to the nature and sequence of any errors and corrections made. Thus, this method allows a quantification of user performance in contrast to the qualitative techniques described above. In addition, these data allow analysis of the pattern of interaction between the user and the system.

System usability and user performance

It is important to appreciate that evaluation of both the system and the user are interwoven; they cannot be considered as independent entities. In the course of improving the interface, the usability of the system will be evaluated. It is therefore necessary to distil the term "usability" into its component parts and to develop a method of measuring system usability and user performance. Once such methods have been developed, it follows that improvement in system usability results in a corresponding improvement in user performance. The work rate metric we propose to use has been suggested by

Whiteside, Jones, Levy and Wixon (Ref. 14), which is :-

$$S = \frac{P\,C}{T}$$

where: S = user's performance score,
P = percentage of task completed,
C = an arbitrary constant based on the fastest completion time for a practised expert,
T = time spent on task.

In principle, such a metric can be applied to speed of work, completion time of a task or set of tasks and number of errors made. The advantage of analysing both speed and accuracy is that it makes possible the examination of speed/accuracy trade-off strategies which the user may adopt.

The four elicitation methods described above, together with the performance metric, will be used throughout the course of the project. Therefore, the process of improvement to the interface will take the form of:
1) Measure current system performance.
2) Make alterations to interface.
3) Measure new performance and compare to value in 1), and so on.

A preliminary study has been performed which evaluated subjects responses to two types of interface display format. The study and analyses of results are briefly described below.

Display format of explanatory text

A user answers questions presented by the system in order to perform a conceptual design. When presented with a question, the user is able to request the display of explanatory text relating to that particular question. All the explanatory text for any one question is presented on screen at one time. The length of explanatory text varies from a single sentence up to three or four paragraphs, each containing perhaps two or three sentences. It was felt that such a display format might be improved upon by displaying the text in an alternative format. The purpose of the first experiment was to compare users' opinions of the present display format to an alternative one.

Experiment 1

For all questions contained within the system, the present display format of associated explanatory text comprises one and only one paragraph. The proposed new display format contains re-wording where appropriate and satisfies the following criteria:

Short length; Explanatory text comprises one paragraph of up to about four lines.
Medium length; Explanatory text comprises up to two paragraphs.
Long length; Explanatory text comprises more than two paragraphs.

Twenty-four subjects viewed printouts of 36 question explanations associated with 18 different questions (18 of the explanations using the present format, 18 using the proposed new format). On each page there were two printouts; one of each format; both relating to the same particular question. Subjects therefore viewed two formats simultaneously and made decisions about both. Ratings of the formats were marked on a 5-point scale ranging from "extremely ..." to "not at all...". The three questions asked were:
1) How pleasant is the design to look at?
2) How clear is the explanation contained within the text?
3) How obvious is the instruction contained within each screen to the user? NB The instruction referred to was "Press any key to continue".

Subjects were also prompted on each page for their comments concerning each pair of screen printouts. Responses were scored according to their position on the 5-point scale, ranging from "extremely.." scored as 5 to "not at all.." scored as 1. Analyses of variance were performed; the main effects being type of format and length of explanation. Format was found to be highly significant for the pleasantness, clarity of explanation, and clarity of instruction ratings ($p<0.0001$). Length of explanation was significant only for the clarity of instruction ratings ($p<0.05$). The interaction between format and length was highly significant for both pleasantness and clarity of explanation ($p<0.001$).

The effects of format were anticipated. Dividing the original single paragraphs into smaller, separate paragraphs (a line space between each paragraph) produces a much more pleasant display. Re-wording of the text where appropriate was performed to improve the clarity of explanation and analysis of the data confirms this expectation. Separating and capitalising the instruction in the new format leads to the anticipated improvement in its perceived clarity to the user.

Analyses of the interaction between format and length shows that for the present format, the longer the text, the lower the perceived pleasantness. This is to be expected because if the content of a design is perceived as unpleasant, then increasing the content is likely to lead to an even lower perception of pleasantness. For the new format, longer text leads to increased perceived pleasantness. This finding is encouraging as it means that the length of text does not necessarily determine its perceived pleasantness. Similar reasoning can be applied to the interaction effects found for the clarity of explanation. If the semantic clarity of text is poor, increasing the amount of text is likely to reduce perceived clarity. The interaction effects for clarity of next instruction approached significance (p=0.06). For both formats, it is generally the case that the greater the amount of text there is preceding the instruction, then the less obvious (visually) the instruction becomes. Design recommendations suggested by these findings will be incorporated into the next version of the interface design to be tested.

It is anticipated that experiments in the near future will examine various menu designs as most of the interaction with the present system is conducted via menus. In the long term, full account will be taken of the approaches to interface design suggested by psychologists and computer scientists in order to arrive at a professionally produced product. Thus, whilst supporting the quote at the head of the paper, we provide the addition, "Consideration of the user must be paramount."

REFERENCES

1. Bannon, L.J. "Issues in Design: Some Notes." In D.A. Norman & S.W. Draper (Eds.), User Centered System Design, Lawrence Erlbaum Associates, Hillsdale, N.J., 1986.

2. Bikson, T.K. and Gutek, B.A. "Advanced Office Systems: An empirical look at use and satisfaction." Proc., AFIPS Nat. Comp. Conf., 52, 319-328, 1983.

3. Boddy, D. and Buchanan, D. "Information technology and the experience of work." In L. Bannon, U. Barry, & O. Holst (Eds.), Information technology: Impact on the way of life (pp. 144-157), Tycooly International, Dun Laoghaire, Dublin, 1982.

4. Card, S.K., Moran, T.P., & Newell, A. The psychology of human-computer interaction, Lawrence Erlbaum Associates, Hillsdale, N.J., 1983.

5. Gould, J.D. and Lewis, C. "Designing for usability: Key principles and what designers think." In B. Shackel (Ed.), INTERACT '84: First conference on human-computer interaction. Amsterdam, North-Holland, 1985.

6. Hooper, K. "Architectural design: An analogy." In D.A. Norman & S.W.Draper(Eds.), User Centered System Design, Lawrence Erlbaum Associates, Hillsdale, N.J., 1986.

7. Johnson-Laird, P.N. Mental models, Harvard Univ. Press, Cambridge, M.A., 1983.

8. Lewis, C. In D.A. Norman & S.W. Draper (Eds.), User Centered System Design, Lawrence Erlbaum Associates, Hillsdale, N.J., 1986.

9. Miles, C., Miles, J.C., & Brandon, J.A. "The utility of a knowledge based system in engineering design". SERC/ESRC grant no. GR/F 89657, 1990.

10. Norman, D.A. "Cognitive engineering." In D.A. Norman & S.W. Draper (Eds.), User Centered System Design, Lawrence Erlbaum Associates, Hillsdale, N.J., 1986.

11. Norman, D.A. & Draper, S.W. (Eds.). User Centered System Design, Lawrence Erlbaum Associates, Hillsdale, N.J., 1986.

12. Reason, J.T. "Lapses of attention." In W. Parasuraman, R. Davies, and J. Beatty (Eds.), Varieties of attention, Academic Press, New York and London, 1982.

13. Shneiderman, B. Designing the user interface: Strategies for effective human-computer interaction, Addison-Wesley, Reading, M.A., 1987.

14. Whiteside, J., Jones, S., Levy, P.S., & Wixon, D. "User performance with command, menu and iconic interfaces." In L. Borman & W. Curtis (Eds.), Human factors in computer systems-11: Proc., CHI '85 conf., San Francisco. Amsterdam: North-Holland, 1985.

4 STRUCTURAL ENGINEERING DESIGN

REASONING ABOUT CONSTRUCTIBILITY: REPRESENTING CONSTRUCTION KNOWLEDGE AND PROJECT DATA

M. Fischer

Center for Integrated Facility Engineering,
Stanford University, Stanford, United States of America

Due to the fragmentation in the construction industry, structural engineers rarely use explicit constructibility knowledge when making decisions about the layout and dimensioning of structural elements. An on-line construction expert offers the opportunity to allow use of efficient construction methods on a project by interfacing with a CAD representation of the design and by providing decision support in the form of construction knowledge. I developed such an expert system called COKE (COnstruction Knowledge Expert) by acquiring constructibility knowledge from field experts, formalizing and organizing this knowledge base in a way suitable for input to the design process, specifying a symbolic model of a reinforced concrete structure that supports constructibility reasoning, and implementing a software prototype that links a CAD system to an expert system. When providing constructibility feedback, COKE first uses high level application heuristics to determine whether a construction method is applicable or not. It then compares the project data in the symbolic model with specific constraints about the layout and dimensioning of structural elements, and alerts the designer about mismatches between these constraints and the structure. This research shows that software technology can be used to integrate downstream project constraints into the work environment of design engineers. Object-oriented programming offers a computing paradigm to support this cross-functional integration. Once implemented, such integrated intelligent tools offer the opportunity to solve problems in early project stages at a fraction of the cost of making changes in the construction or operation phases of a facility.

INTRODUCTION

An on-line construction expert offers the opportunity to improve the constructibility of a project from the very beginning. By interfacing directly with a CAD representation of the design and by providing construction knowledge and decision support for the layout and dimensioning of structural elements, an electronic construction expert enables designers to create more constructible structures. In a research project sponsored by Stanford's Center for Integrated Facility Engineering (CIFE), I developed such an expert system called COKE (COnstruction Knowledge Expert) by

- acquiring a constructibility knowledge base;
- formalizing and organizing this knowledge in a way suitable for input to the design process;
- developing an underlying model of a reinforced concrete structure[1] that supports constructibility reasoning;
- and implementing a software prototype that links a CAD system to an expert system.
- and implementing a software prototype that links a CAD system to an expert system.

This paper describes this prototype system. It first summarizes the purpose of the constructibility adviser system and then presents COKE's system architecture including the main components of the CAD system and the expert system and the link between the two systems. It continues by describing the representation of structural reinforced concrete elements in the CAD system. It also explains how a designer models a structure in the CAD system and itemizes all the data entered and stored in the CAD database. It then describes the project data that are transferred from the CAD system to the expert system and illustrates how project data are aggregated into a symbolic model and represented in the expert system. It then outlines the representation of the constructibility knowledge base in the expert system and explains the reasoning processes required for constructibility feedback. It also presents examples of reasoning functions and discusses tradeoffs between the amount of project information represented in the symbolic model and the complexity of the reasoning functions. It finally summarizes the user interface and explains the overall functioning of COKE. It ends with conclusions about the development and use of such intelligent systems that foster cross-functional integration.

[1] I selected the domain of reinforced concrete structures because it offers an interesting range of labor-intensive and labor-saving construction methods.

PURPOSE OF CONSTRUCTIBILITY ADVISER SYSTEM

The purpose of the constructibility adviser system called COKE (COnstruction Knowledge Expert) is to provide constructibility feedback for the layout and dimensioning of reinforced concrete structural elements. A major goal for the development of COKE was to automate this feedback as much as possible. This was achieved by creating a link between a CAD system (representing the project) and an expert system (containing the constructibility knowledge base). All the project data that serve as the object of constructibility reasoning are taken directly from the CAD database. In terms of the framework outlined in [1] COKE is a system of type (3), i.e., object-oriented with additional attributes; with an indirect link between the CAD and the expert system; capable of all three types of reasoning for its domain; and offering advice only (reasoning depth 1).

SYSTEM ARCHITECTURE

Figure 1 summarizes COKE's system architecture. COKE was developed using CIFECAD as the CAD package [5] and KAPPA as the expert system shell.

CIFECAD is a modified version of AutoCAD. Menus and functions programmed in AutoLISP make AutoCAD an object-based CAD system that can capture the project data necessary for constructibility reasoning. These menu-functions allow a designer to model structural elements without drawing the individual lines of each object. The user also inputs connection information with the mouse (e.g., when entering a slab, he/she clicks on all the columns that support the slab). CAD data can then be written to an ASCII file that can be interpreted by KAPPA in order to build a symbolic (geometrical and topological) model of the project. KAPPA also contains the constructibility knowledge base and reasoning functions that compare the constructibility knowledge to the project data in order to give a designer feedback about the constructibility of his/her design.

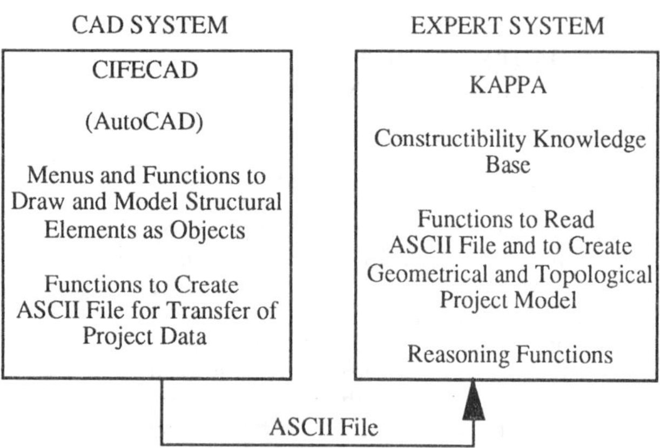

Figure 1. System Architecture

The reasons for this architecture are threefold:

(1) Given the current state of these tools, it is impractical to use an expert system shell to model a structure or a CAD system to reason about constructibility [1]. This necessitates two separate systems, one to represent the project data, and one to store the constructibility knowledge base and to perform the reasoning function.

(2) Project data are needed to initiate the constructibility reasoning process. In order to avoid the tedious manual input of these data to COKE, they have to be transferable from the CAD system to the expert system. However, current CAD systems (e.g., AutoCAD) do not run in a multi-tasking environment on PCs[2]. This makes a batch data transfer necessary—shown with the ASCII file arrow in Figure 1—because data cannot be exchanged dynamically between the CAD system and the expert system.

(3) While state-of-the-art CAD systems allow their users to build a three-dimensional model of a structure, they usually do not know what the lines stored in the CAD file represent[3]. This makes the extraction and transfer of CAD data drawn in a traditional way (i.e., drawing lines) extremely hard. Most systems, however, provide a 'block' function by which the user can group lines to represent objects and attach additional attributes to the block. This simplifies the extraction of CAD data since block data are more readily accessible than vector data in the original CAD database. This, in turn, requires the creation of menu-functions that insert structural elements into a CAD model.

The next sections describe each of the components of this system architecture in more detail.

[2] On workstations running under the UNIX operating system, CAD systems can exchange data dynamically through UNIX pipes (e.g., Design++ built on top of KEE and AutoCAD [7]).

[3] Some CAD systems use solids as primitives for representation. Although the CAD system does not necessarily know what each solid represents, the attachment of type and other information is easier. However, even these tools generally lack topological modeling functions.

CAD REPRESENTATION

The CIFECAD model used with COKE includes five structural elements: columns, beams, walls, slabs, and drop caps. All the data needed for constructibility reasoning are attached to these elements in CIFECAD.

General Description

An analysis of design-relevant constructibility knowledge [2] shows that geometrical and topological models of the project are necessary to support constructibility reasoning. Thus, when modeling a structure in CIFECAD, the user has to input data about the types of structural elements and about their location, dimensions, and connections to other elements [5].

At the beginning of the modeling process, a designer inputs general project information, data about the number of floors including floor-to-floor heights, and grid lines. Then he/she starts modeling the project by inserting structural elements into the CAD model. For every element, he/she enters the data in the following sequence:

(1) Select the type of structural element from the CIFECAD menu. Options include: column, beam, wall, slab, and drop cap. CIFECAD then creates a block with a name that includes the element type and a number assigned by CIFECAD (e.g., column18).
(2) With the mouse, indicate a point or object on which the new component is to be placed.
(3) Enter shape information if needed. For columns, the user has the choice between rectangular and round columns; for beams, between V-shaped and rectangular cross-sections.
(4) Input dimensions, concrete strength, and reinforcement ratio.

This is the general procedure to build a three-dimensional geometrical and topological CAD model. All data items are attached to the object's block in the AutoCAD database. Table 1 summarizes the data stored in the CAD system.

Data Type	Columns	Beams	Walls	Slabs	Drop Caps
Location	*x, y, z coord.*	*x, y, z coord.*	*x, y, z coord.*	*x, y, z coord.*	*x, y, z coord.*
Dimensions	*side x side y (diameter)* height	*top width bottom width depth* length	*thick-ness* **height length**	*thickness*	*side x side y thick-ness*
Connections	*column below*	*column left column right*	*column left column right*	*boundary and suppor-ting columns*	*column*
Addtl Attr.	**name level** *concr. strength rebar ratio orientation round*	**name level** *concr. strength rebar ratio*	**name level** *concr. strength rebar ratio orienta-tion*	**name level area** *concr. strength rebar ratio*	**name level** *concr. strength orienta-tion type*

Table 1. Data stored with structural Elements in CIFECAD (*Italic* data items are entered by the user. **Bold** items are derived by the system.)

The sections below explain the data required for each structural element in more detail. Had the project been modeled in a traditional way, i.e., using 'draw line' and other original AutoCAD functions, the CAD system would not have any notion of the structural elements it contains, nor of their dimensions, locations, and connections. CIFECAD enables a designer to model a structure quickly, and to extract project data easily. However, it limits the designer to only the elements that the system knows about. Nothing prevents a designer from drawing lines just as in a plain geometrical CAD system (using basic AutoCAD functions), but the system would not know what these lines represent and would not be able to pass on any information about these lines.

Columns

The 'x, y, z coordinates' point to a base insertion point at the center of the bottom face of the column. For rectangular columns, the system stores the dimensions of the two sides ('side x' and 'side y'), while for round columns it stores the diameter. It keeps a flag called 'round' that can be set to 'yes' or 'no'. This tells the system whether 'side x' is the side of a rectangular column or the diameter of a round column. The 'height' is always equal to the story height. Since it is assumed that the user models the structure from bottom to top, CIFECAD stores a pointer only to the 'column below' and not to the column above. If the column is not supported by any other column, the value of the 'column below' field is zero. Otherwise, the value is the name ('id') of the supporting column (Figure 2).

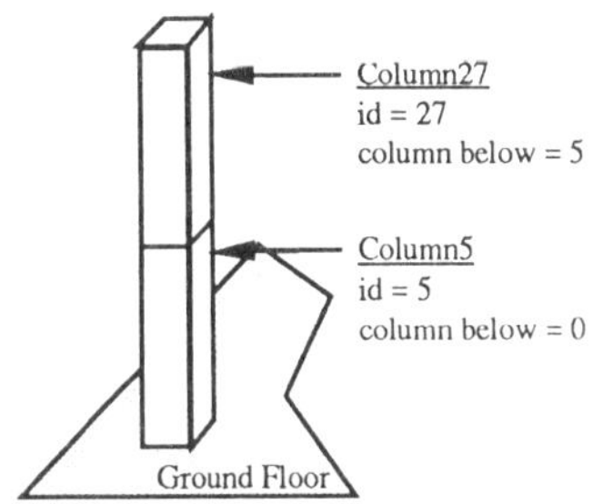

Figure 2. Connection Modeling in CIFECAD

The information described above models a column geometrically and topologically. In addition, CIFECAD automatically stores an 'id' and the corresponding level with each element. Furthermore, the user can enter the 'concrete strength' (in psi) and the 'reinforcement ratio' (in percentage of the gross cross-section) used for a particular element. Finally, the 'orientation' field stores an angle 'Phi' of the column relative to the grid lines (Figure 3).

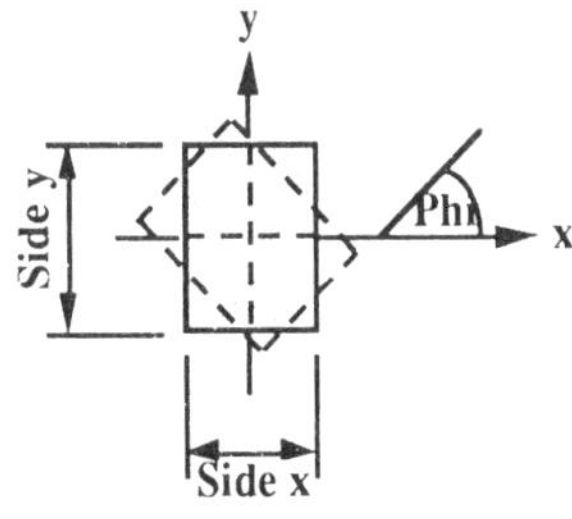

Figure 3. Orientation Convention for Columns

Beams

The 'x, y, z coordinates' point to a base insertion point at the center of the top face of the beam. For each beam, the user specifies whether the beam is rectangular or V-shaped (i.e., the 'top width' is larger than the 'bottom width'). CIFECAD automatically computes the beam 'length' from the x and y coordinates of the supporting columns; i.e., this length is based on center line dimensions. The user enters the two supporting columns by mouse-clicking on them. In CIFECAD, a beam always spans between two columns; i.e., multiple supported beams must be entered as multiple beams. The 'orientation' convention is the same as for columns.

Walls

The 'x, y, z coordinates' point to a base insertion point at the center of the bottom face of the wall. Each wall has one 'thickness'; i.e., wall segments with different thicknesses must be modeled as different walls. The 'height' is again equal to the story height including one slab thickness; i.e., walls that are on top of each other must be modeled separately. The 'length' is calculated in the same way as for beams. For the CAD representation in this research, a wall is inserted between two boundary columns ('column left' and 'column right'). The convention for the 'orientation' is the same as for columns.

Slabs

The 'x, y, z coordinates' point to an arbitrary point picked by the user within the boundaries of the slab. It is a dummy point used by CIFECAD to visualize the slab on the computer screen. Each slab has one 'thickness'; i.e., slab sections with different thicknesses must be modeled separately. CIFECAD can model any polygonal slab supported by columns. The user defines the shape of the slab by entering the 'boundary columns' (in order) or corner points. He/she then adds the interior 'supporting columns' to complete the topological information.

Drop Caps

Drop caps are a thickened portion of the slab. They provide additional shear capacity around columns. Their 'x, y, z coordinates' point to a base insertion point at the center of the top face, whose location relative to the column depends on the 'type' of drop cap used. CIFECAD can represent three types of drop caps with different locations (Figure 4). The 'orientation' convention is the same as for columns.

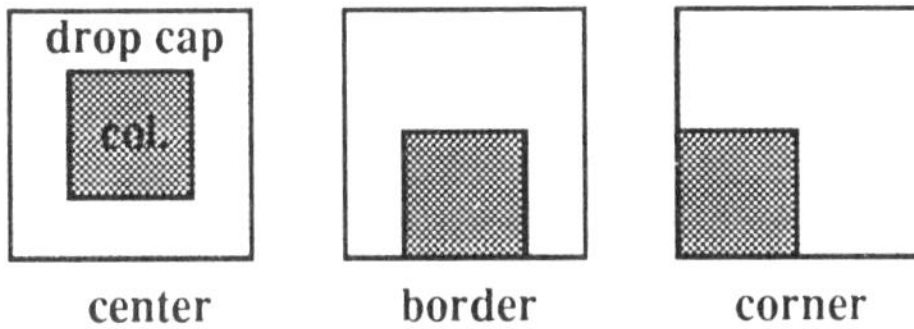

Figure 4. Possible Location of Drop Caps Relative to Columns

Additional Objects Represented in CIFECAD

CIFECAD can also represent openings in walls (windows and doors) and foundations (single footings and strip and mat foundations). No constructibility knowledge related to foundations was identified in this research. Therefore, I do not use the foundation data stored in CIFECAD in any way, and a designer does not have to include foundations in the CAD model in order to receive constructibility feedback. However, wall openings can have a significant impact on the constructibility of a wall. Therefore, the design engineer should include wall openings in the CAD model. The data required to model a door and a window are almost the same. For both the user picks the wall that contains the opening and inputs the base insertion point (center of the bottom of the opening) and the dimensions (width, height) of the opening. For windows, the user also specifies the elevation above the floor.

Discussion of CAD Representation

The previous description of the project data required to support constructibility reasoning shows that—in order to transfer the data—a significant programming effort is needed to represent these data in a CAD system in use today. This is mainly necessary because current CAD systems use points, lines, and arcs as CAD primitives and have a proprietary system architecture. Future CAD systems are likely to be object-based and will allow dynamic access to their data.

When specifying the data needed to model a structure topologically and geometrically, I tried to keep the user input to a minimum. I feel that, considering the current state of CAD representation, the user effort required to input the necessary information during the design and drawing process is acceptable. He/she has to enter information regarding the type, location, size, and connections of structural elements. It would be impossible or difficult for a system to derive these data[4].

Furthermore, within the limitations of CIFECAD, the designer can place and size structural elements any way he/she wants, thus giving him/her maximum design flexibility. Some of the limitations of CIFECAD are:

- no split levels or slanted slabs,
- no stairs, upturn spandrel beams,
- no openings in slabs, and
- limited connection possibilities (e.g., a beam supported by a beam is not possible, non-perpendicular connections of columns and beams are not modeled in detail).

A fully practical system would have to provide these and other options.

DATA TRANSFER

For transfer to the expert system, the project data described in the previous section are written to an ASCII file. Although this ASCII file contains all the information necessary for the checking of design-relevant constructibility knowledge, I decided to include additional topological data in order to simplify the functions that perform constructibility reasoning. In the original ASCII file, each object contains all the data necessary to describe its own geometry and location. However, since in CIFECAD the project is modeled from the bottom to the top, each structural element contains pointers only to the elements it is supported by and not to the elements it itself supports. This "supports" information is the topological data that is added to the original ASCII file.

Connection information plays an important role in reasoning about constructibility. Therefore, it is more convenient to have both "supported-by" and "supports" data available for every structural element. This avoids having to infer the "supports" data from the "supported-by" data every time they are needed in the reasoning process. Although the expert system could derive these data after the data transfer, it is more efficient to treat the ASCII file with a procedural language than with a symbolic language. C-procedures simply loop through all the "supported-by" information available from CIFECAD (and written to the original ASCII file) in order to deduce the "supports" information for every element and to include it with each object (in the second ASCII file). For example, for every beam, CIFECAD knows (and transfers) the supporting columns, but it does not know what beams a column might support. The C-procedures mentioned above look at all the beams and add a pointer to the columns that support a beam. This is done for all structural elements. Table 2 summarizes the data content of this enriched ASCII file and highlights the additions to the original file shown in Table 1. This enriched file is used as input to the expert system. The following discussion explains the differences between the two ASCII files for every structural element.

Columns

Since in CIFECAD everything is initially supported by columns, the increase in topological information is particularly large for columns. Back-pointers to beams, walls, and slabs are added. For supported beams and walls, the enriched ASCII file contains the following new data:

- number of beams and walls supported,
- 'id' of each beam and wall, and
- angles at which a beam, wall, or slab are connected to the column.

[4] With the CAD systems in use today, a designer has to input location and size information. The input of type and connection information is typically lacking. In CIFECAD, this additional input has been simplified through the use of menus and mouse-sensitive elements on the screen.

Data Type	Columns	Beams	Walls	Slabs	Drop Caps
Location	x, y, z coord	x, y, z coord.	x, y, z coord.	x, y, z coord.	x, y, z coord.
Dimensions	side x side y (diameter) height	top width bottom width depth length *clear span*	thickness height length	thickness	side x side y thickness
Connections	column below *column above* *drop cap* *walls* *beams* *slabs*	column left column right *slab left* *slab right*	column left column right *slab left* *slab right* *wall above* *wall below* *windows* *doors*	boundary and supporting columns *supporting beams and walls*	column
Addtl. Attr.	name concr. strength rebar ratio level orientation round	name concr. strength rebar ratio level	name concr. strength rebar ratio level orientation	name concr. strength rebar ratio level area	name concr. strength level orientation type

Table 2. Enriched Data Models for Structural Elements (Additions in *italic*)

For example, for the beams supported by col24 (Figure 5a), the original ASCII file contained a pointer from a beam to its two supporting columns. The enriched ASCII file includes back-pointers from supporting elements (in this instance, columns) to supported elements (e.g., beams). These data are added in the following form[5]:

```
2, 9, Phi1, 12, Phi2
```

This means that column number 24 supports two beams, the first being beam number 9 framing into the column at an angle of Phi1, and the second being beam number 12 framing into the column at an angle of Phi2. Although CIFECAD can represent beams and walls that frame into columns at an angle other than 90°, it does not model non-perpendicular column-beam (column-wall) connections in detail.

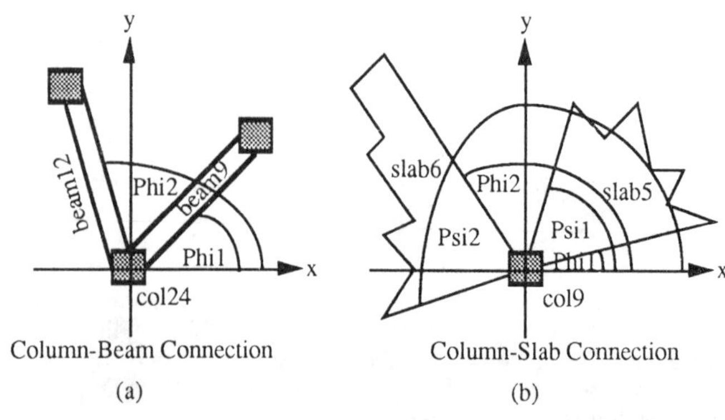

Figure 5. Connection of Columns to Other Elements

For supported slabs, the enriched ASCII file looks like this (using Figure 5b as an example):

[5] The connection to walls is modeled in the same fashion.

```
2, 5, Phil, Psil, 6, Phi2, Psi2
```

This means that column number 9 support two slabs, the first being slab number 5 starting at an angle of Phi1 and ending at an angle of Psi1, and the second being slab number 6 starting at an angle of Phi2 and ending at an angle of Psi2. This is a simple yet powerful way of modeling polygonal slabs with connections to columns. Any polygonal slab with any number of supporting columns at arbitrary positions can be modeled. Other applications have so far modeled only linear elements and simple two-dimensional elements. This way of modeling allows reasoning about edges and corners of slabs and about corner and edge columns.

Beams

For beams, the original file contains only the span between the column center-lines. The C-function calculates the clear span (i.e., from column face to column face) and adds it to the ASCII file. This value, however, is only useful if a beam frames into a column at an angle of 90°. The C-function also adds data about the slabs a beam supports, allowing for the three cases shown in Figure 6.

Figure 6. Beam-Slab Connection Options in CIFECAD

Walls

For walls, pointers are added to the walls above and below and to the supported slabs (same possibilities as for beams; see Figure 6). Furthermore, information about doors and windows that are part of a wall is also included.

Slabs

Since slabs are the last structural elements added to the project model built in CIFECAD, most topological data are already available in the first ASCII file. Pointers to the beams and walls that support a slab are added to the second file.

Discussion of Transfer File

This transfer file significantly enriches the topological project information available for input to the expert system. This simplifies constructibility reasoning because the expert system does not have to derive the data that were added to the second ASCII file. One could therefore imagine even richer data models. For example, what columns align with each other can be an important fact for reasoning but is not stored explicitly with each column. It is rather implicit in the CAD drawing. Data models that already contain all the data ever needed for reasoning would reduce even complex reasoning tasks to a straightforward comparison of two values. However, the effort needed to create such rich data models seems prohibitive: each object class would contain a very large number of attributes and this would make their management difficult. Therefore, one faces a tradeoff between the amount of data and knowledge contained in a data model and the complexity of the reasoning processes needed to provide constructibility feedback (see also below).

The data contained in the enriched ASCII file appear to provide a good balance. The file provides all the data required for geometrical and topological modeling, but is not cluttered with too many additional attributes that would make the data management at the object level difficult.

REPRESENTATION OF PROJECT DATA IN COKE

Since, on PCs, direct and dynamic access from other software to CAD data is not possible with current CAD systems, the data transfer described above is necessary. The data in the transfer file could be represented in various forms to support constructibility reasoning. The project data could be stored in a relational database [6], in a symbolic project model [3], or in a flat ASCII file [9]. I chose to use a symbolic model containing geometrical and topological data because such models can be quickly implemented in frame-based expert system shells. No further data transfer is needed during reasoning because the project data are right where the knowledge is (the knowledge representation in the expert system is described below).

I selected the expert system shell KAPPA as the tool for the expert system part of this research. I implemented functions in KAPPA that read the ASCII file described above. These functions first create a class for every level (floor) of the project under every class of structural elements (Figure 7). Then, they create an object (instance) for each structural element at the corresponding level under the appropriate object (structural element) class. Each class contains a slot for every data item in the ASCII transfer file (Table 2). After creating a new instance for a new structural element, the function in KAPPA that reads the ASCII file fills all the slots of this new instance with the corresponding values (see Figure 7 for the slot names and examples of slot values corresponding to the data items in Table 2). This process works well, but is highly application dependent since the functions that read the ASCII file need to mirror exactly the sequence of the data items in the ASCII file. In addition, they need to know the slot names that correspond to each ASCII data item[6].

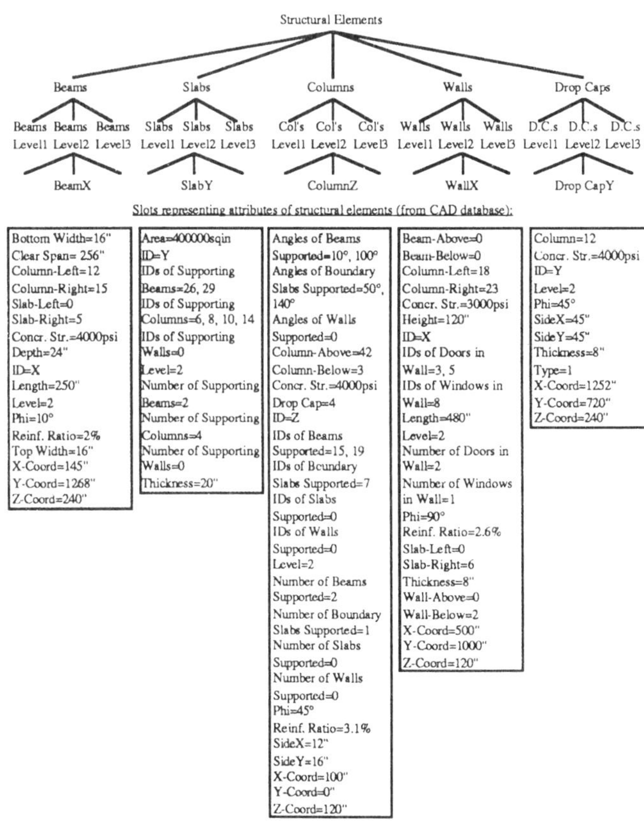

Figure 7. Symbolic Representation of Project Data

---

[6] Research is currently under way at Stanford's Center for Integrated Facility Engineering (CIFE) to develop a general engineering query language usable for engineers and computer programs to access project information [8]. The availability and implementation of such a language would render the somewhat hardwired link used for COKE obsolete.

This hierarchical project model facilitates reasoning about object classes (structural elements) and about the layout of structural elements on one floor. It also makes reasoning about slab boundaries easy because supporting columns are separated into boundary and interior columns (This also simplified data input to CIFECAD).
Figure 7 illustrates the use of single and multiple value slots for the modeling of geometrical, topological, and additional project data. These slots contain either numerical or object values. Geometrical and additional project data are modeled with numerical single value slots (e.g., thickness of slabs, concrete strength). Topological data are modeled with single and multiple object value slots, i.e., the slots store one (or a list of) pointer(s) in order to indicate connection information (e.g., beam(s) supported for columns). These topological data are refined with numerical multiple value slots (e.g., angles of beams supported for columns).

This symbolic representation of project data shows some of the tradeoffs between the richness of the data model and the complexity of the reasoning processes. For example, the project model contains a multiple value slot that describes the angles of the beams supported by a column. The expert system could have also calculated these data from the coordinates of the column it is currently reasoning about and from the coordinates of the other supporting column of a beam (each column contains a pointer to the beams it supports; each beam then contains a back-pointer to this column and a pointer to the other column it is supported by). By directly providing the angles of the beams framing into a column, this data model simplifies reasoning about constructibility. What additional data items are useful is, however, highly application-dependent. Therefore, it is important to develop prototype applications like COKE in order to find out what project data items should be represented explicitly (i.e., what data items are used frequently or are difficult to derive) and what data items should remain implicit in the project (CAD) representation.

REPRESENTATION OF KNOWLEDGE IN COKE

I investigated the use of rules and frames for knowledge representation, and the use of rules, message-passing, and functions for reasoning. I found that frames work well when combined with functions. In the expert system, each construction method is a frame with design-relevant constructibility knowledge items as slots (attributes) including the appropriate values (Figure 8). Each slot therefore represents one item of constructibility knowledge (refer to [2] for a full description of all knowledge items). For example, the knowledge base contains a frame "Flying Forms" with a slot "Maximum Floor to Floor Height" which is filled with the value "20 feet."

The slots in the constructibility knowledge base mirror the slots in the symbolic project model, i.e., constructibility reasoning consists of comparing the value of a slot in the knowledge base with a value in the project model. Functions are used for this comparison. For example, if constructibility feedback for flying forms is required, a function compares all floor-to-floor heights with the value stored in the "Maximum Floor-to-Floor Height" slot in the knowledge base. Other functions perform the reasoning for different constructibility items. The next section explains these reasoning functions in more detail.

Not all constructibility knowledge can be represented in this object-attribute-value format. Spatial knowledge, such as constraints regarding the fit of flying form tables, involves many objects and attributes and does often not lend itself to such a representation. This type of knowledge is represented in functions, thus making it implicit in the reasoning process. As discussed previously, one faces the tradeoff between rich (but complex and application-dependent) or simple (but general) data models. A rich data model simplifies the reasoning functions (allowing for more simple pattern matching functions); a simple data model requires more complex reasoning functions because more data (such as the angles of beams) need to be deduced.

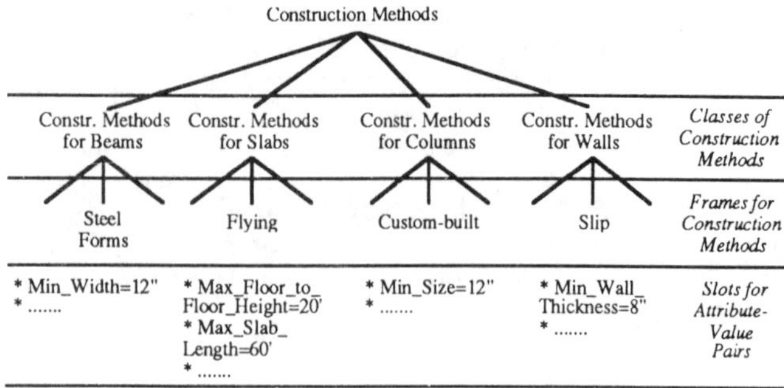

Figure 8. Representation of Constructibility Knowledge

REASONING

Reasoning in COKE is done by functions. Each function performs the necessary tests to check whether the structure at hand fulfills all the necessary requirements for one knowledge item. These functions are grouped by construction methods in order to allow the user to check easily all the requirements for one method. COKE performs the types of reasoning explained below. The reasoning complexity and the data requirements to support reasoning increase significantly from type (1) to type (3). Figure 9 shows an example for each type.

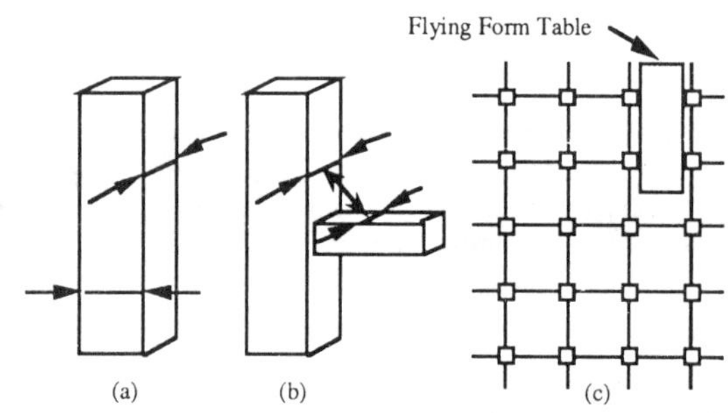

Figure 9. Examples of the Three Types of Reasoning

Reasoning about Attributes of Objects

This is the most straightforward type of reasoning. It simply involves taking the attribute of an object and comparing it to some value from the knowledge base. For example, the width of a beam (an item of project data) is compared with available formwork sizes (a knowledge item) in order to conclude whether the beam can be built with available formwork (Figure 9a). In COKE, such a reasoning function looks like this:

```
ForAll [ beam|Beams ]
  If Not(Member?(SteelBeam:AvailableBottomWidths,
    beam:BottomWidth ) )
  Then {PostMessage( beam # " is " #
    beam:BottomWidth # " inches wide at the
    bottom. Steel beam forms of that size is not
    available." );
      If ( PostMenu( "Do you want to see a list
        with the available sizes?", Yes, No ) #=
        Yes )
      Then PostMessage( "An available bottom width
        is: " # width # " inches." ) ); };
```

This function compares the bottom widths of all the beams with a list of the available bottom width that is acceptable for steel beam forms. If the bottom width of a beam is not part of that list, the function alerts the user with a message.

Reasoning about Relationships between Attributes of Objects

This is a slightly more complicated type of reasoning. It involves taking the attribute of an object and propagating its influence on attributes of a different object. For example, when Cunningham steel forms are used to form beams, the columns should be four inches wider than the beams (Figure 9b). In COKE, this looks like this:

```
ForAll [ beam|Beams ]
  If { ( ( beam:Phi - beam:ColumnLeft:Phi ) ==
    90.0 ) Or( ( beam:Phi - beam:ColumnLeft:Phi )
    == 270.0 ) And ( ( beam:ColumnLeft:SizeX -
    SteelBeam:AddtlColumnWidthReq ) <
    beam:TopWidth ); }
  Then PostMessage( "For steel beam forms, " #
    beam:ColumnLeft # " should be " #
    SteelBeam:AddtlColumnWidthReq # " inches wider
    than the top width of " # beam # " ." )
  Else
    If { ( ( beam:Phi - beam:ColumnLeft:Phi ) == 0
      ) Or ( ( beam:Phi - beam:ColumnLeft:Phi ) ==
      180 ) And ( ( beam:ColumnLeft:SizeY -
      SteelBeam:AddtlColumnWidthReq ) <
      beam:TopWidth ); }
    Then PostMessage( "For steel beam forms, " #
      beam:ColumnLeft # " should be " #
      SteelBeam:AddtlColumnWidthReq # " inches
      wider than the top width of " # beam # " ."
      ) );
```

This function performs some spatial reasoning before comparing a constructibility knowledge item with a project value. First, it figures out from which side a beam frames into a column and then compares the top width of the beam to the corresponding side of the column. The column side should be at least 4 inches wider than the top with of the beam, which is the value stored in the slot "AddtlColumnWidthReq" of the Steel Beam construction method. This example also illustrates the use of slots to make constructibility explicit. In a typical rule-based system, the value "4 inches" for the additionally required column width would be buried in a rule. Here it is made explicit in a frame, accessible to any function, and easily changed by a user, in the case of a change in construction practice, without him/her having to change the reasoning function. This type of reasoning can get difficult if the relationships among different attributes are complicated or ill-defined.

Spatial Reasoning

This is the most challenging type of reasoning because the quantity of data and objects involved is often very large. Humans can easily reason spatially by looking at a plan and recognizing patterns and relationships among objects. No formal models describing how to implement spatial reasoning in a software system are readily available. An example of spatial reasoning is the task of figuring out whether a flying form table will fit into a column layout (Figure 9c).

```
EnumSubClasses( Rectangular, x, ForAll [ col1|x
  col2|x ] { If ( col1:XCoord == col2:XCoord )
  Then AppendToList( col1:ColumnsAlignedX, col2 );
} );
ClearList( Global:TrashList );
ForAll [ rect|Rectangular ]
  If Not( Member?(Global:TrashList, rect:XCoord ))
  Then AppendToList(Global:TrashList,rect:XCoord);
  Global:Counter1 = LengthList(Global:TrashList );
  While  (( Global:Counter1 > 1 )) {
    Global:Counter4 = Min( Global:TrashList );
    RemoveFromList(Global:TrashList,
    Global:Counter4 );
    Global:Counter5 = Min( Global:TrashList );
    EnumSubClasses( Rectangular, x, ForAll [
      col1|x col2|x] {
      If { ( col1:XCoord == Global:Counter4 ) And
        ( col2:XCoord == Global:Counter5 ); }
      Then AppendToList( col1:NextColumnsAlignedX,
        col2 );});
      Global:Counter1 += -1; };
          EnumSubClasses( Rectangular, x,
            ForAll [ rect|x ] {
            ClearList( Global:TrashList );
            EnumList( rect:ColumnsAlignedX, col1,
            EnumList( rect:NextColumnsAlignedX, col2,
              { Global:Counter4 = ( col2:XCoord - (
              col2:SizeX / 2 ) - col1:XCoord - (
              col1:SizeX / 2 ) ); If Not( Member?(
              Global:TrashList, Global:Counter4 ) )
              Then AppendToList( Global:TrashList,
              Global:Counter4 ); } ) );
           If ( LengthList( Global:TrashList ) > 1 )
           Then PostMessage( rect # " is not aligned
           perfectly with the next column line." );}
);} );
```

This function demonstrates the extra effort required to deduce spatial information that is implicit in the geometrical and topological project model. The function above determines whether a flying form table fits between columns in the Y-direction (a similar function is used for the X-direction). For each column at every level, it first determines the columns that are aligned with each other by checking whether two columns have the same X-coordinate or not. It then creates a list of the X-coordinates for all columns. This list is used to determine the next column(s) for each column. Then, the function checks whether the clear distances between the columns are the same, and if they are not, it alerts the designer that a certain column is not aligned perfectly with other columns and therefore hinders the use of flying forms.

Discussion of Reasoning Functions

The types of reasoning required in a given application depend on the knowledge, the project data, and their representation. A rich project data model (i.e., one that associates a lot of data with each object) is likely to make reasoning easier but might require a significant effort for data input. Ultimately, the inference engine compares a constraint value with a project value. This is why reasoning types (2) and (3) are often more difficult. Before the inference engine can conclude whether a certain constraint is violated or not, it often needs to derive the data required for reasoning from the available project data. For example, what columns align with each other can be an important fact for reasoning but is usually not stored explicitly with each column. It is rather implicit in the CAD drawing and has to be derived each time it is needed. One could imagine objects with data models that already contain all the data ever needed for reasoning. This would reduce even spatial reasoning to a straightforward comparison of two values. However, these data models would be highly application-dependent. Therefore, one faces the tradeoff between the amount of information contained in a data model and the complexity of the reasoning processes.

USER INTERFACE

In order to receive constructibility feedback from COKE, a designer must draw the structure in CIFECAD. As described above, he/she can model a structure using pre-programmed menu-functions that help place and connect structural elements. Whenever he/she wishes to check the constructibility of a structure, he/she gives a command in CIFECAD that creates an ASCII file from the CAD database. This file contains all the data necessary for constructibility reasoning. After creating the data transfer file, the designer has to quit CIFECAD. This is necessary because AutoCAD's release 10 does not work in a multi-tasking environment on PCs. He/she then starts KAPPA and then COKE. In COKE, he/she first clicks on the "Load Project Data" button[7]. This activates a function that builds a symbolic model of the project from the previously created ASCII file. As described above, this symbolic model represents the structure of the project with all the data required to perform constructibility reasoning.

---

[7] All commands are given via menus or buttons.

Before starting the actual constructibility reasoning process, the designer can specify the construction methods for which he/she would like feedback. COKE then uses application heuristics to check the applicability of the selected methods. If the designer does not specify any construction methods, the system uses application heuristics to rule out some construction methods. For example, if there are no walls in the CAD database, all construction methods for walls are excluded, or if the building is only five stories high, flying and slip forming are disregarded[8].

After this, the system compares the data in the symbolic model with the constructibility knowledge for the applicable construction methods (layout and dimensioning knowledge), and gives the designer feedback about the constructibility of the structure whenever it spots a mismatch between constructibility requirements and the structure at hand[9]. The designer can then disregard the constructibility advice if other constraints, such as usability constraints, make the consideration of the advice impossible, or he/she can incorporate the necessary changes in the CAD model.

CONCLUSIONS

This paper described an integrated software system that provides designers with construction input during the preliminary design process. I believe that such tools will be part of the engineering environment of the future because they support the inter-disciplinary transfer of knowledge. These tools will enable companies to increase the quality of their engineering because

- a designer starts a project from a higher level (the computer system automatically keeps track of constraints);
- company-wide standards can be implemented and distributed efficiently;
- engineers can be trained more quickly; and
- better decision support data and knowledge will be available.

These are some benefits of computer-enabled integration that should contribute to improve the overall project delivery process. However, to reap these benefits, we still need to

- devise and implement formal methods to compile knowledge bases; and
- develop better product models of the components used on construction projects.

In this research project, I compiled and formalized a significant body of design-relevant constructibility knowledge. This shows that such knowledge exists and that it can benefit a project when linked to individual design decisions. However, a significant effort was required to collect this knowledge at the appropriate level of detail. Companies should develop and utilize more formal methods to acquire and transfer requirements of construction, operation, and maintenance. The experience from this project can serve as a starting point for the identification and classification of such knowledge.

Integrated object-oriented CAD/expert systems are starting to emerge in universities and in industry. In the near future (within the next five years) tools will be available that integrate many of the different functions necessary for designing, constructing, and maintaining a project. They will automatically solve many problems that are now solved by engineers. The object-oriented approach offers the paradigm necessary to achieve this level of integration because knowledge can be put at the object (e.g., structural element) level. Each object can be told how to size itself, how to draw itself, how to calculate its cost, how to interact with other objects, etc. This will provide engineers with powerful design tools that will enable them to either do routine design much faster or to explore more alternatives. This in turn requires the development of objects that represent actual project components and become part of a project model [4]. This research provided the data model for the construction view of reinforced concrete elements. On a larger scale, it provided the groundwork for the development of computer applications that help plan, design, construct, and maintain facilities over their entire life.

ACKNOWLEDGMENTS

I gratefully acknowledge the support of Stanford's Center for Integrated Facility Engineering (CIFE) and of its member companies. I also thank all the experts in reinforced concrete design and construction. Without their continued contributions this research would not have been possible.

BIBILIOGRAPHY

1. Fischer, Martin, "Reasoning about CAD Data," *Proceedings of the Computer Integrated Design and Construction Symposium*, ASCE Construction Congress '91, Cambridge, April 13-16, 1991.

2. Fischer, Martin, "Using Construction Knowledge during Preliminary Design of Reinforced Concrete Structures," *Ph.D. Thesis*, Department of Civil Engineering, Stanford University, June, 1991.

3. Ito, Kenji, Law, Kincho H., and Levitt, Raymond E., "PMAPM: An Object Oriented Project Model for A/E/C Process with Multiple Views," *CIFE Technical Report*, No. 34, July, 1990.

4. Howard, H. C., Abdalla, J. A., and Phan, D. H., "A Primitive-Composite Approach for Structural Data Modelling," Draft, To appear in *ASCE Journal of Computing in Civil Engineering*, Vol. 6, No. 1, January, 1992.

5. Kolountzakis, Mihail and Fischer, Martin, "CIFECAD: A User's Manual," *CIFE Working Paper*, Center for Integrated Facility Engineering, Stanford University, 1991 (forthcoming).

6. Law, Kincho H., Barsalou, Thierry, and Wiederhold, Gio, "Management of Complex Structural Engineering Objects in a Relational Framework," *CIFE Technical Report*, No. 19, September, 1989.

7. Levitt, R., Axworthy, A., and Kattajamäki, M., "Automating Engineering Design with Design++," *Nikkei Artificial Intelligence Journal*, Winter, 1991, 118-127.

8. Nedzel, Jared L., "SAGA-Spatial and Geometric Analyzer," *Dissertation Proposal,* Department of Civil Engineering, Stanford University, March, 1990.

9. Voeller, John, "Implied Engineering," *CIFE Symposium Proceedings, Technical Report*, No. 24, March, 1990.

[8] The designer has the option to receive detailed constructibility feedback even if the application heuristics for this method are not satisfied.

[9] One should note that COKE does not give cost information. It uses only heuristics about minimum technical requirements for the application of a certain construction method to give constructibility advice.

ACCORD: DEVELOPING CONCEPTS TOWARDS INTEGRATION OF ANALYSIS AND DESIGN - THE ROLE OF ASSET IN THE DESIGN ASSURANCE PROCESS

R.G. Parker†, N.H.W. Stobbs†, A.T. Humphrey†, M.A. Pearce†,
J.P. Patureau‡ and A. Azarian‡
†GEC Marconi Research Centre, Great Baddow, Chelmsford, England,
‡Bertin et Cie,Zone Industrielle des Gatines, Plaisir, France

This paper gives an overview of the work accomplished in the ACCORD sub-project ASSET. ASSET's objective was to demonstrate the benefits of database management, knowledge based assistance and concurrent processing within the field of design assurance. This objective has been achieved in the implementation of the ASSET demonstrator.

The ASSET demonstrator integrates three engineering disciplines viz Reliability, Cost and Thermal Analysis, under a common management system. The analytical tools within these domains have the capability to communicate with each other through the development of a novel specification of the product hardware. Further, computational speed up through concurrent processing and new design assurance aids has been achieved.

All the software packages within the ASSET demonstrator can be used either as separate tools or in any one of a number of combined configurations.

INTRODUCTION

ACCORD was a 4 year ESPRIT-1 demonstrator project, which ended in August 1990. It was comprised of two sub-projects, ASSET (ACCORD Suite of System Engineering Tools) and APPEAL (ACCORD Parallel Processing Engineering Analysis Library). A broad overview of ACCORD is given by PATUREAU et al (1989). This paper concentrates on the ASSET sub-project.

ASSET's prime objective was to improve Computer Aided Engineering by making design assurance an integral part of design; (this is now recognised as being part of the 'Concurrent Engineering' Arena). ASSET aimed to achieve this objective through the encapsulation of engineering expertise and the integration of design assurance tools. This would enhance the effectiveness and scope of product management by providing the facilities to monitor the design and design assurance activities more readily. Further benefits should accrue from providing an enlarged system engineering expertise base, thereby freeing highly qualified personnel from the more mundane tasks associated with design assurance and plan preparation.

New technologies and design methods have been introduced into the ASSET system. The benefits arising from these innovations include speed up in the areas of design assurance analyses and the capability for data exchange between analytical tools. Experiments within the Reliability and Thermal domains have shown that the main effort and time required for certain tasks have been reduced by an order of magnitude.

PROGRESS

The work accomplished in the ACCORD sub-project ASSET is encapsulated in the ASSET demonstrator. The ASSET demonstrator is a database application for integrated design assurance analyses (encompassing the engineering disciplines Thermal, Reliability and Cost). A major component of the demonstrator is the Knowledge based element known as ADVISE (ACCORD Design Verification Intelligent Support Environment). The ASSET demonstrator is still only a prototype system which shows the potential and can provide the basis for future exploitation. Here, its main purpose has been to investigate the feasibility of integration between the diverse range of design assurance tools and also to identify the problems associated with it.

Structure of the ASSET Demonstrator

The structure of the ASSET demonstrator, depicted in figure 1, meets the original requirements given in the functional specification. The main elements are:

- an Executive Manager
- a Domain Manager
- libraries of technical knowledge and data
- a selection of analysis tools
- a product analysis database.

These elements are described in turn below:

The Executive Manager provides two separate but linked functions. The first function is concerned with project management, that is the creation and deletion of projects within the ASSET system. The second function facilitates product specification which entails the creation of an electro-mechanical breakdown of the system under analysis. The user enters the engineering domain of his choice through the analyse function.

The Domain Manager is responsible for the organisation of task within an engineering domain. This includes establishing the status of a project for an analysis (ie if sufficient data exists) and providing a common interface to the domain's tools.

Libraries of technical knowledge and data provide the user with a wealth of information at his finger tips. The libraries have been established to help the user to obtain the best predictions out of the tools which have been made available.

The selection of tools provides the analysis capability within the engineering domain.

The product analysis database manages the storage and retrieval of all project related information for each domain.

The demonstrator consists of three engineering domains, viz Thermal, Reliability and Cost (see figure 1). Each of the domains are linked to the others at two key levels:

- at the top level, known as the Executive Manager, for project management

- at the bottom level for product data communication through a unique database managed by a relational database management system (INGRES in our case).

Features of the ASSET Demonstrator

Common user interfaces to the database and to the knowledge-based components have been developed. This provides ASSET with an integrated user friendly environment and also has a beneficial impact on learning curves.

Extensibility has been a major design criterion. For instance, both the Cost Model library and the Thermal Template library may be easily extended. Within ADVISE, domain knowledge may be prepared by an engineer without requiring any knowledge of the underlying software. User interfaces may also be tailored to suit different users.

Another feature similar to extensibility is adaptability. ASSET provides a framework for design assurance disciplines and tools. New engineering domains and/or tools can be readily interfaced with the system.

Portability has featured largely in the development of ASSET. Consequently, an appropriate set of language and programming standards have been adopted. The system currently resides on VAX VMS but should be portable to SUNs and other machines (the ADVISE sub-system already runs under SUN UNIX).

ASSET has been developed in a modular fashion. This means that packages ranging from individual tools up to the complete ASSET suite can be readily configured for future exploitation.

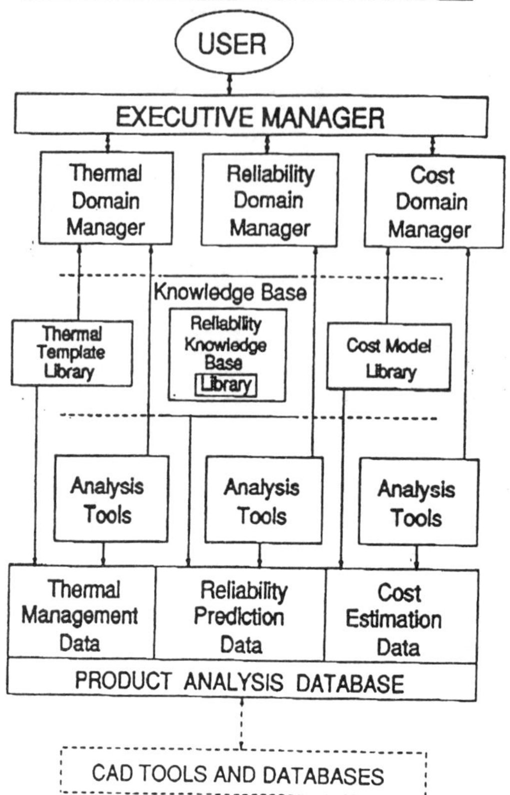

Figure 1

EVALUATION OF ASSET'S OBJECTIVES

Review of Initial Objectives

The overall objective of ASSET was to integrate and rationalise design assurance process, concentrating on the domains of Reliability, Cost (ie Whole Life Cycle Costing) and Thermal analysis (Stress and Vibration analysis was also considered at an early stage, but had to be postponed due to lack of resources). This was to be addressed by the exploitation of three technologies:

Firstly, a set of existing analytical tools and their data would be integrated under a common environment using a Database Management system (DBMS). The adoption of a DBMS would provide centralised control, which would offer the benefits of data sharing between tools, data consistency and reduced redundancy; these in turn could be expected to yield advantages with regard to the development and integration of prototype applications. For the end user, the consistent interfaces generated with the aid of the fourth generation language would reduce learning time, while avoidance of repetitive data entry would lead to acceleration of the design process.

Secondly, a range of knowledge-based facilities for assisting with project and task planning was to be developed using software, built on an Artificial Intelligence toolkit. This would exploit the expressive power of object-based and rule-based representations, together with the reasoning facilities provided by the toolkit, in order to capture and make available both theoretical and experiential expertise with regard to the domains. The use of such a toolkit would, like that of DBMS, be highly conductive to the rapid prototyping of applications. This aspect of the project would include the development of a number of intelligent front ends to existing tools.

Thirdly, the benefits of computational speed-up through parallel processing methods were to be investigated. This aspect was to be addressed through the exploitation of the results of the sister project, APPEAL.

In order to integrate the design assurance domains there was a need to establish a common view of the product specification. This product specification could then be assessed by the design assurance tools and be used as a focal point for inter-domain communication. In addition, it was recognised that if the software system produced were to be readily exploited, it would have to be user-friendly, portable and extendible.

Although the project was largely concerned with the integration of existing tools, there was little in the way of suitable software in the Cost domain. Additional goals for this area were to evaluate existing and emerging Whole Life Cycle Costing (WLCC) techniques, to incorporate a WLCC package into ASSET and to obtain recognition of the importance of WLCC in the design process (at the completion of the design phase 85% of the life cycle is committed).

Review of Achievements

At ESPRIT Week 89 the project was able to demonstrate the exploitation of the above technologies to enhance the design assurance process:

The Reliability, Thermal and Cost domains were successfully integrated under INGRESS DBMS; this allowed consistent user interfaces to the tools to be implemented (but see later discussion). A common representation for describing the product hardware specification, referred to as the System Breakdown Structure (SBS), was developed. The three domains are accessible through an executive Manager, which is responsible for project management and SBS version control. An SBS is issued to the design assurance domains once it has been created in the ASSET Executive Manager, and then provides the basis for transferring information between tools (primarily within the Reliability and Cost domains).

A general environment for the knowledge-based component of ASSET (the ADVISE system), whose central feature is a basic User Interface Management System (UIMS), was developed. This was built using the AI toolkit KERIS, whose roots lie in the ESPRIT-1 PCTE project. This provided the foundation for implementing a broad range of knowledge-based design assurance support modules for the analysis domains.

Finally, a set of parallel processing software modules generated by the APPEAL project was successfully incorporated within the Thermal domain to provide significant computational speed-up on complex routines.

Within the individual domains, the main advances have been the following:

Reliability

A common representation, the System Breakdown Structure (SBS), was developed using the INGRESS DBMS. This provided the basis for integrating tools addressing a range of aspects of Reliability analysis.

An existing Reliability Prediction tool for determining the failure rate and mean time between failures of a system and its components was integrated into the software. This gave a more flexible environment for entering the necessary component values with the facility to build up a library of component data specific to the working environment, thereby reducing User input in the future.

A Maintainability Prediction tool which made two maintainability assessments available was integrated. The first assesses the optimum size for the maintenance team required to support the system during its working life, and the second calculates the time for which the system is likely to be unavailable due to repair and the probable time needed for preventive maintenance and repair in a given period. This uses failure rates from the same SBS module structure and adds maintainability data to the existing reliability data.

The Fault Tree Analysis tool (FTA) is one of the powerful reliability analysis tools developed for determining the causes of observed faults in a system using the standard SBS. Therefore, the FTA tool could be considered as a necessary complement to FMECA and it is incorporated within ANALYSE function. The FTA tool performs at first a qualitative analysis by producing the minimal cut sets and all the implicant families in terms of basic events. Then, it indicates the importance of the basic events and cut sets as well as the probability of the top event (quantitative analysis). Different possibilities are offered to the user in order to:

- select the gate events (options: PROCESS & ALL),
- specify the methodology (options: NELSON, MSUP...),
- measure the basic events according to
  - BIRNBAUM
  - CRITICITY
  - FUSSEL-VESELY
  - BARLOW PORCHAW

A limited interface between the SBS (and hence the tools) and the knowledge-based modules in KERIS was developed. This allows structural models and failure rates to be passed to the knowledge-based system.

The ADVISE sub-system provided the basis for integrating a range of knowledge-based facilities for supporting Reliability Analysis. The most significant of these (all of which, like the UIMS at the heart of ADVISE, exploit fully the KERIS object representation facilities) are:

A fairly powerful graphics-oriented facility for Reliability Modelling and apportionment (using an overall requirement for system availability or failure rate to determine lower level goals). A variety of apportionment strategies are supported and rapid what-if analyses facilitated. Models (or sub-models) may be saved in a library and later imported at any level. The modeller has been applied to a design for a communication system containing several thousand units; the two week time-scale for analysing it from scratch suggests a productivity enhancement of at least one order of magnitude over conventional tools.

A facility for generating largely complete Reliability Programme Plans for delivery to the customer. These are created automatically from a combination of project-specific data entered by the user and a set of generalised, extensible templates of text derived from standard documents. The plans thus generated may be tailored for individual customers and stored for reuse. This facility has also been successfully put to use on a project.

A suite of task scheduling facilities. This incorporates a knowledge base containing information about analytical and management tasks for Reliability and/or Maintainability studies. Knowledge about the effects of various system attributes on given tasks is used to generate timescales for each task, taking into account the characteristics of the system as specified by the user. This information (or a set of user-specified timescales) may then be used in conjunction with knowledge about inter-task relationships, and constraints such as project milestones and available effort, to produce overall task schedules; these may be presented textually and/or graphically.

Cost

The cost application has achieved its initial objectives in establishing a WLCC package for ASSET. This includes a fully functional Development and Production Costing facility as well as a Whole Life Cycle Cost facility, which together are integrated into the ASSET environment.

Within the cost domain, the user is advised initially to undertake a development and production cost analysis. This will predict the development engineering and prototype costs and production costs for electrical boards and electrical or mechanical modules identified within the project's System Breakdown Structure. The data requirements for these predictions include specifications of the environmental conditions and quality grades - the same information as required for the Reliability Prediction tool.

The development and production costs can be incorporated within the Whole Life Cycle cost facility. This facility is used to predict the total cost of ownership of the system from its cradle to its grave. The model for predicting the cost of ownership is based on the concept of a Life Cycle Cost Breakdown Structure (LCCBS). The LCCBS gives an ordered breakdown of the cost defined by the project phase or function, and/or the system construction and/or the resource requirements. The cost estimates for these elements of the LCCBS are evaluated from a selection of cost models which are supplied by ASSET's Cost Model Library.

The Whole Life Cycle Cost facility currently has two tools, viz a Cost Prediction tool and a Sensitivity Analysis tool. The Cost Prediction tool allows the user to modify the initial data prior to undertaking a fundamental WLCC analysis, whilst the Sensitivity Analysis tool evaluates the sensitivity of the LCCBS cost elements with respect to user designated parameters.

Thermal

The project aimed to develop an environment that would support thermo fluid analysis tools capable of being used in trade-off studies to optimise the enveloping packaging design. This is achieved via the provision of a prototype thermal domain package made of two components:

- the thermal domain manager: set of tools enabling the user to manage analysis data and results through a study/task reference

- The modelling base manager that provides access through a standard interface to a range of software tools, modelling aids and also data on cooling processes including natural and forced convection. The facility also contains handbook procedures and data useful in preliminary design studies and a range of programs for assessing junction temperatures for components mounted on printed circuit boards.

It also allows the handling of complete documentation sets for each tool (underlying hypothesis, bibliography, user manual, sketches). Both managers use an enhanced multi-windows, multi-users interface offering various access means to data, by means of keywords, catalogs. Data transfer between the two managers automated, thus permitting ready storage and retrieval of data and results of a previous analysis.

The thermal package includes a third component that realizes the interface with other analysis domains, thus allowing, for instance, result data of PCB thermal analysis tool to be made available via the common standard SBS for use of the MIL HDKB 217 tool for the prediction of failure rates. Tradeoff studies on the effectiveness of the cooling design can thus be based on actual (projected) conditions arising within the unit or enclosure including dissipation levels and distributions.

Studies at system level are completed using software tools that enable detailed assessment to be made of the airflow conditions that arise within the enclosure. The facility is based on the specification of the enclosure geometry including the internal arrangement of ducts and baffles and the location and power of any cooling fans. Thus, a complete assessment can be made of alternative design options during the process of system evaluation.

Software tools/templates incorporated within the current version of ASSET can be considered in the following categories:

Thermo physical data
Handbook cases for various fin/plate configurations
ESATAN thermal network analyser
ESATAN modelling templates
FLOTHERM airflow analyser
PCB thermal analyser.

Use has been made of the APPEAL parallel processing routines to enhance the computationally intensive arithmetic components of the equation solvers used within the PCB thermal analyser. The rapidity with which the data can be generated and the speed of solution permits the packaging engineer to explore a wider range of alternative layouts.

Review of Problems Encountered

In the early stages, the project suffered from incomplete and divergent view of the product specification. In consequence:

- Inter-domain views of Executive Manager and the Electro-Mechanical System Breakdown Structure differed widely. These differences were not fully recognised and consequently had not been resolved by the time the domain designs had been specified

- There was no common Analysis Methodology. A global domain framework should have been given major priority early in the design phase to provide a common engineering approach to be applied across the ASSET system with common terminology and common views of screens and menu keys

- Attempts to match differing representational schema brought about integration difficulties. In particular, the intended project-wide representation of the product under analysis, proved to be of limited relevance to the thermal domain; the complex and varied schema on which thermal modelling was based did not correspond with the tightly defined System Breakdown Structure. As a result, it was not possible to provide more than a superficial integration with the other domains, and data exchange was limited.

Although it was intended that knowledge-based techniques should be exploited, where applicable, on a project-wide basis, this objective was curtailed by budget constraints. Ideas were generated for all domains, many of which reached the design stage, but implementation of knowledge based support was restricted to the reliability domain. ESATAN, a Thermal analysis tool which is widely adopted, powerful, but difficult to use, would have benefitted greatly from an Intelligent Front End (IFE), but this undertaking would have been beyond the resources available.

Some of the tools proved unsuitable for incorporation, as they were out of step with current user expectations. For instance, while the currently available FMECA tool (an old-fashioned batch-driven suite) could have benefited from an IFE, the tool itself did not offer conformity with the generally adopted standard. (In the end the tool was rewritten in KERIS, with a form-based interface. While this is flexible and user friendly, it has yet to be tested on a major project). To a lesser extent, the MIL 217 tool also fell into this category, being batch-driven and rather too slow for the expectations which an interactive environment generates.

Although both the INGRESS-based and the KERIS-based elements of ASSET provide internally consistent interfaces, the two differ significantly. To some extent this is unavoidable, since the INGRES environment is of a traditional single-window, keyboard-oriented nature, while that offered by KERIS is aimed at the provision of multi-window, mouse-driven interfaces. This problem will be reduced when the next version of INGRES becomes available. Although the KERIS-based ADVISE user interface offers a high degree of flexibility in the interaction, it is potentially confusing to inexperienced users. ADVISE uses textual interface specification which may be tailored for individual users; greater flexibility, however, would be offered if some of the general characteristics of user interfaces, such as the maximum permitted number of windows on the screen at any given time, were dynamically adaptable to different classes of user.

The current version of the interface between KERIS and INGRES which has been developed is too limited and slow; the fact that INGRES can communicate with a LISP (and hence a KERIS) application only via C, compounded by constraints on the degree of generalisation possible using the in-built LISP-C interface, tends to suggest that a large amount of low-level C programming would be needed to provide a stronger interface.

One problem arising from the above deficiency is that a module intended to provide remedial advice when analytical goals are not met is of little practical value to the analyst in its current form. At present, it asks a number of questions about the results, then uses the answers to generate a list of plausible corrective actions. This, however, requires considerable work on the part of the analyst and offers only general advice. Ideally it should examine the results directly and give context-specific advice based on inferences made therefrom. Another result is that the FMECA tool is not yet robustly integrated with the Reliability Prediction tool.

FUTURE AIMS

Design and design assurance are recognisable different activities with the former concerned primarily with the creation of a new design concept and the latter in ensuring that the equipment system is fit for purpose when operating within the spectrum of operational environments. It is highly desirable that these activities are carried out concurrently. Design management ensures the progressive advance of the design through the various stages of the process. Close collaboration is required between the various design and development departments each with differing aims and objectives but collectively concerned with the development of an integrated engineering system. Consequently, it is ASSET's ultimate aim to provide an interface to other CAD systems.

The development of an acceptable paradigm upon which to base systematic procedures is not readily achievable largely because there is no single mechanism for defining the system under consideration. Progress can be made if recognition is given to the requirement for a concurrent assembly of models each pertaining to the performance aspects of one or a number of system features. The need for a common view of the evolving product system is largely unmet at present although the so called system breakdown structure goes some way to providing a universally acceptable definition of the system/sub-system/component assembly. Modelling schema pertaining to the mathematical description of some aspect aligned to system functionality provides a further acceptable description. Certain of the schema will refer to a computerised model of the hardware components (solid model) that will map closely the proposed hardware assembly. Similar links between a CAD or drawing representations and certain modelling schema have been devised by CAD vendors to enhance the capability of their software products.

The prototype ASSET system needs to address the application of security restrictions. Limited access should be provided to some functions; for example project creation and deletion in the Executive Manager should be password controlled. Also access to the domains could be monitored by means of passwords.

ASSET's Executive Manager could also benefit from enhancements to its analysis capability. Currently, what-if analysis can only be achieved at a domain level; this capability should be made available at the executive level together with a semi-automatic report generator.

Knowledge based techniques, which are currently restricted to the Reliability Domain, should be applied to support the above enhancements; in addition, Intelligent Front Ends should be provided for all of the domains, and to complex tools such as ESATAN, with a view to providing assistance with the selection of analysis modules and suitable levels of modelling refinement in defining the data, and the planning of trade-off studies.

Investigations into recent advances in technology (eg Object-Oriented databases) will enable ASSET to address current problem areas such as to KERIS/INGRES interface. This would allow the FMECA tool to be properly integrated with the reliability prediction tool, and would pave the way for a robust 'goal achievement expert'.

Within the existing domains, the following enhancements are envisaged:

Reliability

The Reliability and Maintainability tools should be enhanced, in particular extending the tools and methods to include an analysis of mechanical modules (development has concentrated largely upon assessing the reliability of electrical equipment up to now). The Fault-Tree Analysis tool would benefit greatly from a graphical interface.

The Reliability Modelling facility should be extended; although it provides a fast mechanism for analysis, its effective use still depends on the skill of the analyst (garbage in, garbage out). The embedding of domain knowledge, such as typical failure rate ranges for given types of unit (taking into account environmental conditions where necessary), and information regarding different types of redundancy, would extend the range of potential users.

The Reliability Programme Plan generation facility should incorporate more robust and complete tailoring and editing facilities, while the remedial module, (or 'goal achievement expert') should provide context sensitive advice on how goals might be achieved. The latter facility will share some of the knowledge encoded for the modeller, but will also require component level information in order to draw inferences from the raw data provided by the tools.

Thermal

Further development of the thermal domain will involve enhancement of the mechanism for the characterisation of the product form primarily through the introduction of a solid modelling tool and the introduction of further software tools and modelling aids. Bearing in mind the close interaction between the hardware configuration and the mechanical performance, software tools addressing the problems arising from vibration will be introduced.

Special emphasis has been placed on providing the designer with a range of simple to use design checking tools helpful in the early stages when design concepts are still evolving. These handbook cases are to be expanded to include further data including information on materials.

The printed circuit board presents a special range of problems. The more advanced packaging concepts and component forms require improved computing algorithms and software. The programs under development by Pacific Numerix address the problems to a greater degree than those developed in the early phase of the project. The latest program has a facility for addressing the vibrational characteristics of a PCB. These programs are approaching industry standards and are having increasing usage. Advantageously, they interface directly with a range of CAD systems for PCB design, including placement and routing. The incorporation of this code is an immediate project objective.

The incorporation of the thermal analysis solver ESATAN and the computational fluid dynamics program FLOTHERM has provided an advanced facility for thermal analysis studies. At present, both of these programs are operating essentially in stand-alone mode. The basis for an enhanced level of integration devolves upon the incorporation of a solid modelling tool. Future work will see the closer integration of these facilities.

Cost

The work already undertaken in ASSET provides a demonstratable methodology for performing Whole Life Cost studies within the prescribed framework. It is intended to enhance this facility into a marketable product. To achieve this objective three main areas of work have been identified for future development:

- The suite of prime equipment cost estimating models
- The site of in-service cost estimating models
- An Optimal Repair Level Analysis Facility.

The cost estimating models will be designed to cost specific activities, which together form the development, production or in-service phases of equipment. Different parts of the model will be exercised by equipment developed to meet various specification levels. It is intended that greater integration will be achieved through the selection of the cost model input parameters.

The various cost model elements will be parametric in nature and aimed primarily at personnel who have a need to perform comparative evaluations of completing equipment or system designs. The integrity of the cost model algorithms will, however, be such as to allow a user to perform absolute cost predictions, the accuracy of which will depend upon the time the user invests in calibrating the model elements more particularly to reflect his products and business methods. Absolute cost estimating will demand more equipment parameters to be input than purely comparative cost analyses.

As a major utility to the in-service section of the WLCC model an Optimal Repair Level Analysis tool will be developed. Although many customers specify the maintenance scenario(s) that will be employed to support their system, there are often occasions when this information is required from the equipment supplier, with documented justification for the recommendations. For these latter situations it will be necessary to evaluate various parameters against a variety of maintenance concepts. Consequently it is intended to develop a tool to assist the user in these matters.

CONCLUSIONS

The ASSET demonstrator shows that the encapsulation of knowledge and the integration of tools is desirable in that it does improve the efficiency of the design assurance process. This aside, in conclusion, we wish to address the main innovative segments of ASSET. Here, these are recalled as being:

(i) A knowledge based interface for the Reliability domain, which lowers the skill threshold of the end user

(ii) Some new adapted computational fluid dynamic tools which enable the packaging designer to optimise the hardware to improve the effectiveness of the cooling process. This in turn, leads to improved life expectancy of the equipment

(iii) A cost estimating package which can predict the development engineering and production costs together with the whole life cycle costs for the system design under evaluation

(iv) An increase by several orders of magnitude in the speed of computationally intensive design assurance tools which has been accomplished through the incorporation of the algorithms from ASSET's sister project, APPEAL.

With the emergency of concurrent engineering, the future of ASSET is promising. It is, however, clear that (as a preliminary study) a reappraisal of the modelling methodologies and their inter-relationships is necessary for the advances in the process of design management to be realised. This should include investigations into the greater use of knowledge based interfaces applicable to the various classes of analysis tools, thereby facilitating the more effective utilisation of engineers over a wider spectrum of design disciplines. The study should also evaluate the potential for the introduction of object oriented database system for improving communication between the various design domain interests. Nevertheless, in the course of this ESPRIT-1 project, the foundations for commercial exploitation have been firmly laid.

REFERENCES

Patureau J P, Trowbridge W, Bryant C, Stobbs N H W, Parker R G and Denniss S J, 'ACCORD: Developing some Concepts towards Integration of Analysis and Design', Procs of 6th ESPRIT Conf, December, 1989

Deliverable D7, 8, 9. ACCORD/2.2, 2.3, 2.4/DEL/005/7.1.88/SJD, Functional Requirements, Software Architecture and Design, February 1988, Edited by GEC.

Watziels J L, Buchanan R and Azarian A, 'Specifications of FTA computer code, January 1990, ACCORD/WP2/BC/WD/025/4.01.90/AAn.

Pearce M A, 'Maintainability Prediction Design Specification', ACCORD/WP2/GEC/TN/033/3.8.89/MAP.

Deliverable D34, ACCORD/2.4, 2.5/GEC/DEL/014/26.06.89/SJD, Design and Implementation, July 1989, Edited by GEC.

RAMZES: A KNOWLEDGE BASED SYSTEM FOR STRUCTURAL CONCEPTS EVALUATION

G.H. Arafat†, B. Goodman‡ and T. Arciszewski†

†Department of Civil Engineering,

‡Department of Management and Organization Sciences,

Wayne State University,

Detroit, Michigan, United States of America

ABSTRACT

The objective of this paper is to present RAMZES, a computer system that can be used as a decision support tool for structural concepts evaluation. The system is based on a knowledge-based approach and on Multiattribute Utility Theory The knowledge base is used to provide designers with domain-specific information for describing concepts, whil Multiattribute Utility Theory is used for numerically assessing the overall worth of these concepts. The system' architecture, assumptions and operations are presented in this paper. It includes the initial results of evaluation experiment conducted using the system.

1. INTRODUCTION

Design is considered a two-phase process [Ref. 4]. It includes a conceptual design phase, and a detailed or numerical design phase. The objective of the conceptual design phase is to develop a class of concepts satisfying logical and selected numerical constraints, where a concept is defined as a solution in its most general form described by a particular set of quantitative and/or qualitative attributes. The object of the detailed design phase is to conduct analysis and optimization of the proposed concepts in order to arrive at a single best solution satisfying all the imposed constraints.

The conceptual design phase consists of four stages [Ref. 13]: 1) problem formulation, 2) concept generation, 3) component sizing, analysis and optimization, and 4) concept evaluation. The problem formulation is the stage where the design knowledge, needs and requirements are formulated as a formal problem specification. This specification is used in the second and third stages as input to generate concepts. Structural concepts are evaluated in the fourth stage.

The present trend in the computerization of conceptual design is toward the development of formal models for each of the stages and subsequent development of computer tools based on these models. Since the input to the concept evaluation stage is often a very large number of concepts, there is a need for both a systematic and logically defensible evaluation methodology and its computerized implementation. In the proposed system, concept evaluation is a multistage process resulting in the assignment of numerical values, through formal procedures, to individual concepts. These assigned values represent the overall worth of these concepts to the decision maker, and can be used as the basis for a rational choice among them. There are several evaluation methods in engineering design [Refs. 3,6,7,8,10], but most of these methods are not readily translatable into a computer-based decision support tool for concept evaluation.

The object of this paper is to describe RAMZES, a decision support tool for concept evaluation. This description includes first the basic assumptions and the procedure used to develop the evaluation model implemented on RAMZES, and second RAMZES' architecture and operations. In addition, the initial results of evaluation experiments conducted with this methodology are presented.

2. EVALUATION MODEL

RAMZES is based on an evaluation model using both a knowledge-based approach and Multiattribute Utility Theory (MAUT). A knowledge base is developed to provide designers with domain-specific data on the proposed concepts. It includes the specification of each of the evaluation attributes along with its respective range. Attributes are variables or factors that differentiate between the worth of the concepts of the structural system being developed. MAUT is the methodology used for developing the overall numerical assessments of worth of these concepts, based on the defined attributes. Following are the assumptions and a description of the evaluation model.

2.1 Assumptions

The proposed evaluation model is based on three groups of assumptions: methodological, domain, and mathematical [Ref. 2]. The methodological assumptions are those necessary assumptions regarding the general form of the proposed model of the evaluation process. These assumptions are presented throughout the description of the model. Domain assumptions specify: 1) the knowledge base of a given structural system, and 2) the source for determining this knowledge base. The details for these assumptions must be determined for each given domain, but several guidelines should be followed:

- An engineering system is described by a set of qualitative or logical and quantitative or numerical attributes. Qualitative attributes describe the subjective factors of an engineering concept. Quantitative attributes are numerical dimensions, e.g., weight, cost, etc.
- The specific set of quantitative and qualitative attributes must be specified. Candidates for this set can be determined from many sources, including an examination of the relevant literature, working with domain experts, user specifications, etc. [Ref. 14].
- The candidate set must then be organized, pruned, and studied for missing attributes. This is an iterative process and a difficult one.
- The final set of attributes, along with the feasible ranges of the attributes, comprise the system's knowledge base.

Mathematical assumptions based on MAUT present a formal mathematical foundation for the model. The main assumption used in the proposed model is the additive integration rule, which can be used to determine the value of a concept [Ref. 12].

Specifically $u(A_j) = \sum_{i=1}^{n} w_i u(x_i)$,

where $\sum_{i=1}^{n} w_i = 1$ and each w_i represent the importance or weight of the X_i attribute,

$u(x_i)$ is a conditional utility function on X_i such that u_i maps X_i onto [0,1] and

the set of concepts to be evaluated, $A = \{A_j\}, j = 1......n$, is characterized by a set $X = \{X_i\}, i = 1......n$, of attributes, where $A_j = (x_1,....,x_n)$ and $x_i \in X_i$.

2.2 Description

The application of this model to a particular engineering domain is conducted in two phases: the preparatory phase and the standard use phase. The preparatory phase includes the process of building a knowledge base for the engineering system, preparation of data, evaluation of the selected concepts, and presentation of the evaluation results. In the standard use phase the evaluation system developed in the preparatory phase is used directly in evaluating given or additional concepts of the structural system.

2.2.1 Preparatory Phase:

Three main stages have been delineated and are shown in the flowchart presented in Figure 1: 1. Building of Knowledge Base, 2. Evaluation, 3. Analysis and Graphical Representation of Evaluation Results.

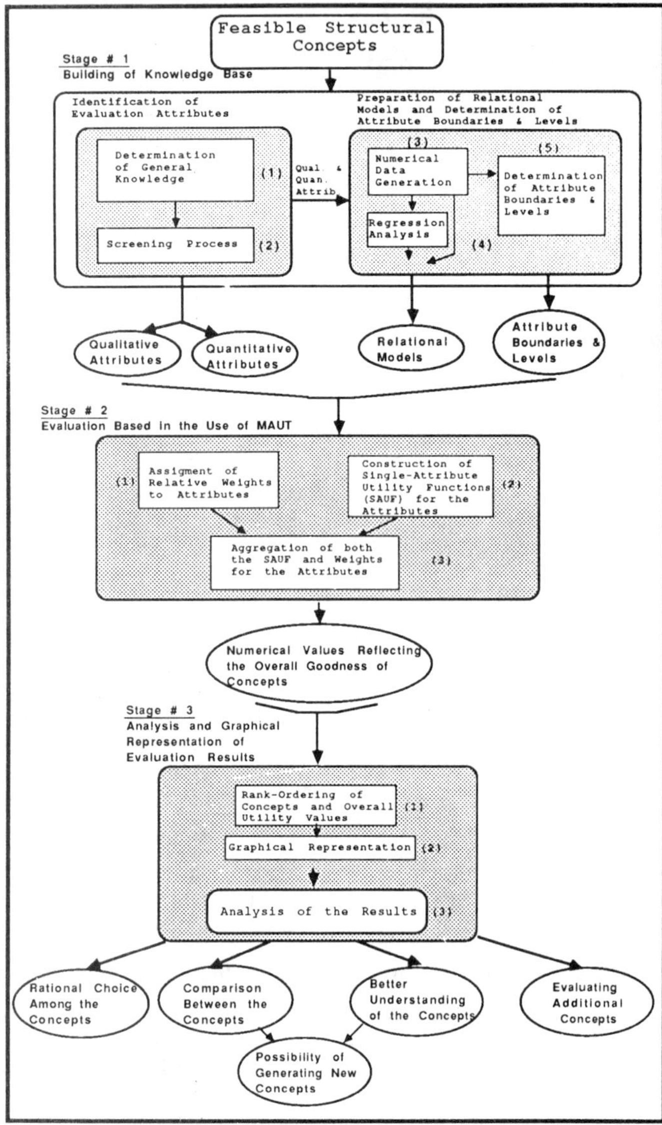

Figure 1 Flowchart of the Evaluation Method (The Preparatory Phase)

1. Building of Knowledge Base:

The objective of this stage is to identify a set of qualitative and quantitative attributes describing the concepts, determine the attribute boundaries or ranges, and prepare relational models. The steps required to complete this stage are:

1)Determination of the general related knowledge about the concepts by referring to domain experts, examining relevant literature, and analytically studying the problem.

2)Application of an organizing and screening process to this general related knowledge to identify a reasonable number of quantitative and qualitative attributes that satisfy a specific set of criteria: completeness, operationality, decomposability, nonredundancy, and minimal size [Ref. 14].

3)Generation of the numerical data which provide the relationships among the design parameters of the engineering system.

4)Preparation of the relational models using regression analysis. Relational models provide designers with the relationships between the engineering system's variables.

5)Determination of quantitative attribute boundaries and qualitative attribute levels. Quantitative attribute boundaries represent the maximum and minimum values of these attributes. The qualitative attribute levels are determined according to the type of attribute and the nature of the set of concepts to be evaluated.

2. Evaluation:

The main objective of this stage is to assign values to each of the concepts using MAUT. These values reflect the overall worth of the concepts. Three main steps are used in applying this theory:

1)Assigning relative weights to each attribute.

2)Constructing the single attribute utility values and functions (SAUF) of both qualitative and quantitative attributes.

3)Determining the MAUT score for each concept as a weighted sum of each single attribute score times its respective importance weight [Ref. 12].

3. Analysis and Graphical Representation of Evaluation Results:

The objective of this stage is to analyze and study the evaluation results in order to 1) compare the concepts, 2) gain a better understanding of them, 3) select the most promising concepts, and 4) generate new ones. This stage includes three steps:

1)Rank-order the concepts based on their overall utilities.

2)Build several graphs based on the above rank-ordering and the overall utilities of the concepts. In each graph, the overall utility values of all concepts are represented on the Y-axis by values ranging from 0 to 1, while the X-axis represents the concepts themselves.

3)Analyze the results based on the rank-ordering and on the graphical representation of their overall utility values.

2.2.2 Standard use phase

In the standard use phase the evaluation system is used to evaluate any set of concepts of the engineering system. A computer evaluation system for practical use has been prepared to aid in this task. The main difference between the different stages of the preparatory phase and standard use phase is in stage 1, the building of the knowledge base. In the preparatory phase, the main objective of stage 1 is the building of the knowledge base for the given engineering concepts. In the standard use phase, the knowledge base is available and the main objective is to use the knowledge base in evaluating the engineering concepts. Input to the evaluation process in the standard use phase includes an appropriate range for each of the design variables, the relative importance of attributes, and the single-attribute utility functions. The output from the evaluation process is the overall utility or worth of each concept. Relating the input to the output of the evaluation method enables the study of the influence of specific attributes on the overall utility of concepts.

3 RAMZES: A COMPUTER SYSTEM FOR THE CONCEPT EVALUATION

3.1 Components of the System

RAMZES is a computer system designed to facilitate both the preparatory and standard use phases of the concept evaluation stage of the conceptual design process. It consists of three components: 1) the knowledge processor, 2) the evaluator, and 3) the modifier. The first two components form the major part of this system. Each of these components consists of several subcomponents, each with a specific task assignment (Figure 2). The first component, the knowledge processor, is a knowledge base shell used for storing the data needed by the system. The second component, the evaluator, is an interactive component used for determining the overall worth of each concept. It uses the stored knowledge base and MAUT methodology. The third component, the modifier, is used to make the necessary changes in the first two components, based on the designer's analysis of the preliminary results of the system.

3.2 Relationships Between the System's Components

The relationships among the different components and subcomponents of RAMZES are shown in Figure 2. The user begins the operation by entering data describing the concepts into the knowledge processor. This data is analyzed and structured into a hierarchical system of quantitative and qualitative attributes that completely capture the essential features of the structural system being considered. The structuring is done during the building of the knowledge base in the preparatory phase. This is the most difficult and subjective aspect of concept evaluation.

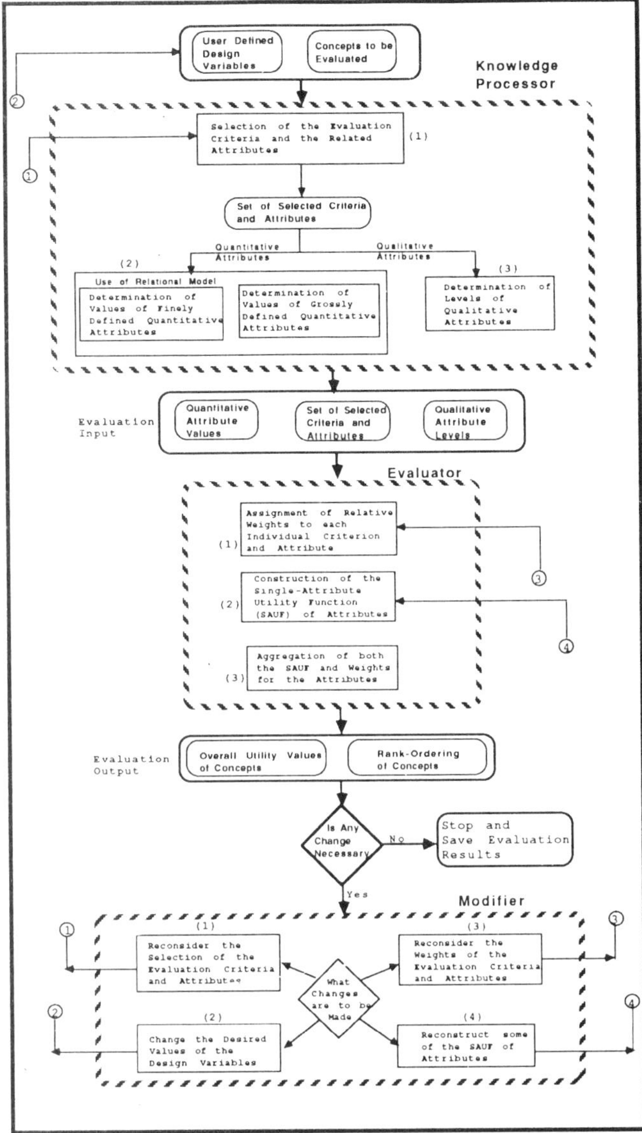

Figure 2 Computer System for Concept Evaluation: Component Relationships

The knowledge processor is used as a support tool during the structuring process. The output of the knowledge processor is a redefining of the basis for the identification of the set of concepts consisting of the necessary quantitative and qualitative attributes, along with their respective ranges or levels.

The main object of the second component, the evaluator, is to determine the overall worth of each of the selected concepts. Three subcomponents perform the MAUT procedures. The first subcomponent aids the decision maker in determining the relative importance of the evaluation criteria and attributes. The second subcomponent is a tool used in the construction of the Single Attribute Utility Functions (SAUF) for the evaluation attributes.

Finally, through the third subcomponent, the resulting weights and SAUFs for the attributes are aggregated into overall utility values for the concepts.

After completing the operations of the second component and obtaining the utility values and if desired, analyzing the results, the user may terminate the evaluation or make changes anywhere in the process. If the decision maker wants to make change, this is done through the third component of RAMZES, the modifier. The modifier allows the user to alter either the knowledge processor or the evaluator components. Then the system can be used to reevaluate existing concepts or to evaluate new concepts. The output of RAMZES, or the concept evaluation stage, is a small set of the best concepts which will then serve as the input to the detailed design phase.

4. EVALUATION EXPERIMENTS

4.1 Description

The feasibility of the evaluation model and the computer system RAMZES have been demonstrated in the evaluation of steel roof truss designs for industrial buildings. Twenty types of roof trusses were considered for evaluation. They are shown in Figure 3. The information needed for this application was gathered from engineers with local construction companies, from engineering faculty, from the relevant literature, and from the analytical study of roof trusses. This information along with a large number of examples generated for minimum weight design of these roof trusses, assisted in building the knowledge base needed for evaluation. The knowledge base of the roof trusses includes two parts: 1) a complete system of evaluation criteria and attributes describing the roof trusses and 2) the possible ranges of the evaluation attributes. The final set of these criteria and attributes is shown in Figure 4. The second part of the knowledge base was prepared by using relational models developed for this purpose, and by analyzing the set of attributes and concepts considered for evaluation. The knowledge base is stored in the computer system.

The evaluation process for the roof trusses was conducted with the cooperation of five people representing three engineering companies, and of civil engineering faculty. The results of these experiments, one associated with each person, are presented in the next section.

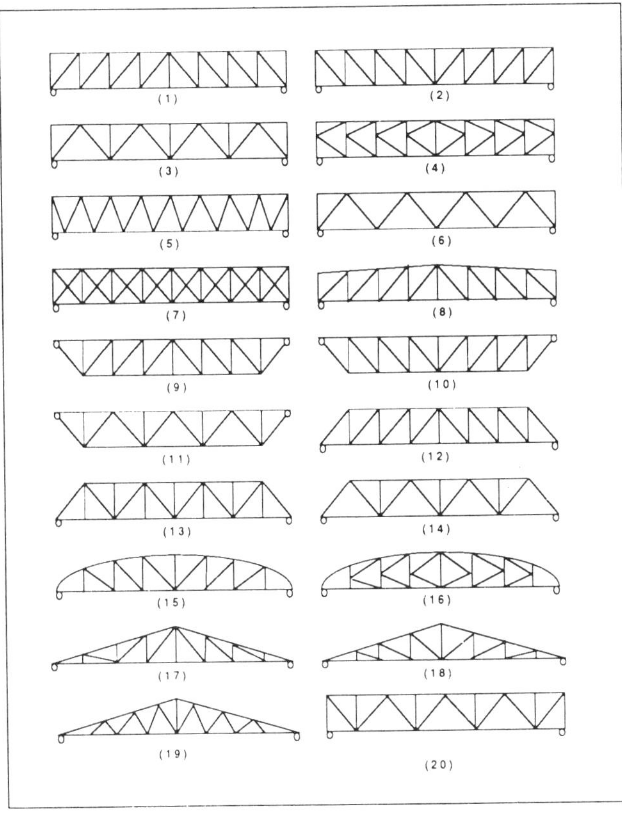

Figure 3 The Selected Types of Roof Trusses for Evaluation

4.2 Results

The process of evaluating roof trusses using this evaluation model and its computer implementation, RAMZES, was conducted considering three cases for the ranges of the design variables span length, overall height, steel grade, and number of panels. Case I represents roof trusses with span length of 72 ft, overall height of 8 ft, steel grade 36 Ksi, and six panels. Case II represents roof trusses with span length of 72 ft, overall height of 8 ft, steel grade 36 Ksi, and eight panels. Case III represents roof trusses with span length of 72 ft, overall height of 11 ft, steel grade 50 Ksi, and six panels. The purpose of changing some values of the design variables is to study their effect on the overall utility values and rank-ordering of roof trusses. A sample of the evaluation results for the five experiments is presented next. These evaluation results were prepared for the purpose of verifying the feasibility of the proposed evaluation model.

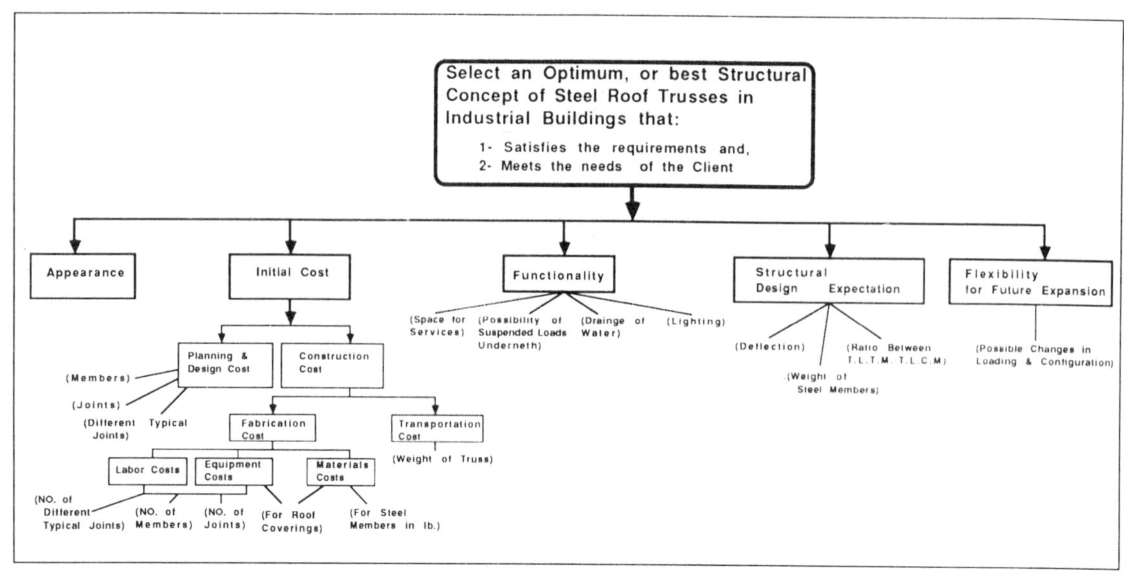

Figure 4 The Evalution Criteria and Attributes of Roof Trusses

In Figure 5, the overall utility values and rank-ordering of the selected roof trusses are presented for experiment 1. The rank-ordering of roof trusses in Figure 5 is based on the overall utility values for case I. The overall utility values of cases II and III are presented in the same graph. This enables the user of the model to study and compare the evaluation results of roof trusses for different evaluation inputs.

A study of Figure 5 shows that:

- For the most part, roof truss types 13 and 14 are the most valued types.
- For the most part, roof types 16, 19, and 7 are the least valued types.
- For most roof trusses, case II utility values are smaller than those of case I. The reason is related directly to the number of members, number of joints, and space available for services (case II, with eight-panel roof trusses, gives a higher value for the number of members and joints, and a lower value for the space available for services.
- For most roof trusses, case III utility values are greater than those of case I. The reason is related directly to the weight and deflection of the roof trusses (case III, with steel grade 50 KSI, and an overall height of 11 ft, produces lighter trusses with lesser deflection).

Another set of evaluation results is shown in Table 1. The top five concepts of roof trusses using the evaluation model for the five experiments are presented. A study was conducted to determine how much agreement exists among structural engineers about five most valued types of roof trusses. The results, expressed in percentages, showed that there is 85 percent agreement about the top five roof trusses. The same study was conducted to determine how much agreement there is about the five least valued concepts of roof trusses. The results showed 95 percent agreement. This supports the validity of this method in the evaluation of engineering concepts.

4. CONCLUSIONS

The initial results obtained from applying and testing the model and its computer implementation for the evaluation of the roof trusses have confirmed the following:

1. A knowledge-based approach and Multiattribute Utility Theory (MAUT) are feasible for a formal evaluation method applicable to the conceptual design stage.
2. The proposed method can be used to determine the order of concepts based on their worth measured by utility, which allows selection and comparison among the evaluated concepts.
3. Evaluation knowledge for a given structural domain can be stored in a computer system for evaluation. This makes the knowledge instantly accessible for evaluation purposes.
4. The developed computer system reduces the effort and time needed to evaluate engineering concepts. It can be used for

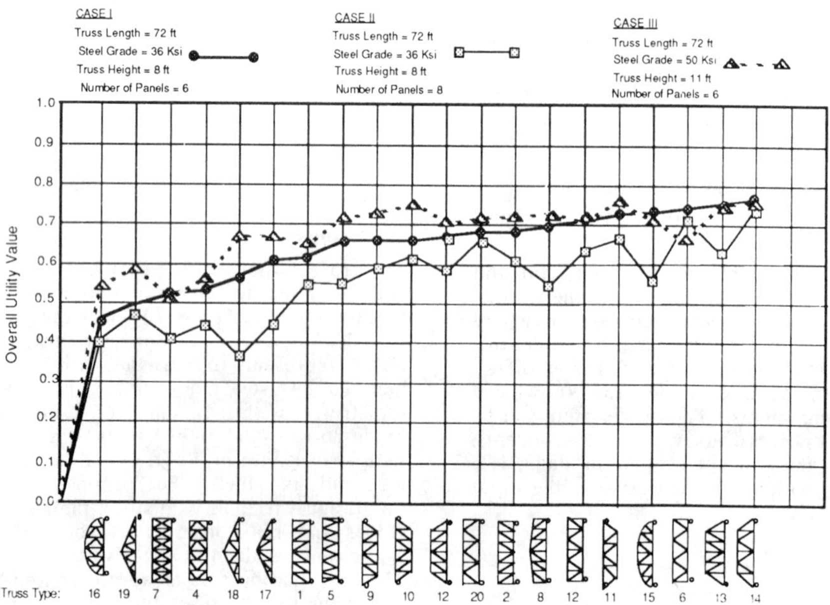

Figure 5 Overall Utility Values and Rank-Ordering of Roof Trusses (Exper. #1)

Rank-Ordering	Truss Type Exper. #1	Truss Type Exper. #2	Truss Type Exper. #3	Truss Type Exper. #4	Truss Type Exper. #5
CASE I					
1	14	13	14	14	6
2	13	14	6	13	14
3	6	12	11	6	10
4	15	10	13	11	11
5	11	2	10	15	13
CASE II					
1	13	13	14	13	10
2	14	14	6	14	11
3	11	18	3	11	14
4	10	12	11	12	13
5	8	10	20	10	9
CASE III					
1	14	14	14	14	5
2	6	6	11	13	14
3	3	13	13	11	11
4	11	3	3	6	3
5	20	2	10	12	10

Case I:
Span Length = 72 ft.
Overall Height = 8 ft.
Steel Grade = 36 Ksi
No. of Panels = 6

Case II:
Span Length = 72 ft.
Overall Height = 8 ft.
Steel Grade = 36 Ksi
No. of Panels = 8

Case III:
Span Length = 72 ft.
Overall Height=11 ft.
Steel Grade = 50 Ksi
No. of Panels = 6

Table 1 The Top Five Concepts of Roof Truss Types for all Experiments

both practical design and research design purposes.

5. The ability to modify the evaluation data input in the computer system enables the user to analyze concepts while changing in the ranges of the design variables, the various sets of evaluation criteria and attributes, and different weights or the SAUFs for the evaluation attributes.

6. The computer system for concept evaluation represents an attempt to computerize conceptual design, which is one of the least investigated, understood, or computerized stages of the design process.

5. REFERENCES

1. Ang, A., Tang, W., Probability Concepts in Engineering Planing and Design, John Wiley & Sons, Inc. 1984.
2. Arafat, G., Arciszewski, T., Goodman, B., "Concept Evaluation: A Knowledge-Based Approach," Proceedings of the Tenth European Meeting on Cybernetics and Systems Research, University of Vienna, Austria, 1990.
3. Arciszewski T., Brodka J., "Technical Quality of Structural Solution, Appraisal Method," in Polish, Journal Inventiveness and Rationalization, Vol. 1, 1978.
4. Arciszewski, T., "Stochastic Form Optimization," Journal of Engineering Optimization, Vol. 13, 1988.
5. Bascaran, E., Bannerot, R., Mistree, F., "Hierarchical Selection Decision Support Problems in Conceptual Design," Journal Engineering Optimization, July 1988.
6. Daru, R., Vangils, A., "Decision Graphics (ADG): Criteria for Development and Evaluation," Preprints CAPE '86 Copenhagen, 1986.
7. D.E. Boer, Ir., "Selection Techniques in Methodical Design," Proceeding of the 1987 International Conference on Engineering Design, Boston, 1987.
8. Diaz, A., Kikuchi, N., "Finite Element and Optimization Methods in the Design Process, "Proceeding of the 1987 International Conference on Engineering Design, Boston, 1987.
9. Dixon, J.R., "On Research Methodology Towards a Scientific Theory of Engineering Design", Artificial Intelligent for Engineering Design, Analysis, and Manufacturing, AT EDAM, Vol. 1, No. 3, June 1988.
10. Dong, Zhongyuan, "Evaluating Design Alternatives and Fuzzy Operations," Proceeding of the 1987 International Conference on Engineering Design, Boston, 1987.
11. Edwards, W., Newman, J., Multiattribute Evaluation, Sage Publications, Beverly Hills, California, 1982.
12. French,, S., Decision Theory: An introduction to the mathematics of rationality, Halsted Press, New York, 1986.
13. Hajdo, P., Arciszewski, T., "Computer Generation of Structural Concepts: A Knowledge-Based Approach," Proceedings of the Seven Conference on Computing in Civil Engineering and Symposium on Data Bases, Washington D.C, May 6-8, 1991.
14. Keeney, R., Raiffa, H., Decisions with Multiple Objectives: Preferences and value tradeoffs, John Wiley & Sons, Inc., 1976.
15. McCormac, J., Structural Steel Design, Harper & Row, Publisher, New York, 1981.
16. Melaragno, M., Simplified Truss Design, Van Nostrand Reinhold, New York, 1981.
17. Merret, F., Building Construction Handbook, McGraw-Hill Book Company, Inc., new York, 1958.
18. Pitz, G., McKillip, J., Decision Analysis for Program Evaluators, Sage Publications, Beverly Hills, California, 1984.
19. Parker, H., Simplified Design of Building Trusses for Architects and Builders, Wiley In. Publication, 1982.
20. Saaty, T., A Scaling Method for Priorities in Hierarchical Structures, Journal of Mathematical Psychology, Vol. 15, 1977.
21. von Winterfeldt, D., And Edwards, W. (1986). Decision Analysis and Behavioral Research, Cambridge U.N. Press.

EXPERT SYSTEMS AS AN INTEGRATING TOOL IN CIVIL ENGINEERING DESIGN

W.J. Spencer and P. Podlaha
School of Civil Engineering and Building,
Swinburne Institute of Technology, Hawthorn, Australia

Design engineers have now accepted the computer as an aid to engineering design offering a wide range of tools to assist in the design process. However the increasing sophistication and power of computer facilities has been matched by increasing complexities in the design task brought about by an expansion of knowledge. Codes of practice that were initially slim volumes of a few pages now often approach the size of modest text books and environmental and social issues are more prominent. Matching the required knowledge to the design task has become a problem since both an awareness and management of information is required. The paper addresses this problem and suggests that real advances in the design process are possible through the judicious use of expert systems coupled with a full utilisation of the power of modern computers. In addition to being seen as another option for use as a design tool in engineering, it is contended that expert systems have a particular role to play in the overall management of the design. Two aspects are considered, firstly the role of expert systems at the level of conceptual design and secondly at the level of detailed design. The concept suggested is one of providing a coordinated computing environment with the designer in overall control of the process. A model showing how expert systems can be used as an integrating tool is described in some detail.

INTRODUCTION

'Now the engine, from its capability of performing by itself all these purely material operations, spares intellectual labour, which may be more profitably employed....'

'......Again, who can foresee the consequences of such an invention?' (Charles Babbage 1842)

Who indeed? The rapid growth in computer technology over the last two or three decades has probably only been matched by the increase in potential knowledge available to the design engineer. Anyone who has been fortunate enough to have been involved in computing over that period will readily testify to exponential growth rates in whatever measure is taken except physical size. The increasing sophistication of computer hardware and software has also been matched by increasing sophistication of codes of practice, building regulations and complexities surrounding the design task such as social and environmental issues. The complexities mean that, unless the design task is properly managed, the designer is unlikely to come up with a better solution to the problem today, than that that might have been achieved ten years ago. The dilemma can be expressed by an analogy to an urban traffic route between two growing population centres. A common experience is that progressive upgrades along the route simply approximately keep pace with the increased traffic demands. The nett result is that it is unlikely that any particular increase in benifit to the user occurs over a period of time.

The problem is one of managing resources that will assist with the specific design task in hand. Knowledge is available but it is often not known to the design engineer confronted with the task of coming up with a solution in a given time frame. The designer may seek to identify all the sources of knowledge available, but often falls back on the depth of previous experience on that particular task. Frequently, and for a variety of reasons, not all the relevant knowledge is identified largely because there is an enormous amount of knowledge even in a specific domain. This may or may not be a serious problem. At one level it may simply mean that the solution is less efficient than it might otherwise have been, but at another level it might be the cause of a failure that could have been avoided. The difficulty is to know what to do about it at a local level in the office of the design engineer. There is now some prospect that the speed and capacities of microcomputers will outstrip the required knowledge access in a specified domain, making it possible for real advances to be made in the outcomes of the design process. This paper takes up that theme with the view that expert systems offer a role as an integrating tool providing the environment to oversee the design. Further, by considering what is involved in the design process and the expectations of the designer, it is possible to suggest the characteristics required of the expert system environment.

Lansdown (Ref. 1) introduced an interesting matrix representation of knowledge sources, as shown in Figure 1, in presenting a concept of the designers' information environment. As may be seen, Lansdown has proposed that the designers' basic education and training, while covering the full range of knowledge types from global knowledge through to task orientated knowledge, only meets the designers needs from the coping to the helpful level. Beyond that other resources are necessary and this is widely recognised. Research engineers and academics may be familar with procedures for literature reviews and using textual databases to research a particular topic. However the design engineer does not have the time for a leisurely review of the knowledge specific to the task. Expert systems appropriately developed can overcome this problem.

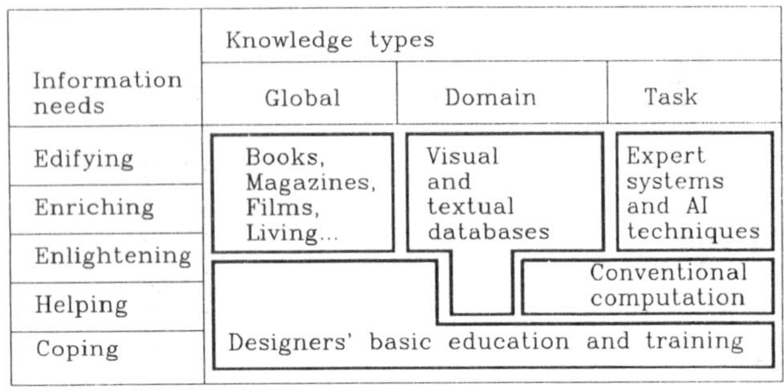

Fig. 1 A knowledge needs - sources matrix

THE DESIGN PROCESS

Two major aspects of the design process are considered in the context of this paper. The first aspect is that of conceptual design and the second is detailed design. It is probably helpful to have a specific area in mind and in this case the authors' experience relates to structural engineering although there is little doubt that the argument is valid for a wide range of design activities.

Conceptual Design

Typically the design process starts with instructions from a client and the preparation of a design brief. The principal design engineers are called in and it is their experience and local knowledge that is brought to bear on the conceptual design. There is a definite emphasis here on past experience and those involved in the conceptual design phase are likely to be senior engineers in the practice. Hopefully innovation and new solutions will not be overlooked and the more successful practices will have senior people who do keep up with the relevant knowledge. Previous solutions to similar problems will often form a starting point although a difficulty in this regard is a suitable system of recall. With effectively nothing in place, little or no effort may be made to consider this aspect and it is likely that time and again the same problem is re-solved. This is particularly true of what might be described as routine projects rather than major structures likely to attract the communities wider attention. The point being made here can be put into context by relating it to the problem of in-house computer program development. Enthusiastic engineer-programmers have had a long history of difficulties with documentation and program maintenance. Consequently many design office practices have experienced the problem of lapsed software for specific tasks and the same problem being solved over and over. Re-inventing the wheel is alive and well at a number of levels.

There is little doubt that the more efficient design offices have this matter in hand with a systems approach to the problem. The office manager will be likely to have a set of procedures in place ensuring that the project goes through appropriate stages. However it is contended that the problem can be more effectively addressed by an expert system designed to illicit a problem definition from the user and direct the designer to previous solutions and appropriate knowledge sources. While alternative systems could approach the problem, expert systems have a number of characteristics that make them more suitable. The first of these has already been mentioned. That is, the user can be asked to define the nature of the problem and be directed accordingly. Previous solutions could be stored as text files, in a frame representation of knowledge or as external data bases all of which can be accessed by the system. Part of the office practice could be a review of the project with recommendations for future projects of a similar nature. Such notes and advice can be readily incorporated and maintained in the expert system operating at this stage essentially as an information retrieval system. The expert system also has the advantage of being able to incorporate local knowledge and local conditions pertinent to the project. This can be quite far reaching, ranging from a restriction on material sizes or specifications to noting sensitive social or environmental issues. It is interesting to note a brief report in a paper by Ashley and Levitt (Ref. 2) of a similar concept being used in the development of a knowledge base for repeating construction project success. While the project aims at isolating the factors for success, it uses a data base of previous project data. The task of implementing such a scheme may appear at first to be a formidable one, demanding of time and resources, but engineers must recognise the increasing capabilities and sophistication of the knowledge engineering tools and the computer hardware that are available to meet this end.

Detailed Design

While conceptual and detailed design processes are interrelated, it is convenient to consider each aspect separately. An early stage of conceptual design may suggest a look at some aspect of detailed design and the converse is also true. That is, in examining some design detail it may be appropriate to review the design concept. Adequate design of the engineers design environment should ensure that such processes are encouraged.

Civil engineering design is characterised by guidelines in codes of practice, regulatory procedures, analytical processes and expert knowledge. Invariably many of the processes involve computer applications but rarely are these continuous on-line processes, in spite of the various attempts that have been made for fully computer-integrated design. Few projects of any substance could be effectively carried out without the aid of a computer somewhere between client instruction and project completion. In the face of increased computer involvement, it remains important to maintain a sense of perspective and not lose sight of the objective of the overall task. The creative aspect of engineering should always be maintained and a sense of achievement should be available in a given task. It is for these reasons that the engineering design process should not become an automated computer orientated one, rather it should move to become a computer managed process with the designer in full control. The emphasis here is on management, direction and advice, clearly a role for expert systems. Coupled with all of the other computer applications such as databases, conventional analytical programs, CAD packages and codes of practice expressed as knowledge bases, an expert system can have a monitoring role. There is some recognition of this in trends in the development of expert system shells and databases. In some cases expert system shells have a form of database as part of the software, using an extension of the frames representation of knowledge. On the other hand, intelligent databases are emerging using rules of logic in the search process.

The proposed development expressed here can already be seen in the development of conventional programming languages. Initially, complete development of a program involved the use of a number of separate, but related, entities. These included a text editor, a compiler, a linker to incorporate library subroutines and the direct use of an operating system to manage all the operations in a limited way including program execution. While the essential elements remain, programmers are now offered a programming environment exercising complete control over all the operations with mouse or keyboard driven pull down menus and windows. The result is increased efficiency of operation and generally a greater sense of control to the user. Once initiated, an entire programming exercise, from setting up the source code to debugged execution, can be managed from the programming environment. This is not to say that the previous functions are not used of course, since indeed they are.

The discussion presented to date could be summed up by a revision of Figure 1 expressing a much broader role for expert systems. While the Lansdown matrix does not attempt to show information flow or control, Figure 2 attempts to do this with a rearrangement of the components. In Figure 2, the designer, with basic education and training, is seen to be in control aided by an expert system managing a range of tools.

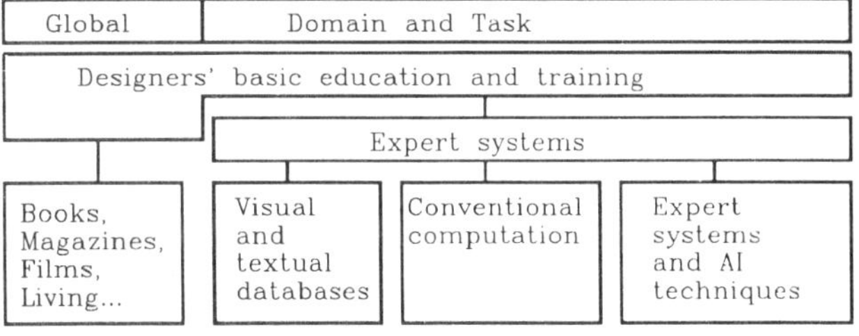

Fig. 2 Integration of activity by expert systems

A MODEL FOR INTEGRATION

The feasibility of using an expert system as a master integrating tool or computer manager has been demonstrated by Spencer et al (Ref. 3) in an application which can be regarded as a subset of the structural engineering design process.

In the development of an expert system for the preliminary design of highway bridges (RCABE), the various design processes were established and the possible solutions identified as the design space. In addition the criteria against which success was to be judged were defined. Against this background a multiple set of knowledge bases was developed. While each knowledge base was capable of independent execution, each was also capable of both contributing to, and accessing a common knowledge pool. The integrating tool in this case was the master knowledge base directing access as required to the specific knowledge base. The general configuration is shown in Figure 3. Since it is proposed that the model used for the development of RCABE represents a general one for civil engineering design, the model will be discussed in some detail.

Requirements of the Model

The engineering design process is an iterative one demanding a 'what if?' option. Any system which purports to support this process must have a great deal of flexibility with an ability to retain previous results for comparative study. It is also desirable that the system should be capable of:

(i) multiple runs related to the one design problem

(ii) the ability to decline the system's offer of advice on a particular design task; and

(iii) the ability to repeatedly access certain knowledge sources during the run.

Frequently design tasks are initiated in the absence of all the relevant data and some gross assumptions may have to be made. It is important for any system to store the design data and allow a user to subsequently return to continue the design or to review the previous tasks. The advisory nature of the system is emphasised in the ability of the user to input information in lieu of the advice of one or more knowledge sources. This is an important aspect of the system design that ensures that the user remains in control of the design process. Often a particular degree of flexibility may be required in a part of the system where the advice from knowledge sources is highly interdependent. In the example under consideration, a special linking was required of the knowledge bases relating to the bridge superstructure, abutments, piers and foundations. This group was referred to collectively as the bridge module. The design tasks associated with the selection of each individual part of a bridge were tied together through a menu called from the master knowledge base. During a particular run of the system, the user has the option to repeatedly seek advice on any particular part of a bridge. In this way the user can investigate the effects of changes say in the pier foundation type, on the type of pier and the type of superstructure.

Generally speaking expert system shells have the characteristics to meet the requirements outlined. Specific characteristics that are exploited in the model include the ability to call external programs including databases and other knowledge bases, and the ability to write and read ASCII files as a basis for exchanging data. Some expert system shells have developed this latter theme in what is generally known as blackboard architecture (Refs. 4 and 5) where the exchange of data between knowledge bases is a design feature of the shell. The expert system shell EXSYS (Ref. 6) for example, has a datalist command where information on a variable is specified by its number or abbreviated name and its value. In essence, this becomes a defined format for an ASCII file which is known. When a knowledge base is linked to a specified datalist, that knowledge base will

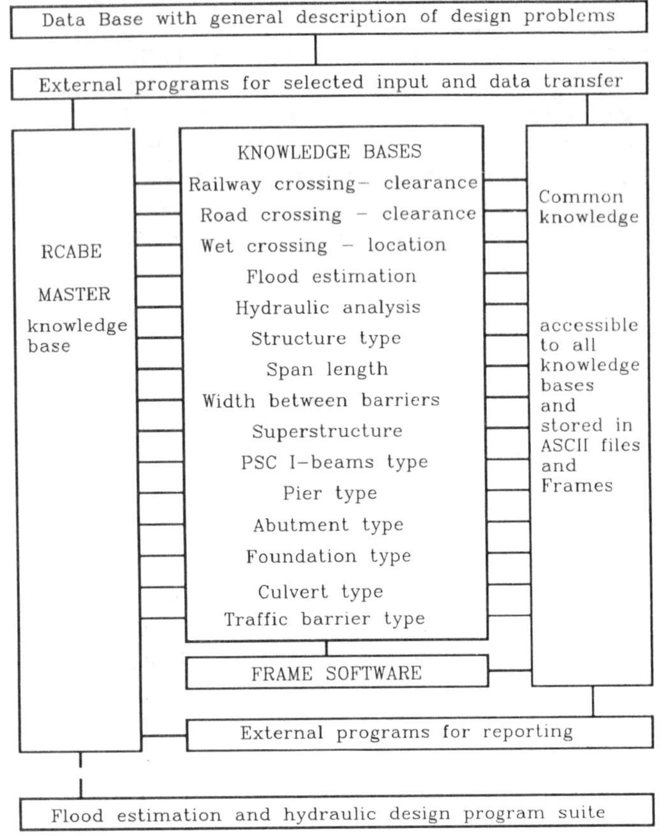

Fig. 3 The RCABE model

automatically check the datalist and read or write data as required. Since the format is known, external programs may also read and write to the datalist. It is apparent that black-boarding is an aptly named facility. The analogy is one of writing or reading information on a blackboard which is available for all to see and chose to act accordingly.

The RCABE Model

With reference to Figure 3 it is possible to recognise the following major functional blocks.

- (i) master knowledge base
- (ii) specific knowledge bases
- (iii) common pool of knowledge accessible to all knowledge bases
- (iv) external programs
- (v) data base
- (vi) external program for system reporting
- (vii) frame software

Any design project can be broken down into a series of design tasks each with their own goals. This was the basis for the development of the specific knowledge bases shown in the model. While they were developed as independent systems each was integrated into the overall system through the master knowledge base and the links to the common pool of information.

In contrast, the master knowledge base contains mainly rules which decide if and when a specific knowledge base or conventional program or data base is to be called. The master knowledge base contains strategic knowledge relevant, in this case, to preliminary bridge design. While the details are not particularly relevant to this paper, it is important to note that it was essential to develop an extensive control structure behind the master knowledge base, reflecting the way in which the expert would proceed through the various design steps.

Particular reference should be made to the external programs in the RCABE model which are in two categories. The first are those programs which were written specifically to meet some shortcomings in the operational functions of the shell. While this might not be particularly desirable, the fact that it was possible demonstrates the flexibility and integrating power of expert systems. More interest here however centres on the external programs representing conventional computational techniques. In this case they are represented by the Flood estimation and hydraulic design program suite. This latter group is a collection of conventional analytical programs well known in water engineering and used in the estimation of design discharge and design flood levels. Under varying circumstances, one program is more suitable than another, presenting a classical case for an expert system to advise on the most appropriate selection. In fact an illustration of this has been reported in the literature (Ref. 7).

In the RCABE model, two knowledge bases were developed to fulfill this function and integrated into the system, although the suite of analytical programs were not directly connected. The system simply advises the user as to what program should be used and allows the user to leave the system at that point, carry out the computation and subsequently return with the necessary data. Far from detracting from the integrating role of an expert system, this approach underscores it. The managerial role of the expert system is more apparent, the user has a greater sense of being in control and the total system is made more flexible. Of course if it is appropriate, a conventional program can simply take its place alongside the specific knowledge bases as a fully integrated part of the system.

The remaining components are fairly self evident and do not need further comment. However there is one other aspect to the development of the model that is worth noting. One of the advantages cited for the use of expert systems is that the system is readily expandable. This lies in both the characteristics of an expert system and the way in which such a system is built. A model such as RCABE clearly has the same capability providing some thought is given to this prospect in the planning stages. The advantages of these are clear. Firstly there is the prospect of staged development with a useful tool being available at an early stage. Secondly the opportunity exists for upgrades to be made with the use of improved knowledge engineering tools. Again, rapid changes are occuring in software development and new versions of propriety software regularly become available. The development time of RCABE was approximately eighteen months during which time there were two upgrades of the original shell selected for the project. More significantly, a Frame (Ref. 8) extension to the otherwise rule-based shell, EXSYS, became available towards the end of the project. It was found that it was relatively easy to incorporate its use into the system.

Any integrated system such as the one suggested here will

never be complete in the sense that both maintenance and upgrades will always be a consideration. However this is not an additional requirement in design office practice since any knowledge base, be it human or machine, must be maintained and developed. Expert systems, because of their tendency towards natural language processing, are easier to maintain than programs written in conventional languages. An experienced user who is not necessarily the developer of the system can quickly suggest and implement changes.

Reflection on Figure 3 will show that it includes the relevant components of Figure 2 and it is therefore a practical illustration of how Figure 2 can be implemented providing integration of design activities through an expert system. A realistic model must be domain specific, but nevertheless the domain could be quite broad. For example RCABE could be expanded to present the design, as opposed to the preliminary design, of short to medium span highway bridges. Another system could embrace the design of industrial and low rise commercial steel buildings. The point here is that the domain needs to be defined so that there are some practical limits and an overall strategy for development can be mapped out.

On a smaller scale work is well in hand on a project to develop an expert system as a Steel Design Tutor, offering guidance to the Australian Standard AS4100 (Ref. 9), the steel structures code. The same modular approach to the task has been taken, although within each specific knowledge base more attention has had to focus on the development of an extensive menu system forming the framework for development. The major difference lies in the fact that it is not so much a strategy for the design process that is driving the system, as it is the requirements of the user. Basically the project aims at guiding the user through various parts of a complex modern code. To do this it is chapter orientated in line with the code with the user having the options to select the essentially independent chapters, represented by the specific knowledge bases, by menu. During development some limitations of the shell being used (Ref. 10) have become apparent. Essentially these relate to the need, in engineering design systems, for flexibility and looping and resetting of variables. The problems are not insurmountable but it is up to the profession to express its needs to the software developers.

CONCLUSION

That expert systems have a role as part of the computer based resources of the design engineer is probably not now in dispute. The propositions of this paper however, are much broader than that, promoting expert systems to an information management role while ensuring that the design engineer remains in control of the task. It will be recognised that the approach suggested here is going to require, in the long run, large amounts of computer memory and high speed access. However time will solve that problem, if it hasn't already, and the question will remain as whether or not civil engineers will be ready to take advantage of the technological advances of computing power. A recent announcement by Intel talks of the release of a 66 MHz i486 chip rated at 50 MIPS. At the beginning of this decade microprocessor clock speeds were around 2 MHz at a time when civil engineers didn't even know what a MIP was. Babbage was a visionary of the nineteenth century and there is still scope for visionaries at the end of the twentieth century.

Computer engineers and scientists are perhaps leading the way at the moment. The danger is that civil design engineers will be led instead of leading. Already some application packages in structural analysis and design are so sophisicated that only 'experts' can use them confidently. There are even examples of expert systems being written to advise on the use of such packages. If the design engineer does not take a stand others will take over the task and managers will ask whether or not the engineer is needed.

The answer lies in the judicious use of expert systems in the manner outlined in this paper. The feasibility of this has been demonstrated and it can be effected by careful planning in an integrated approach to the design task in a given domain. There must be a clear understanding of what engineers want in a design engineers environment and management must be encouraged to support this.

REFERENCES

1. Lansdown, J. "The Designers' Information Environment: Tools for Design Knowledge Manipulation", Proc. of Symp. on Knowledge Based Systems in Civil Engineering, Monash Univ., 1988

2. Ashley,D.B and Levitt,R.E. "Expert Systems in Construction: Work in Progress", Journal of Comp. in Civil Eng., ASCE, Vol 1, No. 4, October, 1987, pp. 303-311

3. Spencer,W.J., Atkins,R.M. and Podlaha,P. "Development of an Expert System for the Preliminary Design of Bridges", Civil Engineering Systems, Vol 6, No.1-2, pp. 51-57, 1989

4. Harmon,P. and King,D. "Expert Systems and Artificial Intelligence in Business", John Wiley and Sons, New York, 1985

5. Waterman,D.A. "A Guide to Expert Systems" Addison-Wesley, Reading Mass. 1986

6. Exsys Professional: "Expert System Development Package", Exsys Inc. Albuquerque, NM, 1988

7. Fayegh,D. and Russell,S.O. "An Expert System for Flood Estimation", Proceedings Symposium on Expert Systems in

8. Minsky,M. "A Framework for Representing Knowledge", in "The Psychology of Computer Vision, ed. Winston,P. McGraw-Hill, 1975

9. Standards Association of Australia AS4100-1990, "Steel Structures", SAA, Sydney, 1990

10. Crystal: "The Expert System Builder", Intelligent Environments, Surrey, 1987

A CONCEPTUAL APPROACH OF FUZZY DECISION FOR SYSTEMS BY "DEEP KNOWLEDGE" AND "DEEP DATA"

Lin Shaopei, Guo Xiuling and Wu Yufei
Civil and Architectural Engineering Deparemnt
Shanghai Jiao Tong University, Shanghai, PR China

It is obvious nowadays the problem that the conventional Data Base (DB) and Knowledge Base (KB) technologies are not satisfied to practical decision support. The decision maker used to have puzzled with the "shallow" data provided by conventional decision suporting system (DSS), which can not be fully argumented in the decision process. Thus the development of intelligent technologies capable to provide "Deep Knowledge" and "Deep Data" seems necessary. From unified fuzzy set approach, this paper deals with the reasoning and decision processes which are represented by fuzzy nearness degree and fuzzy similarity degree in a decision state space. Some developments in this regard are presented based on hierarchical subspaces of state parameters of the decision event. An example of decision making for bidding submitted by project contractor and supported by an intelligent data base is presented for the purpose of illustrating the techniques of this paper, in which the intelligent DB is inferred by fuzzy approach. Finally, the possibility of extending the method to machine learning is also discussed.

INTRODUCTION

It is an ever challenge promoted by the development of science and technology and competitive environment of decision making, a system which can provide data, information as well as knowledge in an efficient way (thorough-reaching implicit information) is urgently needed. The gap between what is expected in real world decision making and what is the infor -mation technology can provide is widening by the fact that there is a misunderstanding which leads the computer science people pay their attention in extending a DB system capability of manipulating data forms and organizations explicitly rather than in improving its quality. It seems the time for rectifying the tecdency and common efforts should be paid by inter-disciplinary cooperation for impro -ving these situations.

It is understandable that different attributes of data are required by people in different post. A professional staff concerns only those data which are directly related to his (her) department. However, the department head will concern data much in the inter-departmental nature. Eventually, for the General Manager of a firm, who are responsible to the strategical planning will pay his (her) concerns on more integrated information for decision making combined with his (her) personal experience and judgement. Certainly, the conventional,trivial and tedious "shallow data" and "shallow knowledge" will not be satisfied in this case.

What is needed for an efficient decision suporting system (DSS) can be listed as below:

* Capacity of information re-production supported to the decision making;
* Availability of existing DB and KB technologies for the DSS with friendly users' interfaces;
* Potentiality of manipulating data-information-knowledge as a whole based on unified theoretical basis for enhancing and facilitating system capacity for decision making support.

As a matter of fact, the essentials of above mention -ed efficient DSS is a problem of fully utilizing the techniques of artificial intelligence into the system for obtaining deep data and deep knowledge on which a sufficient and rational decision can be made. There are two points to be cardinally concerned, namely: method of representation and inference strategy. The previous one relates to the expression of a complicated causal event. However, the latter one pertains to problem solving by indirect searching or indirect reasoning. Therefore, the fine essence of AI can be likely concluded as to simulate the human intelligence for solving complicated real world problems through rational reasoning process.

From the point of view of space of state, based on the inference (searching) strategy the initial state --- a known condition --- is being changed into a series of intermediate states. Eventually, the inference strategy of AI is to find a simplest approach from the initial state to the final objective state in the space of state. Though the objective state is not necessarily to be unique, however, among those feasible objective states, each one represents a fea -sible solution to the problem. What we're concerned is that the aims of AI strategy is to manipulate the movement of the intermediate states in order to achieve an optimum path of inference.

In general, the system,aimed by its system goal, can be modeled by a set of factor constitutively structu -red by certain links between them to form a factorial network which represents the essential of system behaviour. The nodes of which represent the factors concerned. The weight distribution between factors located at the same level determined by the eigenvalue problem of "pair comparison" relation matrix (refs. 1, 2). Successively manipulating the weight distribution of the factor in each level until the fuzzy assessment of system response for decision making consultation is made.

Another problem pertaining to the "intelligentization" of the system is machine learning. A fuzzy learning model with fuzzy input and output is presen -ted, which is essentially a combination of AI technique with the information and system engineering methodologies based on unified mathematical approach.

In order to explain the reasoning and learning processes in a complicated decision system, the hierarcical space of state is introduced for representing the internal relationship of the event in different level and in different aspect. Thus, the theoretical manipulation of reasoning could be performed in an unified approach of mathematical explicity. By means of hierarchicak space of state the network of fuzzy reasoning both for assessment and learninf can be formulated on a rather reasonable basis in its development.

REPRESENTATION OF HIERARCHICAL SPACE OF STATE

For a complicated system, the mathematical model is based on the classification of event with different attributes, which behave in different structural models such as tree structure and network structure etc. The representation of knowledge or information of an event, no matter in production rule, semantic network or other framework, needs a media of its state description. The compositional hierarchy of the event implies the rationality of introducing the hierarchical space for representing and manipulating the inter-relationship of the sub-events in a hierarchical structure of the complicated event. Fig. 1 shows a two-layered hierarchical space, where

$\mathcal{S}=\{t_i\}$ represents top level space of state with $i=1,2,\dots,n$

$\mathcal{S}_1=\{t_1\} = l_{1j},\dots \; \mathcal{S}_i=\{t_i\}=l_{ik},\dots \; \mathcal{S}_n=\{t_n\}=l_{np},\dots$ represent lower level spaces of state with $j=1,2,\dots n_{m1}$, $k=1,2,\dots n_{mi}$ and $p=1,2,\dots n_{mn}$ respectively.

Therefore, $\mathcal{S}=\{t_1,t_2,\dots t_n\} = (l_{11},\dots l_{1j},\dots l_{m1}),\dots (l_{i1},\dots l_{ik},\dots l_{imi}),\dots (l_{n1},\dots l_{np},\dots l_{nmn})$

with total dimension of

$$\sum_{h=1}^{n} m_h \tag{1}$$

The relation between top and lower states can be expressed by

$$\mathcal{S}_1 = f_1(l_{11},\dots l_{1j},\dots l_{1m1})$$
$$\mathcal{S}_2 = f_2(l_{21},\dots l_{2c},\dots l_{2m2})$$
$$\dots\dots\dots\dots$$
$$\mathcal{S}_i = f_i(l_{i1},\dots l_{ik},\dots l_{imi})$$
$$\dots\dots\dots\dots$$
$$\mathcal{S}_n = f_n(l_{n1},\dots l_{np},\dots l_{nmn})$$

or

$$\mathcal{F} = (f_1, f_2, \dots, f_i, \dots f_n) \tag{2}$$

In state expression, for top space event

$$(I_t, O_t, G_t) \text{ i.e. } I_t \xrightarrow{O_t} G_t \tag{3}$$

for lower space sub-event

$$(I_l^{(i)}, O_l^{(i)}, G_l^{(i)}),\; I_l^{(i)} \xrightarrow{O_l^{(i)}} G_l^{(i)} \tag{4}$$

where

I_t, G_t are initial state and objective state of lower level sub-event

From (2) $\mathcal{F}$ is inter-level transform function for studying the operational relationship between top level operation O_t and the i-th lower level operation of the subevent $O_t^{(i)}$, here we understand $i=1,2,\dots n$. It is possible to chart Fig.1 as a network structure rather than a tree or having multi-layered hierarchy space of state if necessary.

MATHEMATICAL DISTANCE IN SPACE

Based on the space of state, the different between two events A and B in the same attributive subspace

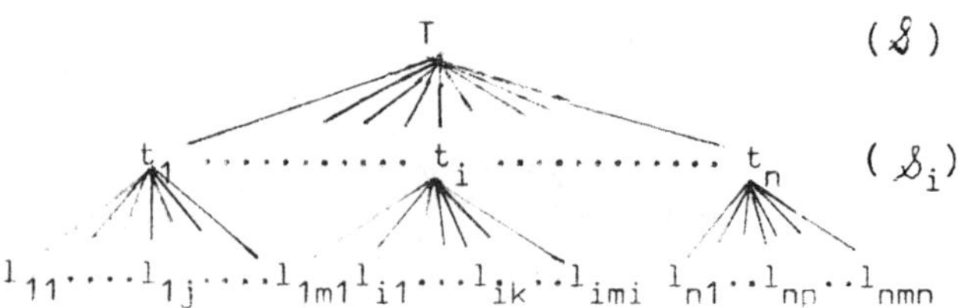

Fig. 1

can be measured by the generalized mathematical distance $\mathcal{D}$. Define

$$\mathcal{D}(A,B) = \sum_{i=1}^{m} w_i(d_i)\, d_i(A,B) \tag{5}$$

where m, the dimension of (sub) space

w_i, influence coefficient of i-th factor (or i-th attribute)

d_i, difference between A and B events in i-th attribute

It is obvious that

$$\mathcal{D}(A,B) = \mathcal{D}(B,A)$$
$$\mathcal{D}(A,A) = \mathcal{D}(B,B) = 0 \tag{6}$$

In addition, we can extend the measurement of difference in (hierarchy) space to uncertain events. For instance

$$d_i(A,B) = \mu_{\underset{\sim}{A}}(v_i) - \mu_{\underset{\sim}{B}}(v_i) \tag{7}$$

where $\mu_A(v_i)$ and $\mu_B(v_i)$ are the membership function of fuzzy events $\underset{\sim}{A}$ and $\underset{\sim}{B}$ with respect to its i-th attribute v_i in the same (hierarchy) space.

FUZZY HIERARCHY REASONING APPROACH

We now discuss the decision making process under uncertain environment. The decision is made taking consideration of several fuzzy attributes, which are both uncertain in quatity and its implicitness of relationship.

If the decisive solution T (Fig.1) is a particular state in its attribute space $\mathcal{S}$, then according to (1) $\mathcal{S}$ is determined by its attribute state $\{t_i\}$ which can be further expressed by corresponding sub-attributes' states in l_{ik}. This situstion holds true even in fuzzy representation. Thus, the process of decision making is transformed by a series of fuzzy state operation.

Space Chart Analysis

Each attribute affected to decision solution T is presented in its own subspace $\mathcal{S}_i$ and is hierarchically through (2) influenced to the state of T in space $\mathcal{S}$. The relation between $\mathcal{S}$ and $\mathcal{S}_i$ is shown in the space chart. The nodes of which represent the essential of the concerned factor subspaces associated with the state of the sttributes (Fig.2).

Fuzzy State Assessment

Fuzzy state assessment for each attribute in corresponding subspace with respect to its root attribute is performed by evaluation matrix through "pair comparison" based on empirical prediction directly related to fuzzy reasoning process. Define five grade scales for "pair comparison" of state:

"1" represents that the state is slightly important to the system goal of decision; "2" represents moderate important; "3" represents obviously important; "4", seriously important and "5", extremely important. This 1-5 grade scales are feasible to quantize the conceptual thinking of professional experts to make judgement for evaluating the state status of attributes and formulating evaluation matrix

$$A = (a_{tij}) = (\zeta_{1i})/(\zeta_{1j}) \tag{8}$$

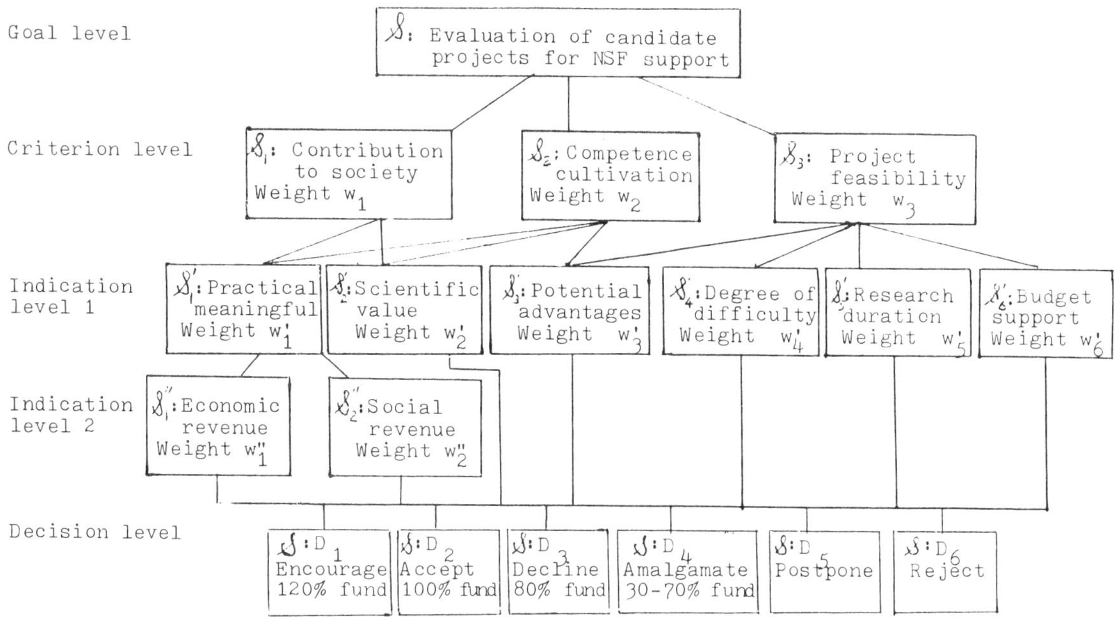

Fig. 2

where i,j are the attributive subspace order of low-er level attributes; t,l mean the abbreviations of top and lower. ζ_{li} and ζ_{lj} mean the membershin functions of i-th and j-th subspaces of lower level with respect to its root space. For instance, in the indication level 1 of Fig.2,

$$(a_{45}) = (\zeta_{\delta 4})/(\zeta_{\delta 5}) = 3/1 = 3$$

it means that the importance of attribute 4 (degree of difficulty) and 5 (reasearch duration) with respect to its root (project feasibility) is obvious important "3" and slightly important "1".

Weight Distribution Between Attributes

It is necessary to weight the attributes in the same level. Simulating this problem as sequencing physical weights of a set of body without a scale, we may get it if the total weight of body 1,2,....,n as we well as each of its weight ratio $r_{ij} = w_i/w_j$ $(i,j = 1,2,....,n)$ are known. Suppose $w_1, w_2, ..., w_n$ are the weight of body 1,2,....,n respectively, then we can set up a weight ratio matrix A, where we have

$$\begin{aligned} A &= (a_{ij}) \\ a_{ij} &= 1/a_{ji} \\ a_{ii} &= a_{jj} = 1 \\ a_{ij} &= a_{ik}/a_{kj} \end{aligned} \tag{9}$$

or

$$A = \begin{bmatrix} w_1/w_1 & w_1/w_2 & \cdots & w_1/w_n \\ w_2/w_1 & w_2/w_2 & \cdots & w_2/w_n \\ \cdots & & & \\ w_n/w_1 & w_n/w_2 & \cdots & w_n/w_n \end{bmatrix} = (a_{ij}) \tag{10}$$

Multiplying weight vector W

$$AW = \{nw_1, nw, \ldots, nw_n\}^T = nW \tag{11}$$

$$(A - nI) = 0 \tag{12}$$

Equation (12) means the weight vector is the eigen-value problem of "pair comparison" matrix A with respect to eigen value n, the eigen value reflects the weight distribution. i.e., the sequencing of weight distribution among attributes can be handled as an eigen-value problem of (12).

Illustrative Example

We have defined the space chart of the event to be decided and have assessed by fuzzy approach the state of its affected attributes as well as its weight distribution with the root sttribute. The fuzzy reasoning process can be carried out level by level. An example of evaluating candidate research projects for fund suppo t is studied hereby and the factor chart corresponding to space chart is presented in Fig.2 , which is comprised of 5 levels. Goal level space also serves for the decisive solution $T = T_i$ $(i=1,2,\ldots,6)$. There are sub-spaces in each level as indicated in Fig.2 and athe attributes associated with its weight are also noted in the block. Since the state of any attribute is described in its own subspace interacted with the others upon its root attribute in the root space, therefore, the evaluation matrix A of each root attribute actually reflect the heuristic prediction of human knowledge and constitute a part of fuzzy reasoning process. The evaluation matrix $A_{\delta i}$ of each node in Fig.2 with weight distribu-tion are:

$$A_{\delta 1} = \begin{array}{c} (\delta'_1) \quad (\delta'_2) \\ \begin{bmatrix} 1 & 3 \\ 0.33 & 1 \end{bmatrix} \end{array} \begin{array}{cc} & w_i \\ (\delta'_1) & 0.75 \\ (\delta'_2) & 0.25 \end{array} \tag{13}$$

$$A_{\delta 2} = \begin{array}{c} (\delta'_1) \quad (\delta'_2) \quad (\delta'_3) \\ \begin{bmatrix} 1 & 0.2 & 0.33 \\ 5 & 1 & 3 \\ 3 & 0.33 & 1 \end{bmatrix} \end{array} \begin{array}{cc} & w_i \\ (\delta'_1) & 0.10 \\ (\delta'_2) & 0.64 \\ (\delta'_3) & 0.2[illegible] \end{array} \tag{14}$$

$$A_{\delta 3} = \begin{array}{c} (\delta'_3) \quad (\delta'_4) \quad (\delta'_5) \quad (\delta'_6) \\ \begin{bmatrix} 1 & 1 & 3 & 2 \\ 1 & 1 & 3 & 2 \\ 0.33 & 0.33 & 1 & 0.5 \\ 0.5 & 0.5 & 2 & 1 \end{bmatrix} \end{array} \begin{array}{cc} & w_i \\ (\delta'_3) & 0.33 \\ (\delta'_4) & 0.33 \\ (\delta'_5) & 0.10 \\ (\delta'_6) & 0.24 \end{array} \tag{15}$$

The weight sequencing for subspaces δ'_i

	(δ'_1) $w_1=.43$	(δ'_2) $w_2=.14$	(δ'_3) $w_3=.43$	w_i
(δ'_1)	0.75	0.10	0	0.34
(δ'_2)	0.25	0.64	0	0.20
(δ'_3)	0	0.26	0.33	0.1[illegible]

(16)

(δ'_4)	0	0	0.33	0.14	
(δ'_5)	0	0	0.10	0.04	(16)
(δ'_6)	0	0	0.24	0.10	

The final decision can be achieved by fuzzy evaluation through fuzzy reasoning.

Fuzzy Decision Making

The fuzzy decision making is based on evaluation of state of individual attribute in its own subspace δ_i. The global decision solution (state) is presented by several alternative solution D_i $(i=1,2,....n)$ in the decision space δ. In Fig. 2, six alternatives are taken: D_1, encouraged project with 120% support funding; D_2, accepted project with 100% full funding; D_3, declined project with 80% support funding; D_4, amalgamated project with 30-70% supports; D_5, project postpond by certain reasons and D_6, insufficient project being rejected.

Membership function of individual attributes in δ'_i subspace can be obtained by sequencing the weight distribution w'_i with respect to D_i (see Fig.2). For which fuzzy evaluation matrix $A_{\delta i}$ (as (13) to (15)) should be formulated.as

$$A_{\delta k} = (\zeta_{ik}/\zeta_{jk}) \quad \begin{matrix}(i,j=D_1,D_2,D_3,D_4,D_5,D_6)\\(k=w'_1,w'_2,w'_3,w'_4,w'_5,w'_6)\end{matrix} \qquad (17)$$

ζ_{ik}/ζ_{jk} represents the ratio of belonging of the solution D_i and D_j with respect to k-th attribute in δ_k space.

Substituting (17) to (12). a set of eigen-vector can be obtained

$$\mu_k = \{\mu_{1k}, \mu_{2k}, \ldots\ldots \mu_{6k}\} \qquad (k=w'_1,w'_2,\ldots\ldots w'_6) \qquad (18)$$

Eq. (18) is belonging vector of the attribute k to the decision solution D_i, which exactly is the membership function of k to D_i. i.e.,

$$\underset{\sim}{\mu_k}(D_i) = \frac{\mu_{1k}}{D_1} + \frac{\mu_{2k}}{D_2} + \ldots + \frac{\mu_{6k}}{D_6} \qquad (19)$$

The fuzzy decision shouldd be made by D_m satisfying

$$\mu_m = \text{Max.}\left(\sum_{k=1}^{6} \mu_{ik}\right) \quad \begin{pmatrix}i=1,2,\ldots\ldots,6\\ 1 \le m \le 6\end{pmatrix} \qquad (20)$$

$$\sum_{i=1}^{6} (\mu_{ik}) = 1.0 \qquad (21)$$

where (20) means the maximum of summation of membership function of all attributes in δ'_k $(k=1,2,...,6)$ subspaces with respect to decisive solution D_i; and (21) means the summation of membership function of attribute k in δ'_k subspace with respect to all decision solution D_i $(i=1,2,...,6)$ equals to unity. Eqs. (12), (18), (19), (20) and (21) represent the essentials of fuzzy reasoning for decision.

FUZZY DECISION IN BIDDING

In the construction market, the clients are offering tender and calling for bid. In responding to the call the contractor will submit the bidding proposal for the project with quotation offer. As a matter of fact , the offering cost is a key issue in total bidding activity. Since the higher the cost quoted, the less competitive the proposal will be, on the contrary, the lower the cost offered, the more the risk of the project will face. One should as soon as possible to estimate the cost before the deadline date of the bid on the basis of available data from data base, in which the attributes and costs of existing buildings are stored. The involving of fuzzy reasoning to conventional DB represents a new era of incorporating artificial intelligenc (AI) ith DB technologies enabling further deepening the available data and knowledge. In other word, it implies a new scenerio of intelligent DB conceptuation, in which the deep data (or deep knowledge) can be drawn from the conventional DB by fuzzy inference machine.

Mathematical Modeling

Supposing the there are n engineering project samples $A_1, A_2, \ldots\ldots, A_n$, and factor attribute set G has m elements

$$G = \{g_1, g_2, \ldots\ldots, g_m\} \qquad (22)$$

As the tall building is concerned, where

- g_1 represents the type of the structure
- g_2 represents the characteristics of structure
- g_3, stories of the building
- g_4, storey height
- g_5, architectural combination
- g_6, type of interior decoration
- g_7, floor area in total
- ....

If $\underset{\sim}{G_i}$ is the fuzzy subset of i-th building to the attribute set G

$$\underset{\sim}{G_i} = g_{i1}/g_1 + g_{i2}/g_2 + \ldots + g_{im}/g_m \qquad (23)$$

where g_{ij} is the j-th membership component of i-th building to subset G and the sign "+" is by no mean of plus, but the meaning of "collection".

The fuzzy subset of building to be quoted A_* is G^*_A

$$G^*_A = g^*_1/g_1 + g^*_2/g_2 + \ldots\ldots + g^*_m/g_m \qquad (24)$$

where g^*_j is the j-th membership component of A_* to subset G. Now the question is focused on fuzzy estimation vector e^*_A of the attributeof the building to be quoted, which can be expressed by

$$e^*_A = M.\lambda[\alpha_1 E_1 + \alpha_2 E_2(1-\alpha_1) + \alpha_3 E_3\ (1-\alpha_1)(1-\alpha_2) + 1/3\ (E_1 + E_2 + E_3\ )(1-\alpha_1)(1-\alpha_2)(1-\alpha_3)] \qquad (25)$$

where M, the total floor areas

λ, empirical coefficient of rectification, which influences to the base price of the quotation very much

α_1, α_2, and α_3 , the top three similarity degree (as shown in eq. (28)) or the top nearest three mathematical distance (as shown in eq. (5)) among existing building samples $\alpha_1, \alpha_2, \ldots, \alpha_n$. According to the principle of priority, it will serve as the basis of latter estimation. we have $\alpha_1 \geqslant \alpha_2 \geqslant \alpha_3$

E_1, E_2 and E_3 are the corresponding estimation vectors of building project related to α_1 α_2 and α_3 respectively.

In practical calculation, λ is suggested as 0.9~1.1. Actually, λ is a multi-factor attributed value and is fluctuated by the condition of individual contractor/firms, market price of material and equipment, the prescribed goal of bidding and site environmental conditions etc. Usually, λ is determined by extensive statistical data processing and studies based on the available data of existing buildings, which serve as samples.

Solving Procedures

(1) In the first step, one should identify the

necessary attributes which reflect the overall characteristics of the building to be quoted.

(2) Determining the approximate membership function of the building to be quoted to the attribution fuzzy set by subjective empirical prediction of by statistical data processing. The strategy is that, taking one or two existing buildings as sampling object, the membership function of which is consider-ed as the normalized standard 1.0, then the other components could be estimated by subjective information. This procedure is going back-and-forth through judgement by (24) until a comparative fuzzy relation between expected and sampling buildings can be established as shown in Tab.1, in which the nece-ssary information about the fuzzy set of the expect-ed building.

(3) Based on Tab.1, the similarity degree of two fuzzy set in the universe X, $\underset{\sim}{A}$ and $\underset{\sim}{B}$ can be expressed by

$$(\underset{\sim}{A}, \underset{\sim}{B}) = \frac{1}{2}[\underset{\sim}{A} \otimes \underset{\sim}{B} + (1 - \underset{\sim}{A} \odot \underset{\sim}{B})] \qquad (26)$$

where $\underset{\sim}{A} \otimes \underset{\sim}{B} = \bigvee_{x \in X} [\mu_{\underset{\sim}{A}}(x) \wedge \mu_{\underset{\sim}{B}}(x)]$ and

$$\underset{\sim}{A} \odot \underset{\sim}{B} = \bigwedge_{x \in X} [\mu_{\underset{\sim}{A}}(x) \vee \mu_{\underset{\sim}{B}}(x)]$$

represent the interior and exterior products of the fuzzy sets $\underset{\sim}{A}$ and $\underset{\sim}{B}$. Select the top three similarity degree under consideration with the corresponding estimations E_1, E_2, and E_3.

(4) The base price for quotation in bidding proposal can be estimated through eq. (25).

Case Studies

Case 1: The quotation of a reinforced concrete frame-work building for bidding.

(1) The building characteristic attribution set

$$G = \{g_1, g_2, g_3, \ldots\ldots, g_8\}$$

where g_1 represents structural type;

g_2 represents foundation type;

g_3 represents the condition of exterior decoration;

g_4 represents the number of storey;

g_5 represents the height of storey;

g_6, the situation of architectural combination;

g_7 represents the material of floor;

g_8 represents the total floor areas.

(2) Select the basic component element in fuzzy set as a calibration standard, assigning the membership function as 1.0, then establish fuzzy relationship by trial-and-error method shown in Tab.1. Define the attributes of fuzzy set of sampling building as above and then the membership function of sampling building with respect to G is sequentially assigned in Tab.1.

(3) The fuzzy subset of sampling building to G are:

$$\underset{\sim}{C}_1 = (0.9/g_1)+(0.6/g_2)+(0.5/g_3)+(0.75/g_4)+(0.5/g_5)+(0.5/g_6)+(0.7/g_7)+(1.0/g_8)$$

$$\underset{\sim}{C}_2 = (1.0/g_1)+(1.0/g_2)+(1.0/g_3)+(0.5/g_4)+(0.3/g_5)+(1.0/g_6)+(0.4/g_7)$$

$$\underset{\sim}{C}_3 = (0.6/g_1)+(0.8/g_2)+(0.8/g_3)+(1.0/g_4)+(1.0/g_5)+(0.8/g_6)+(0.3/g_7)$$

$$\underset{\sim}{C}_4 = (0.8/g_1)+(0.5/g_2)+(0.5/g_3)+(0.55/g_4)+(0.8/g_5)+(1.0/g_6)+(0.3/g_7)$$

....

The fuzzy set of expected building with characteristic attribution set G is

$$A^* = (1.0/g_1)+(0.7/g_2)+(0.5/g_3)+(0.7/g_4)+(0.9/g_5)+(1.0/g_6)+(0.9/g_7)$$

(4) Define the similarity degree of $\underset{\sim}{A}^*$ to C_1, C_2, C_3 and C_4, we take the similarity degree of $\underset{\sim}{A}^*$ and C_1 first:

$$\underset{\sim}{A}^* \otimes C_1 = (1 \wedge 0.9) \vee (0.7 \wedge 0.6) \vee (0.5 \wedge 0.5) \vee (0.7 \wedge 0.75) \vee (0.9 \wedge 0.5) \vee (1.0 \wedge 0.7) \vee (0.9 \wedge 1.0) = \ldots\ldots\ldots = 0.9$$

$$\underset{\sim}{A}^* \odot C = (1.0 \vee 0.9) \wedge (0.7 \vee 0.6) \wedge (0.5 \vee 0.5) \wedge (0.7 \vee 0.75) \wedge (0.9 \vee 0.5) \wedge (1.0 \vee 0.7) \wedge (0.9 \vee 1.0) = \ldots\ldots\ldots = 0.5$$

The similarity degree

$$(\underset{\sim}{A}^*, \underset{\sim}{C}_1) = \frac{1}{2}(0.9+(1-0.5) = 0.7$$

Similarly

$$(A^*, \underset{\sim}{C}_2) = 0.65; \quad (\underset{\sim}{A}^*, \underset{\sim}{C}_3) = 0.55; \quad (\underset{\sim}{A}^*, \underset{\sim}{C}_4) = 0.75$$

....

(5) As mentioned above, take the top three similarity degree as $\alpha_1(=0.75) > \alpha_2(=0.7) > \alpha_3(=0.65)$ and the corresponding cost $E_1=198$; $E_2=294.21$ and $E_3=223.07$ (Yuan/square floorarea)

(6) The expected building is costed by eq. (25), the

$$e^*_A = M \cdot \lambda[\alpha_1 E_1 + \alpha_2 E_2 (1-\alpha_1) + \alpha_3 E_3 (1-\alpha_1)(1-\alpha_2) + \frac{1}{3}(E_1+E_2+E_3)(1-\alpha_1)(1-\alpha_2)(1-\alpha_3)$$

$$= 4420.8 \times 1.0 [0.75 \times 198 + 0.7 \times 294.21 \times (1-0.75) + 0.65 \times 233.07 \times (1-0.75) \times (1-0.7) + 1/3(198+294.21+233.+233.07) \times (1-0.75) \times (1-0.7) \times (1-0.65)] = \ldots\ldots$$

$$= 217.06 \times 4420.8 = 959{,}578.8 \text{ (Yuan)}$$

Practical finalized cost of expected building is provided latter on by contractor as 210x4420.8 = 928,368 (Yuan). The error percentage is

$$\frac{959{,}578.8 - 928{,}368}{928{,}368} \times 100\% = 3.36\,\% \;(\;5\%)$$

Case 2: The quotation of a bridge based on fuzzy reasoning of attributes from data base of existing bridges.

The attributes of existing bridges ZSMRD, DXRB, ... are listed in Tab.2 with its unit price and costs for different individual parts. The fuzzy membership function of each structural component parts are respectively represented by the relational coefficients k as presented in Tab.1 for different items of the structural characteristics. In our analysis, k is regarded as the ratio of the cost of individual structural part with respect to the "standardized" cost of the same rank structural part, since the membership relation can also be evaluated by the comparative cost of the bridge to be examined with the "standardized" one. In addition, k is manipulated for different structural component part, which will further improve the accuracy of fuzzy estimation.

Tab.2 presents the cost data of 8 R.C. bridges with the amount of concrete (M^3), Same operation as the case 1 is carried out for expected bridge ERCB. In 9-th row it shows that ERCB is quoted as 1345.3, yet the real cost is 1407.7, the error rate is

$$(1407.7 - 1345.3)/1345.3 = 0.046 \text{ or } 4.6\%$$

Tab. 1

EXISTING ENGINEERING BUILDING ATTRIBUTIVE DATA AND MEMBERSHIP RELATION COEFFICIENTS

Items	Structural Characteris -tics		Foundation Situation		Exterior Decoration		Storey Number Storey Height		Architectural Combination		Floor Material		Total Floor Areas		Total Structural Cost
	Descrip -tions	Reltn Coeff	Descrip -tions	Reltn. Coeff.	Descrip -tions	Reltn Coeff	Descrip -tions	Reltn. Coeff.	Descrip -tion	Reltn. Coeff.	Descrip -tions	Reltn Coeff	Descrip -tions	Reltn Coeff	(RMB Yuans) per m^2
Existing Building C_1	FCS SPF Semi-Basement	0.9	PFP Deep Foundation	0.6	CSM	0.5	NS=8 HS=3.6m	0.75	ECC	0.5	FAB & FCB	0.7	$4193m^2$	1.0	294.21
Existing Building C_2	FCS SPF	1.0	ICF	1.0	Lime-Stone Mounting	0.5	NS=12 HS=2.7m	0.5	File & Document Warehouse	0.3	FAB	1.0	$12045m^2$	0.4	223.07
Existing Building C_3	FCS SPF	0.6	ICF LGF	0.8	Stone Plate	1.0	NS=7 HS=3.6m	1.0	CAL	1.0	FAB & FCB	0.8	$9856m^2$	0.3	244.43
Existing Building C_4	FCS SPF & SCS	0.8	RCRF	0.5	CSM	0.55	Side NS = 10 Mid. NS = 13 HS=3.3m	0.55	IECC	0.8	FAB	1.0	$16200m^2$	0.3	198.0
....															
Estimated Building A^*	FCS SPF	1.0	PFP	0.5	CSM	0.7	NS=10 HS=3.6m	0.7	ICC	0.9	FAB	1.0	$4420.8m^2$	0.9	To be determined

FAB — Floor of Artificial Brick
NS — Number of Storey
FCS — Framework Casting on Site
ICF — Independent Column Foundation
RCRF — Reinforced Concrete Raft Foundation
CSM — Ceramic Surface Mounting
FCB — Floor of Ceramic Brick
HS — Height of Storey
SCS — Slabs Casting on Site
PFP — Pre-Fabricated Pile
EC — Exterior Corridor
CAL — Classroom and Laboratory
IECC — Inner/Exterior Corridor Compartment
ECC — Exterior Corridor Compartment
SPF — Slabs Pre-Fabricated
IC — Inner Corridor
LGF — Local Grillage Foundation
ICC — Inner Corridor Compartment

CONCLUSIONAL REMARKS

It is recognized that the conceptual approach of fuzzy decision for systems by "Deep knowledge" and "Deep data" presented in this paper can be successfully applied for obtaining "deep data" and "deep knowledge" of cost for quotation offer to bid. The results show that the prediction of cost is approximately less than 5% of engineering tolerance and is compared with the latter on actual cost in favourable agreement.

The strategies of fuzzy reasoning can also be used for the theoretical basis of inference machine of the intelligent data base, which is expected as the powerful supporting tool to a decision system.

It is hoped that the decision process can be described as to define a decision vector in the decision hierarchical space, where the attributes (state) of the event can be represented in hierarchical space of state, upon which the fuzzy reasoning and its mathematical modeling is found.

Tab.2 EXISTING R.C.BRIDGE ATTRIBUTIVES DATA AND MEMBERSHIP RELATION COEFFICIENTS

No.	NOTATION OF BRIDGE	UNIT PRICE (Yuan/m^3)	TOTAL AMOUNT (M^3)	FOUNDATION		PIER		DECK GRILLAGE		DECK PLATES		NOTES
				COST (Yuan)	RELTN. COEFF.	COST (Yuan)	RELTN. COEFF.	COST (Yuan)	RELTN. COEFF.	COST (Yuan)	RELTN. COEFF.	
1	ZSMRD	1406.6	17270.0	239.0	0.48	239.0	0.57	604.8	0.53	98.50	0.39	
2	DXRB	2092.1	12505.0	690.4	0.71	355.0	0.85	899.6	0.79	146.0	0.58	
3	XQRB	563.5	2115.0	186.0	0.19	95.8	0.23	242.3	0.21	39.4	0.16	
4	TDRB	2285.5	11012.0	754.2	0.77	388.5	0.93	982.8	0.86	160.0	0.64	
5	KJRB	825.9	642.0	300.0	0.31	123.9	0.30	349.0	0.31	53.0	0.21	
6	JTRB	2790.0	3161.80	976.5	1.0	418.50	1.0	1143.90	1.0	251.1	1.0	
7	SYJRB	944.0	1584.0	330.4	0.34	141.6	0.34	396.5	0.35	75.5	0.30	
8	JGRB	1456.0	21584.0	480.5	0.49	247.5	0.59	626.1	0.55	101.9	0.41	
9	ERCB			444.0	0.46	228.7	0.55	578.4	0.51	94.2	0.38	Total cost 1345.3

REFERENCES

1. Saaty, T.L., "The analytic hierarchy process", McGraw Hill, Inc., 1980

2. Lin, S.P. and Yang X.F.,"Fuzzy decision for sa--fety assessment of offshore jacket platforms", Proc. first Asia/Pacific Symp. of offshore mechanics, June, 1990, Seoul, Korea

3. Zao,H.Z. et al., "Hierarchical analysis, a simple new method for decision making", Science Press, Beijing, 1986 (in Chinese)

4. Lin, S.P. and Zhou, G.J.,"Economic analysis and decision in tendering and bidding", SJTU Press, 1990, Shanghai, China (in Chinese)

5. Wang, X.Z.,"The application of fuzzy mathematics in bidding for civil engineering projects", J. of civil engineering, Vol.9, No.1, 1986 (in Chinese)

6. Lin, S.P.,"Fuzzy reasoning method -- Its theoretical adaptability and code compatibility to design of earthquake-resistant structure", Res. Report (I) under contract NCEER-88-1006, School of civil and environmental engineering, Cornell University, Ithaca, NY,U.S.A., July 1990

7. Lin, S.P. and Yang J.P.,"Machine Learning in design expert systems", Proc. 2-nd Intnl. Conf. on the application of artificial intelligence techniques to civil and structural engineering, Sept. 1991, Oxford, England

* This work is supported by the National Natural Science Foundation of China.

5 INTEGRATED STRUCTURAL ENGINEERING DESIGN

SOFTWARE INTEGRATION FOR STRUCTURAL DESIGN USING MICROCOMPUTERS

I.A. MacLeod and F.H. Kor
Department of Civil Engineering,
University of Strathclyde, Glasgow, Scotland

Synopsis

The paper discusses how in-house integrated systems for structural design may be developed for a microcomputer environment. An experimental system of this type is described which incorporates a product model for a reinforced concrete beam. The concept of using one subsystem to synthesise part of the design and a separate system to check this is demonstrated. It is suggested that an integrated system should develop from a basic configuration with intelligent resources added.

Introduction

A logical consequence of the increase in the proportion of design tasks being computer processed is that the corresponding software modules should be linked, i.e. that data should be transferred electronically between them. This paper describes some experiments carried out in relation to how in-house integration strategies might develop.

Advantages of an Integrated System

An increasing proportion of design tasks is being computer processed and, for those firms who wish to embrace this technology, linking of the processing modules is an obvious goal. The main plus factor is the reduction of effort and increased accuracy in passing data. As data interchange becomes easier to handle, design processing strategies can develop. These include:

- Better checking by running the same data on different software modules. This concept is demonstrated here for sizing of reinforced concrete beams. It would also be useful for checking of analysis results.
- The linking of graphical and performance data.
- The use of software for assessment at the concept stage of design.
- The assessment of a greater number of design alternatives.

The general objective of integrating software is to improve the quality of designs.

Integrated systems take advantage of modern software strategies such as artificial intelligence.

Disadvantages of an Integrated System

Some commercial software for structural design does provide data links between modules (e.g. transfer of data from analysis to member sizing programs). However the type of integrated system being considered here has, or has the potential to have, any software item needed in the design process incorporated into the system under the control of a common interface. Commercial systems presently available to do this for structural design tend to be expensive and tend to be inflexible in that user additions to the system are not feasible.

For an integrated system, a project database must be maintained. In this is stored all data relevant to the project. Problems arise in the project data with consistency where the same feature is defined more than once in the database and a change in one location should result in a corresponding change at other locations. The responsibility for maintaining consistency must be clear. This problem is not unique to computer processing, being pervasive in design. In structural design, the drawings normally represent the standard against which other representations (e.g. a value of a beam depth in a sizing module) can be compared. This is likely to be the case for integrated software systems in the near future at least. The concept of a unique database representation of the product is presently being investigated but standards for doing this are slow to develop (Refs 1,2).

Another problem with the project database is the logging of changes. Well developed integrated systems would provide a means of identifying changes as to person and date. This adds to the complexity of the interface.

Designers who use integrated systems will need to be trained in the use of modern computing techniques.

Structure of an Integrated System

Figure 1 shows the conceptual structure of the proposed integrated system.

The functionality of the components of the system is as follows:

A **Subsystem** is a software module written for a specific

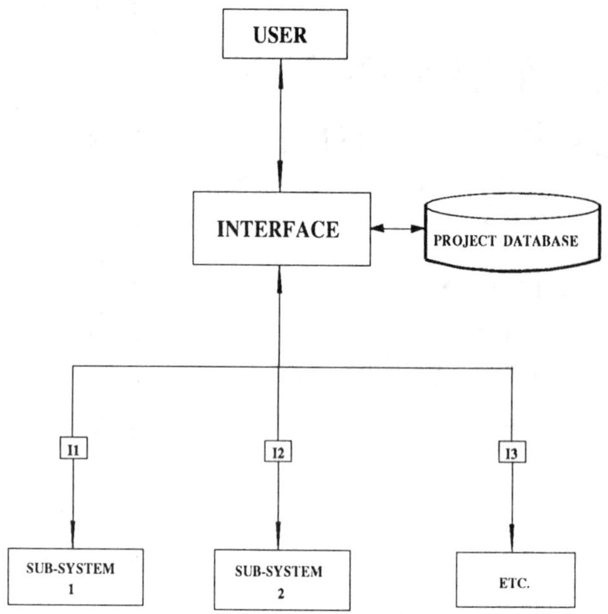

Fig. 1 Structure Of An Integrated System

USER

MENU OPERATING SYSTEM
PRODUCT DATA MANAGER
DATA EDITOR

PRODUCT MODEL
SUB-SYSTEM FILES

I1
I2
I3

STRUCTURAL ANALYSIS MODULE (SAM)

MEMBER SYNTHESIS MODULE (MSM)

MEMBER ASSESSMENT MODULE (MAM)

Fig. 2 Functional Representation Of ISS

task or set of tasks. These modules may be standard software items or may be specially written for the system.

The **Project Database** is a set of files which hold all information about the product being designed that needs to be retained.

Two types of information in the project database are:

- Subsystem files which the subsystems access directly.
- A 'Product Model' file which is a set of data related to a particular product, having a specific representation and being able to support the information needs of a range of activities.

The **Interpreters** transfer corresponding data items from subsystem files to product model files (and vice-versa) making the necessary format conversions.

The **Interface** provides access for the user to the subsystems and to the project database and performs data transfers and data management as appropriate.

An Experimental Integrated System

To investigate the feasibility of creating integrated systems for structural design, some standard software items have been linked in a microcomputer environment based on the structure of Figure 1. The system is called ISS (Integrated Structural System).

The generic task considered is the structural design of a reinforced concrete frame.

Figure 2 shows the structure of ISS.

The **Subsystems** of ISS are:

- Structural Analysis Module (SAM). This is the NL-STRESS program which forms part of the SAND system marketed by Fitzroy Systems Ltd. It is a frame analysis program with linear or non-linear capability. It is used here for plane frame analysis to provide actions for design of the frame members.
- Member Synthesis Module (MSM). This is a spreadsheet module specially written for concrete beam sizing (Ref 3).
- Member Assessment Module (MAM). This is a Scale Proforma for concrete beam sizing which also forms part of the SAND suite of programs. The version used has no direct link with NL-STRESS. It is used in this context to provide an independent check in the beam details generated by MSM.

The **Interpreters** of ISS are as follows:

I1 converts specific data items values from the product model tables to NL Stress data format via an intermediate ASCII format.

I2 converts specific data items values from MSM database to product model format via an intermediate comma separated value (CSV) format.

I3 converts specific data items values from the product model tables to Scale Proforma variables of MAM.

Note that the reverse of the above transfers has yet to be implemented in the interpreters.

The **Project Database** incorporates a product model for reinforced concrete beams. The product model is implemented in DBF format for processing by DBaseIV. Three of the relational tables used have field names as follows:

Field Names	Descriptions
BEAM_TYPE	Concrete beam type
SUPP_TYPE	Beam support type
BREADTH	Beam breadth
DEPTH	Beam depth
LENGTH	Beam length
D-COMP	Cover to compression steel
COVER	Cover to link

TABLE 1: Beam Type

Field Names	Descriptions
MATERIAL	Concrete material type
FCU	Concrete characteristic strength
FYc	Main steel characteristic strength
FYV	Link characteristic strength
E	Young's modulus of material
G	Shear modulus
DENSITY	Concrete density

TABLE 2: Material Type

Field Names	Descriptions
BEAM_MARK	Beam type
AXIAL_P	Beam axial force
SHEAR	Beam shear force
MOMENT	Beam bending moment
DIA_AST	Diameter of tension bar
DIA_ASC	Diameter of compression bar
DIA_LINK	Diameter of link
SHEAR_BARS	Number of link bars
AST_REQS	Area of tension bars required
AST_NO	Number of tension bars provided
AST_PERCEN	Percentage of tension bars
ASC_REQD	Area of compression bars required
ASC_NO	Number of compression bars provided
ASC_PERCEN	Percentage of compression bars

TABLE 3: Concrete Beams

This is not proposed as a standard for concrete beams since it is clear that many more fields would be needed to fully describe such a beam. One strategy for defining a standard format for a concrete beam would be to base it on the format of a reinforced concrete detailing package. This might be the basis of a general strategy for in-house systems for defining product model objects. That is to use the format from the subsystem which gives the most detailed description of the object.

The Interface

The Interface has an operating system written using the public domain software Micro Menu. From the main menu the following submenus can be called:

- Structural Analysis Module (SAM).
- Member Synthesis Module (MSM).
- Member Assessment Module (MAM).
- Data Editor.
- Project Data Manager.

Figure 3 shows the main and submenus of the system.

The Data Editor was specially written in Arity Prolog. This is a modification of the editor for DESIGNER-M (Ref 4). The editor has the ability to handle several files at one time by putting them in separate buffers. Contents of files in the buffers can be 'cut and pasted' from one to each other. Other facilities provided by this editor are file merging, printing, word finding and replacing, and temporary exiting to the DOS environment. Data files for an analysis program usually have a standard format and it would be quicker and easier for a designer to create a data file if he could refer and obtain data from previously created files. The editor developed has five buffers in which the designer can use four to store previously created files for reference.

An editor of this type is very useful for data transfer between subsystem files which are not linked via the product model.

The task of the Product Data Manager is to create or edit the data in the product model tables.

Operation of the System

The first user action is to call the Product Data Manager to insert or edit values in the Beam Type and Material Type tables of the product model. I1 is then called from the SAM menu and the analysis carried out using the SAM utility menu commands. The Member Synthesis Module is then used via the MSM menu, followed by data transfer to the product model via I2. Finally the member assessment module is used via the MAM menu after transfer of data by I3.

The above process describes a normal order for calling the subsystems. The ordering of the process is however entirely under the control of the user. A task controller as part of the interface is an obvious addition although the provision of full control of the process by the user without task control software should always be available. This would allow practitioner mode of computer use (Ref 5) which requires that users have full control over the process of design.

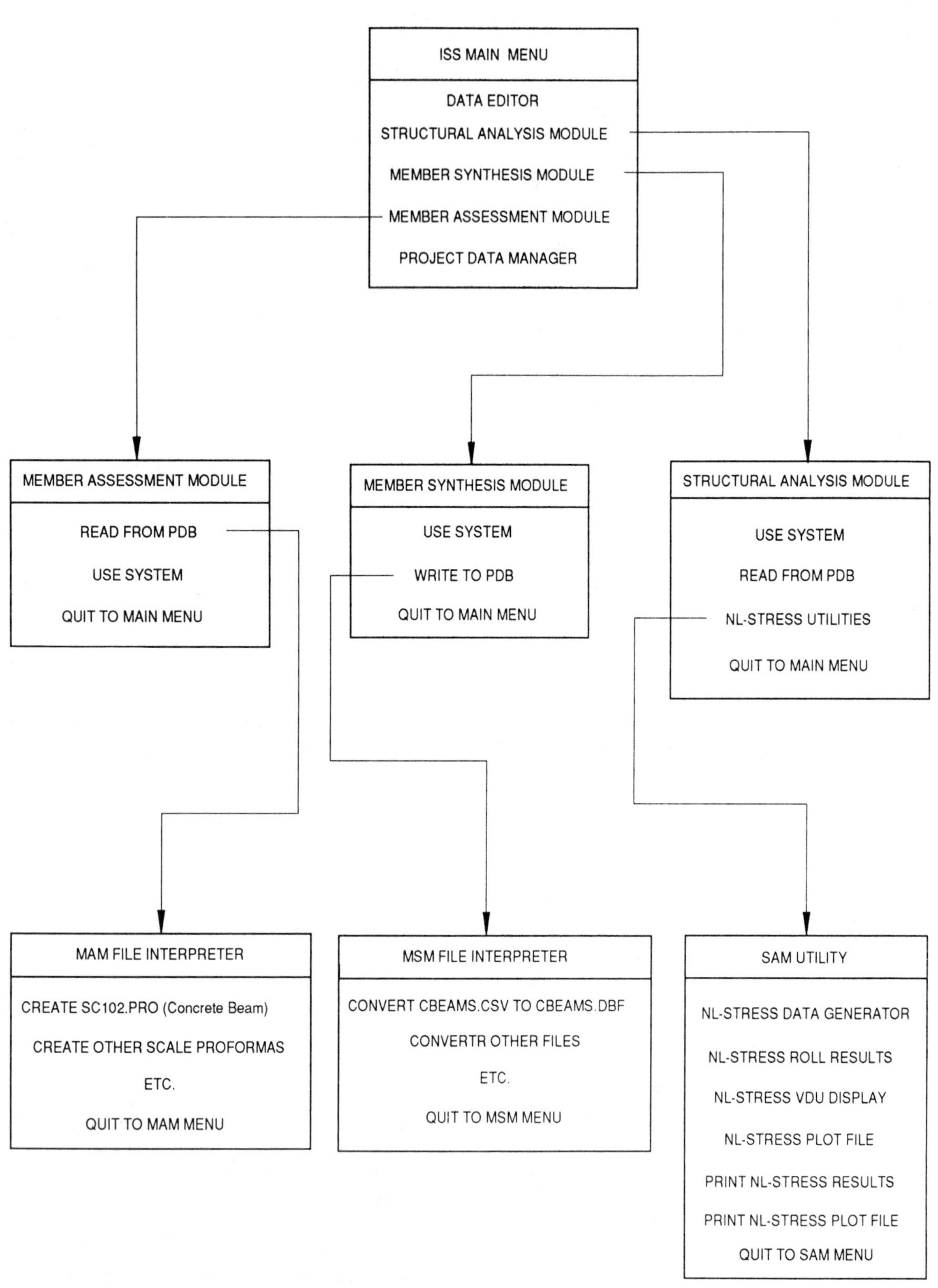

Fig. 3 Menu Structure Of ISS

Proposed Configuration of an Integrated System for Structural Design

Basic assumptions of an integrated system would be:

- A set of objects for the product model defined in a standard format. The relational format described in this paper has been found to be suitable but an object oriented approach is likely to be more advantageous.
- A set of subsystems.
- Interpreters to convert subsytem data to product model data.
- An interface which allows a flexible call structure for the subsystems, editing and database management facilities.

Desirable components of an integrated system would be:

- An intelligent task controller to assist the user in managing the design process.
- Software which simplifies the creation of interpreters and additions of subsystems to the interfaces.
- A consistency checker for information in the project database.

It seems likely that the more effort made to add 'intelligence' to the system, the more effort will be required to add subsystems. Possibly subsystems can be added which would be independent of the intelligent resources. However the possibility of providing such additions must be incorporated into the architecture of the system otherwise the user will tend to use only what is available in the system. The design choices should be constrained by the user and not by the computing system.

Conclusions

To set up an in-house integrated design system requires consideration of two competing factors. Firstly, one wants to be able to insert new subsystems without significant programming effort. To achieve this one wants to keep the system as simple and as uncoupled as is practicable. Secondly, one wants the system to operate in a way that users find easy, without requiring extensive training. For this one needs a more intelligent system which provides a degree of automation in the transfer of data, for task control etc. The system described in this paper is based on the former philosophy, allowing maximum user control but requiring significant user knowledge of the operation of the subsystems.

It is suggested that this is the best way to initiate an integrated system and then to add intelligent resources.

The work done to date has led us to believe that the concept of microcomputer-based integrated systems is worthy of further experimentation.

References

1. CAD and Other Information Exchange in the Building Industry Working Party, "Information Transfer in Building", Technical Report, National Economic Development Office, Feb, 1990.

2. "Four Years of Product Modelling - Collected Papers", TNO Report, Netherlands Organisation for Applied Scientific Research, TNO Institute for Building Materials and Structures, Aug, 1989.

3. KOR, F.H., "Structural Design using Information Technology", PhD thesis, University of Strathclyde, 1991.

4. ADES, N. and MACLEOD, I.A., "Designer-M: A processor for numerical rules", User Instructions, Department of Civil Engineering, University of Strathclyde, 1991.

AN OBJECT-ORIENTED ARCHITECTURE AND CONCEPT FOR AN INTEGRATED STRUCTURAL ENGINEERING SYSTEM

J.A. Abdalla

Center for Integrated Facility Engineering, Stanford University, Stanford, United States of America

The development of an integrated engineering system of broad scope is an enormous task that can only be achieved, with an elaborate study, over a long period of time. Also, an important prerequisite for the development of such a system is a solid foundation of theory, principles and techniques. At present this foundation is weak. This paper presents an overview of part of the research that have been conducted to investigate some of the principles and techniques needed for developing integrated engineering systems, using object-oriented concept. The paper will, of course, address only small part of the overall research effort. Many details have been omitted for brevity.

During the last three decades there have been several attempts to develop integrated engineering systems using different architectures and various techniques. Recently, there has been a growing interest in developing integrated systems using object-oriented techniques. In this paper, a number of object-oriented principles and techniques are explored which are believed to be useful for developing integrated engineering systems. An overall object-oriented architecture of an integrated system and its essential elements is proposed, with (a) a central data base containing data common to all application areas; (b) a number of application programs performing application-specific operations; and (c) an application data base for each program, containing application-specific data. The central data base is accessed through a data base management system, while the application data bases are accessed through a simpler object manager. In addition, the paper will briefly address other issues that affect the development of integrated engineering systems.

Key Words

Object-Oriented Approach, Integrated Systems, System Architecture, Engineering Design Objects, Design Object Versions, Communication Channels, Communication Protocols, Communication Mediums.

Introduction

An integrated software system for structural engineering must include application program modules for many different tasks. These tasks include geometric modeling, conceptual design, preliminary design, structural analysis, detailed design, and drafting. Although significant progress has been made in developing software for individual tasks, integrating all of them into a single system has not been achieved successfully. Therefore, the development of a fully integrated system, that integrate all the activities outlined above is still an open research issue in computer integrated civil and structural engineering applications [Ref. 1]. Also, the design and implementation of integrated structural engineering software are important topics in both the academic and commercial communities. A number of engineering and CAD software companies have developed software tools to facilitate the development of integrated systems. Most of these tools, however, are patched together, in an ad hoc basis, rather than truly integrated [Ref. 2]. True integration requires a substantial development of theory and foundation on which the integrated system should be based.

In developing early integrated systems several different architectures and various techniques have been used. In recent years, there has been a growing interest in developing integrated systems using object-oriented techniques. In this paper, a number of object-oriented principles and techniques are explored which are believed to be useful for developing integrated engineering systems. An overall object-oriented architecture of an integrated system and its essential elements is proposed. Issues affecting the development of integrated engineering systems, mainly (1) representation of engineering design objects [Ref. 3], (2) management of design objects and design object versions [Ref. 2, Ref. 4], and (3) communication channels, mediums and protocols for design objects [Ref. 5] will be briefly discussed. The paper will, of course, address only small part of the overall integration problems and issues. The main objectives, as indicated, have been to contribute to the theoretical foundation needed for the development of a truly integrated structural engineering software system. The results, however, are believed to be applicable to larger scale integration across other civil engineering tasks and disciplines.

In the following sections, object-oriented concepts and terminologies will be introduced, several reasons that made integration in structural engineering domain difficult, will be stated. The needs and benefits of integration in structural and civil engineering environment will then be presented. Some previously proposed means and architectures for integrated systems will be outlined together with some of their advantages and disadvantages. A framework based on object-oriented concepts and a central data base architecture, for developing integrated systems, is then proposed. Brief descriptions of the elements of the integrated systems, their features and functionalities, are then presented. The paper concludes by summarizing the work done, and identifying several problems and issues that need further study in order to successfully achieve a fully integrated engineering system.

Object-Oriented Concepts and Terminologies

The object-oriented paradigm has emerged recently as one of the most unifying concept in developing and integrating programs from various fields. The approach has already gained popularity in a number of fields in computer science and engineering. These fields include: programming languages, software engineering, database applications, data modeling, artificial intelligence, knowledge representation, office automation, and CAD/CAM [Ref. 6, Ref. 7, Ref. 8]. The object-oriented approach has several features that made it attractive to these fields. In this section, only a brief summary of the object-oriented concepts and terminology will be presented.

In object-oriented systems, all physical and conceptual entities are represented by *objects*. Each object has a number of *attributes* that define its state and a number of *methods* that describe its behavior. An attribute may has a simple integer, real or string value, or may be a complex data structure. A method is a procedure that manipulates, sets or returns the state of the object. Attributes are usually defined private to the object. This aspect is known as *encapsulation*. Objects interact by sending *messages* to each other. Messages constitute the public interface of an object. An object reacts to a message by executing one or more *methods* and by returning a response. Objects that have similar attributes and methods are grouped together into *classes*. A class describes the form of the attributes and the behavior for all objects that are members of the class. *Inheritance* refers to creating a new class as a specialization of an existing class. The new class is termed a *subclass* of the old class which is called a *superclass*. The subclass usually inherits some or all the properties and behavior of its superclass. A class may inherit properties from only one class (*single inheritance*) or from several superclasses (*multiple inheritance*).

Integration Difficulties in Structural Engineering

Several attempts have been made to integrate engineering disciplines of constructed facilities and the activities involve in each discipline. These disciplines include architecture, structural engineering, geotechnical engineering, mechanical engineering, electrical engineering, and construction management. Although significant progress has been made in integrating the activities in individual disciplines, integrating all of them into a single system has not been achieved successfully. One reason is simply that the task is enormous. There are, however, several other reasons that make integration difficult, as follows:

1. Each discipline has its own distinct view of the structure being designed. For example, the architect sees the structure from the point of view of its functions for office space, residential space, parking, etc., and its components as rooms, balconies, partitions, etc. The structural engineer, however, sees the structure in terms of its ability to carry gravity and lateral loads, and the components are viewed as moment resisting members, axial load resisting members and so on. This multiplicity of views leads to a lack of uniformity, and therefore to complexity and conflicting demands.

2. The work in each discipline may involve several phases or sub-disciplines, each with different needs, particularly with respect to the type of information and the level of detail that must be considered. For example, structural design involves conceptual, preliminary and detailed design, requiring progressively increasing levels of detail. Also, for design the structure will be viewed as beams, columns, slabs, etc., and the connections between them; for analysis the structure will be viewed in terms of elements, nodes and degrees of freedom; and for drafting the concern is with elevations, cross-sections and connection details.

3. The software modules for each discipline not only require different views of the structure, but are likely to be developed by different people. As a result, the user interfaces, data structures and other details are likely to be different.

4. Some phases, particularly the design phases, are characterized by ill-structured sets of procedures and activities. This typically results in a diversity of styles among different designers, and ill-defined decision-making processes that involve intuition and judgement. These factors make some phases not only difficult to automate but also challenging to integrate with phases in which the processes are better defined and more algorithmic in nature. This is another aspect of lack of uniformity.

5. Some disciplines can experience frequent changes or modifications in the governing standards or provisions, with the result that frequent changes need to be incorporated into the software. If the software for these disciplines is not developed with this provision in mind, managing modifications and upgrades can be very difficult.

Integration Benefits and Needs

In spite of these difficulties, there are clearly major advantages and benefits to be gained, and also urgent needs, for developing and using integrated systems, for both the user and the developer of the system.

A. From a user (designer) viewpoint:

(1) An integrated system can improve reliability. Since design data is characterized by complexity, variety and large size, transfer of data by hand tends to be error prone. An integrated system can provide for automatic transfer of data between application program modules, and hence can help reduce errors.

(2) An integrated system can enhance productivity. Since large design projects usually involve a team of designers working simultaneously on different aspects of the project, having an integrated environment where information can be communicated easily is crucial. Communicating information through drawings and text documents, as is typically done, could be largely eliminated by having users share data through a common source. This also improves consistency, by eliminating the risk of keeping multiple copies of the same data at different locations.

(3) A truly integrated system is capable of unifying application program modules with different views of the system being designed. Although each module may have its own view of the data that represents the structure to be designed, an integrated system can automatically provide each module with the view of the data it needs. This enhances integration by making the system appear as one integral unit in spite of the diversity of the views and activities of its modules.

B. From a developer (programmer) viewpoint:

(1) A truly integrated system can enforce uniformity. Uniform methods and development standards allow new modules to be developed at different times by different developers, without later need for major restructuring of the system or major reprogramming effort.

(2) A truly integrated system can enforce modularity. Modularity makes a system easier to develop and maintain, since modifications and changes in one

module will not spread widely into others. Ideally, each module can be developed and tested independently, before being integrated into the complete system.

Integration Means and Architectures

General

To fully achieved integration, a thorough investigation of the key elements involved in developing an integrated system is necessary. Over the years, a number of attempts have been made to develop integrated systems. These attempts tend, however, to have been made without a solid theoretical foundation on which to base the design of the system. One approach, which is a necessary one for practical reasons, has been to integrate modules that were developed without the notion of integration in mind. This approach results, however, in systems which are only loosely integrated, and which tend to be specialized for relatively narrow applications. To achieve the goal of developing a truly integrated system, a solid theory covering all aspects of integrated system development needs to be devised.

There are many aspects of integration that need to be addressed. Two particularly important ones are: (1) approaches for data and information exchange, and (2) overall system architecture, particularly as it relates to data and information exchange. In the following sections, a number of approaches for information exchange are first reviewed. These include the Translator, Neutral File and Central Data Base (CDB) approaches. It is concluded that the CDB approach is clearly the desired goal. A number of different architectures that might be used with the CDB approach are then identified, and the data base centered architecture is chosen as the most promising. Several secondary aspects related to integration are then reviewed.

Direct Translator Approach

In the *direct translator* approach, the output data from one application program module is directly translated into the format required for input to another module (Figure 1). The process may involve the processing of multiple output files to generate the desired input file, since the output from several modules may need to be combined into one input file. For each pair of modules that must communicate, a special purpose translator is needed to convert from one format to the other.

The obvious disadvantage of this approach is the large number of translators that must be developed and maintained. However, if the individual application programs are stable, with little or no modification expected, this approach can be effective because it provides direct, highly specialized, and fast communication links between modules.

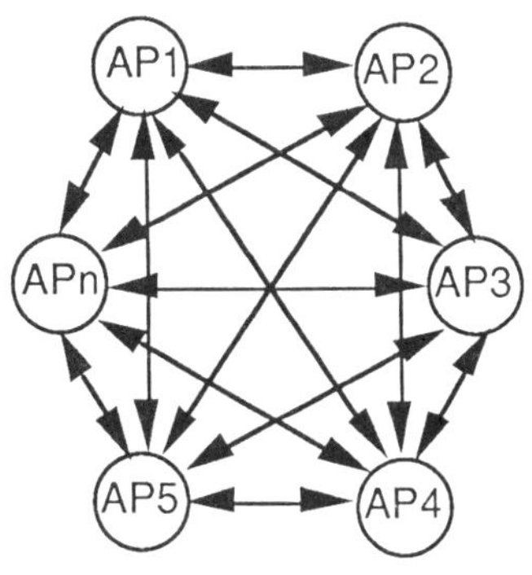

Figure 1: Direct Translator Approach

Standard Exchange Format (Neutral File) Approach

In the *standard exchange format* approach, all application program modules read input from and write output to files in a standard format (called the neutral file) that is understood by all modules (Figure 2). Although there is no accepted standard exchange format for structural engineering, this approach has been used in a number of other fields. These fields include: IGES (Initial Graphic Exchange Standard) [Ref. 9] for exchanging graphics data, PDXI (Process Data eXchange Institute) [Ref. 10] for exchanging process data, EDIF (Electronic Data Interchange File) [Ref. 11] for exchanging electronic component and circuit data, PDES (Product Data Exchange using STEP) [Ref. 12] and STEP (Standard for the Exchange of Product Model Data) [Ref. 13] both for exchanging product model data (including architecture, engineering and construction data). Details of these and other standard formats, such as SET (Standard d'Exchange et de Transfert), can be found in [Ref. 14, Ref. 15].

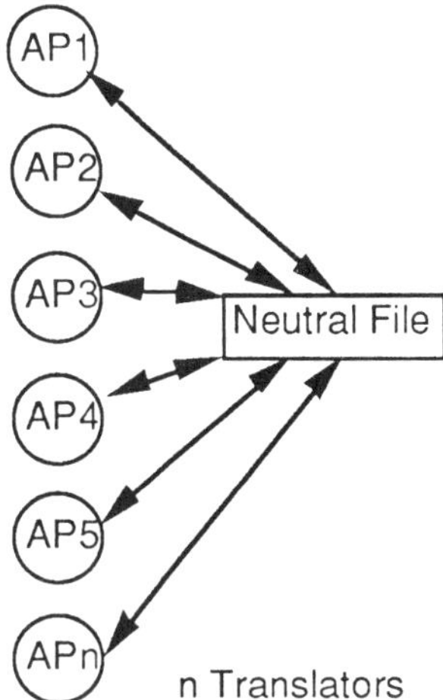

Figure 2: Standard Exchange Format Approach

Advantages

The advantages of the standard exchange format approach over the direct translator approach are as follows:

(1) A uniform neutral file is used to transfer data among all application program modules. This introduces uniformity to the system.
(2) If a new application program module is to be added to an existing system that uses a neutral file, only the new application program module needs to be modified to comply with the neutral file format.
(3) Removal of an existing application program module will not affect the neutral file or other application program modules.

Disadvantages

While different neutral file formats have been developed for different CAD systems and other application domains, each suffers from some limitations, these include:

(1) A great deal of time and effort must be spent to develop a neutral file format that supports any given application domain. Although there is substantial agreement on formats for graphics, solid and geometric modeling and VLSI CAD, there is no such agreement in civil engineering, and it is likely to take years before agreement is reached.
(2) In the neutral file format each application program keeps its own copy of the data, even if some or all the data is common among these programs. Such duplication of common data can result in consistency problems.

Central Data Base Approach

The *central data base* approach is one in which the data common to all applications is stored in a central data base and accessed through a Data Base Management System (DBMS). This allows several application programs to share data among themselves as indicated in Figure 3. The approach has several advantages and disadvantages, as outlined below.

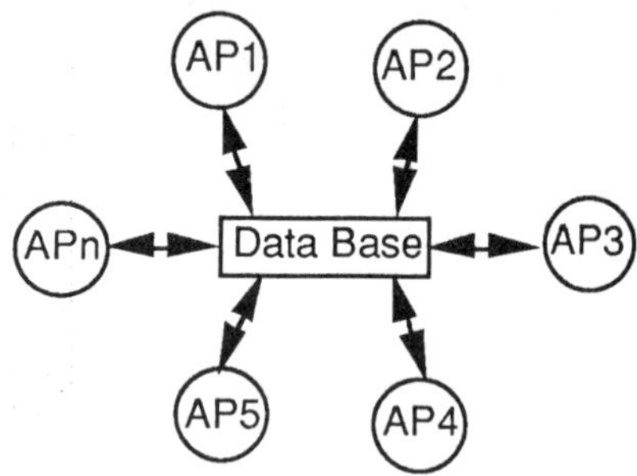

Figure 3: Central Data Base Approach

Advantages

The advantages of the central data base approach over other approaches are as follows:

(1) There is a single data repository for all application program modules, so that data common to all modules can be maintained in one place.

(2) The DBMS provides a unified set of access utilities to the data base, through which different application program modules with different needs access the common data.

(3) Addition or removal of an application program module does not affect other modules or the data base.

(4) Each application program module can request information from the data base in the form that matches its view of the problem. Although this requires powerful DBMS capabilities, it provides more flexibility than the standard exchange format approach.

Disadvantages

The disadvantages of the central data base approach are as follows:

(1) The design and implementation of a suitable data base and DBMS are even more difficult than the design of a neutral file format. This is especially true for integration across disciplines and organizations, since it is very difficult to design a data base that supports the many views required by varied organizations.

(2) The application program modules must be programmed to work directly with the data base, rather than only with their own data.

(3) The efficiency of access to the data is likely to be lower than if the data is accessed directly, because of the DBMS overhead.

Clearly, however, over the long term the central data base approach has advantages over other data and information exchange approaches. Therefore, this approach is the one adopted in this investigation.

Control of Information Exchange

For any data and information exchange approach, issues associated with overall control must be addressed, as well as the details of the approach. Only those control issues associated with the central data base approach are addressed here. A number of different control architectures has been proposed, each of which differs in the way it controls the process, in the degree of control imposed, and in the distribution of duties between the DBMS and other control modules. Examples of these architectures are as follows.

(a) The *Database-Centered* architecture, in which the data base is central to all applications, and in which access to the central data base is controlled by the DBMS. The DBMS manages access to the data in the central data base. Each application program module has its own user interface, and there is no central module that controls the activities of these modules. There are no direct communication links among the various application program modules (i.e., all data and information are communicated through the central data base).

(b) The *Executive-Centered* architecture [Ref. 16, Ref. 17], in which the data base, the application program modules and the user interface are all centered about an executive. In this architecture the role of the DBMS is reduced, as some of its duties are shifted to the executive. The executive is a central control module that coordinates application program access to the data base as well as to the user interface utilities.

(c) The *Data-Driven* or *Blackboard* architecture, in which control information is conveyed through a "blackboard" [Ref. 18]. In this architecture, the modules communicate information among themselves both through the blackboard and through the data base that holds information common to all applications.

Although these architectures have substantial differences, the concepts are similar, since all are centered around a data base that holds data common to all applications. In this research only the database-centered architecture is considered further.

Levels of Integration

The benefits to be gained from integration depend on the type of activities that are integrated and the level of integration considered. The following three broad levels of integration can be identified:

(1) Integration Across Organizations, in which organizations of different natures performing different activities are integrated. For example, organizations may include Developers, Financial Institutions, Government, Designers, Contractors and Facility Managers, as shown in Figure 4.

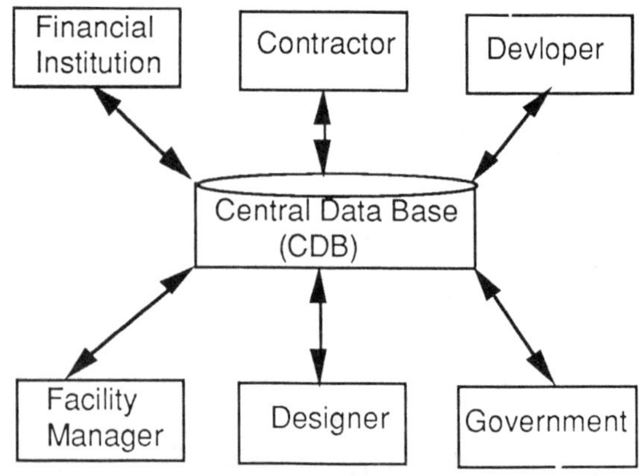

Figure 4: Integration Across Organizations

(2) Integration Across Disciplines in a certain organization, in which the activities of all disciplines in one organization are integrated. For example, disciplines of a design organization may include Architectural Design, Structural Design, Foundation Design, Mechanical System Design, Electrical System Design and Communications System Design, as shown in Figure 5.

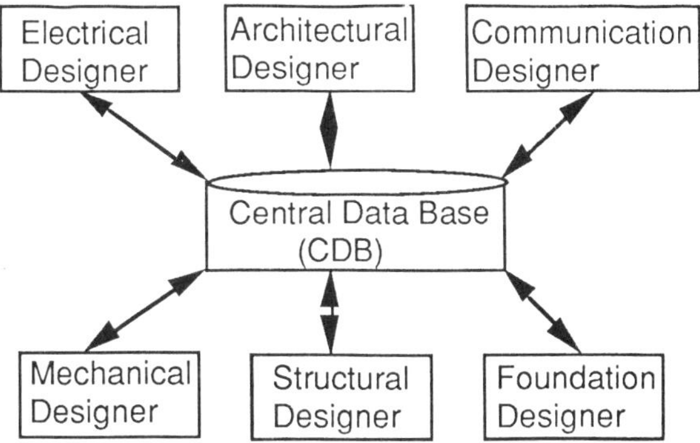

Figure 5: Integration Across Disciplines

(3) Integration Across Tasks, in which all the activities in one discipline are integrated. For example, Design tasks may include Geometric Modeling, Conceptual Design, Preliminary Design, Structural Analysis, Detailed Design, Drafting and Specification Writing, as shown in Figure 6. Detailed Design can be divided into more specialized modules.

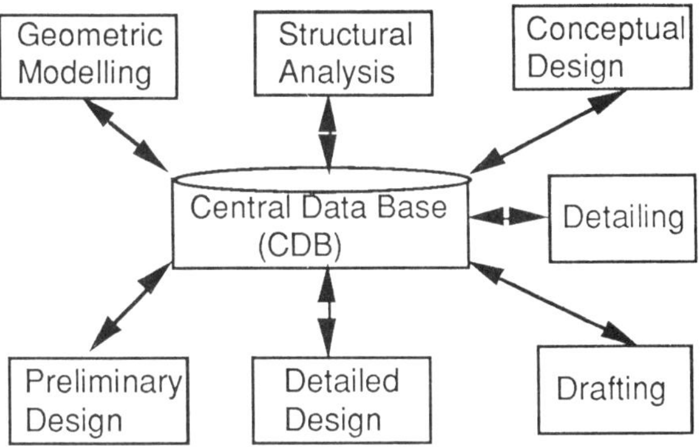

Figure 6: Integration Across Tasks

The creation of a truly integrated system that crosses the boundaries of organizations and/or disciplines is a truly enormous goal, which can only be reached over the long term and with involved research and development. A shorter term goal is integration involving only the structural design and analysis tasks, including such activities as preliminary design, code compliance checking, detailing, and drafting. Even this, however, is also ambitious, and is well beyond the scope of the research that is described here. In this research, it has been necessary to focus on a small number of narrow aspects of the complete problem. In the following sections, some of the broad issues associated with integration of the structural engineering design activity are first discussed.

Structural Design Activities

Since this research is concerned primarily with structural design, it is useful, therefore, to review the general tasks that make up the structural design process, and to consider the types of computing activities involved in detailed design. Structural design is a complex process, and the discussion of a detailed model is outside the scope of this research. A discussion of a model that is particularly promising can be found in [Ref. 19, Ref. 20]. In the following section only a brief discussion of the process is presented.

The design process consists of a number of goal-directed activities in which decisions are taken about the physical form of the structure and its components, in order to ensure their fitness for the intended purposes. The designed object must satisfy some functional specification, which is usually expressed in terms of design rules and criteria. The structure must achieve the desired performance with a high degree of confidence, and the design must be consistent with the available resources.

The design process is complex, involving a variety of strategies, activities and tools. In a simplified sense, the structural design process for a reinforced concrete frame may be viewed as a sequence of stages [Ref. 21], which include preliminary design, structural analysis, detailed design, detailing, and drafting.

In the preliminary design stage, the designer employs intuition, experience and some design rules to produce preliminary design parameters. This involves the synthesis of a potential structural configuration and the consideration of some design constraints. This stage may be subdivided into substages such as: (1) definition of the geometrical parameters that define a feasible structural configuration, (2) formulation of design constraints applicable to the configuration, (3) preliminary selection of the geometrical parameters, and (4) definition of the material properties. There have been some attempts to automate this stage, using expert system techniques [Ref. 21], but preliminary design generally requires a lot of human experience.

Following preliminary design, structural analysis can be used to estimate the demands on the structural components due to the applied loads. Analysis involves: (1) transformation of the structural description into a mathematical model, (2) definitions of the gravity load, live load, wind load, etc., (3) calculation of the bending moments, shear forces, axial forces and deflections, and (4) interpretation of the results.

Using the structural analysis results and design constraints, the structural components can be designed in detail. This involves selection and proportioning of the structural components (specifying section dimensions and other details), and checking the conformity of the designed components with the code requirements. The information generated at this stage is then used to produce detailed drawings of the structural components and detailed construction specifications.

The design process is, of course, usually iterative, as indicated in Figure 7. At any stage, if the design fails to meet the design requirements, new geometrical parameters and/or material properties must be chosen, and all or part of the process repeated. As shown in Figure 7, several design tasks share common data stored in the central data base.

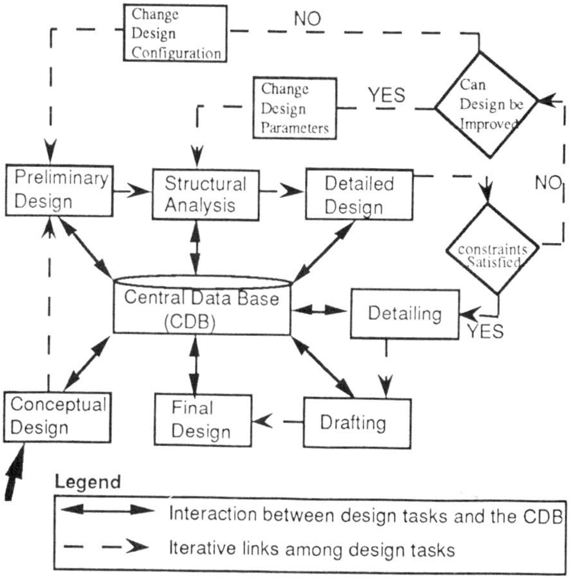

Figure 7: Structural Design Tasks

Problems Associated with Integrated Systems

The use of computers in structural analysis is well established, but there has been much less use in structural design, largely because of the ill-conditioned nature of the design process.

Over the years, many stand-alone programs covering specific structural design problems have been developed. Some attempts had been made to integrate such programs together, and the effort had led to the development of some integrated systems. These include ICES (Integrated Civil Engineering System) [Ref. 22]; STRUDL (STRUctural Design Language) [Ref. 23] and its derivatives GTSRUDL [Ref. 24] and McAUTO STRUDL [Ref. 25]; and GENESYS (GENeral Engineering SYStem) [Ref. 26]. More recently, substantial effort has been devoted to integrated system development by commercial vendors such as Intergraph and Impell, and by Architect-Engineers such as Skidmore Owings & Merrill [Ref. 27], Bechtel [Ref. 28], and Stone & Webster. All of these systems tend, however, to be rather loosely integrated, and lack many of the desirable features identified in this paper. A major reason is that there is no solid theoretical basis for the development of truly integrated systems. There are many detailed problems that need to be addressed before such a basis can be developed. Some of these are as follows.

(1) Problems associated with the approach to information transfer and the control architecture. The central data base approach with the database-centered architecture appears to be the most promising, but many complex details need to be addressed.

(2) Problems associated with the representation of design and control entities. Design objects, blackboards, etc., are complex entities. The development of a uniform and powerful scheme or data model to represent the various entity types is necessary.

(3) Problems associated with entity-to-entity communication. Entities of many different types need to communicate with each other. The communication channels among entities need to be identified more clearly, and communication protocols and mediums need to be defined.

(4) Problems associated with the detailed planning and organization of the representations of specific entities. Even after a uniform representation scheme or data model has been defined, a great deal of work remains to define the details of different entity types.

(5) Problems associated with data base management. It is well known that the technologies developed for business databases are not ideal for representing and manipulating the entities encountered in engineering systems.

(6) Problems associated with the management of design object versions. During design, several versions of a single design object are likely to be explored. More powerful techniques for managing versions need to be developed.

Overall Architecture of the Integrated System

To identify system architecture needs, it is necessary to discuss some of the broad problems associated with overall system architecture. Several benefits of an integrated system, from both the user and developer points of view, were presented in a previous section. To gain these benefits, however, many problems need to be addressed. These problems concern both the overall architecture of the system and the details of organizing and implementing the system. Before these problems can be solved they must be clearly defined and stated.

As noted previously, the Central Data Base (CDB) architecture has major potential advantages, and it is this architecture that will be explored in this research. In this architecture, the CDB is shared by a variety of applications, which places severe demands on the CDB and its Data Base Management System. Some particular problems are as follows.

(1) Since the Application Programs perform many different activities, the CDB must be able to support many different views of the data.

(2) The DBMS must manage transactions from Application Programs that need to access the same data simultaneously. This can be difficult because the transactions in engineering applications are often of long duration.

(3) The Application Programs communicate among themselves only through the CDB. If certain programs are intimately related and need to communicate frequently, the fact that communication is possible only through the CDB may be an obstacle rather than a help.

The first of these problems can be particularly important if an object-oriented database is used, because objects combine both data and procedures. In a typical passive database, the procedures required to create and operate on application-specific views of the data are contained in the Application Programs, and the DBMS contains procedures that perform only data management functions. An object-oriented DBMS can, however, be more active, and in principle could contain many (or even all) of the procedures that might normally be in the Application Programs. Clearly, however, it is not feasible for the CDB and its DBMS to contain the data and procedures for all the operations in all application programs. Rather, it appears that only the data common to all applications should be kept in

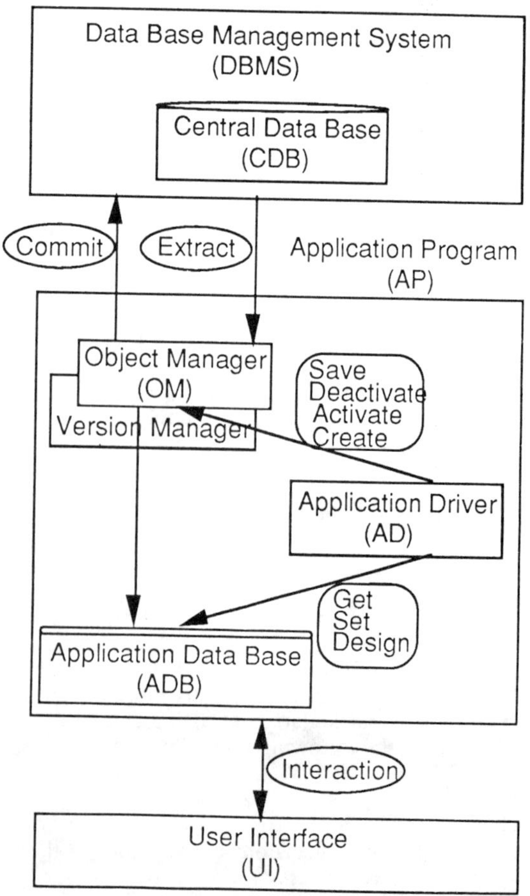

Figure 8: Overall Architecture of the System

the CDB, and that the DBMS should include only the operations needed (a) to help the application programs create appropriate views of the data, and (b) to ensure consistency of the data. An object-oriented CDB can thus be more active than a typical CDB, but still can not be expected to perform application-specific operations. Features required of an object-oriented data base can be found in [Ref. 29 and Ref. 30]. Each Application Program will continue to maintain and operate on its own view of the data. For an object-oriented system this application-specific view will be termed an Application Data Base (ADB). For this report, the overall architecture of the system is thus as shown in Figure 8. Details of the elements of the system are given in the following sections.

Other Aspects of Integration

Besides the uniformity in management of data that is provided by the central data base architecture, there are other key requirements that must be satisfied in order to achieve a high level of uniformity in an integrated system. These requirements include: (a) uniform means of representing entities, (b) uniform means of communication between entities, and (c) a uniform user interface.

(a) Unification of the *representation* of data. Although entities in different disciplines vary considerably, a data model that represents all entities uniformly and completely is needed. For example in an integrated civil engineering system, the architectural designer, the structural designer, the construction planner, etc., all have different views of the entities that make up a project. A representational model (data model) that is capable of supporting the views of all participants while preserving uniformity is required.

(b) Unification of the *communication Aspects (protocols, mediums and channels).* All application program modules, whether from different organizations, different disciplines, or different tasks, should ideally use a uniform protocol and medium for managing communications between entities, following well defined communication channels.

(c) Uniformity and consistency of the *user interface.* Although each application module performs a different set of activities, the user interface to each module should ideally have a similar interaction flavor - the same "look and feel."

Elements of the Integrated Systems

General

This section will briefly discuss the elements of the integrated system shown in Figure 8. It is important to note that, most of the details of these elements is not presented here, such details can be found in [Ref. 1].

Central and Application Data Bases (CDB)

The CDB (shown in Figure 8), which is accessed through and managed by the DBMS, holds the data that is shared by all applications. The CDB is thus the repository for permanent design data, and its data must be stringently controlled to ensure consistency. The details of the design and implementation of the CDB and the DBMS are outside the scope of this research. Work related to this problem can be found, for example, in [Ref. 31, Ref. 32].

Each ADB typically contains a subset of the objects in the CDB, corresponding to the design objects that are currently being operated on, plus objects defining loads, material properties, design requirements, and other data needed to perform the design. There is always a one-to-one correspondence between design objects in the ADB and those in the CDB. However the operations on the objects in the ADB will support application-specific activities such as sizing and proportioning, performing design constraint checks, and detailing reinforcement layouts. The objects in the ADB are thus instances of different object classes than those in the CDB, with application-specific methods (present in the Application Program) rather than strictly data management and consistency methods (contained in the DBMS). Also objects in the ADB are temporary, and require less stringent consistency control than corresponding objects in the CDB. Since consistency is a less important issue in the ABD, there is no need for a formal DBMS, and instead an Object Manager (OM), which is much less complicated than a DBMS, can be used to manage the objects.

It may be noted that an ADB could be considered as a separate part of the CDB (a single complex object), created by an Application Program and accessed through the central DBMS. However having an ADB for each Application Program allows the applications to executed independently, with no need to make use of the DBMS except for the "extraction" and "commitment" phases as described later.

Data Base Management System (DBMS)

The DBMS is responsible for managing common data stored in the CDB. Because of the need for stringent control over this common data, and to provide for concurrent access by several Application Programs, severe demands are placed on the design and implementation of the DBMS. As already noted, details of the DBMS are outside the scope of this research. However, the role of the DMBS in the extracting and committing of object data to the CDB is noted later.

Object Manager (OM)

The OM is responsible for creating objects in the ADB, storing them as persistent objects on disk, and keeping track of their addresses when they are in memory. It also coordinates the activity of extracting objects and committing them back to the CDB, as shown in Figure 8.

The creation of an object in the ADB involves sending a message to the appropriate object class, to create a new object with mainly undefined instance variables. The extraction process then involves extracting data for the corresponding object from the CDB, and using this data to initialize the instance variables of the ADB object. The commitment process involves informing the DBMS about changes in the attributes of an object in the ADB, and requesting that it updates the corresponding object in the CDB. There are several operations supported by the OM. Details of a proposed OM and its operations are beyond the scope of this paper and can be found in [Ref. 1].

Version Manager (VM)

As a design evolves, it may be necessary to keep several different versions of a single design object, until a decision can be made on the most appropriate version. Such versions need to be managed during the design process. A Version Manager needed to perform this task. There are several operations supported by the VM. Details of a proposed VM and its operations are beyond the scope of this paper and can be found in [Ref. 4].

Application Program (AP)

Each AP uses the methods supported by its application-specific object classes to carry out design and other activities. As shown in Figure 8, the AP makes use of the OM in accessing the objects in the ADB. For example, an AP for reinforced concrete (R/C) design might consist of modules for sizing, proportioning and detailing of structural components; for checking design constraints; and for browsing through the objects using graphical and/or textual displays. A prototype application program for data translation facility has been developed, using this architecture [Ref. 33]

User Interface (UI)

The User Interface provides a set of interface utilities for interaction between the user and the Application Programs. These utilities should provide an easy to use environment for display of the structure, its components and its connection details, and to allow the user to control the design activities. This is an ongoing research and development issue [Ref. 34, Ref. 32].

This section has described the general architecture that is being explored, and has outlined the main features of the components of the system. Details of the Central Data Base, Object Manager, the Application Data Base, the Version Manager, and the object communication aspects are outside the scope of this paper.

SUMMARY AND FUTURE WORK

This research has investigated some of the tools and techniques needed for developing a truly integrated software system for structural engineering, using object-oriented principles. It is recognized that the development of an integrated system is an enormous task that can only be achieved over time. This research has addressed only a small part of the overall problem, and many issues have been left for future work. The following section summarizes the principles and techniques that have been contributed by the research, and identifies several problems that need to be investigated in more detail. Remember that most of the details of the research have not been presented in this paper.

Summary and Contributions

In this study a number of principles and techniques have been proposed which may be helpful for developing truly integrated software systems. The topics that have been addressed are as follows.

1. An overall architecture has been proposed, with (a) a central data base containing common data; (b) a number of application programs performing application-specific operations; and (c) an application data base for each program, containing application-specific data. The central data base is accessed through a data base management system, while the application data bases are accessed through a simple object manager.

2. The needs for having a formal data model for uniformly representing design objects have been recognized. Features required of the model have been identified.

3. The need for standardized protocols, mediums and channels for object-to-object communication has been recognized. In the study, several protocols, mediums and channels have been proposed to facilitate object-to-object communication.

4. The features required for an Object Manager that manages objects in the Application Data Base have been identified. Several specific operations for managing design objects have been defined.

5. A scheme for defining and representing design object versions has been proposed. The features required for a Version Manager which managers design object versions have been identified, and several specific operations for managing version have been defined.

Future Work and Directions

While it is believed that this research has made significant contributions, there are obviously a large number of research issues that remain to be tackled. Some of these are as follows.

1. The investigation has addressed mainly issues concerning the local integration of structural engineering tasks, and even in this limited area has not developed complete solutions. To solve problems of global integration, across civil engineering and other disciplines, will require a great deal of work and ingenuity. The development of a truly integrated system is a complex task that will require development of a rigorous theory before computer code can be developed.

2. The communication protocols suggested in this study is based on the use of certain mediums to convey arguments and responses. Although it provides a uniform mean of communication, it involves substantial computational overhead. More study of this, and possibly other protocols, are needed.

3. The capabilities of the object and version managers described in the investigation are limited. Work is needed to develop managers with broader capabilities. In particular, the version manager should have the ability to manage object alternatives as well as versions.

4. The preceding points all follow aspects that have been studied as part of this research. There are many other aspects that need to be addressed before a truly integrated system can be implemented. In particular, the user interface is a critical ingredient. The design and implementation of an interface for civil engineering applications are a major task that needs to be addressed.

Acknowledgement

The research described in this paper has been supported in part by a grant from University of Khartoum, Sudan, and in part by National Science Foundation Grant No. DMC-8517338. The Center for Integrated Facility Engineering at Stanford University has provided the facilities and environment for producing this paper. The supports of these organizations and that of my thesis advisor, professor Graham H. Powell of the University of California at Berkeley, are gratefully acknowledged.

References

[1] Abdalla, J. A. *Object-Oriented Principles and Techniques for Computer Integrated Design.* Unpublished Ph.D. Dissertation, Department of Civil Engineering, University of California at Berkeley, Berkeley, CA, USA, 1989.

[2] Powell, G. H. and Abdalla, J. A. "Object Management in an Integrated Design System," *Proceedings, ASCE Sixth Conference on Computing in Civil Engineering*, Atlanta, pp. 556-563, 1989.

[3] Abdalla, J. A. and Powell, G. H. "Object Design Framework and Methodology for Structural Engineering," draft paper under revision, to be submitted to the Journal of Engineering with Computers, 1991.

[4] Abdalla, J. A. and Powell, G. H. "Version Management Needs for Computer Integrated Structural Design," in the Journal of *Engineering with Computers, an International Journal for Computer-aided Mechanical and Structural Engineering*. Vol. 7, No. 3, Summer 1991.

[5] Abdalla, J. A. and Powell, G. H. "Design Object Communication: Protocols, Channels and Mediums," draft paper under revision, to be submitted to the Journal of Engineering with Computers, 1991.

[6] Proceedings of the ACM Conference on Object-Oriented Programming, Systems, Languages, and Applications, (1986-1990).

[7] Spooner, D. L., Milicia, M. A., and Faatz, D. B., "Modeling Mechanical CAD Data with Data Abstraction and Object-Oriented Techniques," *IEEE International Conference on Data Engineering*, Los Angeles, CA, pp. 416-424, 1986.

[8] Zdonik, Stanely B., and Maier, David Readings in Object-Oriented Database Systems Morgan- Kaufman Publishers, Inc, San Mateo, California, USA, 1990.

[9] The Initial Graphics Exchange Specification (IGES) Version 5.0, U.S. Department of Commerce, National Institute of Standards and Technology, Center for Building Research, Gaithersburg, MD 20899, USA, 1990.

[10] Process Data Exchange Institute (PDXI), American Institute of Chemical Engineers, 345 East 47th Street, New York, New York 10017, USA, 1990.

[11] Personal contact, Prof. R. Newton, EECS Department, University of California, Berkeley.

[12] *IGES/PDES Organization Reference Manual*, National Computer Graphics Association (NCGA), 2722 Merrilee Dr., Suite 200, Fairfax, VA, 22031, USA, 1991.

[13] Wilson, P.R., and Kennicott, P.R., *PDES/STEP Integrated Product Information Model (IPIM)*, ISO TC 184/SC4/WG1, Doc. 4.1.2., Working Draft, September 1988.

[14] Wilson, P. R. "A Short History of CAD Data Transfer Standards," *Computer Graphics & Applications*, 1987.

[15] Encarnacao, R. S., and Voge, E.*Product Data Interfaces in CAD/CAM Applications*, Springer-Verlag, 1986.

[16] Staley, S. M., and Anderson, D. C. "Functional Specification for CAD Databases" *Computer-Aided Design*, Vol. 18, No. 3 PP. 132-138, April, 1986.

[17] Howard, H. C., and Rehak, H. C. "KADBASE: A Prototype Expert System-Database Interface for Engineering Systems." *IEEE Expert*, 4(3), 65-76, 1989.

[18] Fenves, S. J., Flemming, U., Hendrickson, C., Maher, M. L., and Schmitt, G. "An Integrated Software Environment for Building Design and Construction," *Proceedings of the Fifth ASCE Conference on Computing in Civil Engineering: Microcomputers to Supercomputers*, PP. 21-32, Alexandria, VA, March, 1988.

[19] Sause Jr, R. *A Model of the Design Process for Computer Integrated Structural Engineering*. Unpublished Ph.D. Dissertation, University of California at Berkeley, Berkeley, CA, USA, 1989.

[20] Deepak, J., Luth, G. P., Krawinkler, H., and Law, K. H. "A Formal Approach to Automating Conceptual Structural Design," *Technical Report No. 31,* Center for Integrated Facility Engineering, Stanford University, 1990.

[21] Maher, M. L., and Fenves, S. J., HI-RISE: A Knowledge-Based Expert System for the Preliminary Design of High Rise Building, Report No. R-85-146, Department of Civil Engineering, Carnegie-Mellon University, Pittsburgh, PA, 1985.

[22] *ICES: Programmers Reference Manual*, Dept. of Civil Engineering, MIT 1971.

[23] Logcher, R. D. and G. M. Sturman " STRUDL - A Computer System for Structural Design," *Journal of the Structural Division, Proceeding of the ASCE* Vol.92 No.st6 December 1966.

[24] Swanger, M. H. and Kahn, L. F. "Computer-Aided Design of Reinforced Concrete Structures," *Proceedings of the Second Conference on Computing in Civil Engineering*, 1981. PP. 302-317, 1981.

[25] DasGupta, D. "MCAUTO STRUDL RECON- A Reinforced Concrete Frame Design Software" *Concrete International*, PP. 37-42, November, 1982.

[26] Stahl, F. I. "The Standards Interface for Computer-Aided Design: An Overview of Some Technical Problems Associated with Automated Design Checking" PP. 560-567, April, 1984.

[27] Thomas, V. C. "Mechanical Engineering Design Computer Programs for Building," *Center for Integrated Facility Engineering Symposium Proceedings*, Stanford University, 1991.

[28] Borrows, K., Bechtel Group of Companies: The AIM Group, personal contact, 1991.

[29] Atkinson, M. Bancilhon, F., DeWitt, D., Dittrich, K., Maier, D. and Zdonik, S., "The Object-Oriented Database System Manifesto," *First International Conference on Deductive and Object-Oriented Databases*, Kyoto, Japan, December 1989.

[30] Kim, W., "Object-Oriented Databases: Definition and Research Directions," IEEE Transactions on Knowledge and Data Engineering, Vol. 2, No. 3 September 1990.

[31] Bhateja, R. *A Database Design for Structural Engineering*. Unpublished Ph. D Dissertation, University of California, Berkeley, CA, USA, 1989.

[32] Abdalla, J. A. and An-Nashif, H. N. "An Object-Oriented Data Base for a Structural Engineering System," *The Fourth International Conference on Computing in Civil and Building Engineering*, Tokyo, Japan July 29-31, 1991.

[33] Abdalla, J. A. and C. John Yoon "An Object-Oriented Design and Implementation of Finite Element and Graphics Data Translation Facility," to appear in the American Society of Civil Engineers Journal of Computing in Civil Engineering, Special Issue on Object-Oriented Programming, 1992.

[34] Howard, C. S. and Howard, H. C., "Issues in User Interfaces for Engineering Databases," *ASCE Sixth Conference on Computing in Civil Engineering*, Atlanta, GA, USA, 1989.

USING KNOWLEDGE ISLANDS IN AN OBJECT ORIENTED FRAMEWORK FOR INTEGRATED STRUCTURAL DESIGN

Z. Turk and J. Duhovnik

Department of Civil Engineering,
University of Ljubljana, Slovenia

An important part of computer integrated design environment is knowledge. This can be coded proceduraly or declaratively. Traditional knowledge based systems separate knowledge (data) from inference engines (procedures) and knowledge from contexts. Traditional object oriented environments that are essential for modelling of building products group data and methods into an object. In the paper architecture of an integrated design environment will be proposed that enables integration of existing tools and programs and utilises a combination of traditional procedural and artificial intelligence (AI) techniques and is based on an object oriented (OO) data model. In this preliminary communication, abstract and data models will be suggested where knowledge becomes the organic part of objects on all levels of complexity forming usually small knowledge islands, hanging from a tree of objects on methods.

Key words: integrated design, object oriented, knowledge based, knowledge islands, hybrid tools.

Introduction

Computer based tools assist a human engineer in a growing range of tasks that have to be performed during the process of designing a civil engineering product. The computerisation has been particularly intense during the phase of structural analysis and design. In today practice software tools used are still discrete and stand-alone. Fenves /FENS91/ writes about the islands of automation in a largely manual design environment. The civil engineer is becoming an intelligent interface between largely incompatible software. His activities seem intelligent when compared to a computer, but are routine and not creative in comparison to the engineering tasks he was trained for. The AEC industry is still exchanging information on paper /HOWA90/.

The main task of civil engineering software development is the integration of various stand alone computer tools that would enable computer integrated construction process of the next century /BJOR89/. The primary source of problems is the prototype nature of buildings, its complexity, the large number of components and participants and the interaction of multiple technologies /SRIR90/. While looking forward to STEP/PDES, AIS, SDAIS ... compliant software products, we have to consider the installed base of in-house written programs that need to be integrated here and now.

Model of Computer Integrated Design Environment

Computers and Design

In reference to the integration of computers into the design process we see three main stages of development:

- COMPUTER ASSISTED/AIDED DESIGN - computer based tools are used for some tasks or phases of the design process.
- COMPUTER INTEGRATED DESIGN - computers are used for the integration of software, hardware, engineers and processes involved in the design process.
- COMPUTER AUTOMATED DESIGN - asymptotically the computers are approaching the state, where they would be able to produce the entire design with no help (instructions) from the engineer. The emphasis is on the asymptotic approach meaning that the state will never be entirely reached (Fig. 1).

Due to the pragmatic and economic reasons the transition from one stage to the other is expected to be evolutionary, meaning that the existing solutions will be employed whenever feasible. The possibility of turning over some portions of control and decisions to the computer must be taken into account during the development of the computer integrated environments as well. The possible integration of software must be considered during the development of the new stand alone tools and during the adaptation of older ones.

We can distinguish two main approaches to implementing a computer integrated design environment (CIDE):

- PIPELINE ARCHITECTURE - various tools are connected through some interfaces with one another. The approach is giving efficient

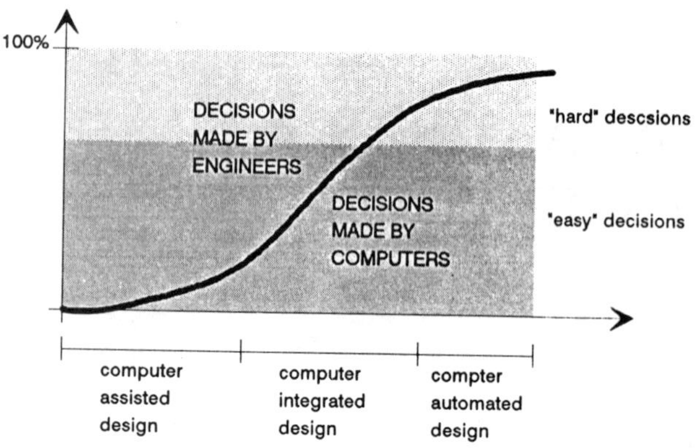

Fig 1. Stages of usage of computers during the design process. Most of practical applications are in the first level, while a lot of efforts of various researchers is centred on porting or writing applications for the second level. Elements enabling some computer automated design are expected by the end of the century.

short term results, particularly when interfacing a smaller number of programs using similar abstract models. If there are N programs involved in the design process, between N-1 and N2/2 interfaces are needed to enable the flow of the data.

- CENTRALIZED ARCHITECTURE - we define a common concept that serves as an integration instrument. On the physical level, this can be a central database, an expert system shell, an operating system tool ... or a combination. On the conceptual level, it can be a common data model type, compatible abstract models or building product models. There are 2*N interfaces needed to connect N existing programs to a central data base.

The two architectures are schematically shown in Fig. 2.

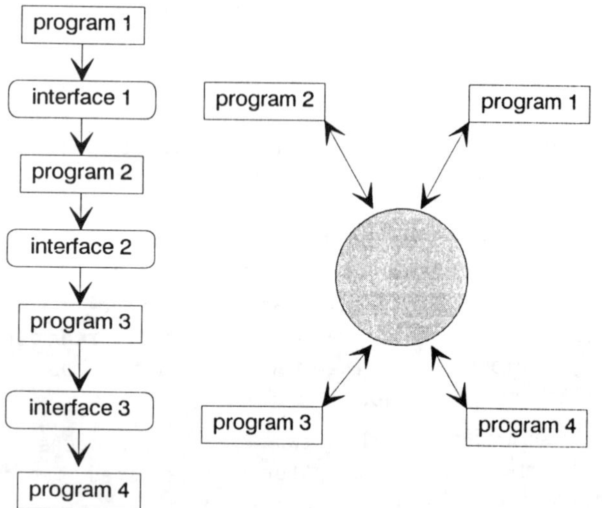

Fig 2. Pipeline (left) and centralized architecture (right) of a computer integrated design environment.

Detailed abstract modelling or conceptual modelling (or systems analysis) of all the models that are either implicitly or explicitly present in the programs participating is one of the most important steps in developing a computer integrated design environment.

Our approach to integrated (and later automated) design is based on the following assumptions:

- The development should be evolutionary, not revolutionary.
- Traditional software like FEM analysis programs and CAD packages are tested reliable tools. There is no need to write special versions for the integrated design.
- Evolutionary approach with integration of already available packages enables gradual transition from "aided" to "integrated" design and later to "automated" design. The approach to integration that admits the existence of other software simplifies integration of tools on various hardware platforms.
- OO, AI and traditional techniques have its own advantages and drawbacks. The solution should be found in the synthesis of the three.

<u>Model of Computer Integrated Design Environment</u>

In Fig. 3 there is a model of a computer integrated design environment that has been divided into five sub-models.

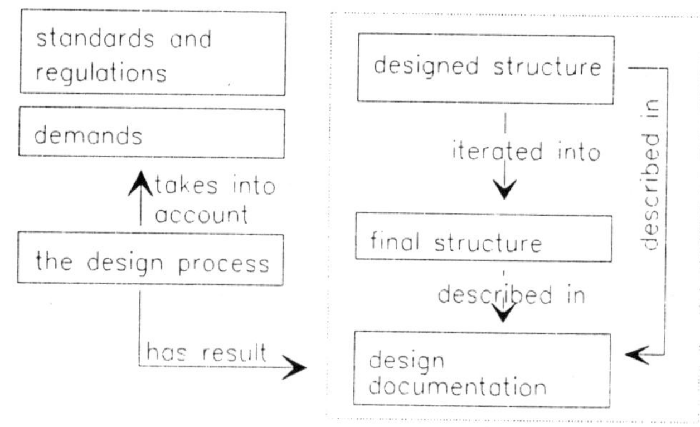

Fig 3. Semantic network of models participating in computer integrated design.

The relation between the models is clear from their function. During the *design process*, various *alternative designs* and calculations are made in order to be able to finally produce and refine all the necessary *plans and documentation* for the *designed structure*.

- STANDARDS, REGULATIONS and DEMANDS define the boundary conditions that the design should fulfill. Some are enforced by the state authorities, some by the investor of the building. Their semantic and syntax is often similar. The investor might want a building at least 20 storeys high, the state regulations may require all buildings lower than 10 storeys because of the unstable ground.

- THE DESIGN PROCESS models the actions and participants of the design process. The actions like analysis, synthesis and evaluation ... are preformed by the participants (designers, programs, computers ...).
- THE DESIGNED STRUCTURE models the design of the building. It is an extension of the final structure and also contains information of various versions of the design, design decisions and other information regarding the design that was used in the design process. It is sometimes referred to as a "process model" /CORN91/.
- THE FINAL STRUCTURE models the building as it will be built. It describes the final, chosen design or a "static model". Most building product models being developed describe this model.
- THE DESIGN DOCUMENTATION models the interface with non-computer users of the design. For some time to come, final designs will have to be communicated in a form that will be used without computers.

The five models are closely related. They are all based on the static (final design) model, being their extensions or translations (process model, documentation models) or being implicitly referred to in standards, regulations and the design process /TURK91/.

Most of the literature we had the opportunity to study deals with the model of the designed structure, a discuss the design process. The documentation of the design is often seen merely as an output of the design structure database. The lack of a special model of the structure in the process of design complicates both the model of the structure and the model of the design process. But, unfortunately, many programs particularly in the field of analysis, work on models that only remotely correspond to the central (usually geometric) model of the structure. There are many sets of results for some parts of the structure and no results at all for others. Some results are considered valuable, some are deleted. The structure changes during the design process, is being refined, some results get obsolete, some are dependent on each other ...

Object oriented data model

As we have shown, the building product models are the key for integrated design. Unfortunately, these product models are very complex. In a description of a building several hundred and maybe thousand of various data item types are used. In order to organize these vast amounts of data types (and even larger amounts of data items) it has been recognised by many researchers that object oriented approach has many advantages /FENG90/, /POWE88/. It provides hierarchical order of data types, encapsulation and inheritance. Very similar data model type can be seen present under the STEP standard and the EXPRESS language.

An object can be a compound object (consisting of one or more member-objects which can again have members) or a simple object which can be one of the following:

- integer number (stores an integer value);
- real number (stores a real value);
- string (stores text);
- object reference (stores a pointer to an object);
- object (stores a member object);
- method (stores functionality or knowledge about the object).

Object oriented abstract data models provide three built-in relations:

- is_kind_of defines that a class of objects is derived from another class of objects ("sedan" is_kind_of "car");
- is_type_of defines that an object belongs to a certain class ("Ford Sierra" is_type_of "car");
- is_member_of (is_client_of) defines that an object has another object as a member. "length" is_member_of "car", "car" is_client_of "length").

Unfortunately, no other relations among data items are built into the OO paradigm. We too have been tempted to use one of the above relations to describe something else, i.e. connections, part/whole relations, but we believe that all such misuses revenge later. We should not attach any additional semantics to the predefined relations.

The other inconvenience with object oriented programming is that an object is a collection of data and methods that manipulate that data. In an object data and methods (procedures) are one.

This is in contrast with the general idea of most knowledge based systems that are characterized by the separation of the data (knowledge base) from the procedures (inference engine), and more, by the separation of the knowledge based part from the algorithmic part.

The main motives for separation were:

- Existing engineering programs were very complex and adding additional knowledge seemed not feasible. Separated knowledge based modules lowered complexity. Object orientation provides other means to fight complexity.

- It would be easier to maintain, add and customize the knowledge, if it was separated from the rest of the software. It turned out that manipulating knowledge base requires trained knowledge engineers and usually can not be done by end users.
- It would be possible to make plug compatible products that would enable to plug different knowledge modules into the same engine, making it possible, for instance, to apply different national codes to the same analysis program. The feasibility of this does not depend only on the architecture of knowledge based system, but mostly on compatibility of codes.

We feel that the two technologies, AI and OO should be merged. An attempt will be made in this paper.

Models, views and controllers

The Smalltalk environment introduced the Modell-View-Controller paradigm /GOLD83/. We like to see

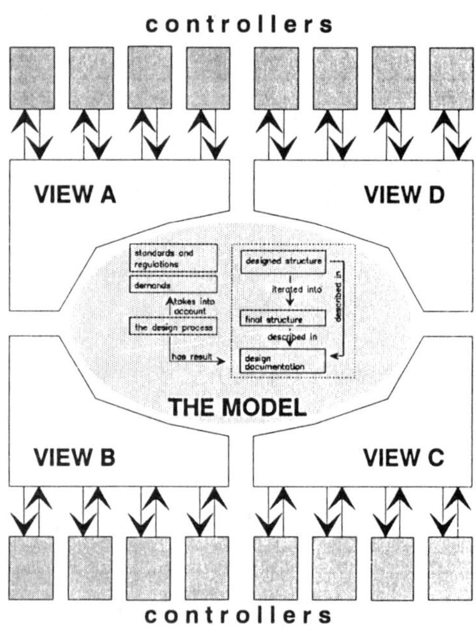

Fig 4: MVC paradigm applied to integrated design environment.

our integrated design environment in a similar way. It is shown in Fig. 4.

The model is implemented as an object oriented interface over a relational database. OO expert system shell running under MS Windows is used as a data modelling tool and OO database. There is a one to one mapping between objects in an expert system shell and some of records stored in data base files. The data base is storing values of objects, expert system shell is storing most of the knowledge. Some procedural knowledge is coded in specialized programs. The solution is temporary and will be replaced with an object data base (ODB).

There are several views of that model. Geometric modellers would see one view of the model, FEM analysis programs another (each), programs for reinforcement design perhaps a combination of the two. Usually, one view is shared by a few applications.

The controllers are several programs through which either software or humans manipulate the model (not the view but "seeing" the view only).

Although not as extensive, the architecture of such a system is similar to the one that would use Application Interface Specification (AIS). The views are playing the role of the AIS as presented in /MAGL90/.

Knowledge Islands

The knowledge island architecture is based on the premise that knowledge is about objects and that it should be tightly connected with the objects.

Types of knowledge

After the introduction of expert systems the word "knowledge based systems" was often used as a synonym, implying some declarative representation of the knowledge. We feel that in engineering in particular, knowledge should be understood as broadly as possible. Different problems would require different coding of the knowledge. We like to think that *a unit of knowledge is any generic relation between two or more objects* (similar definition is given in /COST91/). We see two types of knowledge:

- procedural knowledge, which defines the value of objects using an algorithm;
- declarative knowledge, which defines the value of objects by querying a data base. The data base can contain rules, facts, or relations between objects and classes ...

On the basis of knowledge usage, two extremity representations of knowledge stored in the computer may be defined:

- Human usable representation that enables mainly interactive evaluation of knowledge. A human is an interface between the knowledge base and the context data. The computer often only provides means for faster and more efficient gathering of relevant information.
- Machine usable representation that enables automated evaluation of knowledge. The knowledge is tightly connected with the context data.

We feel that both representations should be supported and compatible. The first one is much easier to implement and can be delivered to the end user relatively quickly. The second one requires a well designed building product model.

Knowledge and Objects

We have been using OO techniques on two scales. In our data model, everything is an object. In the large scale we see whole programs and program groups with their inputs and outputs as objects, and on the small scale we see every piece of data and functionality in any of the four models as an object.

Knowledge is stored with objects (or, if it is of a more generic nature, with classes). Procedural knowledge is stored as methods that are part of objects. Declarative knowledge is stored in its specialised database and is accessed through a method like "solve", "consult" or "query". What we get as a result are knowledge islands or knowledge fruits that hang from the object tree. The schema is shown in Fig. 5.

We see such knowledge based parts ranging from trivial range checking of some input value, to complex knowledge bases comparing design alternatives. Such an organisation is both a simplification and generalisation of the traditional record/field or frame/slot approach and, at the same time a synthesis of the OO and AI worlds. The implementation of such a scheme is not directly supported by traditional tools (rule based expert shells and OO languages), but we found a good approximation in hybrid expert system shell Kappa that allowed us to extend it with features we wanted.

Encapsulating knowledge into knowledge islands with objects is expected to give the following advantages:

- Knowledge is structured and coded with the object in question. Knowledge on abstract objects has attributes of deep knowledge.
- There are many small, simple knowledge islands each dealing in its own level of abstraction.
- Knowledge is portable with objects.
- The frame like structure itself is considered knowledge based.

Knowledge Island Class

The knowledge islands themselves are modelled as objects. The class KnowledgeIsland is shown in Fig. 6. The slots are listed below:

- RULES is an ordered list of rules that are used for declarative representation of knowledge contained in this island.
- GOAL is an ordered list of goals to trigger the backward chaining or to control end of the forward chaining process.
- RETURN stores what the consultation returned as a result. Usually this would be a Boolean value, but it can be an object of any type as well.

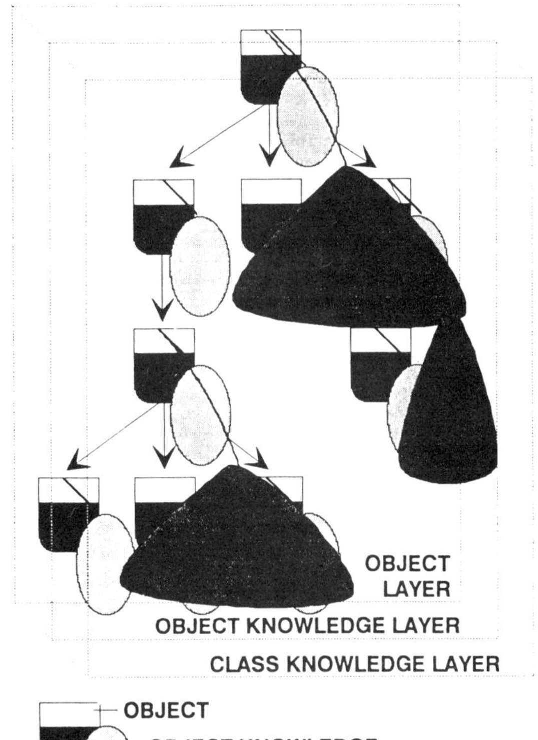

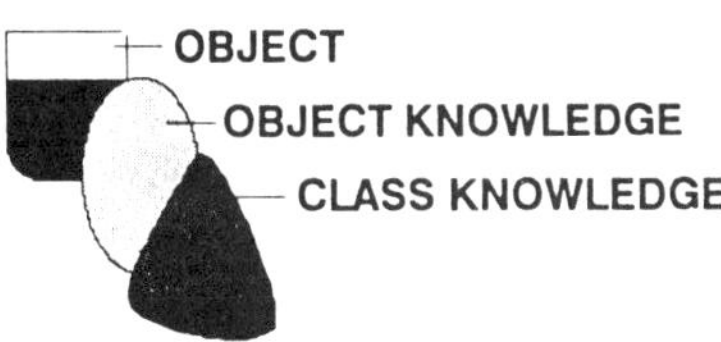

Figure 5: Knowledge islands.

- RESULT stores the longer, usually textual and more descriptive result of the consultation. It is generated by the rule consequences in the form suitable for inclusion in the design documentation.
- CONTEX defines the type of the knowledge island which may be object (the context of knowledge island is the object on which the island hangs), class (the context are object that directly or indirectly belong to a certain class) or global (the context is the whole object tree).

The essential methods are:

- CONSULT runs the rule evaluation process.

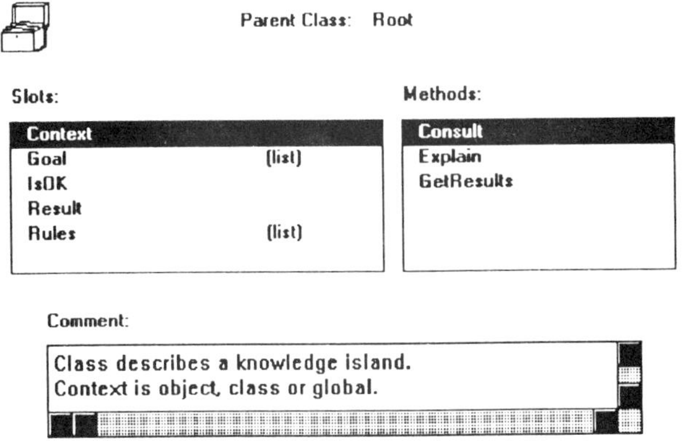

Figure 6: Slots and methods of the KnowledgeIsland class.

- EXPLAIN explains inference process by displaying inference network and/or messages explicitly coded into the rules.
- GET_RESULTS is an interface to Knowledge Island clients and enables them to fetch the data resulting from the consultation.

The usage of knowledge islands is simple. An object or a class of objects includes an instance of knowledge island type as its slot. There can be more than one knowledge island attach to an object simply by having more than one knowledge island instance.

Example: Integrated design environment

In Fig. 8 there is a simplified class tree for the integrated design environment. There are three main branches (controllers, programs and view) and some server classes (only the knowledge island class is displayed).

A *view* can be a real, indirect view like a view of geometry or a particular finite element view of the building, or a file (input, output or both), which needs special methods to communicate with one of the views. A view is manipulated by *controllers* which are grouped according to the view they handle.

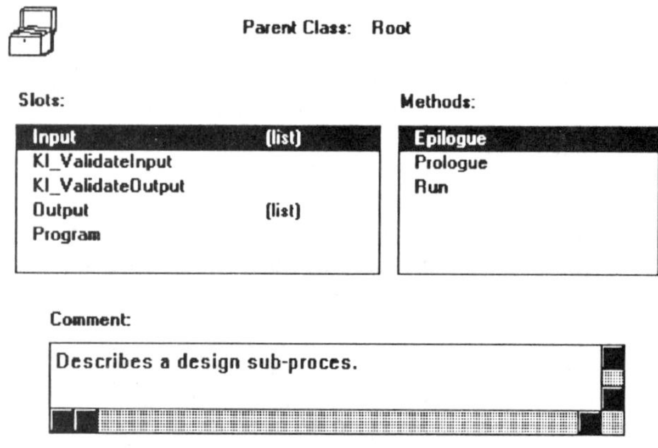

Fig 7. The structure of a generic process object.

A *process* describes one "run" of a program, its inputs and outputs. Its main components are lists of view and controller objects. The generic process class is shown in Fig. 7. It has two "knowledge island slots". One to manipulate input, the other to analyse output. The run method first starts the prologue method, which among other, extract data from the view consults KI_validate input, then the program is run and, finally, the epilogue method is called to analyse the output (through KI_validate output) and the results are translated into the view compatible form. The translation is performed using the methods in the class that describes the output file in question.

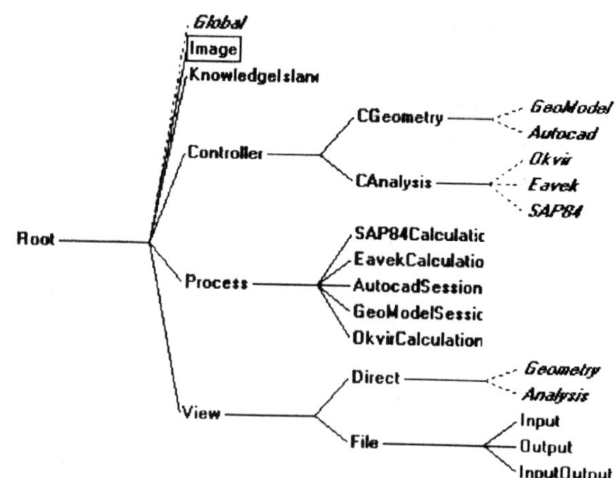

Figure 8: The tree of objects that model existing programs in the computer integrated design environment. Classes and objects are explained in the text.

Conclusions

We have been testing the architecture in a prototype environment developed under a hybrid expert system shell. We found the shell very good for the modelling and the analysis of data, suitable for prototyping, but rather insufficient for implementing the real system. It lacks the data storage capabilities and the efficiency of object oriented databases (ODB) and provides very limited tools for the combination and cooperation or merging of several models and objects taking place.

Using the decomposition of knowledge into islands, we found the inference capabilities sufficient. We would find a simple inference engine a useful add-on to ODBs.

The knowledge island architecture solved the problem of combining knowledge based and algorithmic programs. By knowledge decomposition, we believe larger, portable knowledge bases could be built. On the other hand, knowledge decomposition is causing problems when knowledge can not be decomposed using class/object boundaries but is associated with group of objects related by some user defined relation.

Further work will be focused on developing the prototype into a useful product for the integrated design of hi-rize buildings. Alternative declarative knowledge representations (like facts /TOPP89/) are also being considered.

References

BJOR89 Bjork, B.C., Penttila,H., "A Scenario for the Development and Implementation of a Building Product Model Standard", Adv. Eng. Software, Vol.11, No.4, 1989.

CORN91 Cornick, S.M., Leishman, D.A., Thomas, R.J., "Integrating building codes into design systems" , VTT workshop Computers and Building Standards, Espoo, Finland, 27.-29.5. 1991.

COST91 Costa, F.C., "Treatment of knowledge and its use in construction applications - state of the art and results of a case study", available from the author at Institut de technologia de la construccio de Catalunya - ITEC (1991).

FENG90 Fenves, G.L., "Object oriented programming for engineering software development", Engineering with Computers 6, 1-15 (1990).

FENS91 Fenves, S.J., "Status, Needs and Opportunities in Computer Assisted Design and Analysis", Structural Engineering International, 2/91.

GOLD83 Goldberg, A., Robson, D., "Smalltalk-80: the language and its implementation", Addison Wesley 1983.

HOWA90 Howard, H.C., "Prototype integrated environment for AEC data", proceedings of the ASCE 6th Conference: Computing In Civil Engineering.

MAGL91 Magleby, S., Ranyak, P., Sanford, D., Jackson, D., Bean, R., "Towards a standardized application interface for product modellers: issues, developments and results", proceednings of the Design Productivity International Conf., Honolulu, HA, Feb. 3-9, 1991.

POWE89 Powell, G.H., Bhateja, R., "Data base design for computer-integrated structural engineering", Engineering with Computers 4, 1988.

SRIR90 Sriram, D., Groleau, N., "Object oriented databases for cooperative design", proceedings of the ASCE 6th Conference: Computing In Civil Engineering (1990).

TOPP89 Topping, B.H.V., Kumar, B., "Knowledge Representation and Processing for Structural Engineering Design Codes", Eng. Appli. of AI, Sep. 1989.

TURK91 Turk, Ž., "Building Model Standard as a Foundation for Computer Integrated Design", VTT workshop Computers and Building Standards, Espoo, Finland, 27.-29.5.1991.

6 TECHNIQUES FOR KNOWLEDGE-BASED STRUCTURAL ENGINEERING DESIGN

A DEVELOPMENT ENVIRONMENT FOR KNOWLEDGE BASED SYSTEMS IN ENGINEERING DESIGN

C.S. Krishnamoorthy, S. Rajeev, S. Karimulla Raja and H. Shivakumar

Department of Civil Engineering,
Indian Institute of Technology, Madras, India

Knowledge-Based Systems (KBS) have been developed to deal with problems which are ill-structured and do not present themselves readily for ordinary procedural programming. Conventional KBS development shells provide environments which allow knowledge representation in a single form, such as rules. But in practice, design activities do not always involve similar processes and each lend itself to a different knowledge representation technique. This brings in the idea of a KBS development environment where different classes of tasks are performed by different generic components. A generic component is a program which can reason and arrive at solutions based on knowledge presented to it in appropriate form. This paper presents DEKBASE, a KBS development environment made using generic components and a number of knowledge representation and data abstraction schemes.

INTRODUCTION

Knowledge-Based Systems (KBS) are finding increasing application in various aspects of engineering design, particularly in situations where heuristics play a major role. Many stages in a typical engineering design process are ill-structured and do not present themselves readily for representation on a computer. Such activities generally involve the use of designer's expertise, judgment and very often, thumb rules. Research in knowledge based methodologies has tried to find ways and means of solving such problems on a computer. As a result, a number of knowledge representation schemes and inference mechanisms have evolved for representing various types of knowledge that one comes across in most engineering design problems (Ref. 1). Design problems cannot be completely solved using only the knowledge based problem solving techniques. Conventional programs also have to be used along with the knowledge based ones to complete the design of engineering systems. Hence a problem solving environment for design of engineering systems should be able to address both knowledge based components and conventional program segments individually as well as in combination. In addition it should have a number of knowledge representation schemes and inference mechanisms to handle different kinds of knowledge that are generally encountered.

The steps that generally form the process of engineering design are conceptual design, preliminary design, analysis, detailed design and drafting. Each of these stages can further be reduced into a number of tasks and subtasks. It is observed that, some of the tasks can be grouped together on the basis of the methodology adopted to solve them. In other words, the solution processes for such tasks use a generic methodology on knowledge specific to that task. It is evident that if generalized methodologies are incorporated as components in the development tool, they can be repeatedly used for solving tasks that can be addressed by them. A number of such generic tasks and knowledge representation schemes are identified and are integrated to form a problem solving environment for engineering design. This paper presents such an environment, DEKBASE (Development Environment for Knowledge BAsed Systems in Engineering design), which has been designed and implemented to assist in developing knowledge-based systems.

The motivation behind design and development of DEKBASE was the need to provide a tool for developing Knowledge-Based Systems that would meet all requirements placed on such a tool by engineering design problems. It was felt that the existing tools either did not meet all the requirements or were too costly to make them economically viable. The architecture of DEKBASE was arrived at after detailed evaluation of existing expert system development tools (Ref. 2).

GENERIC TASKS IN ENGINEERING DESIGN

A close examination of various tasks and subtasks in a design process shows that many tasks follow similar methodologies. Hence, it would be appropriate to identify these tasks and use generic components to solve them. This section explores structural design process as a typical case of engineering design to identify tasks which are similar and can be solved using a single methodology.

The conceptual design stage, which is the first stage in structural design, essentially deals with the task of arriving at a suitable structural system for the given functional requirements. The designer uses his knowledge and experience along with standard guidelines to arrive at an appropriate system. This stage is primarily a knowledge-based one, where reasoning is the major activity. Conventional rule-based expert systems are used to arrive at a structural system for the given requirements. There may be a number of solutions which are feasible, but search techniques such as backward chaining, forward chaining etc. through rule bases stop once a solution is reached.

A design synthesis approach (Ref. 3) is ideal in this situation, where all the feasible solutions are generated. These multiple feasible solutions must be evaluated and rated in an order of preference to select the most appropriate one. It may be noted that the design solutions that are generated as a result of synthesis have to be evaluated for criteria pertaining to other domains such as services, constructibility etc. The methodology used for such evaluation is termed as critiquing.

The preliminary design of the selected system which forms the next step in the sequence, is also a predominantly knowledge-based activity. Rule-based reasoning can very well be applied to arrive at sizes and dimensions of various components of the structural system.

The next stage is analysis, which computes the response of the structure under various loading conditions. It may be noted that modelling of the structure to be analyzed is not an easy task. A number of facts should be taken into account to properly model the structure as well as to decide what type of structural analysis has to be carried out. A knowledge-based analysis consultant would help the designer at this stage, which can be implemented using a rule based approach (Ref. 4). On the other hand, the actual process of analysis is highly computational and procedural and is thus better performed by a program written using a procedural programming language. Review of analysis results follows a methodology similar to critiquing. Depending on the results of criticism one may either have to perform a reanalysis or to go ahead with the next task viz., detailed design.

Detailed design consists of the design of connections, joint details, reinforcements etc. It depends on the type and nature of structure. These can be detailed in a number of ways and each has its own merits and demerits depending on a number of factors. A methodology based on design synthesis can be adopted here to generate a number of detailed designs. For instance, in the case of steel structures, a typical member can be made up of either a single angle or a double angle back to back or starred. A knowledge-based synthesis approach would generate a number of such detailed designs based on the system decomposition and feasibility constraints. These solutions can then be criticized based on aspects like constructibility and can be graded. Finally, the most suitable detailed design may be selected based on the result of criticism.

From the above discussion of the structural design process, it can be observed that a number of design tasks follow similar methodologies such as rule base reasoning, synthesis, critiquing etc. Incorporating such generalized methodologies as components in a development tool, enables one to solve the tasks addressed by them. Eventhough the solution methodologies of different tasks are defined independent of the domain knowledge, proper flow of information has to be achieved from one task to another. For instance, some basic information defining the geometric configuration of the structure is required by all the tasks. Information like site conditions, resource availability etc. are required in a number of tasks, where critiquing for constructibility is carried out. Also the solution generated by synthesis has to be accessed by the critiquing modules. This points to the requirement of having an efficient and effective information flow from one module to another. That means, in addition to having methodologies implemented for generic tasks, different knowledge representation schemes and data abstraction schemes have to be made available. This will enable one to systematically organise the data and knowledge in a natural form depending on their type and nature. For instance, the solutions generated by synthesis can be conveniently expressed using frames and a network of frames. Predominantly heuristics-based active knowledge can be represented using rules while procedural knowledge can be represented using functions and procedures in languages like FORTRAN and C. Information such as properties of steel sections, reinforcement bars, cables etc., are better represented in a database form. All this lends credence to the fact that an ideal knowledge based problem solving system should have all the requirements as described above. DEKBASE which was designed and implemented as a problem solving environment has the above mentioned knowledge representation schemes and consists of corresponding generic methodologies (Ref. 5).

ARCHITECTURE OF DEKBASE

DEKBASE is made up of modules that provide three control strategies - the rule base reasoning mechanism, design synthesis and design criticism. In addition, a knowledge acquisition facility is provided, which uses the concept of decision tables to acquire knowledge from the expert. Knowledge representation in the form of rules, frames and functions are provided. A relational database management system forms an effective mechanism to have proper organisation and flow of physical information and data (Ref. 6). Fig.1 shows the overall organisation of DEKBASE, and each component is described in detail in the following paragraphs.

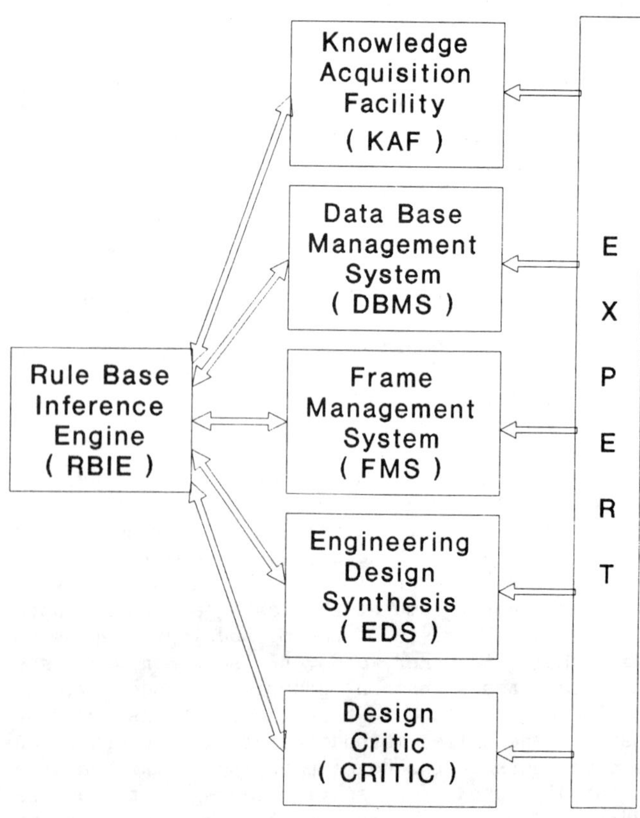

Fig. 1. Organization of DEKBASE

Rule Base Inference Engine (RBIE)

The Rule Base Inference Engine primarily acts on the knowledge represented in the form of rules and provides three reasoning mechanisms, viz., backward chaining, forward chaining and hybrid chaining. In addition, problem information can be stored in frames and data base and RBIE provides facilities for full-fledged interaction between frames, databases and rules. The two generic components -- engineering design synthesizer (EDS) and design criticism (CRITIC), are invoked from the rule base. The information flow from rulebases to these generic components is controlled by the rule base inference engine. The overall organisation of problem solving starts from a higher level rule base, which controls the sequence of tasks that are performed and has the knowledge about the overall design process. It can be said that this rule base at the highest level is at a meta level and has meta rules in it. Since the problem is controlled by this meta level rule base and rules can access frames, databases and external functions, the rule base inference engine handles the overall flow of information from one module to another in an elegant manner.

Knowledge Representation as Rules

A typical rule is of the form

IF <a set of conditions>
THEN <a set of actions>
ELSE <an alternate set of actions>

the else clause being optional. It represents a piece of knowledge which translates to "If the given set of conditions is satisfied, perform the actions given in the THEN clause, otherwise perform the actions given in the ELSE clause, if any".

The information in Rule Base Inference Engine (the problem data, decisions and context) is stored as values of a number of variables. Five classes of variables have been provided to represent five kinds of information. They are: boolean, integer, double, string and group. The group type is defined to handle variables which can have more than one string assigned to them at a time. The antecedents and consequents of various rules primarily try to modify the values of a set of variables in the system based on the values of another, possibly overlapping,set of variables. While antecedents can only check whether a variable has any value or whether it has a particular value, the consequents can perform any of the following functions:

1. Assign values to variables
2. Deassign values from variables
3. Display a piece of information to the user
4. Read from and write to data files
5. Store and access information in frames
6. Store and access information in databases
7. Invoke external functions or programs
8. Invoke another rulebase

RBIE also allows confidence levels to be associated with the values of variables. A confidence level of 0 indicates the least confidence while a level of 100 indicates complete confidence. When a variable is assigned a value based on the values of a set of variables in the rule base, its confidence level is obtained by suitably combining the confidence levels of the set. The knowledge can declare some variables as 'overridable', in which case, RBIE allows the user to override the values assigned to them by the inference mechanism. The knowledge can also declare some variables to be 'volunteerable', in which case, if RBIE is unable to deduce their values from the inference process, it requests the user to supply them.

DEKBASE permits the organisation of the knowledge into a number of rulebases and can invoke one rulebase from another. The variables which are common to more than one rulebase can be declared as global and thus their values can be shared by respective rule lists. Access to frames and databases is provided through a Frame Management System (FMS) and a Data Base Management System (DBMS) respectively. All the frame functions and data base functions supported by the respective systems can be invoked from rules. The details of these functions are given in following sections. Parameters can be passed to and from external programs using the file read/write facility. Eventhough this approach makes the inference process slow, this seems to be the only alternative in a limited memory environment like PCs.

In addition to rules, the knowledge in the rule base also contain information about the possible goal states of the solution. During inference, when the RBIE recognises a valid goal state, it signals the termination of the inference process.

Knowledge Representation as Frames

DEKBASE allows the storage of static information in frames. A frame describes an object (such as beam) by means of a number of attributes (such as depth and breadth) whose values identify the object. The frames in rule base inference engine are stored in a frame base which is managed by a Frame Management System (FMS) (Ref. 7). FMS permits a variety of operations on the frame base as given below:

1. Creation of new frames
2. Deleting existing frames
3. Addition of attributes in existing frames
4. Deletion of attributes from existing frames
5. Assigning values to attributes of frames
6. Removing values of attributes in frames
7. Modifying values of attributes in frames
8. Retrieving value of an attribute in a frame

The values of attributes in frames can be of five types, viz., integer, long integer, single precision floating point, double precision floating point and character strings. FMS also allows the definition of unidirectional links between frames. These links are used to form a hierarchical or network organisation of the frames in the frame base. FMS can assign the values of attributes that do not exist for a frame by inheriting the values from frames to which the current frame is linked. Thus repetition of information common to a class of frames is avoided. In the case of a hierarchical organisation of frames, the path of traversal for inheritance is well defined. But in the case of network organisation, where many-to-many relationships between frames exist, it is required to specify the path of traversal for inheriting attribute values. The following additional functions are used to define links between frames and to specify the kind of links that should be followed for inheriting values for different attributes.

9. Defining links between frames
10. Defining the links to be followed for an attribute.

DEKBASE provides complete interaction between the frame management system and the rule interpreter thus allowing the knowledge based system to access the complete frame base through rules.

Data Base Management System (DBMS)

As described earlier, there can be a number of data items in an engineering design process, which have information on standard items like rolled steel sections, reinforcement bars, cables etc. Storing and accessing such type of information can efficiently be handled following the concepts of relational database management system. DEKBASE provides an interface to a database which is organised as a collection of relations. The database management system in a stand alone mode allows operations such as create relations, insert tuples, delete tuples, retrieve tuples, project attributes of relations, select tuples from relations etc. Since the inference mechanism is expected only to get one attribute value at a time from database and set one attribute value, the database interface allows only two operations from the rulebase. They are:

1. Get attribute of a tuple satisfying a given condition from a relation
2. Set attribute of a tuple satisfying a given condition in a relation

Since all the database management functions are implemented as functions callable from any C program, a complete interaction with database can be achieved indirectly by calling external C programs with database functions called from it. The interaction of rule base inference engine with frames and databases are schematically shown in Fig. 2.

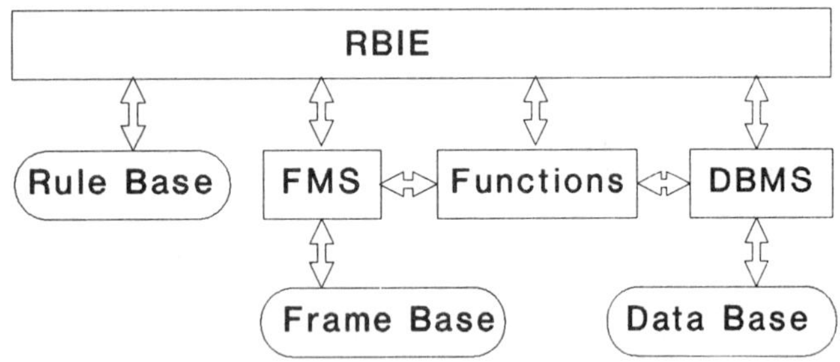

Fig. 2. Interaction Between RBIE & Knowledge Sources

External Programs

Numerical computations during the inference process cannot be avoided. RBIE allows writing simple expressions in the rules itself. But many times large amount of calculations may have to be carried out using values of variables obtained as a result of inference. Also the results of these computations have to be used for further inference. In such cases external programs can be called to do the required computations. DEKBASE permits values of variables to be passed to the external programs through interface files. The file read/write functions provided by the RBIE are used to communicate between the inference engine and the external programs.

Inference Mechanisms

Rule Base Inference Engine provides three inference mechanisms, any of which can be invoked to act on the data and the knowledge to arrive at a solution. They are : backward chaining, forward chaining and hybrid chaining mechanisms.

In backward chaining, the control flows from the goal states to the initial states. Each goal state is considered in turn and the requisites for achieving the goal states are studied. While some of these requisites may be tested for directly, the others may have further sub-requisites which need to be tested. This may go to any depth and a goal state is said to have been reached when all its prerequisites and so on are tested to be true. The chaining mechanism is said to have failed when it cannot achieve any of the given goal states.

In forward chaining the control flows from the initial states to the goal states. Each initial state is considered in turn and its co-acting requisites (which when combined with it result in achieving a subgoal) required to achieve a subgoal are tested. If they fail, inference starts at another initial state. If the requisites are tested to be true, the proved subgoal is treated as the current initial state. Its co-acting requisites are identified and the process goes on recursively until a goal state is achieved or until all initial states have been tried as starting points.

The hybrid chaining mechanism is a combination of the backward and forward chaining mechanisms. The system starts in forward chaining and when a co-acting requisite of an initial state is not defined, instead of looking for an alternate starting point, it tries to prove the requisite by chaining backward. The process terminates when a goal state is reached or the main forward chaining fails. It was found that when the knowledge net is very large and there are many goal states and initial states, hybrid chaining mechanism seems logically to be the appropriate one.

Engineering Design Synthesizer (EDS)

In the earlier discussion on identifying the generic tasks in engineering design process, it was seen that design synthesis methodology can be adopted in a number of tasks. The conventional reasoning mechanisms terminate once a goal state is reached. The synthesizer generates as many feasible solutions as possible satisfying a set of constraints provided. The structure of the knowledge base as well as the method of search in synthesizer differ very much from the rule based approach. The knowledge for this module is in the form of a decomposition tree (a hierarchy of the components of the system) and a set of constraints. The synthesizer carries out a constraint-directed depth-first-search on the decomposition tree. The problem data is expressed as preconditions. Each component in the decomposition may be characterized by a set of attributes. The solution from EDS is as a hierarchy of frames, each frame representing a component complete with the value for its attributes (Ref. 8). In an overall problem solving environment the EDS is invoked from the rulebase and the problem data (as preconditions) is generally obtained from a rule based inference process. The file write functions of the RBIE are used to create a preconditions file before invoking EDS. Thus, tasks requiring synthesis methodology can use EDS by providing the respective knowledge as decomposition trees and constraints. Fig.3 shows the interaction between RBIE and EDS.

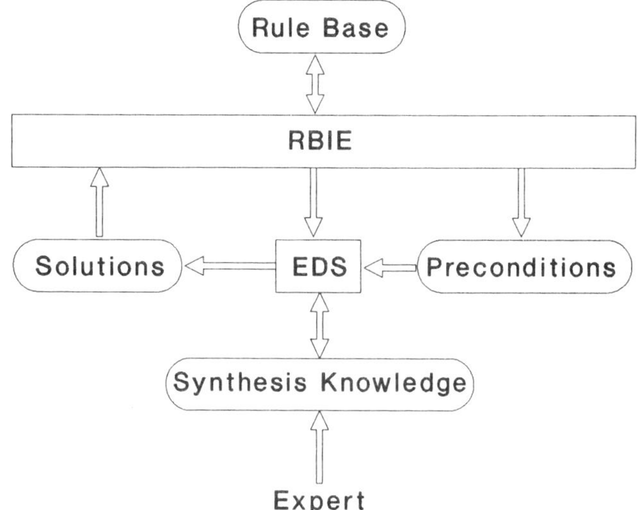

Fig. 3. Interaction between RBIE and EDS

Engineering Design Critiquing System (CRITIC)

The critiquing module of DEKBASE is designed and implemented to carry out either of the following two tasks. First, it can evaluate a given set of solutions using the knowledge contained in the critic knowledge base and grade the solutions in an order of preference. If the solutions are generated using the knowledge in a specific domain, they are generally criticized to test whether the solutions satisfy the design criteria of other domains. For instance, building plans generated by an architect for given functional requirements can be criticized for requirements from domains like structural design, ventilation design, plumbing design, electrical design etc. Based on each criterion, the solutions can be assigned a rating value which reflects the relative goodness of the solution. Secondly, a given solution can be evaluated to specify how good the solution is from a particular domain point of view.

Similar to the design synthesizer EDS, CRITIC knowledge base consists of a decomposition tree where the components of the engineering system to be criticized are organised as a hierarchy. Rating conditions are assigned to components, subsystems or group of components or subsystems. These conditions consider the individual attributes as well as inter-attribute interaction (in a single component or across two or more components) to rate the solution presented to the CRITIC. It has been noted that design criticism is a generic task in engineering design process. Hence it has to be invoked a number of times during the design process. This is done similar to the manner in which synthesizer is invoked. The meta level rule base which contains the knowledge about the problem reduction and design process, invokes the CRITIC as an external program as and when required. The interaction between rulebase inference engine and CRITIC is shown in Fig. 4. The solutions to be criticized should have already been generated in the form of hierarchy of frames, before it is invoked.

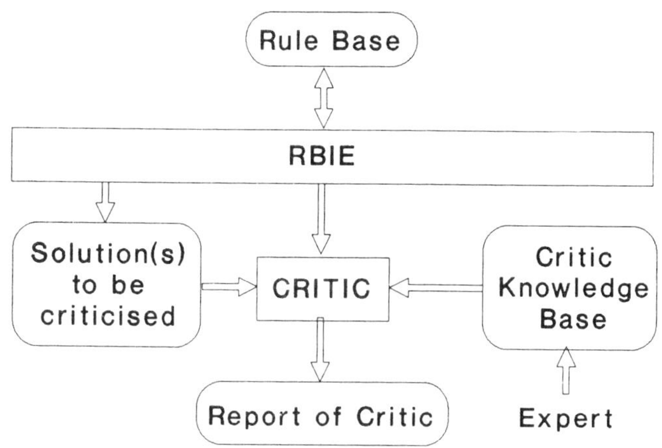

Fig. 4. Interaction between RBIE and CRITIC

Knowledge Acquisition Facility (KAF)

It has been reported in literature that decision tables are a good medium for acquiring knowledge from experts (Ref. 9). The rules written sequentially on a sheet of paper may not give an idea about correctness of the knowledge to the expert during the consultation process. As a first step in knowledge acquisition, the knowledge engineer can put down the knowledge he/she received from the expert in a decision table form. This step helps in determining the inconsistencies and redundancies that may have entered into the knowledge base during the acquisition process due to inherent problems of communication between the expert and the knowledge engineer. In addition, presentation of knowledge in decision tables makes more sense for the expert, since it provides a more comprehensive picture of the knowledge than a rule list. The knowledge acquisition facility provided in DEKBASE is based on decision tables. It is a stand alone module, which is used to create knowledge base consisting of rules. All the facilities provided in rule base such as hierarchical organisation of rule lists, global variables to share information among different rule lists, handling of confidence factors etc. are incorporated in the KAF. It acts mainly as an interface between the expert/knowledge engineer and the rule base.

A typical decision table consists of a condition stub and an action stub on the left hand side and condition and action entries on right hand side (Ref. 9). A decision table editor displays one decision table at a time and entries for conditions and actions are made in it. A decision table processor converts the knowledge represented in the decision table into rules conforming to the syntax of the rulebase compiler of DEKBASE. There is absolutely no difference between the method of knowledge representation using decision tables and using rules. The only difference is in the presentation of the knowledge to the user or the expert. It provides a clearer picture of a larger portion of the knowledge base than would be obtained by viewing a single rule or a set of rules placed one after another. It is easier for the knowledge engineer to use decision tables than the strict syntax which goes into the writing of a rule base. If the knowledge has to be modified to reflect for instance, the difference in design and construction practices at different place it can be done in much easier manner using the decision tables. That means the KAF module of DEKBASE provides a facility to adapt the knowledge for changed environments. Fig. 5 shows the organisation of knowledge acquisition facility of DEKBASE.

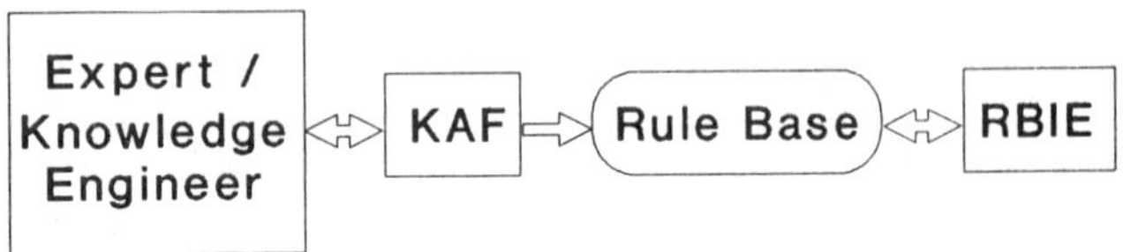

Fig. 5. Knowledge Aquisition Facility of DEKBASE

DEKBASE is formed by integration of these individual components. The strategy used for implementation of the individual components and their integration is described in the next section.

DEKBASE IMPLEMENTATION

A schematic representation of working of DEKBASE is presented in Fig. 6. The components where the knowledge is stored are rule base, frame base, data base, functions and knowledge bases of EDS and CRITIC. The three control strategy components are inference engine, synthesis processor and critic processor. The implementation is designed keeping in mind the fact that one should be able to use the components of DEKBASE individually or in combination. A brief description of the implementation aspects of each of these components is given in the following sections.

The rulebase and the knowledge bases of EDS and CRITIC are created as text files and are then compiled. These compiled knowledge bases are used by the respective interpreters to carry out inference and search. Frame base and database are created using either functions called from C programs or invoking external C functions from rule base.

Rule Base Compiler (RBC)

Rule Base Compiler consists of a lexical analyser which identifies the key words, a parser which identifies the logical sequences of elements and a compilation module, which compiles the information into a compact form. The compiled information is written into a file. The compiler converts the rules into a series of predefined operations called opcodes. It collects and organises the information received from the rule base source code into a form that is then used by the interpreter during the reasoning process. Since DEKBASE supports the organisation of rulebases in a hierarchical form, i.e., calling other rulebases from the rule bases at higher levels, and also allows sharing of variables across rulebases, it is necessary to take care of this aspect during compiling. Each independent knowledge module is created in a separate file which is compiled independently. All the compiled rule bases in a domain are linked together using a linker. The linker writes a compiled rule base file.

Inference Engine

The inference engine of DEKBASE consists of a reading module that reads information from the

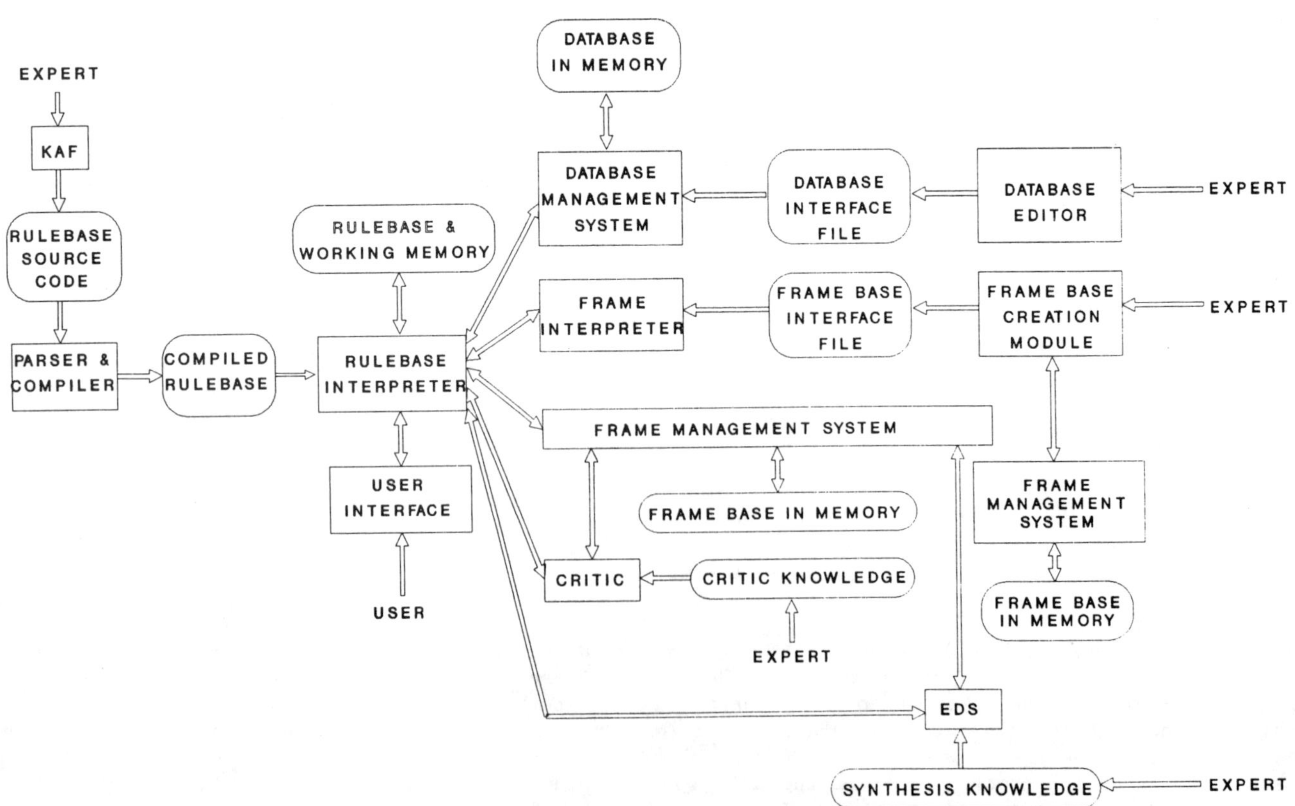

Fig. 6. The schematic representation of working of DEKBASE

compiled rule base file, a control module that monitors the overall flow of control, a chaining module that monitors and executes the inference process and a code interpreter that executes the code generated by the rule base compiler. The chaining module supports backward, forward and hybrid chaining search strategies. The detailed architecture of the DEKBASE inference engine is shown in Fig. 7.

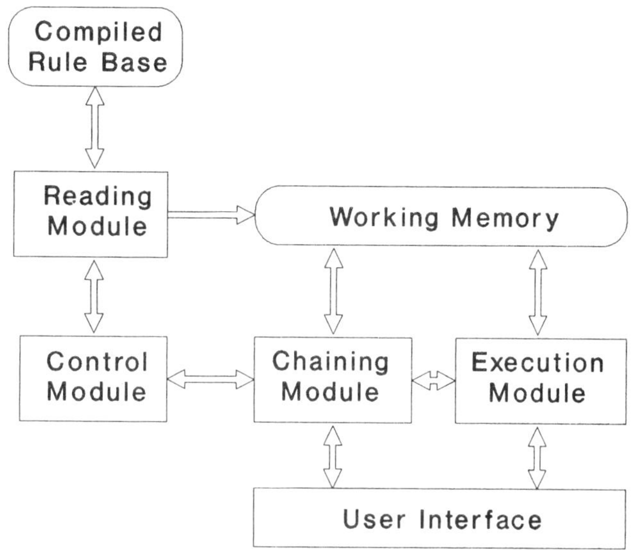

Fig. 7. Architecture of DEKBASE Inference Engine

Frame Management System (FMS)

The Frame Management System (FMS) is a collection of functions callable from any C program as well as from rulebase, to create frames and manipulate the information contained in the frame. The FMS functions are used for creation, deletion and manipulation of various kinds of frames, attributes, values and procedures. A frame base creation module is also provided, which permits one to write the frame functions following a specified syntax, and a compiler converts them into corresponding frame base. This compiler creates a frame interface file which is later used by the frame interpreter during reasoning. To support frame management calls within rules the FMS is linked with the rule base interpreter.

Frame Interpreter

The Frame Interpreter contains a procedure which in turn uses the frame management system functions to execute the code generated for frames. During inference, when the rule interpreter comes across the frame opcodes in the interface file, it invokes the frame interpreter which then invokes the FMS functions to execute the code (Ref. 7).

Implementation of database management system is also done in a manner similar to that of frame management. The details of data structures and other micro level aspects of implementation are not provided in the present paper. The data in the DBMS is stored in files while the control information is stored in memory to permit quick access of information.

Synthesis Module

This module is consists of two parts. The first module, the EDS compiler, processes the decomposition, the preconditions and the constraints provided in a text file and stores them in the memory. It checks for syntactic as well as semantic errors. The attribute evaluation expressions for the various systems are converted into a series of opcodes and operands. After the entire knowledge and the problem data have been loaded into the memory in suitable data structures, the synthesis commences. The synthesis process traverses down the decomposition tree, evaluating the values for the various attributes of the systems. When one set of feasible values for all relevant attributes is obtained, this information is compiled into a hierarchy of frames, with each frame representing a system in the original decomposition tree. For this purpose, the EDS uses the Frame Management System discussed earlier. The frame base is then dumped onto a file. The EDS proceeds to identify the remaining solutions, and each is successively dumped into files as frame bases.

Critiquing Module

This also follows a two stage process concept. The decomposition tree consisting of the evaluation criteria is first compiled into suitable data structures in the memory after a thorough check for syntactic and semantic errors. The critiquing module then reads a solution file written by EDS and constructs a solution frame base in the memory. It then proceeds to evaluate each system successively, based on the relevant criteria. The process proceeds along the tree in a depth-first manner, evaluating the leftmost leaf-node first, and then proceeding upwards. To evaluate a system, all its sub-systems must already be evaluated. Each system is assigned a rating based on how well it satisfies the criteria. A higher level system combines the ratings of its sub-systems in three different ways -- the maximum of the ratings, the minimum of the ratings and the weighted average. Each of these ratings is carried up the tree, until the entire solution has been rated. Thus, each solution is given three ratings evaluated in the three ways. The controlling knowledge uses these ratings for the various solutions in an appropriate manner to choose its 'best' solution.

The implementation of various modules are designed in such a way that, each component can be used either individually or in combination with other components. The implementation aspects are designed with this view in mind. This enables DEKBASE to be suitable even for those applications, where not all components are required.

ILLUSTRATION

In the earlier discussion on engineering design process, it was pointed out that there exist a number of design tasks which have similar methodology but work with different knowledge and data. To illustrate the use of DEKBASE for solving engineering design problems, a typical application in the domain of bridges is taken and an analysis is made on the nature and type of knowledge that is used in various design tasks and how it can be organised and used in the knowledge-based problem solving environment DEKBASE. The major design tasks in a bridge design problem are:

1. Selection of site
2. Selection of structural system
3. Sizing of components
4. Analysis
5. Detailed design
6. Drafting

Eventhough the tasks such as construction and maintenance are also part of bridge engineering, knowledge relating to them is not included in the present study. The nature and type of knowledge in each of the above design tasks and its organisation for an integrated problem solving approach are discussed below:

Site Selection

The objective of this task is to choose an appropriate site from all the alternate sites considered for bridge construction after reconnaissance and hydraulic and geotechnical investigations. There are a number of factors that have to be taken into account such as characteristics of alternate sites, economic benefits of constructing the proposed bridge at alternate locations, existing alignments of roadways and highways etc. This is typically a classification type of problem. The knowledge that goes into the selection process is primarily of the heuristic kind based on the experience of the design engineers. A rule base inference mechanism can be used to select the most appropriate site, where the experience of the design engineers is represented as rules. Another elegant approach is to criticize the alternate sites for considerations such as socio-economic, hydraulic, geotechnical, realignment of highway/railway etc., and grade them in an order of preference. The CRITIC module of DEKBASE provides the facility to represent the knowledge required for critical evaluation of alternate sites from different considerations. The critiquing methodology used in DEKBASE also assigns a rating value to each of the solutions, based on the goodness level of every attribute.

Conceptual Design

The next task in the sequence is providing the most appropriate structural system for the bridge. At the site selected, there can be a number of different structural systems possible. The major factors that go into the selection of structural systems are traffic requirements, nature of loading, environmental conditions, availability of equipment and materials for construction, nature of subsoil and hydraulic characteristics of river. In addition to this, the chosen structural system should meet requirements from constructibility, economy and aesthetics. The objective of this design task is to arrive at a strong, safe and economic structural system from the load transfer point of view. In addition to some general guidelines, the knowledge based on the experience of design engineers plays a major role in this design task. Elicitation and coding of this knowledge in the form of rules is very difficult, because the knowledge contains both procedural as well as declarative components. Frames can be used to represent the declarative portion of the knowledge. Eventhough DEKBASE supports interaction between rules and frames, it is not a very natural way of representing this type of knowledge. A comprehensive study on the nature of this knowledge shows that it has to undergo a large amount of transformation to make it possible to be represented in the form of rules and frames. An experienced design engineer will be able to enumerate all possible structural systems and a set of constraints. The constraints are essentially elimination constraints which specify the incompatible combinations of attributes of structural system, subsystems and components with aspects influencing their selection. The synthesizer module of DEKBASE allows one to specify this type of knowledge in the form of enumerating all the possible structural systems and constraints. In addition to the synthesis knowledge, only a set of problem specific preconditions needs to be given to test the feasibility of different combinations of structural subsystems and components. Synthesis process eliminates all the infeasible structural systems and stores the feasible ones separately. The synthesis module EDS of DEKBASE does exactly this. One major advantage of going in for synthesis is that the representation of synthesis knowledge is done in DEKBASE in a natural manner. Hence the loss of useful information during the process of transforming the expert knowledge to machine readable form is minimized.

Out of all the solutions generated at the synthesis stage, the best one has to be selected. The knowledge that has gone into the process of generating alternate solutions is essentially that pertaining to the structural aspects. Now these solutions are rated by evaluating them from the point of view of constructibility, economy and aesthetics. The critiquing module of DEKBASE allows one to specify the knowledge required for evaluation. Depending on the importance of each subsystem or component, the attribute values being criticized are assigned rating values. These rating values are combined at every node in the decomposition tree containing the critiquing knowledge and carried over up to the root. Thus different rating values are obtained for different solutions. Better the solution, higher is the rating value assigned. The final selection of the structural system is made based on these rating values.

The next task is the sizing of components. A rule based inference approach is generally adopted to arrive at preliminary sizes of components of the structural system. The depth of deck slab, positions and dimensions of the main and the cross girders, dimensions and cross sectional areas of the components of load transferring systems such as trusses, arches, box girders, cables etc. are fixed at this stage. Depending on the structural system selected, the knowledge that is used differs. In most cases a rule based inference approach can be used effectively. Knowledge for sizing subsystems can be organised in different rule lists for better and efficient organisation and use of knowledge. It was noted that in some cases, the component sizing and conceptual design tasks cannot be separated. That means when alternate solutions are generated using synthesis, the dimensions and sizes of components also should be fixed. Then, the critiquing is done on the structural systems with the component sizes fixed and the best one is selected.

Analysis and Detailed Design

Analysis and detailed design is the next task. It consists of computing responses of the selected structural system under actual loading conditions. It is highly computational and procedural. Standard analysis programs can be used to carry out analysis. Depending on the structural system selected, the analysis and detailed design methods vary. For instance, if load transferring subsystem of the deck is made up of framed structures like trusses and beams, the modelling will be simpler. On the other hand, if it is a prestressed concrete box girder, then analysis modelling becomes difficult. If a finite element analysis method is adopted, then the type of elements, properties of elements like aspect ratio, modelling of reinforcement, prestressing cables, anchorages etc., play a major role in the accuracy of the computation of responses. A designer would look for advice from analysis experts on these issues. A knowledge based analysis consultant can be implemented using rule based inference strategy. In many structural design problems, analysis and detailed design stage go together. The reinforcement design, design and placement of prestressing cables, parameters of chosen sections etc. affect the analysis. Hence analysis and detailed design stages are carried out together. The primary objective of these stages is to check whether the design satisfies all the codal provisions or not. Since the codal provisions change from country to country and are also revised periodically, the module that checks for conformance with codes of practice should be implemented in a standard independent manner. A number of studies conducted by many researchers on standards processing show that a rule based inference approach efficiently handles this task. As the codal provisions are arrived at after detailed deliberations by leading professionals, and the manner in which they are presented in the codes makes rule based inference the ideal choice for organising and representing this knowledge. The provisions given in codes for different types of components being separate, it is appropriate to organise the knowledge as independent rule lists for different components which are controlled from a higher level rule base. The facility of global variables in DEKBASE makes it possible to share information that is common to more than one rulelist.

Drafting

Design development or drafting is the next task, which concerns with the preparation of construction drawings. It may be noted that the flow of information from the previous stages to this stage has to be organised very carefully, so that there is no loss or distortion of information. Drafting is a routine process, where the design solution represented in the form of information stored in frames, databases and interface files is transformed into drawings.

The results of different design tasks are represented in frames, database and interface files. Database essentially stores the shared information in addition to details on cross section properties of standard and built-up steel sections, reinforcement bars, prestressing cables etc. Frames store the design solutions in addition to the declarative component of knowledge.

The above discussion on a problem in the domain of bridges essentially illustrates the organisation and use of knowledge in different design tasks. It also illustrates how generic methodology can be used to effectively solve tasks which are similar. DEKBASE provides all the necessary knowledge representation schemes and generic components to solve different tasks of design problems. The organisation of the knowledge for this problem as implemented using DEKBASE is shown in Fig. 8.

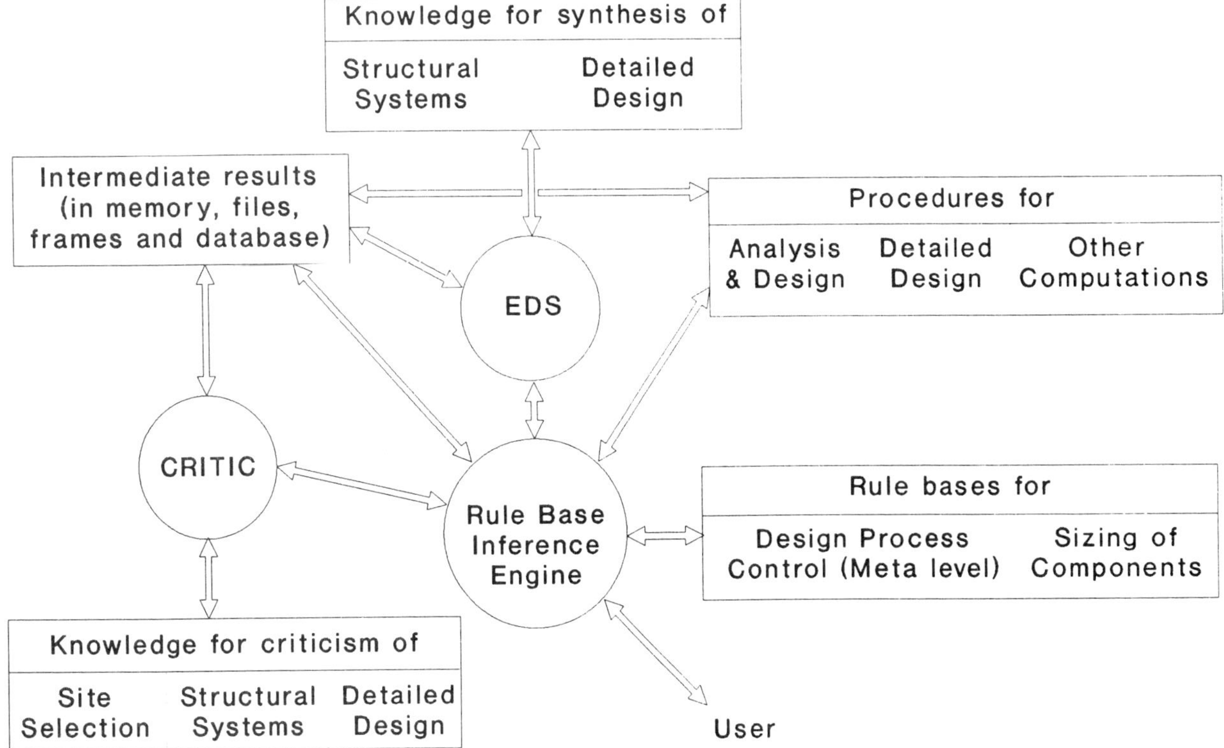

Fig. 8. Organization of Knowledge for Bridge Design using DEKBASE

SUMMARY AND CONCLUSIONS

A knowledge based problem solving concept that makes use of the generic methodologies is presented in this paper. The different knowledge representation and data abstraction schemes required for solving engineering design problems are identified. The methodology used for solving various tasks involved in design process of engineering systems are analysed and similar ones are identified. Based on the study, three control strategies are suggested, which can be used as generic strategies, acting on different knowledge and data at different stages of design process. The generic components that have been identified are: rule base inference, design synthesis and design critic.

Design and implementation aspects of a knowledge based problem solving environment DEKBASE is discussed. DEKBASE has been successfully implemented on a range of machines and is being used for developing integrated prototype knowledge based systems in the area of structural engineering. A detailed analysis of different issues pertaining to the organisation and use of knowledge in design problems is presented for a typical problem in the domain of bridges. It illustrates the use of DEKBASE as a tool for knowledge based design of bridges. Based on the investigation on the aspects of design knowledge and from the experience of design, implementation and use of DEKBASE, the following conclusions have been drawn.

1. Generic methodologies can be conveniently used for solving similar tasks in design problems. The most commonly used generic tasks are rule-based inference, synthesis and critiquing.

2. Different schemes for knowledge representation should be available in a development tool, so that loss of knowledge can be made to minimum during the process of transforming it from the stage of knowledge elicitation to the stage of converting it into machine readable form. The five important knowledge representation schemes identified are rules, frames, functions, decomposition tree and constraints.

3. Different data abstraction schemes improve the effectiveness of flow of information from one module to another. A relational database management system improves the efficiency of information handling.

4. There should be full fledged interaction between various knowledge representation schemes, data abstraction schemes and inference mechanisms.

5. Development environment for knowledge based problem solving with different knowledge representation and abstraction schemes and control strategies can be efficiently used for organising, representing and using knowledge in a near natural manner.

REFERENCES

1. Fenves,S.J., What is an Expert System, in Expert Systems in Civil Engineering, (ed. C.N.Kostem and M.L.Maher), ASCE, Seattle, 1-6, 1986.

2. Suresh, S., Rajeev,S., Krishnamoorthy,C.S., and Fenves,S.J.,State-of-the-Art Report on Knowledge-Based Expert System Development Shells, Research Report No.CE-90-01, Dept. of Civil Engrg., Indian Institute of Technology, Madras, India, 1990.

3. Maher,M.L., EDESYN: An Engineering Design Synthesis Environment, Tech. Rep., EDRC 12-31-89, Carnegie Mellon Univ., USA, 1989.

4. Bennet,J., Creary,L., Englemore,R. and Melosh, R., SACON: A Knowledge-Based Consultant for Structural Analysis, Tech. Rep., STAN-CS-78-699, Stanford Univ., USA, 1978.

5. Rajeev,S., Fenves,S.J. and Krishnamoorthy,C.S., Architecture and Generic Components of Knowledge-Based Expert System for Civil Engineering Design, Research Report No.R-90-186, Dept. of Civil Engrg., Carnegie Mellon University, USA and Dept. of Civil Engrg., Indian Institute of Technology, Madras, India, 1990.

6. Shiva Kumar,H., Suresh,S., Krishnamoorthy,C.S. and Rajeev,S., Development Environment for Knowledge-Based Systems in Engineering Design (DEKBASE) - Users Manual and Implementation Manual, Research Reports No. CE-91-04 and CE-91-05, Dept. of Civil Engrg., Indian Institute of Technology, Madras, India, 1991.

7. Karimulla Raja,S., Rajeev,S. and Krishnamoorthy, C.S., Frame Management System and Interaction with Rules, Research Report No.CE-91-03, Dept. of Civil Engrg., Indian Institute of Technology, Madras, India, 1991.

8. Srinivasa Rao,C., Krishnamoorthy,C.S. and Rajeev,S., Engineering Design Synthesis: Users Manual and Implementation Manual, Research Report No.CE-91-06 and CE-91-07, Dept. of Civil Engrg., Indian Institute of Technology, Madras, India, 1991.

9. Ibarra-Anaya,E. and Fenves,S.J., Eval-KAF: A Knowledge Acquisition Facility for C and OPS83 Programs, Tech. Rep., Dept. of Civil Engrg., Carnegie Mellon Univ., Jan 1990.

ACKNOWLEDGEMENT

The work reported in this paper is based on the Indo-US research project on "KBES for Civil Engineering Design" under NSF Grant nos. INT-8819520 and INT-891386 currently in progress at Indian Institute of Technology, Madras in collaboration with Carnegie Mellon University, Pittsburgh, USA. The authors are grateful to Prof.S.J.Fenves for his valuable advice and suggestions.

DESIGN CRITICISM: A GENERIC COMPONENT IN KNOWLEDGE BASED SYSTEMS FOR ENGINEERING DESIGN

S. Rajeev, S. Suresh and C.S. Krishnamoorthy
Department of Civil Engineering,
Indian Institute of Technology, Madras, India

In this paper, engineering design is viewed as a multi-domain problem in which each domain has a distinct set of requirements. An integrated solution to such a problem therefore involves the simultaneous satisfaction of the requirements of two or more separate knowledge sources. In some cases, the requirements of one source may be in conflict with those of another. This paper presents a methodology for solving such problems. It highlights the role of design criticism which is central to this problem solving methodology. The paper also attempts to establish the generic nature of the criticism task. An architecture for a generic tool for design criticism is proposed. The paper also shows how this tool can be used as a generic evaluator.

INTRODUCTION

The desire to use computers to aid in solving problems in engineering has resulted in the investigation into various problem solving techniques. The limitations of the conventional algorithm based approach and their consequent inability to address problems in design and decision making have already been brought out by researchers. Subsequently, expert system techniques evolved as a result of the need to involve computers in these hitherto unaddressed phases of the engineering problem solving process.

There are several instances in the engineering domain, where the expertise required to solve a design problem may rest with two or more sources. The design of reinforced concrete multistorey buildings is a case in point. An integrated design solution should satisfy not just the structural requirements, but should also be acceptable from the point view of functional, service and project execution considerations. The expertise in these various areas do not necessarily come from the same source. Such an integrated solution, therefore, requires the simultaneous satisfaction of the requirements of more than one domain.

The process of arriving at a design solution can be viewed in terms of generating a description of the system that is being designed, that satisfies the requirements set by the domain. When more than one domain influences the final solution, the system description should correspondingly satisfy the requirements of each of these domains. It is natural to expect that the requirements of one domain may be in conflict with those of another. A final acceptable solution can be said to have been obtained only after each such conflict has been resolved. One way to achieve this is to resolve each and every conflict as and when it arises. In this case, the knowledge sources associated with the competing domains are invoked as and when a conflict arises and the processing proceeds only after the conflict has been resolved. This is the kind of conflict resolution done in the blackboard model.

Another approach is to involve the knowledge sources in a sequential manner. In this approach, one of the knowledge sources proposes a solution to start with. The other knowledge sources affecting the solution, then sequentially evaluate it from the point of view of their requirements and suggest modifications, if any, to the solution. Once the modifications have been incorporated to the satisfaction of all competing knowledge sources, an acceptable solution can be said to have been evolved. This paper presents the salient features of this problem solving model.

THE SYNTHESIZE-CRITICIZE-MODIFY MODEL

From the preceding section, it is clear that this problem solving model consists of essentially three stages. The first stage involves synthesizing a solution based on the needs of one of the domains. Synthesizing a solution involves identifying all the attributes required to completely describe the system being synthesized and then assigning a value to each of these attributes. Once this is done, the solution is separately subjected to criticism by each of the other knowledge sources. Criticism involves the examination of the description of the synthesized system from the point of view of the requirements of the domain that is criticizing it. The result of this criticism is a report in which the critiquing domain presents its opinion of the description. In this report, those aspects of the description that are unacceptable are listed out with reasons for their being unacceptable. Besides, while certain aspects of the synthesized solution may not be entirely unacceptable, they may contribute in some way to inefficiency in performance as far as the critiquing domain is concerned. Such aspects are also listed out along with suggestions for possible improvement. The third and final stage is the incorporation of the modifications to the synthesized solution to make it acceptable from all points of view. The three stages in this process, which shall be termed from now on as the SCM process, are dealt with in greater detail in the succeeding sections.

SYNTHESIS

In the engineering design process, a design goal can be resolved into a set of subgoals. Each subgoal in turn may consist of a further set of subgoals and so on until eventually terminating in making a functional or physical selection. Such a representation is termed as decomposition of the knowledge of the domain. Decomposition, is thus a type of knowledge representation. The entire set of goals and subgoals thus forms a goal hierarchy. A solution to the problem would result if one alternative solution is generated for each node in the hierarchy. This process is termed design synthesis. Synthesis thus involves selecting components to form a system that meets the design objectives while satisfying all the constraints governing the selection. Each object in the decomposition is called a system. Systems describe objects and their relation with other objects. The systems are arranged hierarchically with the first system defining the whole artifact being synthesized. Fig.1 shows a portion of a decomposition tree for a building system. The artifact to be designed is the building system. The vertical load resisting system and the horizontal load resisting system are subsystems. Similarly, the horizontal framing system, which is a subsystem of the vertical load resisting system can be one of flat slab or concrete joists or a beam and slab system or a bearing wall and slab system. The beam and slab system is in turn made of the beam system and slab system. The attributes depth, breadth, support conditions and the span-to-depth ratio define the beam. The decomposition also contains knowledge that helps modify the knowledge based on the context. This is required either when a reordering of the attributes of a system is called for or when several instances of a particular system are to be created. Constraints are used to specify knowledge that describes incompatible assemblage of certain systems. The synthesizer processes this knowledge for a given context to synthesize a solution for the given design problem(Ref. 2,4). In this case, the building system is the artifact being designed. This consists of smaller parts which in turn consists of still smaller parts. A solution for this synthesis problem consists of assigning a definite value or description to all the parts that are in turn part of others. It may not be noted that a part may either be a physical object like a beam or slab or an abstract object like the ratio of span to depth of the beam. The span to depth ratio is an important criterion by which the performance of the beam system can be gauged and the advantage in directly synthesizing an attribute like the span to depth ratio is that, in so doing, the knowledge required to criticize it can be stated easily. It must be noted that the artifact is synthesized based on the requirements of one of the domains and that the decomposition, consequently, encapsulates the knowledge of only one of the sources that influence the design.

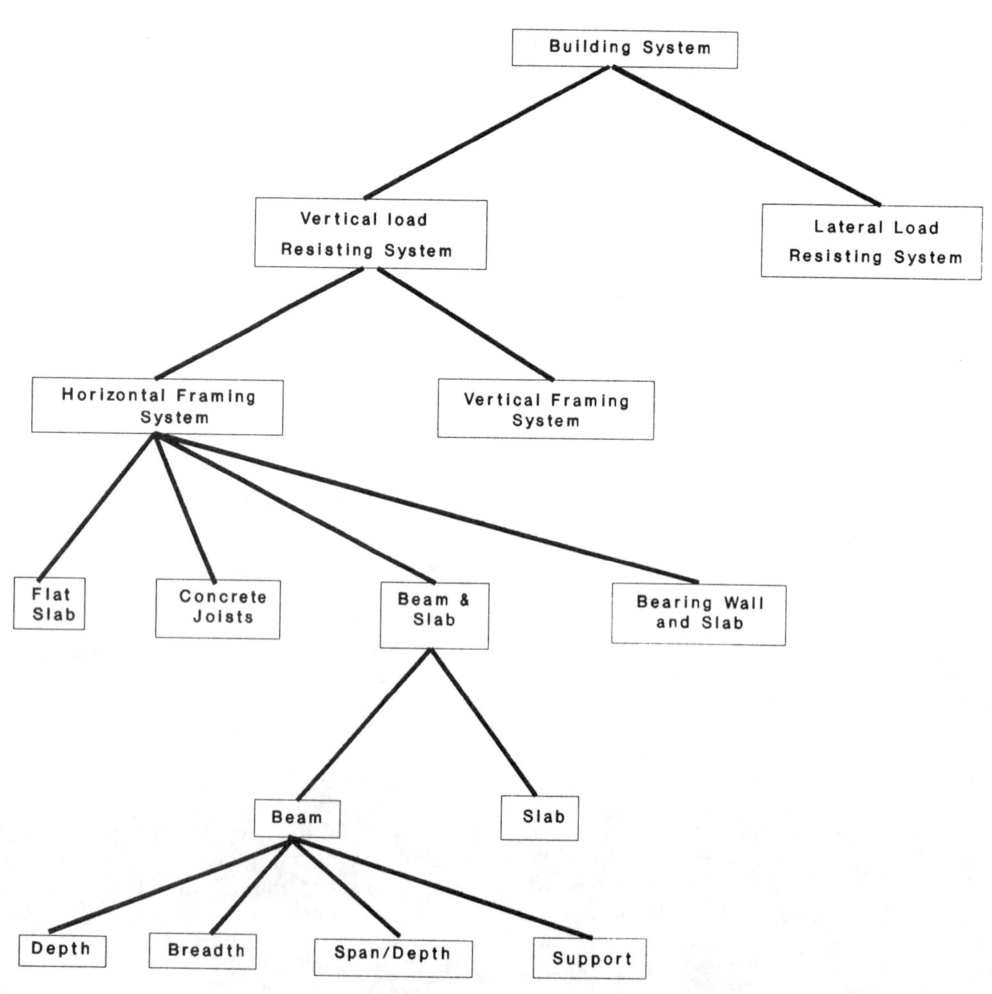

Fig. 1 A portion of a Decomposition Tree
for a Building System

CRITIC

The synthesized solution is separately subjected to criticism by each of the influencing knowledge sources. The input to CRITIC is the solution generated by the synthesizer. During synthesis, a definite value or description is assigned to each attribute of the various systems that together constitute the artifact being designed. The critiquing process involves evaluation of each of these attributes, their interaction with other such attributes within the system and the interaction between the various systems that define the artifact.

The knowledge required to criticize a solution is represented in the form of constraints that direct CRITIC to assign a certain rating to an evaluated attribute or group of attributes that satisfy that set of constraints. For the purpose of evaluation, a generalised attribute group is defined. In its simplest form, an attribute group consists of just one attribute while in its most general sense, it could represent the interaction between attributes from across systems. Corresponding to each such attribute group, the CRITIC knowledge base has a set of constraints that assign a suitable rating (RF), to that group depending on the value of the attributes that constitute the group. In addition to this quantitative estimate, a remark is also posted by CRITIC in text form that represents an English interpretation of the quantitative estimate. The relative importance of the attribute group in the performance of the solution is stated through weighting factors. By default, an attribute group has a weighting factor of one. If a particular group is relatively less important than the other groups its weighting factor is correspondingly less than one. Similarly, a relatively more significant attribute group has a weighting factor that is greater than one. Weighting Factors (WF), are fixed by the expert at the time of representing CRITIC's knowledge base.

For instance, consider the following piece of CRITIC knowledge:

```
SYSTEM floor-system

type:

   IF  (height-restriction == EXISTS)
   AND (additional-stories == REQUIRED)
   AND (floor-system.type == FLAT-SLAB)

RATING :  90

REMARK : Flat slab is the ideal floor  system  for
        these conditions

END-SYSTEM
```

This is an extract from the knowledge that is used to criticize the type of floor system adopted for the particular building system depending on user requirements and local zoning regulations. The name of the system is floor-system and the attribute being criticized is the type of floor-system that has been adopted. This knowledge states that if the zoning ordinances impose height restrictions at the site and additional floors are necessary then a system like flat slab which uses minimum structural depth is the ideal floor system to be adopted.

Similarly consider the following extract:

```
SYSTEM floor-system

type:

   IF  (building-shape == IRREGULAR)
   AND (large-openings == REQUIRED)
   AND (floor-system.type == CON-JOIST)

RATING :  20

REMARK : Concrete joists are  not  suitable  from
         architectural considerations.
END-SYSTEM
```

This is an extract from an architect's critic for criticizing the type of horizontal framing system chosen for the building system. This states that the irregular building shape and the need for large openings do not permit use of concrete joists as an ideal flooring system.

The following is an extract from a foundation engineer's CRITIC.

```
SYSTEM floor-system

type:

   IF  (soil-conditions  == POOR)
   AND (floor-system.type  ==  FLAT-SLAB)

RATING  : 20

REMARK  : Flat slab is not a suitable floor  system
          for the existing soil conditions
END-SYSTEM
```

This piece of knowledge also is for criticizing the type of floor system that has been adopted for the building system that is being designed. It states that a heavy weight structural system like flat slab system is not the ideal flooring system for the existing difficult foundation conditions.

Each of the influencing domains has its own CRITIC knowledge base and the synthesized solution is separately criticized using each of these knowledge sources.

MODIFICATION

Once the critiquing domains have examined the synthesized solution and filed their reports, the suggestions made in the reports have to be incorporated to arrive at a satisfactory solution. The knowledge required to perform these modifications is encapsulated in rules. This knowledge includes information regarding the threshold values that the criticized attribute groups should attain to make them acceptable. In the event of different domains suggesting conflicting modifications, the rule base transfers control back to the synthesizer to come up with an alternative solution for the attributes in dispute. The rule base contains knowledge to tackle situations where no satisfactory solution emerges after all feasible alternatives have been criticized. This knowledge specifies which of the CRITIC's suggestions should prevail in the event of such conflict. This is done by getting the expert to assign priorities to different domains. Different domains are assigned different priorities and the one with a higher priority has thus a greater say in the modification to be performed.

MULTIPLE LEVELS OF SCM

One point to be noted is that the solution proposed by the synthesizer need not be a completely specified design description. Partial descriptions can also be criticized and modified. The design of the artifact can be split up into a number of stages and the three stage process of synthesis, criticism and modification can be applied at each of these stages. This approach in fact has some advantages. In the first place, by criticizing a solution at several stages, it can be ensured that no criticism is too late. For instance, let the artifact to be designed be a reinforced concrete building. If the synthesis proceeds till a detailed description of the structural components of the building is arrived at, before being subjected to criticism by the constructibility critic, it might have become too late to incorporate the modifications suggested. Besides, multiple levels of synthesis, criticism and modification also ensure that the conflict set is kept reasonably small enough for a rule based modifier to perform effectively. Another advantage to be gained out of the multiple level approach is that the criticism carried out at any level can generate constraints for the next level of synthesis.

GENERIC NATURE OF THE CRITICISM TASK

From the preceding discussion, it follows that the design of an artifact can be split up into a series of tasks some of which correspond to synthesis, some to criticism and the rest to modification. Solutions to any of these tasks comprises of using a generic methodology on knowledge specific to that task. This implies that, generic tools can be used to perform these tasks. The generic tool provides the framework to represent the knowledge and has the means to process this knowledge to perform the set task. Such a tool could then be used repeatedly, to perform all tasks that are solved by the same generic methodology. EDS (Engineering Design Synthesizer) (Ref. 5) is one such generic tool that can be used on a given decomposition to synthesize a solution

A generic component for design criticism, similarly, provides the framework for representing the knowledge required to criticize a solution and is also equipped with the means to process this knowledge for a given context and present a report. For each level of the SCM process, each of the influencing domains has a critic. The generic tool provides the syntax in which the knowledge is to be represented. An architecture for such a generic tool for criticism is shown in Fig.2. The knowledge, as stated earlier, consists of a set of constraints for each attribute group that is to be evaluated and a corresponding rating function that assigns a suitable rating to the attribute group, as directed by the constraints. An attribute group when evaluated, gets a rating between 0 and 100. Higher the rating, better is the performance of the attribute group. These ratings are decided by the expert at the time of putting together the critic knowledge base. Ratings at any higher level are obtained by combining the ratings at the preceding level. Weighting factors define the relative importance of the attribute groups in the performance of the artifact. The default value of the weighting factor of an attribute group or system is 1. These factors are again decided at the time of preparing the knowledge base. Remarks, in English text form, give a qualitative idea of the goodness of the attribute group. The core of this design tool is the CRITIC processor that processes the solution that is fed to it and uses the CRITIC knowledge to evaluate it. It is also equipped with the capability to combine the individual attribute group ratings to arrive at an overall rating for the artifact. The CRITIC knowledge base may contain calls to external procedural programs to evaluate the rating of an attribute group based on certain parameters passed to them.

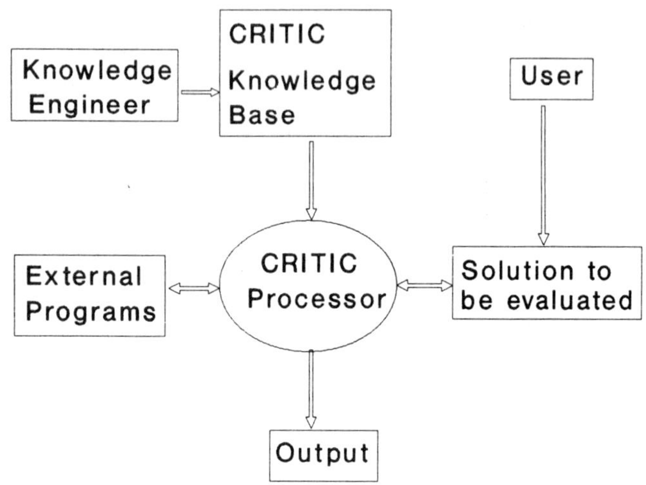

Fig. 2 Architecture of CRITIC

CRITIC can either perform as a stand alone shell in which case a user would directly input a solution to be evaluated or it could be used as a generic component of Knowledge-Based Systems. For instance, an inference engine can call CRITIC from within a production rule to evaluate a solution that has been generated through a similar call to a synthesizer in a previous rule. Control returns to the expert system after CRITIC terminates.

CRITIC AS AN EVALUATOR

One of the features of the engineering problem is that it has no unique solution. Therefore, rather than look for one solution, it might be worthwhile to pursue a finite number of solutions. The end product of design is usually an artifact that has to satisfy a set of constraints placed on it. These constraints can be classified as hard feasibility constraints and soft evaluation constraints. Feasibility constraints are such that they can either be satisfied or violated. Evaluation constraints, on the other hand, can be met with varying degrees of satisfaction. Feasibility constraints can normally be incorporated at synthesis time, so that any synthesized solution would necessarily satisfy the feasibility constraints. Even so, a number of solutions will be synthesized for the given artifact. These solutions will have varying degrees of acceptability and it is the evaluation constraints that decide how acceptable a solution is.

The methodology for evaluation is similar to that of criticism in the sense that here too, a set of constraints are to be evaluated and the resulting rating is to be attached to the attribute group referred to in the constraints. CRITIC can perform as a generic evaluator, as it provides the means to specify these evaluation constraints and also has the capability to rate a solution depending on how well the constraints are satisfied.

The evaluation criteria can range from the cost or the weight of the structure to the stresses and deflections caused during service of the structure. Evidently, depending on the criteria being used to evaluate the structure, the way the individual attribute ratings are to be combined to arrive at the overall rating would vary. For example, if deflections were the evaluation criteria, then the overall rating would only be as good as the worst rating among the individual ratings. If, instead, it is used to evaluate the net effect of a variety of criteria, then, the overall rating would be the weighted average rating of the individual attribute ratings.

The process of arriving at the overall weighted average rating of a solution from individual attribute ratings is illustrated using a typical solution tree shown in Fig. 3.

Node A is the system that represents the designed artifact. B, C, D, I and K are subsystems at various levels in the solution hierarchy. E, F, G, H, J, L, M, N, O and P are leaf level attribute groups.

For the purpose of the illustration, the following notations are used.

RF_x - Rating assigned by CRITIC to the node represented by subscript x.

WF_x - Weighting Factor associated with the node x.

CRITIC's knowledge base contains constraints that direct the processor to assign ratings to the evaluated attribute groups. Evaluation, thus involves a search through the solution tree. It is a constraint directed depth first search, starting with the leftmost node at any level. At any level a system can be evaluated only if all its child nodes have already been rated. The attribute groups are , thus first evaluated and assigned a rating by CRITIC processor as directed by the constraints. Individual attribute groups are assigned a rating between 0 and 100 by the processor. Ratings at any given level are combined to arrive at the ratings at the next higher level. For instance, in the given sample solution tree, ratings of O and P are combined to arrive at the rating of K. Similarly ratings of L,M and N are combined to arrive at the rating of I. The norm for combining is that the rating of the parent is equal to the weighted average rating of the children. The weighted rating of the node K, for instance, is given by

$$RF_K = \frac{WF_O \cdot RF_O + WF_P \cdot RF_P}{WF_O + WF_P}$$

In general, the weighted average rating of a node at any level in the tree, is derived from its child nodes through the following combining formula,

$$RF_x = \frac{\Sigma WF_y \cdot RF_y}{\Sigma WF_y}$$

where, X represents the node whose rating is required and index Y is summed over all the child nodes of the node X.

At the next level, the rating of node D, for instance, is obtained by combining the ratings of J and K in a manner similar to the one stated above. This process of combining is carried over from one level to the next till the root node is arrived at. The resultant rating of the root node will then be the combined weighted average rating of the solution.

Thus, CRITIC can be used to evaluate a series of solutions for predetermined criteria which can then be graded in the order of performance. The user can use this information to either choose one best solution or discard those that fall below a certain threshold rating.

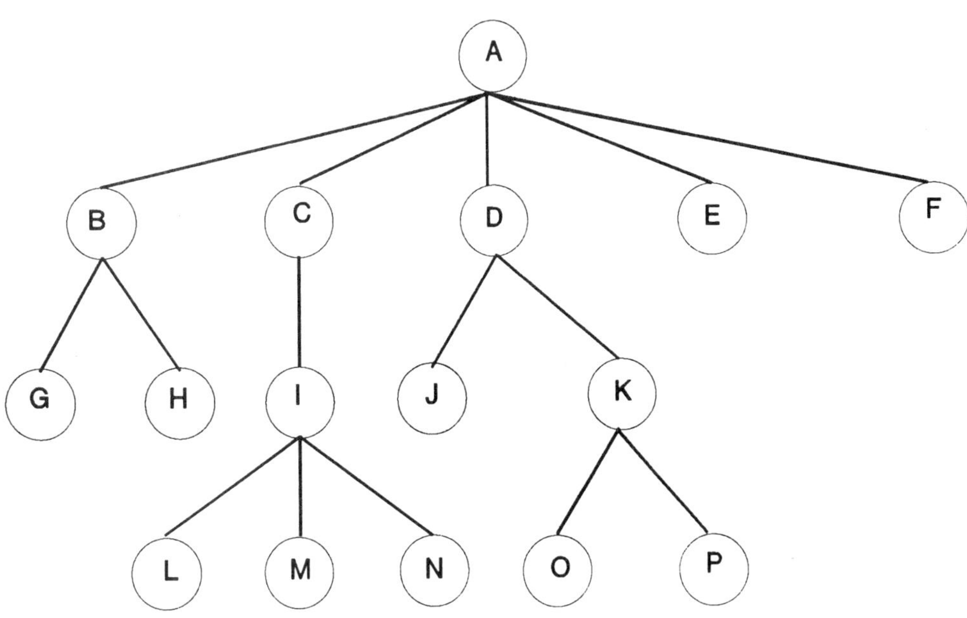

Fig. 3 A Sample Solution Tree to be processed by CRITIC

CONCLUSIONS

A model for solving design problems, that involves simultaneous satisfaction of the requirements set by two or more knowledge sources, has been presented. This three stage process of synthesis, criticism and modification can resolve conflicts that will arise during the solution process. Multiple levels of synthesis, criticism and modification is suggested as a means to improve the efficiency of this process. The generic nature of the SCM tasks has been established and the need to develop generic tools, that provide the framework for knowledge representation and the means to process this knowledge, has been brought out. A framework for design criticism has been suggested and its role in evaluating solutions based on any evaluation criteria has been highlighted.

This paper highlights the role of design criticism in problem solving in the engineering domain. The tasks involved in the design process are found to be generic in nature. The value in identifying such generic tasks and building suitable tools to perform them lies in the richness in knowledge representation and inferencing that each provide and the facility it provides to choose a tool that best applies to a particular task. Design Criticism is a powerful model of problem solving and can be also used to evaluate and rate design solutions using any evaluation criteria.

ACKNOWLEDGEMENTS

The financial support of NSF through Grants INT-8819520 and INT-891386 to the Department of Civil Engineering, IIT, Madras for the Indo-US collaborative Project on Knowledge-Based Expert Systems for Engineering Design, is gratefully acknowledged. The authors would like to express their thanks to Prof. S.J.Fenves of Carnegie-Mellon University, Pittsburgh, for his suggestions and advice on this work.

REFERENCES

1. Maher, M.L, and Fenves, S.J. HI-RISE: A Knowledge-Based Expert System for the Preliminary Structural Design of High-Rise Buildings, Technical Report, R-85-146, Dept. of Civil Engineering, Carnegie-Melon Univ., 1984

2. Lin, T.Y., and Stodesbury, S.D. Structural Concepts and Systems for Architects and Engineers. New York: Wiley, 1981

3. Reinschmidt et al (Editors), Planning and Design of Tall Buildings, Vol la, Tall buildings Systems and Concepts , Proceedings of the International Conference on Planning and Design of Tall Buildings, Lehigh University, Pennsylvania, USA, 1972

4. Maher, M.L. , Engineering Design Synthesis: A Domain-Independent Representation, Artificial Intelligence for Engineering Design, Analysis and Manufacturing 1(3):207-213, 1988a

5. Engineering Design Synthesiser: User's Manual, Research Report No. CE-91-07, Department of Civil Engineering, Indian Institute of Technology, Madras, India, January 1991.

7 DETAILED DESIGN OF STEEL STRUCTURES

AN EXPERT SYSTEM FOR THE DESIGN OF INDUSTRIAL BUILDING STEEL ROOF TRUSSES

V.K. Koumousis and P.G. Georgiou
Institute of Structural Analysis and Aseismic Research,
National Technical University of Athens,
Athens, Greece

An Expert System is presented for the Design of Roofs of Industrial Buildings and Warehouses. These buildings are rectangular in plan and the roof is designed according to Eurocode 3. The expert system is written in Turbo Prolog. The basic scheme followed for the selection of the structural system is that of Generate and Test. The system generates a number of possible designs by using statically determinate trusses to span these buildings. The trusses are triangular or trapezoidal in shape and the nodal loads due to weight, snow and wind loading are deduced automatically by the system. The analysis of the trusses is performed following a logic program. The system determines the minimum weight or minimum cost design for the entire roof.

INTRODUCTION

The process of design is strongly related to Structural Optimization. Structural optimization problems are generally represented as nonlinear mathematical programming problems. In their standard form require the minimization of an objective function, depending on the selected design variables, subject to equality and inequality constraints.

In many situations and certainly in steel design the design variables take values from a prescribed discrete set as for example the cross sectional properties of the standardized profiles. In these cases the nonlinear programming problem depends on both, continuous and discrete design variables (Ref. 10 and 11). As a result the problem becomes difficult and computationally tedious as the number of design variables increases to represent a real design problem.

Frequently the method used treats the discrete variables as continuous and at the end finds the closest values from the given discrete set. However this method is not always a valid technique. The inherent nonlinearities in relation to the non equidistantly distributed discrete values may affect considerably the optimal solution which most of the times is a near optimum solution.

A completely different approach for selecting the near optimum design is to use the brute force strategy of Generate and Test to produce a large number of acceptable designs and select the optimum within this specified design space. This approach can be implemented using heuristic methods that incorporate the knowledge of Experts and new programming methods (Ref. 3 and 4).

In this paper a system is presented that performs a heuristic optimization for the design of steel roofs of industrial buildings. The system developed explains the steps it takes and in that sense represents an expert system written entirely in Prolog. The system is not written with the aid of an expert system shell. The role of the inference engine is played by the Generate and Test scheme.

LOGIC PROGRAMMING

Logic programming is a powerful tool for modeling design knowledge. It can be used for the description, inference, generation and control of possible designs for certain types of structures.

Logic programming is based on an abstract computation model derived by Kowalski from Robinson's resolution principle A logic program is a set of rules that define relations between objects. Computation in the context of logic programming means to prove that a goal statement (rule) is true using the other defined rules. This process is constructive and provides the values for the goal variables which constitute the results of the computation. In other words computation of a logic program is a deduction of consequences of the rules defined in the program. The implementation of logic programming in Prolog is performed by using a control mechanism based on sequential search with backtracking on a restricted class of logical theories, namely Horn clause theories. A Horn clause has the form

$$A \longleftarrow B_1, B_2, \ldots, B_n \qquad n \geq 0 \qquad (1)$$

The important property of Horn clauses is that they can be read both declaratively, saying that A is true if B_1 and B_2 and ... and B_n is true and procedurally saying that to solve problem A, one can solve subproblems B_1 and B_2 and ... and B_n. In order to establish whether A is true, a Prolog program attempts to prove that B_1 and B_2... and B_n are true, starting from left to right. It uses the internal unification routines of the language as a form of a bidirectional pattern matching. It assigns values to the variables of the rule B_1 in order to prove that it is true and subsequently moves to B_2 after leaving a marker in between to remember that up to that point the truth of clause A had been established. This way the program by invoking the internal unification routines, which sequentially search for the right values of the variables from the set that already exists as facts in the program, establishes the truth of B_1 and B_2 and ... B_n and thus the truth of A.

All rules are of the form given in relation (1) where A is referred to as the head of the rule and the conjunction of B_1 and B_2 and ...B_n, as the body of the rule. Every subrule B_i is a predicate or categorim which is either true or false.

Among the most interesting features of Prolog is that of recursion, that is: a rule can be defined in a way that contains itself in its body. Recursion in Prolog is accomplished by the build in control mechanism, which puts the goals and the clauses in an order and keeps them in stacks together with the markers in a way that permits backtracking.

In order to restrict backtracking and thus reduce unnecessary search, the predicate "cut" (!) is used, which in effect deletes the markers for backtracking up to that point and thus works as a barrier that doesn't allow backtracking to the previous subgoals.The predicate "fail" of Prolog is one that makes the rule to fail and thus causes backtracking.

Based on Horn clause theory Prolog programs have both a declarative and procedural meaning. Although Prolog has a close relation with mathematical logic, due to its clause form syntax, it is not necessary for understanding and using Prolog to refer to mathematical logic. The procedural meaning specifies how Prolog satisfies a list of goals. What is important and decisive about the declarative and procedural meaning is that by rearranging the order of clauses or the order of goals in the bodies of clauses the declarative meaning of a program remains the same, but the procedural meaning can change considerably or the program may not work at all. I.Bratko (Ref. 5) addresses these problems and summarizes the existing views. The declarative semantics may cause considerable conceptual problems because it neglects consideration of the flow of control.

A detailed presentation of the Prolog language can be found in references 6, 7, 8 and 9. A friendly version of Prolog for Pc under MS-DOS operating system, that contains also graphics is Turbo Prolog (Ref. 9). Turbo Prolog is not a standard version of Prolog compared to that defined by F.W.Clocksin and C.S.Mellish (Ref. 6). It does not permit operations inside the predicates, nor it allows uninstansiated variables in the predicates, which limits the ability to perform symbolic computations. Moreover its lists are very restricted. Despite its limitations Turbo Prolog offers a good introductory version of Prolog and a friendly environment with graphics which are important for structural analysis problems. The code of this paper was written in Turbo Prolog.

THE DESIGN SPACE

The dimensions of a rectangular in plan industrial building or warehouse are the main given parameters which are considered fixed. The height of the building and the maximum overall height are also parameters which are specified by the user. The number of trusses used for the entire roof is considered as a free variable. Another important design variable is the number of spacings in the inclined upper chord (rafter) of the truss which is related to the selection of the type of purlins. The inclination of the roof is also a free variable. These variables together with the fixed ones are sufficient to determine the layout of the roof, consisting of the trusses, the purlins and the necessary bracings to maintain the integrity of the whole roof. The selection of the type of cladding is also a parameter defined by the user and imposes certain constraints regarding the spacing of the purlins.

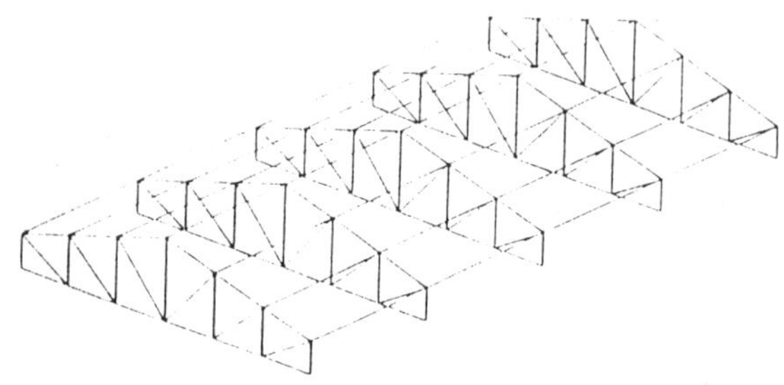

fig. 1 Layout of Steel Roof

These are the overall design variables. All these variables are discrete except the height which is a continuous variable. By changing these variables different sets of descriptions are defined which comprise the design space for this problem. The user can impose certain restrictions to the free variables or he can bound certain variables to specific values. This process restricts the design space and has an accelerating effect on the whole process of design.

For the truss design it is common practice to select statically determinate trusses to span this type of buildings. Among the different types of statically determinate trusses the triangular or trapezoidal in shape trusses are the most frequently used, while the type of internal configuration can vary from V, N type and others. Even though for a specific span there exists a recommended type of truss, in the Expert System the user can select from a set of prescribed types and ask the system to select the best type from the specified range of different configurations. Thus the system searches for the optimum design within the topological space of possible designs determined by the different types of trusses.

In addition to the above variables which characterize the general layout of the design, the most important design variables are the cross sectional properties of all members that constitute the steel roof of the industrial building or warehouse namely the purlins the truss members and the bracings.

The selection of the profiles is performed on the basis of the requirements of Eurocode 3 for steel structures. All members of the truss are grouped in three main groups: the upper chord, the bottom chord and the internal members. Usually the chords are designed using the same section while the internal members are designed usually from the same type of section with every member designed individually. Moreover there exist certain rules which determine the suitability of the combination between these three groups. Both chords are usually designed from the same type of section but with different sizing. These affiliations are default in the system but the user can make his own choice and/or form his own groups of members. These decisions affect considerably the search space and consequently the total time required by the system to come up with the near optimal solution.

THE STEEL ROOF GENERATOR

The strategy followed for the design of the steel roofs is that of Generate and Test. Following this scheme a set of possible designs are generated and subsequently are tested according to the requirements of Eurocode 3. The accepted designs are sorted and finally the minimum weight or cost design is selected as an optimum solution within the prescribed design space.

The truss generator which is part of the generate procedure generates alternative types of statically determinate trusses adjusted to the current values of the problem. From the general parameters of the truss the system defines the shape of the truss as triangular or trapezoidal. Subsequently decides the number of purlins that will be placed on the upper chord of the truss and then by changing the angle of half of the internal members it generates different types of internal configurations.

The truss generator can be augmented by incorporating other types of trusses. In this way the design space is enlarged and the validity of the final design as optimum becomes more reliable.

The relevant predicates are:

```
generate_truss:-                          % trapezoidal
   current_settings(_,TL,SH,FH,F,Segments,Type),
   SH > tolerance,
   select_theta(Type,F,Theta),
   A1=TL/2/Segments/cos(F),
   B=A1*sin(Theta)/(sin(Theta+F)),
   A=TL/2/Segments,DH=(FH-SH)/Segments,
   NN=4*Segments+2,
   assert(number_of_nodes(NN),parameters),
   LoadN=Segments*2+1,
   assert(loading_nodes(LoadN),parameters),
   assert(truss(TL,SH,FH,Segments,DH,A,B,LoadN),
          parameters),
   generate_nodes_trapezoidal(1),
   generate_members_trapezoidal(1),
   generate_supports.

generate_nodes_trapezoidal(K):-
   truss(TL,SH,_,Segments,DH,A,B,LoadN),
   K<=Segments,!,
   N1=K,N2=1000+N1,N3=(LoadN+1)-K,N4=1000+N3,
   XS1=(K-1)*A,XT1=TL-(K-1)*A,
   XS2=(K-1)*B,XT2=TL-(K-1)*B,
   Y=SH+(K-1)*DH,
   assert(node(1,N1,XS1,Y),data),
   assert(node(2,N2,XS2,0),data),
   assert(node(1,N3,XT1,Y),data),
   assert(node(2,N4,XT2,0),data),
   K1=K+1,
   generate_nodes_trapezoidal(K1).

generate_nodes_trapezoidal(_):-
   truss(TL,_,FH,_,_,_,_,LoadN),
   L2=TL/2,N1=(LoadN+1)/2,N2=1000+N1,
   assert(node(1,N1,L2,FH),data),
   assert(node(2,N2,L2,0),data).

generate_members_trapezoidal(R):-
   truss(_,_,_,Segments,_,_,_,LoadN),
   R<=Segments,!,
   MS1=R,MS2=1000+MS1,MS3=2000+MS1,MS4=3000+MS1,
   MT1=LoadN-R,MT2=1000+MT1,
   MT3=2001+MT1,MT4=3000+MT1,
   JS1=R,JS2=1000+JS1,KS1=JS1+1,KS2=JS2+1,
   JT1=(LoadN+1)-R,JT2=1000+JT1,
   KT1=JT1-1,KT2=JT2-1,
   assert(memb(1,MS1,JS1,KS1),data),
   assert(memb(2,MS2,JS2,KS2),data),
   assert(memb(3,MS3,JS2,JS1),data),
   assert(memb(4,MS4,JS1,KS2),data),
   assert(memb(1,MT1,JT1,KT1),data),
   assert(memb(2,MT2,JT2,KT2),data),
   assert(memb(3,MT3,JT2,JT1),data),
   assert(memb(4,MT4,JT1,KT2),data),
   R1=R+1,
   generate_members_trapezoidal(R1).

generate_members_trapezoidal(_):-
   truss(_,_,_,Segments,_,_,_,_),
   N1=Segments+1,N2=1000+N1,M=2000+N1,
   assert(memb(3,M,N2,N1),data).

generate_supports:-
   truss(_,_,_,_,_,_,_,LoadN),
   N1=1+1000,N2=LoadN+1000,
   assert(support(N1,0),data),
   assert(support(N1,1.57),data),
   assert(support(N2,1.57),data).

select_theta(Type,F,Theta):-
   Type = "Vertical",   Theta = pi / 2 - F  or
   Type = "verticalToTheRoof", Theta = pi / 2.
```

Inside the generation procedure another generator generates the appropriate bracings recommended by the codes of practice.

These generators are developed following a logical algorithms that establish the general layout of the configuration adjusted to the current values of the design variables and then following a specific pattern add the required members as to complete the full configuration of the roof.

THE DESIGN PROCESS

A critical parameter for the overall performance of the system is the time required for the completion of its goals. The number of acceptable solutions in relation to the complex design space which consists of various types of discrete continuous and topological variables requires the development of fast algorithms for each particular goal.

The analysis of the truss for the specified loading conditions is performed in a way that implements strategies and rules of Classical structural analysis for the solution of simple trusses (Ref. 2). The time required for this type of solution varies linearly with the number of nodes of the truss while the time required using a direct stiffness method is exponential. Moreover due to the grouping of the members of the upper and lower chord the axial forces required are the max and min values occurred in all members of the group. This approach results in a substantial reduction of time required for the solution.

THE DESIGN CODE IN PROLOG

The truss members are designed following the requirements of EuroCode 3 for the Design of Steel Structures (Ref. 1). The weight, the wind load and the snow per unit area are taken from DIN 18800. The partial safety factors for the ultimate limit states and serviceability limit states are the ones defined in Eurocode 3. The analysis of the truss is performed for three different unit load cases that correspond to weight and wind loading in two different directions. These results are combined according to the Eurocode 3 to determine the axial forces that correspond to a particular limit state loading case.

The tensile members are designed in accordance with paragraph 5.4.3 of the Eurocode 3 and the relevant part of the Prolog program is as follows:

```
compression_requirements(Type,Cnt,_,Min,Length,
          Iarea):-
   section(Type,Name,Height,Width,Thickness,Area,
          Weight,Jx,_),
   chosen_section(_,_,_,WeightOld),
   Cnt * Area >= Iarea,
   Cnt * Weight < WeightOld,
   Tmp = Jx / Area,
   Radius_of_gyration=sqrt(Tmp),
   buckling_curve_axis(Type,Height,Width,
          Thickness,Curve,_),
   imperfection_factor(Curve,Alfa),
   Lamda=Length/Radius_of_gyration,
   Lamda1=pi*(sqrt(m_of_elasticity/y_stress)),
   Lamda2=(Lamda/Lamda1)*sqrt(Alfa),
   F=0.5*(1+Alfa*(Lamda2-0.2)+Lamda2*Lamda2),
   X=1/(F+sqrt(F*F-Lamda2*Lamda2)),
   NcRd=Cnt * Area * yield_stress,
   NbRd=X*NcRd,
   abs(Min) <= NbRd,
   CntArea = Cnt * Area, CntWeight = Cnt*Weight,
   save_chosen_section(Type,Name,CntArea,
          Cntweight),fail.
compression_requirements(_,_,_,_,_,_):- !.
```

The compression members are designed following the requirements of paragraph 5.4.4 and 5.5.1 of Eurocode 3 for buckling resistance. The relevant part of the Prolog program is:

```
tension_requirements(Type,Cnt,Max):-
   section(Type,Name,_,_,_,Area,Weight,_,_),
   chosen_section(_,_,_,WeightOld),

   Cnt * Weight < WeightOld,
   Rnpl= Cnt * Area * yield_stress / gama_M1,
   Max <= Rnpl,
   CntArea = Cnt * Area, CntWeight = Cnt*Weight,
   save_chosen_section(Type,Name,CntArea,
          CntWeight),fail.
tension_requirements(_,_,_):- !.
```

The purlins are designed for bending resistance. They behave as continuous beams of equal span. The moments and shearing forces for these components of the roof are determined from closed form relations. The purlins are designed following paragraph 5.4.5 for bending about one axis or 5.4.7 for bending and shear.

```
bending_and_shear_design(Panels,A,Type,Q,M1,M2):-
   section(Type,Name,Height,_,Thickness,_,
          Weight,Jx,_),
   chosen_purlin(_,_,WeightOld),
   Weight < WeightOld,
   beam(Panels,A,Weight,Q1,_,M11,M12),
   Shear = Q + Q1,
   Moment = M1 + M11,Moment2 = M2 + M12,
   Av=1.04*Height*Thickness,
   RVpl = Av*yield_stress/gama_M1/1.73205,
   Wpl=Jx/(Height/2),
   RMpl=Wpl*yield_stress/gama_M1,
   R=(2*Shear/RVpl-1)*(2*Shear/RVpl-1),
   WplR=Wpl-R*Av*Av/(4*Thickness),
   select_resistance(Shear,RVpl,WplR,RMpl,RMcv),
   Shear <= RVpl,
   Moment <= RMcv,
   Moment2 <= RMpl,
   save_chosen_purlin(Type,Name,Weight),fail.
bending_and_shear_design(_,_,_,_,_,_):- !.
```

GRAPHICAL INTERPRETATION OF DESIGN

The interactive character of the system and the ability to explain its moves are greatly improved by using graphical representations. The system is written in Turbo Prolog which has an efficient graphics library. These features are used to demonstrate the current design, previous designs or after the completion of the process the sorted list of the produced designs. The general form of this graphical representation of a design is presented in Fig.1. The information contained refers to the plan dimensions, the number of purlin and the inclination, while in the report window the face of solution and the current member which is under design is reported.

After the completion of the whole process the system preserves all the produced designs in a database together with all the information that concerns the particular design. This database can be used to retrieve useful information. It can be sorted with respect to the total weight or total cost, or with respect to the weight of cladding, purlins, trusses and bracings.

Various other cross references can be made that relate specific attributes of the database that the user can select. This information can be plotted in curves so as to give a global overview of the alternative designs. The user can select his final design which can be displayed on the screen and then can be directed to produce a graphics file in a DXF format which can be read from almost all graphical packages like AUTOCAD, CLARISCAD e.t.c.

OUTPUT DRAWINGS IN DXF FORMAT

The great advantage of design systems is their ability to incorporate all faces of design into one integrated system. The final stage of a design is the set of drawings and subdrawings necessary for the construction. The option of producing these drawings facilitates the work of an Engineer considerably. This goal is achieved by developing two interfaces. The one that forms a DXF file from the particular objects used in the Prolog program and the other that does the inverse job i.e from a DXF file forms the necessary predicates that express the objects involved in the design which may have been altered by the user inside the CAD program linked to this system.

In the present form the system produces a drawing of the roof that contains the purlins, the trusses and the braces Fig.6. In a subsequent version the system will include also the design and drawings of typical connections.

EXAMPLES

The roof of a building with span 20 m and length 50 m is used to demonstrate the system. The trusses used by the system are triangular or trapezoidal in shape. In fig.2 a triangular truss is presented.

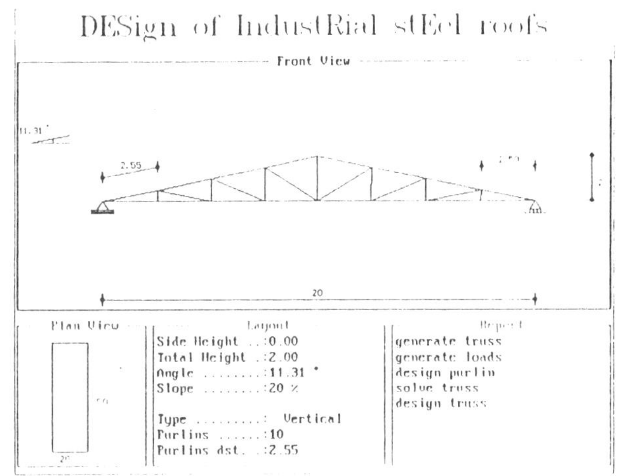

fig.2 A Triangular truss

The trusses used for this example are restricted to trapezoidal trusses with the side and total heights raging from 0.6 m to 1.5 m with step 0.3 m for the side height and 1.5 m to 4.7 m with step 0.4 m for the total height of the truss. For every type of truss the total number of trusses for the whole roof varies from 6 to 26. For every trapezoid different number of purlins is considered that ranges from 10 to 20 purlins. This description of the design space corresponds to 4536 alternative configurations and the total cpu time required for the solution is 120 min on a PC 386/33MHz computer which corresponds to approximately 40 designs per minute. The layout screen for these trusses is presented in fig. 3,

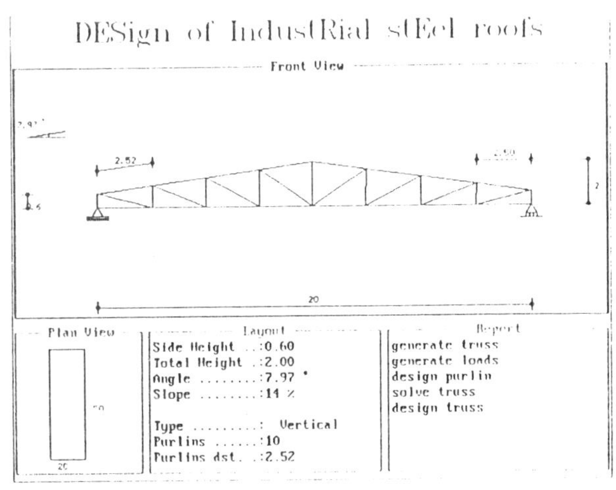

fig. 3 Trapezoidal Trusses

In addition there exists an option to replace the critically compressed members by their counterparts that are critically stressed in tension. In that case the truss takes the form presented in fig.4.

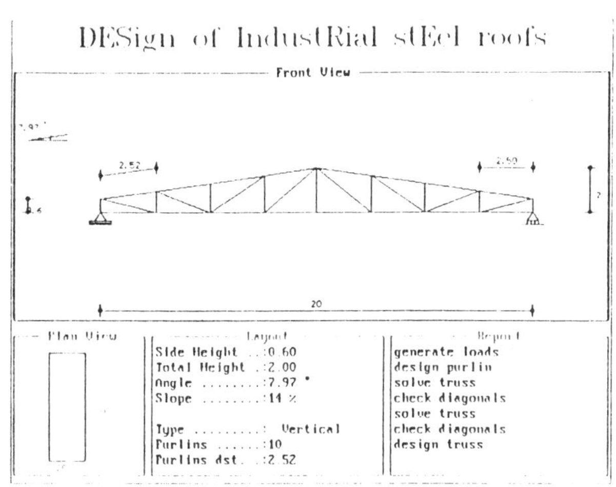

fig.4 Truss with mostly Tensile members

The minimum weight design for the above specified design space is 207.95 kN, 151.91 kN for cladding, 13.70 kN for purlins, 20.38 kN for braces, 21.96 kN for trusses. This occurs for side height=1.5 m, total height=3.1 m and 10 total purlins at equal distances at the top chords and 16 trusses for the whole roof. In table 1, the 15 best designs are given from which is observed that the optimum area is quite smooth for this particular problem and corresponds to a wide range of descriptions.

A T T E M P T S									
HEIGHTS		Number of		Type	WEIGHTS				
Side	Total	Purlins	Trusses		Cladding	Purlins	Braces	Trusses	TOTAL
1.50	3.10	10	16	Vertical	151.91	13.70	20.38	21.96	207.95
1.20	3.10	10	16	Vertical	152.68	13.70	20.16	21.70	208.25
1.50	3.10	12	16	Vertical	151.91	16.44	20.99	19.09	208.43
1.20	3.10	12	16	Vertical	152.68	16.44	20.77	18.63	208.53
1.20	2.70	12	18	Vertical	151.68	16.44	20.63	20.27	209.01
1.50	2.70	12	18	Vertical	151.08	16.44	20.91	20.62	209.04
1.20	2.30	12	15	Vertical	150.90	16.44	20.24	21.53	209.11
1.50	3.50	12	15	Vertical	152.97	16.44	21.18	18.53	209.12
1.50	2.70	10	18	Vertical	151.08	13.70	20.40	23.97	209.15
1.20	2.70	10	18	Vertical	151.68	13.70	20.12	23.70	209.20
1.50	2.70	14	14	Vertical	151.08	19.18	20.60	18.40	209.25
1.50	3.10	10	17	Vertical	151.91	13.70	20.55	23.10	209.25
1.50	2.30	12	15	Vertical	150.48	16.44	20.44	22.09	209.45
1.20	3.10	10	17	Vertical	152.68	13.70	20.30	22.88	209.56
1.20	3.10	12	17	Vertical	152.68	16.44	20.86	19.58	209.56

In fig.5 the variation of the total weight of the roof is plotted with the number of trusses used for the whole roof for the specified optimum truss.

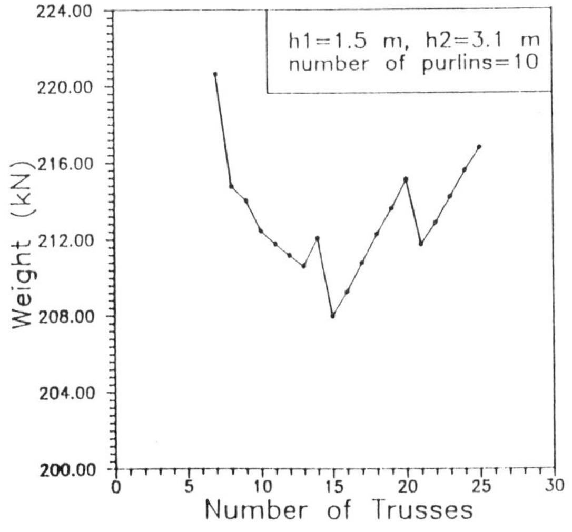

fig.5 Weight with respect to the number of trusses

In fig.6 the variation of the weight of purlins, braces and trusses is plotted with the number of trusses used for fixed side height=0.6 m, total height=2.7 m and 10 purlins.

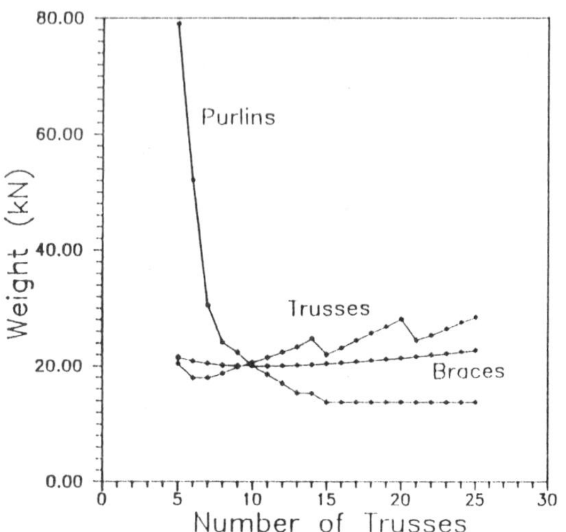

fig.6 Variation of weight with the number of trusses

From fig.6 it is observed that the weight of the purlins has an exponential form and after a specific number of trusses the distance between the trusses is such that corresponds to the minimum design for the purlins. The purlins are designed as continuous beams of equal span. The weight of the bracings is rather constant. the small variations correspond to differences in length. The bracing system utilized in this analysis consists of a roof bracing system that uses braces for the upper chord at the first and last panel. It consists also of vertical and horizontal braces. The bracing system is not designed on the basis of a specific analysis. The cross section used for the described layout is angles 45x5. The weight of the trusses varies in a non smooth manner.

The variation of the weight of a single truss has the form presented in fig.7.

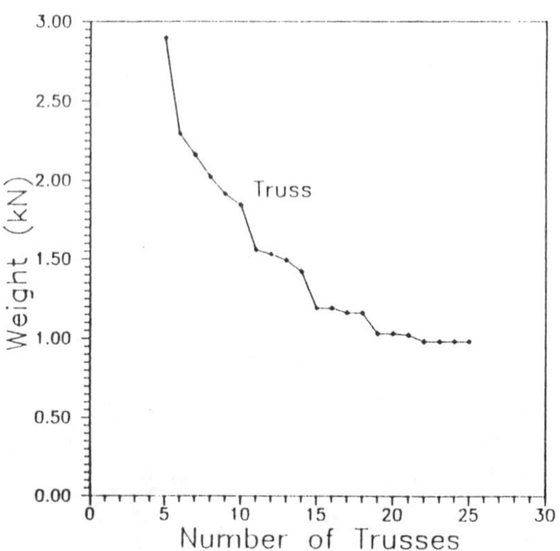

fig.7 Weight of a Truss with the number of total trusses

It is observed that the weight of one truss decreases in an exponential way. The form of the decay is not smooth. Some times the lower stresses can not select smaller sections and thus the weight of the truss remains the same. Following the rule of practice the design of the lower and upper chords is made from the same section on the basis of the most critical bar of the lower and the upper chord respectively. This results in several rapid drops of the weight of a truss shown in fig.7.

In fig.8 the weight of all trusses is plotted with respect to the number of trusses used to cover the whole roof.

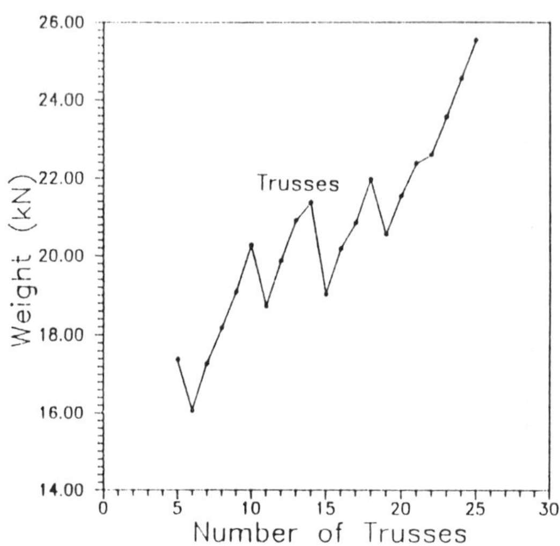

fig.8 Total weight with the number of trusses

It is observed that the form of the total truss weight changes in a discrete manner. Consecutive segments of the same slope correspond to the same single truss design.

In fig.9 the weight of the purlins, braces and trusses is plotted with respect to the total height of the trapezoidal truss of side height=1.2 m with fixed number of purlins=10 and 16 total trusses for the whole roof.

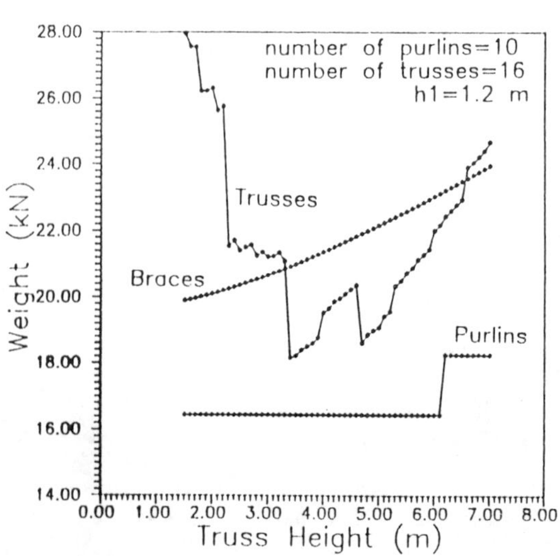

fig.9 Weight with respect to height of truss

From this figure we observe that we are in a region of minimum weight design for the purlins and thus the corresponding curve is flat. The step of this curve correspond to the change of the wind pressure load which after a specific angle of the roof changes sign and stresses more the purlins. The weight of the braces is increased due to the increase of the height of the truss. The most interesting curve is that of the truss weight. It is not smooth and it is characterized by rapid drops which correspond to changes in the chord design. Rules of thump correspond to the flat area around two and three meters but the optimum is to the right.

By relaxing the constraint that all chord members are designed by the same section (different for the upper and lower chords) and leaving every member to be designed individually from the same type of section the variation of the weight with respect to the height of the truss takes the form given in fig.10.

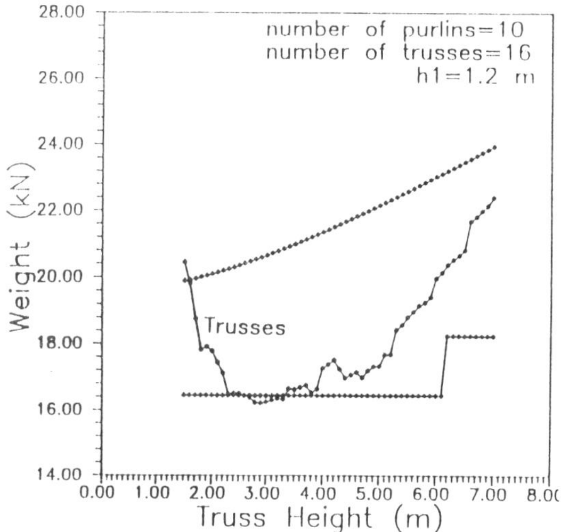

fig.10 Weight with respect to the height

It is observed the the curve that corresponds to the weight of the trusses becomes relatively smoother as compared to the constraint situation. Moreover the total weight is drastically decreased.

CONCLUSIONS

An Expert System for the Design of Steel Roofs of Industrial Buildings and Warehouses is presented. The main characteristics of this Expert System which performs a heuristic optimization with respect to discrete, continuous and layout variables are:

a) The use of the Generate and Test strategy for the solution.

b) The implementation of Prolog language for the development of the system.

c) The graphical interfaces for the description and explanation of its moves.

d) The implementation of Eurocode 3 in Prolog.

e) The use of logic algorithms to solve the statically determinate trusses of the roof.

f) Its overall efficiency in determining the near optimum solution in the specified design space.

g) From the numerical results presented it turns out that the discrete nonlinear nature of the problem can be treated effectively with the method presented.

REFERENCES

1. Eurocode No.3 "Design of Steel Structures" Part 1 General Rules and Rules for Buildings.

2. V.K.Koumousis and P.G.Georgiou "Statically Determinate Trusses Programmed in Logic" ASCE Journal of Computing (submitted for publication)

3. Adeli, H. and Balasubramanyam, K. V. " A Knowledge-based System for Design of Bridge Trusses" Journal of Computing in Civil Engineering, 2, No.1, 1988.

4. Adeli, H. and Al-Rijleh, M. M., "An EXpert System for Design of Roof Trusses", Microcomputers in Civil Engineering, 2, No. 3, 1987.

5. Bratko, I., "PROLOG Programming for Artificial" 2nd Edition, Addison-Wesley Pub. Co. 1990.

6. Clocksin, W.F. and Mellish, C.S., "Programming in Prolog", 3rd Edition, Springer-Verlag, New York,1981.

7. Ross, P., "Advanced Prolog: Techniques and Examples", Addison-Wesley Pub. Co. 1989.

8. Sterling, L. and Shapiro, E., "The Art of Prolog: Advanced Programming Techniques", MIT Press, 1986.

9. Turbo Prolog Manual, Ver. 2.0, Borland Int.

10. John K.V., Ramakrishnan C.V. and Sharma K.G., "Optimum Design of Trusses from Available selections - Use of sequential linear programming with branch and bound algoritm", Engng Optimizn 13, 119-145, (1988).

11. Ringertz U.T. "On methods for discrete structural Optimization", Engng Optimizn 13, 47-64 (1988).

AN EXPERT SYSTEM FOR LATERAL BUCKLING DESIGN

C.T.K. Lai, C.M. Martin, G.J. Hancock, J.P. Papangelis and N.S. Trahair

School of Civil and Mining Engineering, University of Sydney, Australia

The lateral buckling of a structural steel member is a complex phenomenon affected by section type, slenderness, bending moment magnitude and distribution, load height, restraints and several other factors. This paper describes progress in the development of LATXPERT, an expert system for the design of structural steel beams against lateral buckling. Through its user-friendly format and various explanation facilities, the expert system is able to build the confidence and lateral buckling expertise of the design engineer.

The heart of the system is a knowledge-based diagnosis tool which enables the user to determine if there is likely to be a buckling problem, if a simple solution is available, if a buckling analysis computer program is required, or if specialist advice should be sought. The diagnosis tool is integrated with a database of commercial beams and their section properties, and with analysis programs for cross-section properties and lateral buckling. The use of these analysis programs in an expert system environment provides a new and useful tool for the lateral buckling designer.

INTRODUCTION

The designer of a structural steel beam must consider the possibility of lateral buckling failure, involving both lateral (or out-of-plane) deflection and axial twisting of the beam. A beam's resistance to lateral buckling is influenced by many factors, including cross-section shape and size, member slenderness, bending moment magnitude and distribution, load height (relative to the shear centre), and the extent of restraints against lateral deflection, lateral bending, and twisting. While each of these effects has been thoroughly investigated in isolation [Ref. 1], their interactions are rather more difficult to predict accurately. While finite element computer programs [Ref. 2] can account for all of these factors, the restraint conditions must often be idealised, and furthermore, the analysis assumes that purely elastic lateral buckling occurs. In reality, the situation is complicated by the presence of yielding and residual stresses, which cause inelastic buckling, and initial crookedness and twist which increase the stresses and reduce the load capacity below the inelastic buckling load. In formulating rules for design standards, [Refs 3-5] it has been necessary to supplement theoretical work on elastic buckling with extensive experimental investigations of real beam buckling.

There are problems associated with the use of both design standards and computer analysis programs for lateral buckling. The attempts of some recent standards [Refs 3-5] to provide increased guidance on lateral buckling have tended to give a false impression of the accuracy of their formulations, and to hide some of their meaning from designers who do not have a clear understanding of lateral buckling. This leads to a danger that designers will merely insert numbers into formulae, look up tables and consult graphs, without gaining an understanding of the rationale behind the rules. Similarly, there is a risk that a computer analysis program may be used to perform design blindly, without the user being aware of the program's limitations. Such users are unlikely to develop any feel for the behaviour being modelled, or the relative importance of the various factors which influence a beam's lateral buckling and load capacities.

This paper discusses LATXPERT, an expert system approach to lateral buckling design which seeks to overcome many of these difficulties.

THE AIMS OF LATXPERT

An expert system is a computer program which simulates the knowledge and decision making processes of a human expert in a well-defined field. The characteristics of expert systems which distinguish them from other problem-solving programs are summarised in Reference 6. Possibly the most important of these features is the ability of an expert system to explain and justify its requests for input data, as well as explaining the reasoning behind its answers and conclusions. The following features were identified as being essential for the lateral buckling expert system LATXPERT.

- The user should be constantly involved throughout the running of the program.

- The system should identify situations where there is no likelihood of a lateral buckling problem.

- The system should identify unusual problems requiring specialist advice or further consultation, and suggest a course of action.

- All aspects of the system should operate in an organized and user-friendly manner, so that even an inexperienced designer is able to use the program easily and properly.

- Lateral buckling designing or checking should be performed by the expert system itself, at least for a simple class of beam problems.

- More complicated situations requiring the use of finite element computer analysis should be identified, and the user referred to a sub-program of the expert system. These sub-programs should also be user-friendly.

- Use of the program should broaden the user's knowledge of lateral buckling terminology, design methods and the various factors which influence a beam's buckling resistance.

- The system should be able to provide explanations, both for its input requests and its conclusions. In this way the user may gain confidence in the abilities of the program.

KNOWLEDGE ACQUISITION FOR LATXPERT

The process of accumulating knowledge is a time-consuming and vital step in the creation of any expert system, and is formally described as knowledge acquisition (the means of expressing this knowledge in the form of rules and explanatory text is called knowledge representation). The following principal sources contributed to the knowledge about lateral buckling in LATXPERT:

- AS 4100 Steel Structures Code 5 (Section 5: Design of Beams);

- The textbook "The Behaviour and Design of Steel Structures";

- Academic staff at the University of Sydney;

- Computer analysis programs;

- A database of cross-section properties;

- Machine experimentation using a finite element buckling analysis program.

Each of these sources is discussed in detail below.

The Australian Standard AS 4100-1990 Steel Structures [Ref. 5]

Before this new standard was developed, working stress methods were used to design steel structures in Australia. AS4100-1990 introduces the limit state design method for steel, and features many revised design procedures, reflecting advances in research and development since the previous standard AS 1250 [Ref. 7] was first published in 1968. It was seen as essential that the expert system should be based on the new standard, so that designers would gain a feel for the latest approach to buckling design.

AS 4100 rules for designing beams against failure by in-plane yielding, local buckling or lateral buckling. While LATXPERT does not consider local buckling as a possible failure mode, because this limit state is rarely critical in hot-rolled beam design, there are frequent reminders that the user should also check the local buckling adequacy of the design. Beams are otherwise checked or designed in full accordance with the standard, which provides two tiers for beam design. The lower tier method is designed to handle common problems where a high degree of accuracy is not required. The upper tier method is recommended for unusual problems or those where a high degree of accuracy is warranted; it assumes that a buckling analysis computer program is available to the designer.

Lower Tier Design Method in AS4100

Restraints: The first step in the lower tier method is to establish the nature of the restraints acting. Full restraint must provide effective lateral deflection and twisting restraints to the section. A partial restraint provides effective lateral deflection restraint to some point in the cross-section, but only partial restraint against twisting. A lateral restraint is one that only prevents lateral deflection of the critical flange of the section, with twisting not restrained at all.

Division into segments: A segment is a length of the beam which has full or partial restraints at both ends, or full or partial restraint at one end, with the other end free. If lateral restraints are present, then they further subdivide a segment restrained at both ends into subsegments. The aim of these subdivisions is to isolate lengths of the beam which may be considered to buckle independently.

Full lateral support: Any segment restrained at both ends which also has its critical flange continuously restrained against lateral deflection is said to have full lateral support. Such segments cannot undergo lateral buckling, so the design may be based on in-plane yielding alone. Segments without continuous restraint may also be checked to determine if the segment's slenderness ratio falls below a limiting value in which case the segment is also deemed to have full lateral support.

Effective segment length: Each remaining segment must be examined separately, to find which one governs the buckling capacity of the beam as a whole. The length of the segment is adjusted by a series of empirical multipliers to account for, loads acting at the critical flange which are free to move with the flange as it buckles, or restraints against lateral rotation, or partial twisting restraints against end twisting.

Reference buckling moment: The reference buckling moments M_o is the uniform in-plane bending moment which would cause a simply supported beam to buckle elastically [Ref. 8].

Elastic buckling to strength conversion: The capacity M_b of a real beam is approximated by

$$M_b = \alpha_s M_s \tag{1}$$

in which
$$\alpha_s = 0.6\left\{\sqrt{\left[\left(\frac{M_s}{M_o}\right)^2 + 3\right]} - \left(\frac{M_s}{M_o}\right)\right\} \tag{2}$$

and M_s is the section moment capacity. The slenderness reduction factor α_s accounts for the effect of slenderness, initial curvature, twist and residual stresses on the beam strength.

Bending moment distribution: The shape of a beam's bending moment distribution is important in buckling design, because the reference buckling moment M_o is derived for the case of a beam subjected to uniform moment. This effect may be approximated by

$$M_c = \alpha_m M_o \qquad (3)$$

where M_c is the maximum moment at elastic buckling and α_m is a moment modification factor.
The more dangerous loadings (low values of α_m) are those which produce more uniform distributions of major axis bending moment. Where the peak bending moments are restricted to the ends of a segment, much larger values of moment can be tolerated before the same beam buckles.

Three methods for computing the moment modification factor α_m. Firstly, there is a table for common loading patterns. For more unusual bending moment distributions, α_m may be approximated by

$$\alpha_m = \frac{1.7 M_{max}}{\sqrt{(M_2^2 + M_3^2 + M_4^2)}} \qquad (4)$$

where M_2, M_3, and M_4 are the bending moments at the quarter points and mid point of a segment, and M_{max} is the maximum moment in the segment. Finally, a buckling analysis computer program may be used to compute the value of α_m more accurately.

Segment capacity in bending: The nominal bending capacity of the beam segment M_b is determined from:

$$M_b = \alpha_m \alpha_s M_s \leq M_s \qquad (5)$$

Design requirement: AS4100 requires the beam to satisfy the inequality

$$M^* \leq \phi M_b \qquad (6)$$

in which M^* is the design moment determined from the factored nominal loads, and ϕ is a capacity reduction factor of 0.9 which allows for variations in the nominal dimensions and material properties.

Upper Tier Design Method in AS4100

The upper tier method of lateral buckling design requires a buckling analysis computer program. This method is generally more accurate, and can be used for a much wider range of buckling design problems. The computer program gives an accurate prediction of the elastic buckling load, which incorporates all the effects of load height, restraints, bending moment distribution and variations in cross-section.

The computer program gives the bending moment distribution at elastic buckling. This must be converted to an estimate of the real beam bending capacity by a method which is essentially the same as that for the lower tier method. The beam is divided into segments, and each is considered separately to find the critical one for buckling failure.

The Behaviour and Design of Steel Structures [Ref. 1]

This text book, which provides a companion to AS 4100, describes the general background to the clauses in Australian, British and American standards [Refs 3-5], and discusses some special areas such as the lateral buckling of monosymmetric beams and non-uniform beams.

Academic Staff at the University of Sydney

An expert system is most effective when it is based on the knowledge and problem-solving techniques of an actual human expert. In the case of LATXPERT, there were several staff members from the Centre for Advanced Structural Engineering of the University of Sydney who had particular interests in lateral buckling. Their experience in structural design was of assistance in designing the opening part of the system, in compiling a list of computer programs, articles, papers and human experts, and in clarifying the meaning and intent of the new standard's clauses.

Computer Analysis Programs

Many expert systems have been interfaced with conventional algorithmic (non-knowledge based) computer programs. Two such programs, FELB and OPEN, are used in association with the LATXPERT system. These programs are used to perform essentially analytical tasks, and were not programmed in a rule-based format.

FELB (Finite Element Lateral Buckling) is a lateral buckling analysis program [Ref. 2] based on the flexural-torsional buckling finite element method derived in Reference 9.

OPEN is a program for analysing section properties [Refs 10, 11] which is based on the thin-walled section theory derived in References 8 and 12.

Database of Section Properties

The BHP catalogue of the nominal properties [Ref. 13] of Australian hot-rolled sections was converted into a database during the development of another program [Ref. 14] and is used as a source of knowledge by LATXPERT.

Machine Experimentation

By performing many successive runs of FELB, it was possible to simulate the process by which an expert accumulates knowledge from his experience. This process was used [Ref. 15] when developing the rules for the BTXPERT bridge truss design system. The term machine experimentation was applied to this method of learning.

The important result given by FELB is a load factor against elastic buckling. The first step in machine experimentation was to use FELB to analyse a reference case, and to obtain a load factor. A single variable was then altered and the analysis was repeated. A new load factor was obtained, so that it was possible to gauge the effect of the variable change as a multiplying factor on the elastic buckling capacity of the beam.

FELB was used to calculate multiplying factors for the effects on the elastic buckling resistance of, the major axis bending moment distribution, top and bottom flange loading, and lateral rotation restraints.

The first multiplier is the moment modification factor α_m. The latter two factors have been termed the load height multiplier and the lateral rotation restraint multiplier. A multiplier value smaller than one indicates a reduction in buckling resistance compared with the reference case, and vice versa.

A significant advantage of using this multiplier system is that the effect on elastic buckling capacity is immediately evident. In contrast, the series of effective length multipliers used in the lower tier of AS 4100 not nearly as helpful, as the real influence of each effect remains hidden from the designer.

The different effects such as load height and lateral rotation restraint were varied one at a time, in an attempt to separate their influences on the buckling resistance. Only a simple set of beams was investigated, including simply supported beams, cantilevers, propped cantilevers and built-in beams, subject to point and uniformly distributed loads, and segments with constant moment gradients.

Approximately 20,000 automated runs of FELB were performed and the results plotted as graphs. All multipliers showed a fundamental dependence on slenderness, but simply plotting them against L/r_y gave an unacceptable amount of scatter. Instead, it was found that a better variable was a modified slenderness parameter $50(L/r_y)(t/D)$, in which t is the web thickness and D the web depth. This gave rise to much more uniform curves, with all sections falling close to a single line, as shown in Fig. 1 for top and bottom flange loading of a cantilever.

Each set of multipliers was produced by altering a certain effect in isolation, and these were multiplied together, to obtain a combined multiplier which was then applied to the reference buckling moment. This procedure ignores any interaction between the separate effects, which are accounted for when using FELB. Figure 2 illustrates a case where the simplified approach is erroneous for beams of low slenderness.

THE LATXPERT SYSTEM

LATXPERT consists of three distinct sub-programs

- a knowledge-based program for initial diagnosis and solutions to simple buckling problems,
- BEAMFELB, a user-friendly buckling analysis program, and
- BEAMLSC, a program for computer-assisted design to the lower tier of AS 4100,

discussed below.

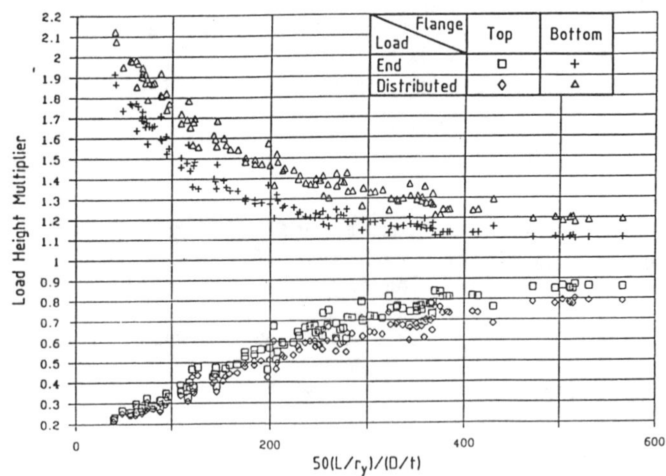

FIG. 1 CANTILEVER LOAD HEIGHT MULTIPLIERS

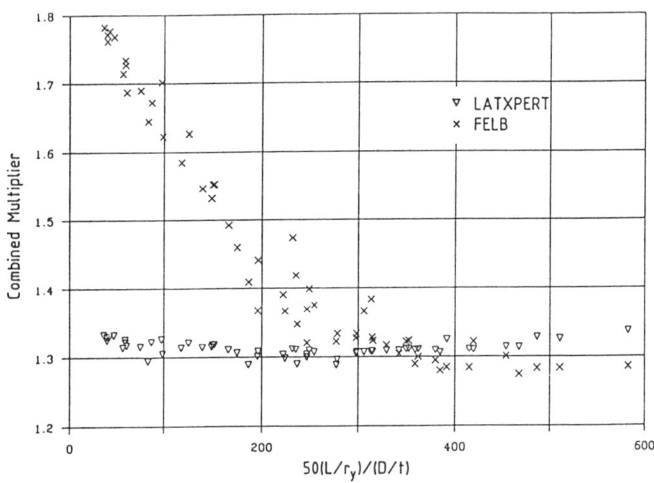

FIG. 2 PROPPED CANTILEVER WITH UNIFORM TOP FLANGE LOAD AND LATERAL ROTATION RESTRAINTS AT BOTH ENDS

Knowledge-Based Program

The VP-EXPERT Shell

A simple expert system shell would be the most appropriate tool for VP-EXPERT [Ref. 18] was selected for developing the knowledge based section of the package. This shell provides all logic and screen display functions for the expert system, and helps the user to create a knowledge base file which customises the VP-EXPERT shell to a specific subject. Each knowledge base file consists of three sections:

1. The ACTIONS block. This consists of one or more FIND commands, which instruct the shell as to what conclusions it is to seek.

2. The RULES. These are IF...THEN rules which contain knowledge, facts, rules of thumb, formulae, etc.

3. A series of ASK statements. These specify the way in which input data is to be sought from the user (usually in the form of a question, followed by a menu).

As an example of how the shell works, Fig. 3 provides an excerpt from the knowledge-based program. The three FIND commands instruct the

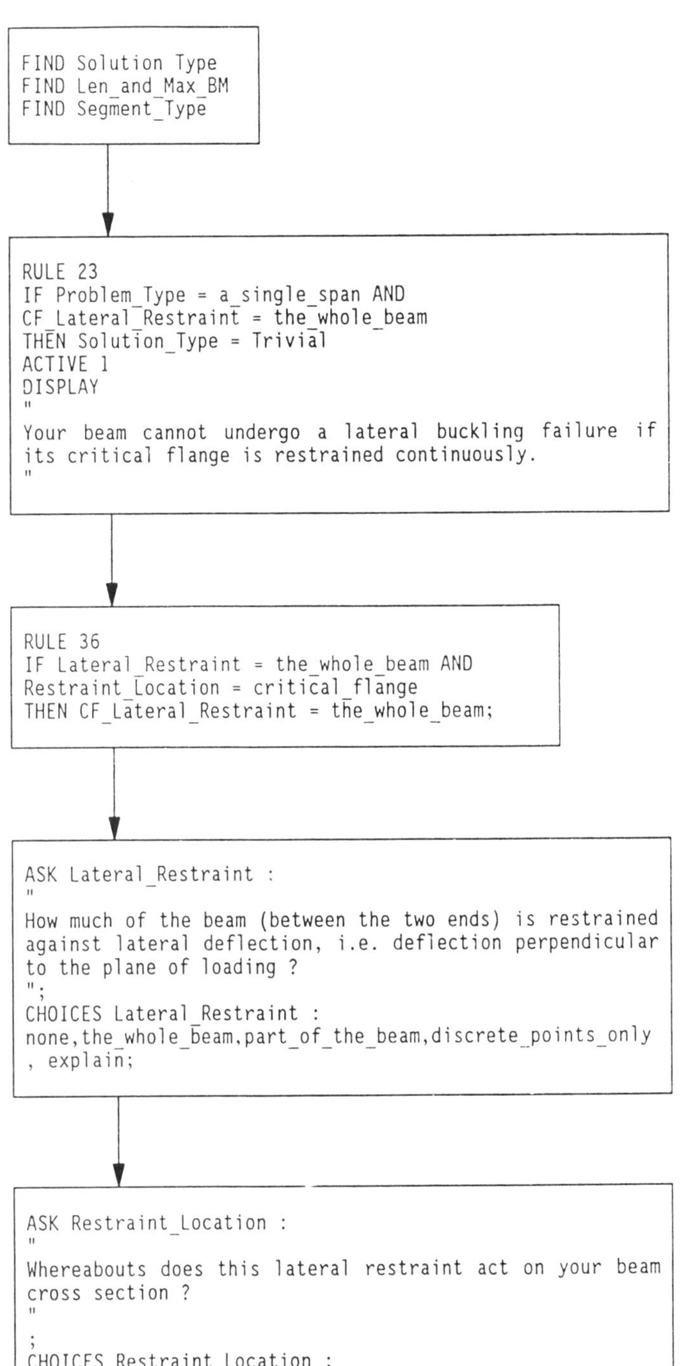

FIG. 3 EXAMPLE OF SHELL ACTION

shell that it is to establish, firstly, the "Solution Type" (simple single span, referral, etc.), then the length of the segment and its maximum bending moment, and then the "Segment Type" (unloaded segment, cantilever with lateral restraint, etc.). To find "Solution Type", the shell looks for a rule which has this variable in its conclusion, such as RULE 23. To see if this rule passes or fails, the IF part must be checked. Assuming that "Problem Type" has been entered as "a single span", "CF Lateral Restraint" must be evaluated. Again the shell chains to a rule with this in its conclusion, such as RULE 36. For this rule to be checked, both the variables "Lateral Restraint" and "Restraint Location" need to be found. These two variables do not appear in the conclusions of any rules, but they are provided with ASK statements. These instruct the shell to seek information from the user through a question and a menu of responses. If the user enters "the whole beam" and "critical flange" for the respective questions, then RULE 36 passes, which means RULE 23 then passes, and "Solution Type" has been found to be "Trivial" (i.e. no buckling problem). An explanatory message is then displayed.

Content and Program Structure

The clauses and formulae of the AS 4100 [Ref. 5] represent a considerable amount of expert knowledge condensed into a practical form. Such knowledge is ready made for programming into an expert system knowledge base, although the actual VP-EXPERT rules vary widely. Some clauses have been adapted directly, such as those dealing with stocky members deemed to have full lateral support. For such members, an explanatory message is displayed to inform the user that a buckling problem is not likely, quoting the clause of the standard from which the conclusion was derived. AS4100 clauses which deal with continuous lateral restraints, requiring the restraint to satisfy several criteria before it may be considered effective, have been programmed into the knowledge base as a number of separate rules, taking advantage of VP-EXPERT's backward chaining ability.

But perhaps more important is the general procedure used to tackle the problem of lateral buckling design, derived from human expertise rather than from AS 4100 itself. With the expert system it is possible to guide users through this design process, using explanation facilities to pass on knowledge of both terminology and the various factors which are important in lateral buckling design.

Figure 4 shows the overall structure of LATXPERT's knowledge based diagnosis program. In all there are 439 IF...THEN rules, which necessitated the divison of the knowledge base into six separate modules which are automatically executed in succession. The operation of the modules is most easily demonstrated with a sample design problem (see Fig. 5).

INTRO Knowledge Base

The first knowledge base to be executed is INTRO. The user is asked whether the problem falls into any of nine unusual categories which are presently beyond the scope of the knowledge-based system. If it does, then an information screen is displayed which refers the user to an appropriate computer program, research paper or other means of solving the problem. Every question asked by the expert system has "explain" as one of the menu options. Selecting "explain" for the question calls up an explanation screen. When this information has been read, the user simply strikes any key, and the question is repeated.

The next piece of information required is the cross-section type being used. The knowledge based diagnosis part of LATXPERT can only handle Channel sections and I-sections, and more sophisticated analysis program BEAMFELB is referred to if the user wishes to design some other section. Subsequent questions firstly determine the general type of problem being tackled, then the relevant quantitative details are sought.

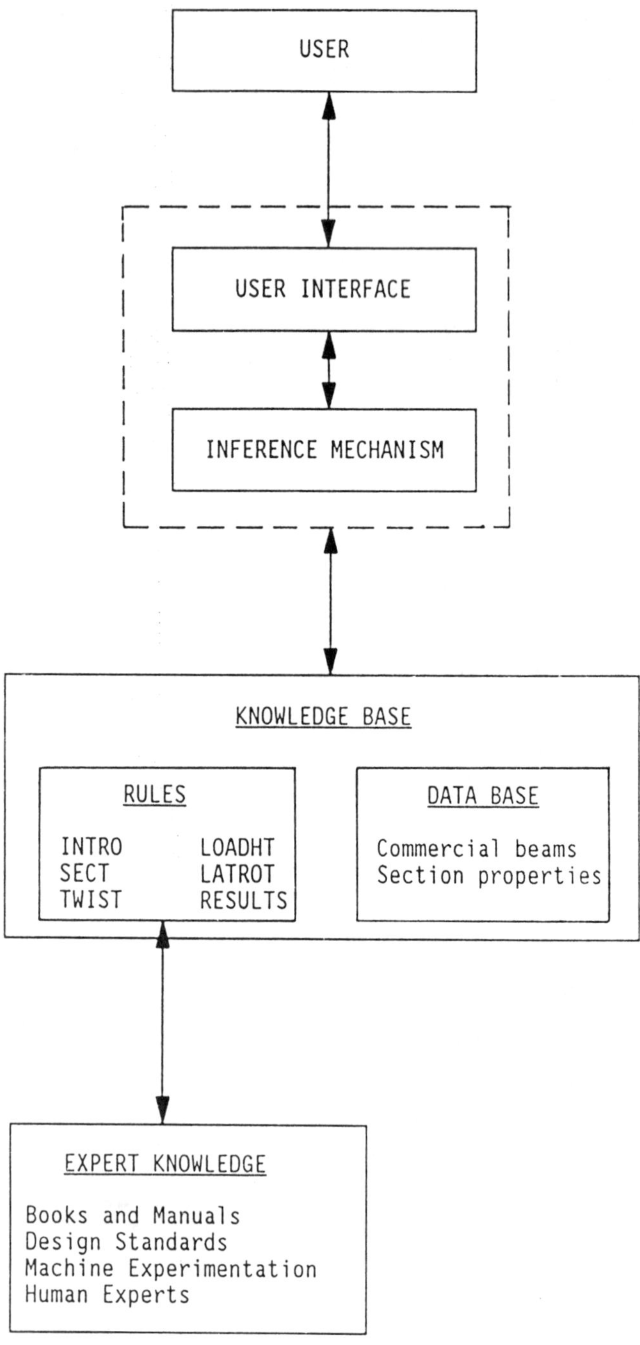

FIG. 4 STRUCTURE OF LATXPERT

The loading configuration is identified by approximating the bending moment distribution by one of those shown to one of a number of standard distributions although BEAMFELB may be used for a more accurate design. The type and extent of lateral restraints acting on the beam are then ascertained. In the case of the example problem (Fig. 5), the continuous lateral restraint is effective, since it acts at the critical flange (the segment is in negative bending). An explanation describes in detail what is meant by critical flange, and the user never has to consult a reference manual or textbook to understand any terminology used by the system. Finally the system requests quantitative data for segment length (6 m) and the maximum design bending moment M^* (195 kNm).

SECT Knowledge Base

The aim of the second knowledge base SECT is to determine a cross-section size to be used for calculating the reference buckling moment and assessing the effects of load height, end twist restraint and lateral rotational restraint.

The user may either select a standard I- or channel section from the manufacturer's range or, if using a welded plate girder, the cross-section dimensions. In either case the expert system advises the user of the smallest size of beam needed to satisfy the in-plane yielding requirement of AS 4100. In a design problem such as the example, the system recommends that this lower bound beam be chosen for the initial buckling check; however this may be over-ridden by a larger section choice if desired. The section properties are passed to the expert system, having been read from disk files (for the standard sections) or computed with formulae (for the user-defined sections).

With these values, the SECT module computes the reference buckling moment M_o. This reference value is then adjusted using a series of multipliers to account for the various important factors influencing buckling resistance. The first of these multipliers is equivalent to the moment modification factor α_m. Machine experimentation was used to determine the α_m factors for the various types of beam. For simply supported beams the α_m factors were found to be independent of slenderness. However, the α_m factors for cantilevers, were found to vary with the slenderness, and so piecewise linear interpolation was employed to represent these curves in the knowledge base.

For unloaded segments of continuous beams, the α_m factor is a function of both the slenderness and the ratio of the applied end moments, but again these factors were investigated by machine experimentation and programmed as rules.

The user is asked for information about the shape of the bending moment distribution (Fig. 6), which allows the expert system to match the bending moment described with the nearest standard case. Explain options with the questions allow the user to appreciate the importance of bending moment distribution to buckling resistance.

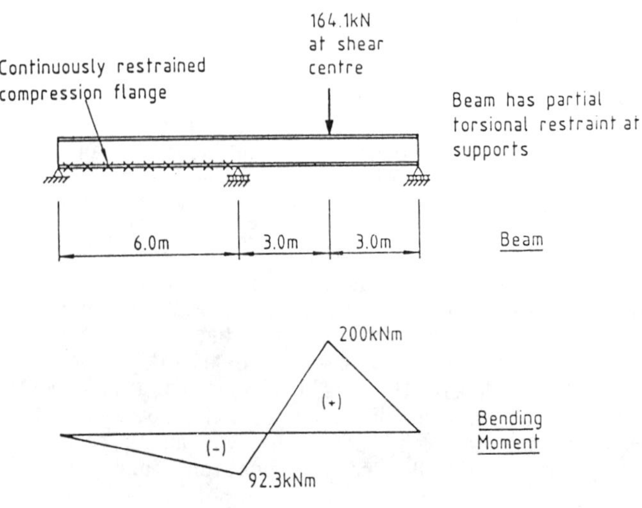

FIG. 5 LATXPERT EXAMPLE

Before leaving the SECT module, a check is made on the stockiness of the beam (or beam segment), to test the AS4100 formulae relating to stocky members deemed to have full lateral support.

TWIST Knowledge Base

The TWIST knowledge base considers the effects of torsional restraints at the ends of a beam on lateral buckling resistance. AS 4100 recognises four degrees of restraint (full, partial, lateral, and unrestrained). By selecting "explain", the user unfamiliar with AS 4100 terminology is given a detailed description of each term.

Machine experimentation was performed on a wide range of beams to obtain a multiplier which is applied to the reference elastic buckling moment. This moment multiplier shows the influence of the twist restraint conditions on the beam's capacity.

LOADHT Knowledge Base

The knowledge base LOADHT considers the important effect of the height of the load relative to the shear centre. Preliminary FELB runs showed that the load height buckling capacity multiplier depended primarily on the slenderness and the bending moment distribution, as well as on the load height itself.

It was possible to derive accurate buckling capacity multipliers for top and bottom flange loading for the full range of slenderness by machine experimentation with FELB.

LATROT Knowledge Base

The LATROT Knowledge Base contains rules for determining the effects on the elastic buckling moment of restraints against lateral rotation which may act at the ends of the beam. The adjustment to the reference buckling moment is achieved by a lateral rotation restraint multiplier, once again evaluated by machine experimentation with the finite element analysis program FELB.

RESULTS Knowledge Base

Finally RESULTS, the last of the six knowledge base modules, is executed, with a choice of output to printer or screen (Fig. 7). First, the main items of input data are echoed for checking, and the slenderness and span-to-depth ratios are displayed for information.

The reference elastic buckling moment from the SECT module is then modified by the series of multipliers for the bending moment distribution, the load height, the twist restraint and the lateral rotation restraint to give the elastic buckling moment. Each of these multipliers is displayed for the user to observe the relative importance of each effect. The elastic buckling moment is then converted to an estimate of real beam capacity, taking into account the interaction of elastic buckling and yielding and the effects of crookedness and twist. Once the design bending capacity M_b is known (Eq. 6), dividing by the actual maximum moment in the beam gives the design load factor against out-of-plane buckling failure. This load factor is compared with the load factor against in-plane yielding failure, allowing the user to compare the relative importance of the modes of failure.

```
Consider the bending moments around the CENTRE of your beam segment.

Is the greatest moment in this central region equal (or close) to Mmax,
or is it substantially less than Mmax ?

 eq or close to Mmax ◄    much less than Mmax      explain


Now consider the bending moments at each END of your segment.

Which of the following is applicable in describing their magnitude ?

 both ends appr Mmax      one end appr Mmax        both ends appr zero ◄
 explain


Is the region of maximum moment localised (e.g. due to a point load), or
is it fairly extensive (e.g. due to a UDL) ?

 [localised]              extensive                explain

↑ ↓ → ←  Enter to select   END to complete   /Q to Quit   ? for Unknown
```

FIG. 6 MOMENT DISTRIBUTION QUESTIONS

```
                    LATXPERT - CONSULTATION OUTPUT


DATA FOR THIS RUN :

Section being used = 360 UB 50.7.
L/ry = 155.00.
Span to depth ratio = 16.87.
Slenderness parameter for calculation of multipliers = 164.19.

Load is acting at : shear centre.
Twisting restraint at ends is : partial partial.
Lateral rotation restraint at : one end.

ELASTIC BUCKLING

The reference elastic buckling moment for your beam is 135.4 kNm.

The multiplying factor for bending moment distribution is 1.36.
The multiplying factor for load height is 1.00.
The multiplying factor for end twisting restraints is 0.90.
The multiplying factor for lateral rotation end restraints is 1.71.

Thus the maximum moment at elastic buckling is estimated at 283.44 kNm.

STRENGTH

The full plastic moment of your beam is 223.8 kNm.

The multiplying factor for bending moment distribution is 1.36.
The multiplying factor incorporating all other effects is 0.57.

Thus the maximum moment at out-of-plane failure is estimated at 156.44 kNm.

A summary of your design :

Load factor against failure by IN-PLANE YIELDING =    1.01.
Load factor against failure OUT-OF-PLANE         =    0.78.

NOTE : both these load factors include the capacity factor φ = 0.9.

For the next run it is suggested that you try a 410 UB 59.7. This will probably
raise your load factor against out-of-plane failure to the required 1.00.


                    END OF BUCKLING CONSULTATION
```

FIG. 7 OUTPUT FOR FIRST LATXPERT CONSULTATION

The knowledge-based problem diagnosis not only provides the user with a pair of load factors, but also gives some further design advice if the out-of-plane buckling load factor is less than 1.0. The formula for M_b (Eq. 5) suggests that simply factoring up the plastic capacity M_s should increase the buckling capacity proportionally. However, experience revealed that when M_s is increased, α_s also increases, so that a proportional increase results in a sometimes considerable overestimate of the new cross-section size required. After some experimentation, an empirical method was derived to estimate the new section size more accurately by finding a correction factor which is applied to the proportional method:

This is achieved by rules in the RESULTS knowledge base, and a BASIC program is used to search for the smallest cross-section with the required plastic capacity. This is returned to the expert system and displayed, so that the user may retry the problem with the new section. If the correct advice has been given, the load factor against lateral buckling should be just over 1.0 with the new section.

The WHAT IF Knowledge Bases

The knowledge-based section of LATXPERT has the ability to carry out a "What If?" evaluation after a run, to observe the effect of changing one of the loading, cross-section or restraint conditions. For example, a user may wish to check the buckling capacity obtained with top flange loading instead of shear centre loading, or the effect of providing full torsional end restraint instead of partial restraint. To carry out a "What If?" run the appropriate one of 4 slightly modified knowledge base modules SECTWI, TWISTWI, LOADHTWI, LATROTWI is consulted.

For the example problem above, the system suggested that a 410UB59.7 be tried in order to raise the lateral buckling safety factor above 1.0. To try this out, it is not necessary to re-run the whole expert system. The user only runs the SECTWI module, and chooses the 410UB59.7. All the ensuing knowledge bases are executed automatically to evaluate new twist restraint, load height, and lateral rotation restraint multipliers (which are a function of slenderness, and therefore depend on the section size). Note that no other input is required by the user, as the initial problem data is retained in a disk file. After a short time the RESULTS module is re-entered, and the new load factors for yielding and buckling are available.
Similarly, other aspects of the problem may be changed using their corresponding "What If?" knowledge bases. For example, to check the effect of critical flange loading instead of shear centre loading in the original problem, the LOADHTWI module would be run. This convenient "What If?" capability is a feature of the rule-based machine experimentation approach - a revised answer is available immediately, whereas an analysis program would have to be completely re-executed.

BEAMFELB

When a problem is too difficult for the knowledge-based program, the user may use BEAMFELB, a finite element analysis tool for solving complex lateral buckling problems. This uses an on-screen editing style of data entry. A main menu allows the user to access any one of the three groups of data required. These are geometry and restraints (Option 2), cross- section properties (Option 3), and loads (Option 4). The use of these are fully illustrated for sample lateral buckling check in Reference 19.

Option 5 in the main menu enables the user to create a problem data file which can be recalled at a later stage for alteration and/or repeat analysis. The elastic buckling analysis is executed by choosing Option 6. The user can view the results using either Option 7 or Option 8. Option 7 gives a screen or printer output of the numerical results, while Option 8 draws 3-D picture of the beam's buckled shape as shown in Fig. 8.
The user may perform "What If?" type investigations simply by changing one or more of the data items previously entered. By comparing the buckling safety factors obtained, the user can quickly appreciate the important factors that affect the elastic lateral buckling resistance of a beam.

A conversion from elastic buckling to design strength is carried out in BEAMFELB using the upper tier design method of AS 4100.

BEAMLSC

When the accuracy of BEAMFELB is not warranted, the program BEAMLSC may be used. This uses the lower tier of the new Limit State Code AS 4100 and is limited to top flange loading, shear centre loading, or bottom flange loading, and the most common restraint conditions. Further details of BEAMLSC are given in Reference 19.

CONCLUSIONS

This paper summarises LATXPERT, which provides an expert system for the design of steel beams against lateral buckling (a more detialed description is given in Reference 19). LATXPERT contains a knowledge-based diagnosis tool which enables the user to determine if there is likely to be a buckling problem, if a simple solution is available, if a buckling analysis computer program is required, or if specialist advice should be sought. The rules of the knowledge base are obtained from textbooks, manuals, design standards, and human experts. This knowledge is supplemented by rules derived by machine experimentation using computer programs for cross-section and elastic buckling analysis. A data base of commercial beams and their section properties is integrated with the rules.

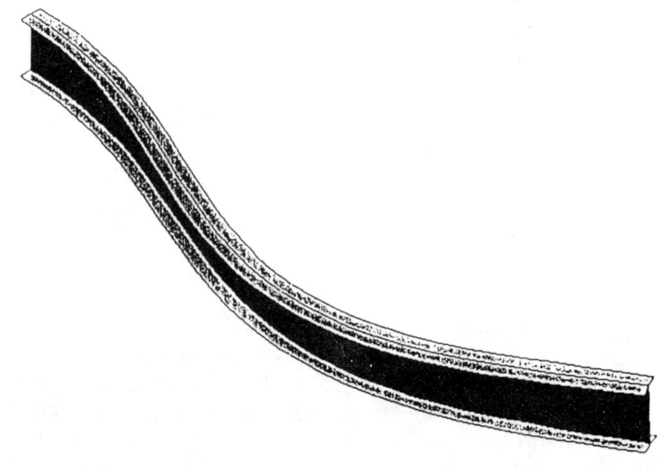

FIG. 8 BUCKLED SHAPE

LATXPERT includes "What If?" facilities which enable the designer to explore the consequences of changing various data items which affect lateral buckling. It is presented in a user-friendly format and with explanation options which help to build the confidence and expertise of the user.

LATXPERT is complemented and extended by two user-friendly computer programs BEAMFELB and BEAMLSC, which perform the elastic buckling analysis of steel beams and determine their design capacities. BEAMFELB allows a wide range of steel beams to be analysed and designed by using the design by buckling analysis method (upper tier) of AS4100, while BEAMLSC performs similar tasks using the less accurate lower tier of AS4100.

It is concluded that the process of designing steel beams against lateral buckling may be simplified by using an expert system such as LATXPERT, which increases the designer's knowledge of the lateral buckling phenomenon, dispels the confusion that may be generated by the complicated rules of some design standards, and avoids the use of these rules as a "black box". It is further concluded that an expert system may be used to offer much wider guidance than that available in design standards, and when combined with user-friendly computer programs may allow the solution of a much wider range of lateral buckling design problems.

REFERENCES

1. Trahair, N.S., and Bradford, M.A., *The Behaviour and Design of Steel Structures*, 2nd ed., Chapman and Hall, London, 1988.

2. Hancock, G.J., and Trahair, N.S., "Finite Element Analysis of the Lateral Buckling of Continuously Restrained Beam-Columns", *Civil Engineering Transactions,* Institution of Engineers, Australia, CE20, (2), 120-7, (1978).

3. British Standards Institution, *Structural Use of Steel in Building, Part 1, Code of Practice for Design in Simple and Continuous Construction: Hot Rolled Sections*, BSI, London, 1985.

4. American Institute of Steel Construction, *Load and Resistance Factor Design Specification for Structural Steel Buildings*, AISC, Chicago, 1986.

5. SA, *AS4100-1990, Steel Structures,* Standards Australia, Sydney, 1990.

6. Rychener, M.D., (ed), *Expert Systems for Engineering Design*, Academic Press, Boston, 1988.

9. Barsoum, R.S., and Gallagher, R.H., "Finite Element Analysis of Torsional and Torsional-Flexural Stability Problems", *International Journal of Numerical Methods in Engineering*, 2, 335-352, (197).

10. Hancock, G.J., and Harrison, H.B., "A General Method of Analysis of Stresses in Thin-Walled Sections with Open and Closed Parts", *Civil Engineering Transactions*, Institution of Engineers, Australia, **CE14**, (2), 181-188, (1972).

11. Hancock, G.J., and Rasmussen, K.J.R., *User's Manual - Programs OPEN4 and GOPEN4*, School of Civil and Mining Engineering, University of Sydney, 1989.

12. Vlasov, V.Z., *Thin-Walled Elastic Beams*, Israel Program for Scientific Translations, Jerusalem, 1961.

13. BHP, *Hot Rolled Products*, BHP Steel, Melbourne, 1989.

14. Trahair, N.S., and Papangelis, J.P., *LIMSTEEL Version 2.0 User's Manual for the Limit States Design of Steel Structures According to AS4100-1990*, Australian Institute of Steel Construction, Sydney, 1990.

15. Adeli, H., and Balasubramanyam, K.V., "A Knowledge-Based System for Design of Bridge Trusses", *Journal of Computing in Civil Engineering*, ASCE, 2, (1), 1-20, (1988).

16. Harrison, H.B., *Structural Analysis and Design - Some Microcomputer Applications*, 2nd ed., Pergamon Press, Oxford, 1990.

17. Sawyer, B., Moose, A., and Shafer, D., *VP-Expert Rule-Based Expert System Development Tool*, Paperback Software International, Berkeley, CA, 1987.

18. Lai, C.T.K., Martin, C.M., Hancock, G.J., Papangelis, J.P., and Trahair, N.S., "An Expert System for Lateral Buckling Design", *Research Report R627*, School of Civil and Mining Engineering, University of Sydney, December 1990.

7. SAA, *AS1250-1981, SAA Steel Structures Code*, Standards Association of Australia, Sydney, 1984.

8. Timoshenko, S.P., and Gere, J.M., *Theory of Elastic Stability*, 2nd ed., McGraw-Hill, New York, 1961.

A KNOWLEDGE REPRESENTATION SCHEME FOR THE DESIGN OF HYBRID STRUCTURAL STEELWORK CONNECTIONS

T.J. McCarthy and Z. Nouas

Department of Civil and Structural Engineering, UMIST, Manchester, England

ABSTRACT

This paper describes a KBS for the design of hybrid connections in structural steelwork. The system is based on an on-going research programme aimed at providing design guidance for those connections. The system is developed using the expert system shell, GOLDWORKS II. The approach is a combination of object-oriented programming and rule-based programming. The main components of the system are described. The paper shows the suitability of such an approach in developing a KBS for an engineering design problem. It also discusses some benefits of developing a KBS alongside a research programme where the domain knowledge is not well established.

INTRODUCTION

Hybrid connections are those where different types of connectors (friction grip bolts and welds) work together in resisting the loads applied to the joint. Their behaviour and analysis is more complicated than conventional joints. Little guidance can be found in the general literature or codes on how to design such connections and therefore, structural designers tend to avoid them. However, situations do arise where these connections can be used. Strengthening an existing bolted connection may be required, due to an increase in the loading, for example, and the use of hybrid connections may be a cost effective solution. Also, in continuous construction, it may be required to have the connection that can develop the full plastic moment capacity of the beam (eg. in the case of earthquake resistant structures). In such cases, with some type of connections, it is not possible to make bolted only connection match the strength of the beam and a hybrid solution is required (Ref. 1).

A research program into the behaviour of combined HSFG bolted and welded connections has been carried out in the Structural Assessment Unit at UMIST over the past ten years. The aim of this programme is to provide guidance for the design of these connections. The KBS described in this paper is based on the findings of this research programme. It is used as a vehicle for implementing the new design guidance and as a test bed for the design recommendations that are made.

BACKGROUND TO KNOWLEDGE REPRESENTATION

Knowledge representation is an important question in developing a knowledge-based system. Several representational schemes have been developed in the past and a central issue for the KBS developer is to decide which scheme to use for a particular application. Unfortunately, there are no simple recipes for selecting the right scheme. The selection of the knowledge representation scheme is often constrained by practical considerations such as availability of hardware and software, programming skills as well as practical experience in knowledge-based system development. Moreover, if the system is developed using a commercial shell , which is generally the case, the developer has no alternative but to use the representation scheme(s) provided by the development tool. However, the decision should take into account both the overall problem solving approach and the domain problem. The scheme should allow for a natural description of the facts, information and entities which constitute the object domain. Also, it should be possible to naturally express the problem solving approach adopted for the problem at hand.

Different problem-solving approaches have been developed for solving various engineering problems. The most popular of these approaches is the rule-based approach. It has been used in the first expert systems such as MYCIN (Ref. 2,3,4). Other problem-solving paradigms such as object oriented approach and blackboard approach have since emerged as powerful methods, showing the progress made in knowledge representation and problem-solving techniques.

Rule and frame based systems

The main components of a typical rule-based system are the knowledge base, the inference engine and the working memory.
The knowledge base holds a representation of the knowledge the expert uses to solve the problem. Whereas the developer has no control over the inferencing mechanism, he or she has more latitude to organise the knowledge base. Basically domain knowledge falls into two categories, passive or static knowledge and active knowledge. A fact an expert knows to be true is considered passive knowledge, while a method he uses to make deductions is known as active knowledge.

Passive knowledge or factual knowledge is a term which stands for all the facts, general information and entities which make up the object domain of the problem. Symbols or list of symbols may be used to represent simple facts and general information. The O.A.V (Object Attribute Value) representation is commonly used to represent such knowledge. However, in real-word engineering problems, domain objects can rarely be successfully represented using only the techniques mentioned above. Indeed, reasoning about an entity of the domain problem often requires accessing the underlying structure of that entity. For example, reasoning about a beam object in structural engineering often requires the description of the beam in terms of the material it is made of, its section properties etc... Frame-based representation has proved very practical in that respect (Ref. 5).

In a frame-based representation, related pieces of knowledge about a typical domain object are grouped together in a single data structure. At the developer level, a frame is perceived as a collection of slots and fillers. Each slot stands for a particular attribute or property of the entity being described. An instance of the frame is created by supplying values for the slots. Each slot of the frame can be further described by facets which can hold more specific information about the slot. Commercial shells, such as GOLDWORKS II, with system defined facets which provide additional functionality to a slot. For example, if the facet :when-accessed of a slot contains the name of a procedure, this later is called each time the slot is accessed. Similarly, the facet :when-modified can point to a procedure which is called when the value of the slot is changed (Ref. 6). These procedures are also referred to as daemons.

Also, using a frame representation the domain object can be organised into a hierarchy. A frame in the hierarchy is taken to represent a class of similar objects. By adding more slots to the initial frame, more specific frames can be formed to represent sub-classes of the initial class. An important feature of the frame-based representation is inheritance. A child frame in the hierarchy inherits all properties of the parent frame, thus making storage of information more efficient.

Active knowledge stands for the kind of information used by an expert to make inferences and draw conclusions. Certain tasks can be procedural by nature and lend themselves to a procedural treatment. For example, in structural design the analysis phase is still treated using a conventional programming approach. Other tasks such as preliminary design, rely mostly upon heuristics and rules of thumb gained from past design experience. These tasks are generally difficult to reduce to simple algorithmic procedures. Rather, the related knowledge is best formulated as if-then rules (Ref. 7,8).

Two basic control regimes are used by the inference engine to process the rules. Under a forward chaining control regime, conditions of rules are matched against the working memory, and their conclusions added to the working memory probably causing further forward chaining. This process continues until a solution is reached or no more rules can be used to advance the solution state. In backward chaining, on the other hand, the rules are scanned and those whose conclusions can assert a particular conclusion (goal) are found. For each of these rules a check is made whether its conditions match against the working memory. If a condition does not match, it is set as a new subgoal to establish. This recursive process continues until subgoals are proved and therefore, the initial goal ascertained. Sometimes, subgoals cannot be proved and the system must resort to querying the user. If multiple rules qualify, a conflict-resolution strategy is applied by the inference engine to select the rule to consider first.

Forward and backward chaining are sometimes combined in what is called mixed-mode chaining. The idea behind this strategy is to find the most efficient inference process by switching from forward to backward chaining and vice versa. An example of mixed-mode chaining is backward reasoning with opportunistic forward chaining or goal directed forward chaining where the availability of data is opportunistically exploited by a forward chaining (Ref. 9,6).

In general, rules are organised into rule sets. Each rule set is specialised for a particular task and meta-rules are used to choose the appropriate rule sets. This approach is preferred to a single rule set for ease of debugging and maintenance and speed of execution.

Blackboard approach

The blackboard approach views problem-solving as the cooperative effort of various knowledge sources or specialists working together to solve the same problem. The knowledge about the problem is distributed among a set of specialists, each of which has a particular area of expertise. The evolving solution of the problem is recorded in a global data base, the blackboard, through which the specialists communicate and which is controlled by an inferencing mechanism. Ideally a specialist can write onto the blackboard and contribute to the problem solving on an opportunistic basis. However, in most blackboard implementations, the inferencing mechanism or one specialist has the control function of guiding the problem solving activity. The blackboard approach was first implemented in the speech recognition system called HEARSAY II (Ref. 10). More recently, the concept was adopted in systems such as DESTINY (Ref. 11) for integrating the various activities of structural design and SPEX, a knowledge-based structural component design system (Ref. 12).

Object-oriented programming (OOP)

The basic principle of object-oriented programming is the encapsulation of data and procedures in prepackaged software components called objects. Traditionally, conventional programming has generally favoured a procedure-oriented approach to problem-solving. This consists in applying procedures to data. Thus a clear distinction is made between passive data and active procedures. No such distinction exists in the object-oriented approach which views problem-solving knowledge as a collection of self-contained objects, each containing both specific data and their corresponding procedures. In object oriented programming, the problem-solving activity occurs through message passing between the objects. When an object receives a message, it executes the corresponding procedure, also called method, which operates on its private data and eventually sends messages to other objects.

As with a frame-based representation, objects in OOPs are organised into a hierarchy and inherit all properties of their ancestors. In addition, objects inherit methods from their parents in the hierarchy.

Another important feature of object-oriented programming is information hiding. The sender of the message does not know how the object processes the message. The same message can therefore be sent to different objects which will process that message in different ways. As a result of information hiding, modification or updating of knowledge about an object is made easier. Any modification of the knowledge about an object is local to that object and will not affect the rest of the program.

Object oriented approaches have been first used in graphics and simulation systems. Relatively few applications in civil and structural engineering problems have been reported. Fenves (Ref. 13) demonstrated the application of the object-oriented approach for structural analysis and design. Varghese (Ref. 14) developed an object-oriented based representation scheme for the design of reinforced concrete beams. Forde (Ref. 15) and Miller (Ref. 16) described programs using the object-oriented approach for structural analysis using the finite element analysis method.

HYCON KNOWLEDGE BASED SYSTEM

HYCON is a knowledge-based system for the design of hybrid connections in structural steelwork. HYCON acts as a design assistant and helps a structural engineer investigate various strengthening alternatives. Given an existing bolted connection, the system creates various solutions for strengthening the connection and presents them after evaluation of their relative merits. However, the user can override the decision of the system.

Different strengthening strategies can be adopted by the user. He can either allow the connectors to develop their full plastic capacities or restrict their behaviour to elastic. Also, initial loading of the joint is permitted before strengthening. This allows the end user to consider strengthening the joint without having to jack the structure up.

Currently, the system deals only with basic construction joints, such as a double angle beam to column connection or a flexible end plate beam to column connection where the load path through the joint is either in plane shear or in plane shear plus moment. Two other types of connections will be added in the future, namely, the extended end plate beam to column connection and the flushed end plate beam to column connection.

The approach.

The approach adopted in HYCON can be described as a combination of object-oriented and rule-based approaches. This was the result of two considerations. Firstly, we needed a system capable of undergoing extensive modifications from time to time. This is due to the fact that we are dealing with an immature problem domain where the knowledge has not yet taken its final form. Preliminary results from the ongoing research on hybrid connections are used by the system and the corresponding knowledge may need to be updated. The object oriented approach seemed suitable in that respect. Also, the object oriented approach was considered because it allows a direct correspondence of the real objects of the problem domain and those object in the system. This proved to be very helpful during the knowledge elicitation phase as well as aiding the validation process. During validation 'faulty' objects were easily identified and their behaviour corrected.

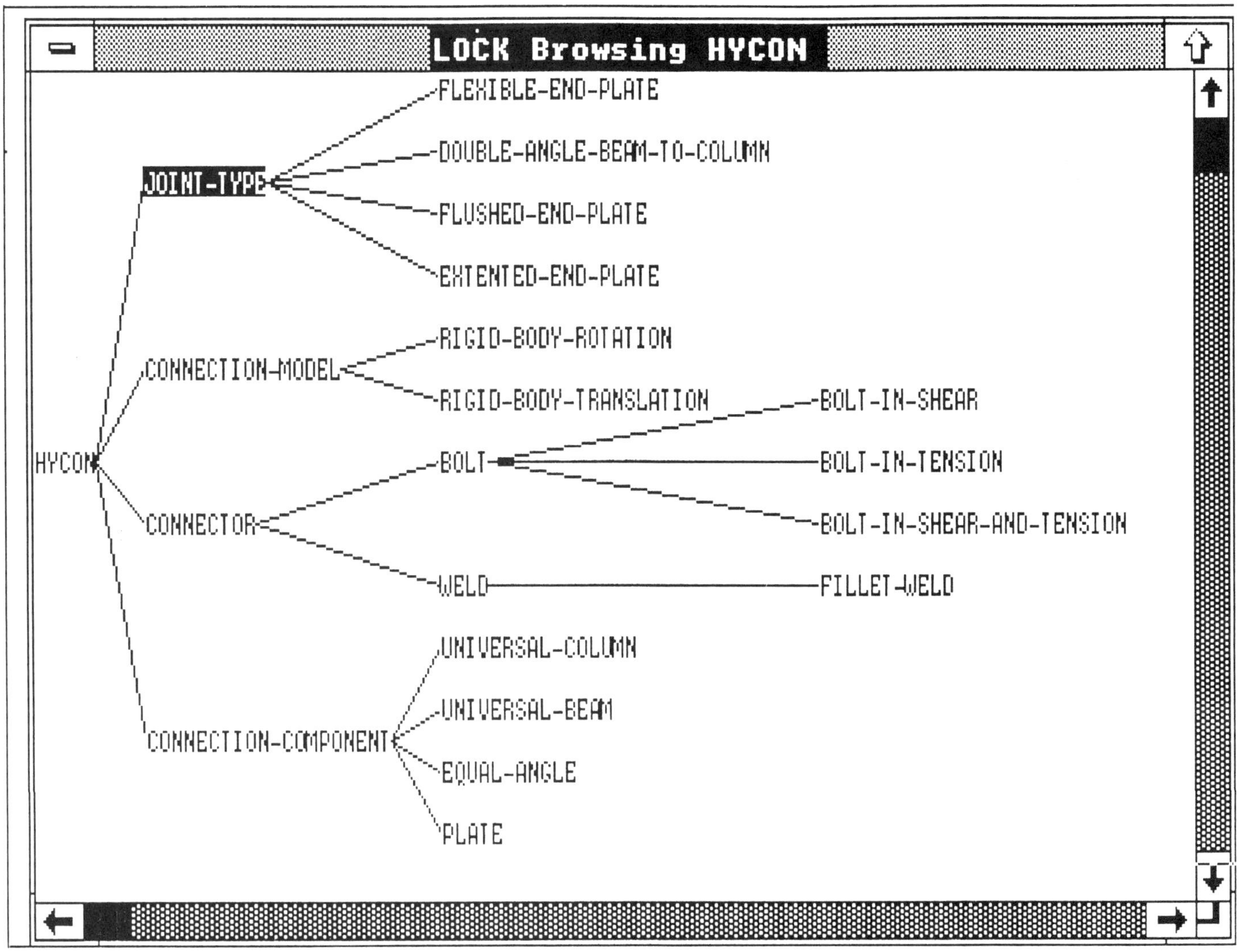

LOCK Browsing HYCON

HYCON
JOINT-TYPE
FLEXIBLE-END-PLATE
DOUBLE-ANGLE-BEAM-TO-COLUMN
FLUSHED-END-PLATE
EXTENTED-END-PLATE
CONNECTION-MODEL
RIGID-BODY-ROTATION
RIGID-BODY-TRANSLATION
CONNECTOR
BOLT
BOLT-IN-SHEAR
BOLT-IN-TENSION
BOLT-IN-SHEAR-AND-TENSION
WELD
FILLET-WELD
CONNECTION-COMPONENT
UNIVERSAL-COLUMN
UNIVERSAL-BEAM
EQUAL-ANGLE
PLATE

Figure 1 Part of HYCON knowledge base

On the other hand, the rule-based approach proved a natural way to formulate the heuristical nature of some design decisions. These include the different strengthening alternatives to consider or evaluating and comparing different solutions. In HYCON, an object can have, in addition to conventional procedures, local rule bases which can be processed using backward or forward chaining. This approach differs from the classical view of domain knowledge divided between a central or several rule sets representing the active knowledge and an object domain represented by an organised collection of passive entities. Rather, this active knowledge is distributed among domain objects.

The knowledge base

Two kinds of objects were identified during the knowledge elicitation phase. Some objects, like a *beam* object, a *column* object or a *plate* object have a direct mapping onto the real world entities they stand for. They are referred to as physical objects. Conceptual objects, on the other hand do not correspond to physical entities of the problem domain. Rather, they represent concepts which have been identified by the expert and are used in solving the problem. An example of a conceptual object is the rigid-body-in-rotation model. Figure 1 shows part of HYCON object domain. The main objects are the joint components, the connectors, the connection models and the joint types.

Joint components

The joint component class of object has sub-classes such as beam, column or plate. These primary objects are used to make up different types of joint. The same primary object, once it has been stored in the knowledge base, is used in various joint configurations. For example, it was found particularly practical to use the beam object in different joint types such as a flexible-end-plate beam to column connection or in a double-angle-beam-to-column connection (Figure 2). The beam object is armed with its own methods. For example, when the local shear capacity of the beam is required, it is simply obtained by sending the message :local-shear-capacity to the beam instance. This latter will use the corresponding method and return its local shear capacity.

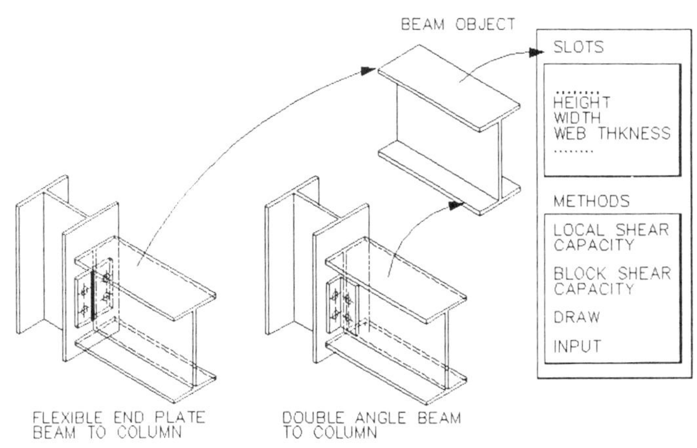

Figure 2 Object reusability

The connectors

HSFG bolts are used in combination with welds to transfer loads between the joined parts of a hybrid connection. The relative deformation capabilities of both types of connectors, welds and bolts, is of primary importance in hybrid connection. It was found particularly useful to store this kind of knowledge along with other basic characteristics of the connector within the connector object itself (Figure 3). For example, the load-deformation curve of an HSFG bolt in shear is attached to the bolt itself. The bolt can therefore answer to requests such as *:displ-ult* (ultimate displacement of the bolt) or *:displ-allow* (allowable displacement of the bolt). The load-deformation curve allows the bolt to determine the force acting on it, given its displacement. Load-deformation curves of both weld and bolts are based on the research carried out by the Structural Assessment Group at UMIST (Ref. 17). Also, since initial loading of a joint before strengthening is permitted, it was found practical to store bolts initial displacements in the bolts themselves.

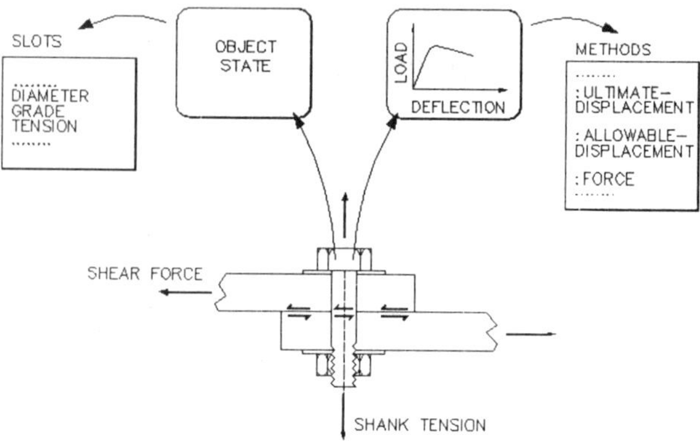

Figure 3 HSFG-bolt-in-shear object

Connection models

Connection models are used to describe the load path within a joint. When a joint is created, instances of connection models are created to describe the load path. Currently, only two connection models are used by the system, i.e. the rigid-body-translation model and the rigid-body-rotation model. The rigid-body-translation conceptual object models the load transfer between two joined parts of the connection when their relative movement is assumed to be a simple in plane translation. The rigid-body-rotation object models the load transfer between two parts when their relative movement is an in plane rotation due to an eccentric shear load. In both models, the connected parts are supposed to move as rigid bodies. The connectors may be bolts, welds or a combination. These two models are depicted in Figure 4.

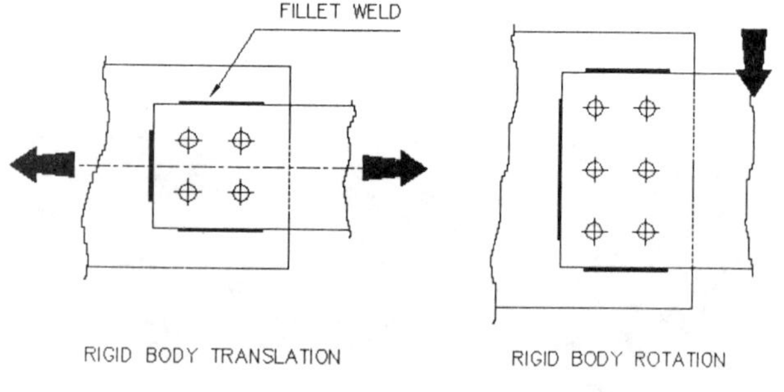

Figure 4 Connection models

These objects have different methods associated with them. For example, the message *:displ* can be sent by a bolt to the appropriate instance of the connection models to have the displacement at a particular location (the position of the bolt for example). Another method associated with both objects is the method *:capacity* which returns the capacity of the particular instance. These two messages are the same for both types of models but they are treated differently.

Joint types

A joint object represents the actual joint the user wants to strengthen. Two joint types are already in the knowledge base, namely a flexible end plate beam to column connection and a double angle beam to column connection.

Several slots are used to describe the joint in terms of the different components it is made of, how the different parts are connected etc... Slots such as column or beam are used to point to a particular instance of column or beam. A slot such as angle-to-beam (in the case of a double angle beam to column connection) points to an instance of the rigid-body-rotation model since the angle to beam connection is supposed to behave as such.

At the creation of an instance of a joint, a rule base attached to the instance is responsible for defining the load path. The rule base creates instances of connection models to describe how different parts of the joint are connected.

Different methods are attached to a joint object. Two important methods are the capacity method and the strengthening method. Basically, the capacity of the joint is obtained by scanning the different joint components as well as the connection model instances that describe the load path and sending the appropriate messages. The strengthening method, on the other hand, uses a rule base associated with the joint to produce strengthening alternatives. The strengthening rule base is composed of different rule sets. The feasibility rule set checks whether strengthening the joint to a given load is feasible. The *hybrid-create* rule set is responsible for the creation of different strengthening alternatives. The *analysis* rule set is responsible for checking the strengthening alternatives proposed by the previous set. Finally, the *refine* rule set modifies the strengthening alternatives until they satisfy the capacity requirement. Eventually some alternatives may be discarded if they prove unfeasible.

HYCON interface

The interface is the component of the system through which the user interacts with the system. The user interface is also developed using the object-oriented approach. It uses system defined objects such as screen-layouts, pop-up menus available in GOLDWORKS II.

The interface relies also on drawing methods associated with almost all the objects. A beam object for example has a *:draw* method responsible for drawing itself. Arguments can be used when sending the message *:draw* specifying the scale, the location, the view etc... A compound object like a joint also has its drawing method. When the joint object receives the message *:draw* it will forward the same message to all its components with the appropriate arguments.
The interface is designed to allow different strengthening strategies to be investigated by the user. This is made possible by saving the current environment to a checkpoint file and restarting later on from that point.

DISCUSSION

An unusual feature of HYCON is the immature nature of the domain knowledge. This is due to the fact that the system is derived from an on-going research programme rather than on a past practical experience. Developing the software alongside the research (but with a time lag) was only possible because of the flexibility of the KBS approach.

By combining the development of the KBS with the ongoing research a number of benefits accrued. One of the benefits was to have the researcher think carefully on how to formulate their knowledge and look at the practical implication of their work. The system was used by them as a simulation tool and let them investigate various approaches.

Presently, HYCON considers the strengthening problem of existing bolted connections only. The authors believe that it can, without great modification, be used to design basic bolted or welded connections.

The approach that was adopted in HYCON benefitted from the development environment provided by GOLDWORKS II. Its hybrid nature allows the combination of the rule-based and the frame-based knowledge representations. Access to the Lisp programming language is possible from within the rules as well as through the daemons. This allows the developer to define new functions as required. Through its frames and their attached methods the shell allows object-oriented programming. The graphic toolkits and the Windows environment permits the efficient development of user-friendly interface. The one weakness of the shell was the relative difficulty in manipulating the explanation facility.

CONCLUSION

This paper has presented a flexible approach to building a KBS. The object oriented approach has been extended to include the attachment of local rule bases to specific objects. This approach has applications in many types of KBS. In particular, it is suited for engineering design problems. The use of objects allows the developer and the expert to concentrate on specific aspects of the problem at any time.

It is feasible to develop an object oriented KBS for domains that are still undergoing research. Indeed, one of the many benefits of undertaking the knowledge elicitation at an early stage is that it helps the researchers to formulate their expertise. In a university research environment, this type of development has added relevance, since most researchers are on three year contracts. Also, as the research programme expands, so too can the KBS.

REFERENCES

1. Burdekin, F.M., Gurtuna, A.G., Rahal, H.M., Abdelfattah, F.A. and Watts, W.S., Combined HSFG Bolted and Welded Connections, Proc., Int. Conf. Welded Structures '90, London, UK, 26-28 November 1990, pp. 128-138., Pub. by The Welding Institute, 1990.

2. Hayes-Roth, F., Waterman, D.A., Lenat, D.B. (eds), Building Expert Systems, Addison-Wesley, 1983.

3. Winston, P.H., Artificial Intelligence, Addison-Wesley, 1984.

4. Buchanan, B.G. and Shortliffe, E.H., Rule-Based Expert Systems, The Mycin Experiments of the Stanford Heuristic Programming Project, Addison-Wesley, 1985

5. Balachandran, M. Gero, J.S., A Knowledge-Based Graphical Interface for Structural Design, in Sriram, D. and Adey, R.A. (Eds) Artificial Intelligence in Engineering: Tools and Techniques, Computational Mechanics Publication, Southampton, 1987.

6. Gold Hill Computers, GOLDWORKS II, Reference Manual, Gold Hill Computers, Inc. Cambridge, MA, USA, 1989.

7. Maher, M.L. and Fenves, J.S., HI-RISE: An Expert System For The Preliminary Design of High Rise Buildings, in Gero, J.S. (Ed.), Knowledge Engineering in Computer-Aided Design, Elsevier Science Publishers B.V. (North-Holland), 1985, pp. 89-117.

8. Adeli, H., An Overview of Expert Systems in Civil Engineering, in Adeli, H. (ed.), Expert Systems in Construction and Structural Engineering, Chapman and Hall, 1988.

9. Creative Logic Limited, LEONARDO, an Expert System Development Tool, Middlessex, England.

10. Reddy, D.R., Erman, L.D. and Nelly, R.B., The HEARSAY Speech Understanding System: an Example of Recognition Process, Proc. of the International Joint Conference on Artificial Intelligence, Tokyo, pp. 185-93, 1973.

11. Sriram, D.,Knowledge-Based Approaches to Structural Design, Ph.D. Thesis, Department of Civil Engineering, Carnegie-Mellon University, Pittsburgh, February 1986.

12. Maher, M.L., Fenves S.J. and Garret, J.H., Expert Systems for Structural Design, in Adeli, H. (ed.), Expert Systems in Construction and Structural Engineering, Chapman and Hall, 1988.

13. Fenves, S,J., Object-Oriented Programming for Engineering Software Development, Engineering with Computers 6, pp. 1-5, 1990.

14. Varghese, K., An Object-Oriented Representation for the Design of Reinforced Concrete Beams, M.Sc. Thesis, The University of Texas at Austin, Austin, May 1988.

15 Forde, B.W.R., Foshi, O.R. and Stiemer, S.F., Object-Oriented Finite Element Analysis, Computers and Structures, Vol. 34, N. 3, pp. 355-374, 1990.

16. Miller, G.R., A lisp-Based Object-Oriented Approach to Structural Analysis, Engineering with Computers, 4, 197-203, 1988.

17. Gurtuna, A.G., Hybrid Joints in Structural Steelwork: Combined Shear and in Plane Moment of HSFG Bolts and Fillet Welds, M.Sc Thesis, Civil and Structural Engineering Department, UMIST, April 1984.

8 DESIGN AND REPAIR OF REINFORCED CONCRETE STRUCTURES

ARCAD386: A KNOWLEDGE-BASED SYSTEM FOR REINFORCEMENT DESIGN

D. Zlajpah and J. Duhovnik,
Department of Civil Engineering,
University of Ljubljana, Slovenia

In the process of the production of technical documentation for reinforcement, more time is usually spent on direct work for technical documentation as for solving technical problems. Computer aided work can shorten this time, but, on the other hand, it might also be a risk. The final user loses the control over the process and has no influence on the course of work. The knowledge-based system ARCAD 386 tries to offer new possibilities for better work: higher level of the user's creativity, quick consideration of all changes in instructions and standards, higher level of reliability, and better control over the process. The system ARCAD 386 is written in FORTRAN and AutoLISP, AutoCAD is used for graphical environment. The possibilities of the ARCAD 386 system's use in the reinforcement design, and in the production of reinforcement drawings are described in the article.

INTRODUCTION

The design of reinforcement concrete (R/C) structures is a very complex problem because concrete as a material allows a large variety of shapes. The aim of reinforcement design is that the reinforcement has to be adapted into this concrete shape without reducing its effectivity. Therefore, this problem is not analytical with only one solution. The designer's task is to choose one solution from the solution space of a construction region /KROP90/. The solution space is first of all limited by the knowledge of the branch and by the standards that must to be respected by all designers. Because of that, these restrictions can be built into the programmes. More difficult is to capture expert's heuristic knowledge and final user's wishes into the programmes.

In the age of algorithmic programmes the heuristic knowledge was fixed into programmes, which output were unique admissible solutions. The development of new programme techniques /COX86/ and the use of artificial intelligence methods /WINS84/ enable the final user to choose the subjective optimal solution (for each case differently), from the suggested admissible solution space (Fig. 1). This manner of the work increases the final user's creativeness and reduces the negative influence of automatization, when the user gives in the programme the input data and then waits on the results without any influence during the process. Creative interactive work means: the computer is used only to control and to help final user with his decisions.

A large number of programmes (each type of element requires its own programme), is also deficient of algorithmic programmes in the field of reinforcement design. These programmes have to be maintained too. Object oriented techniques enable such data structure and methods in a way that different elements' types can be worked out with one programme and different knowledge bases. In the knowledge base all needed knowledge about the current element type is presented. This is very useful for the final user, because the manner of work is always the same, independent of the element type. On the other hand, the knowledge base can be easily changed and adapted to the changes in standards or technology.

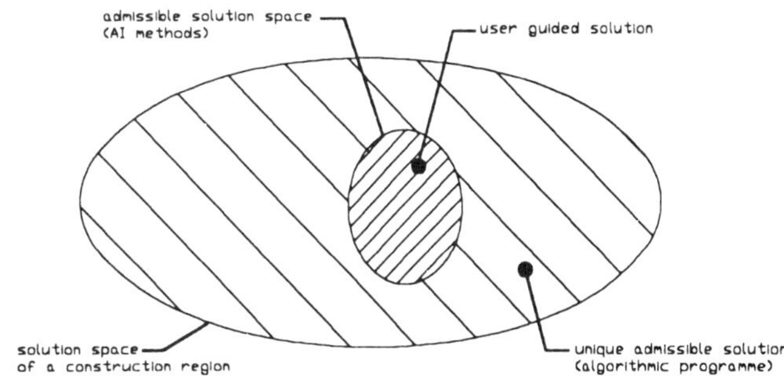

Figure 1 Levels of solution space

The production of technical documentation for reinforcement can be divided into the following three stages:

- reinforcement design (building up the 3D model of the element);
- drawing of the reinforcement plan; and
- production of the reinforcement schedule.

At the first stage the calculated reinforcement must be arranged into the concrete element according to standards, knowledge of branch and final user's wishes. The result of this phase is the 3D model of the element, containing concrete shape and arranged reinforcement. The representation of this 3D model as a reinforcement plan is a result of the second stage. The reinforcement plan contains different views, cross-sections and details, equipped with standardized signs and marks /DIN/, /JUS/. Reinforcement schedules can be presented in a form of tables or sketches, where type, size and weight of reinforcement are stored. Reinforcement schedules are made for single elements as well as for the whole construction.

THE CONCEPT OF THE PROGRAMME SYSTEM ARCAD 386

The programme system ARCAD 386 (written in FORTRAN and AutoLISP) is used for computer aided reinforcement design and is based on a three-dimensional (3D) model of concrete element's shape and reinforcement positions in this shape. The system is based on a number of modules (KONAVT, KONSPP, GA, PCRAR, PCRIS, SEZNAM ...) and interfaces (RAR_DXF, RISACAD). Such structure of the system (Fig. 2) allows different ways of reinforcement design (automatized /DUHO84/, interactive /KNIF90/ and knowledge based /ŽLAJ88/), dependent on the current case. The main characteristics of the programme system ARCAD 386 are:

- the possibility of working on various types of elements;
- the simple correction of knowledge base;
- the minimal number of the input data in the current case;
- the informing of the final user during the working; and
- the process of the work is always the same for the final user and independent of the complexity of the problem.

For efficient use of the programme system ARCAD 386 the following minimal hardware equipment is recommended:

- personal computer IBM 386 or compatibles;
- math coprocessor;
- 4MB RAM;
- VGA video board;
- 16" color monitor;
- printer (laser); and
- plotter (format A1).

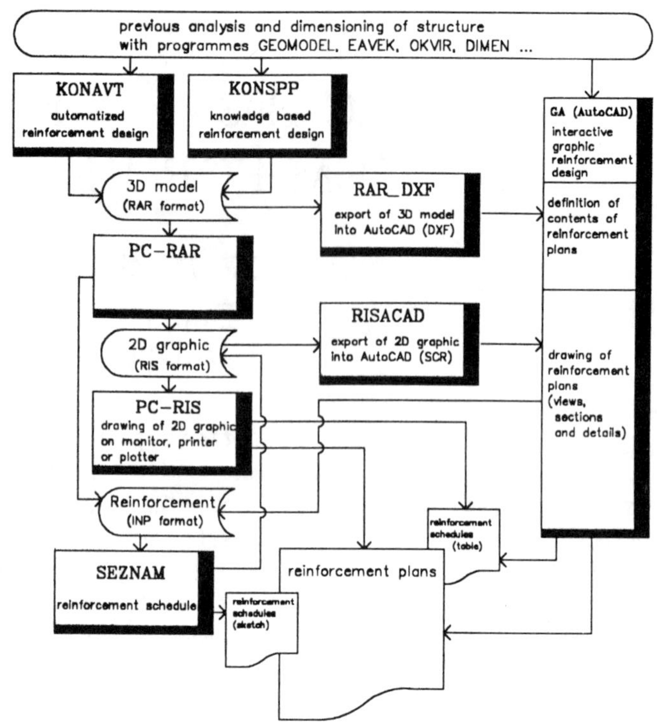

Figure 2 Scheme of the programme system ARCAD 386

3D model of the element

The 3D model of the element (Fig. 3) is the starting point for further drawing of reinforcement plans and the production of reinforcement schedules. All geometry and reinforcement data are stored in the 3D model. Geometry (Fig. 4) is represented by a wire model, where element is composed by several subelements, which are defined by initial and final cross-section (the same number of points in both cross-sections is required). The section is defined by set of successive points and Every point is defined by three coordinates (X_i, Y_i, Z_i) in the global coordinate system of the element.

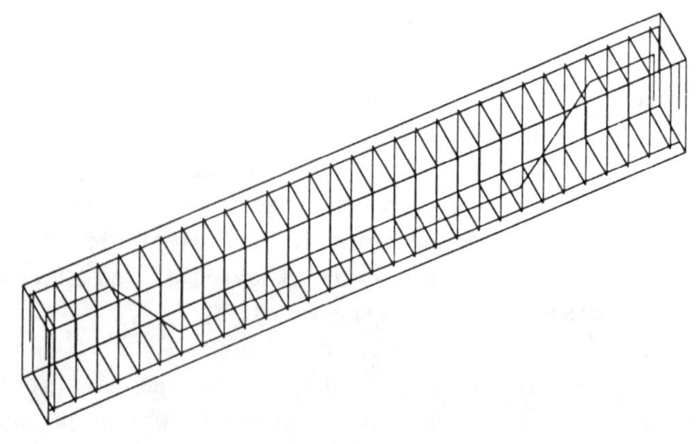

Figure 3 Representation of the 3D model

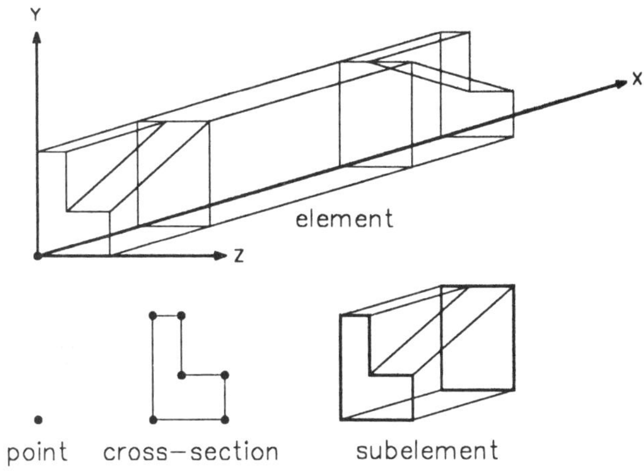

Figure 4 Geometry of the element

Reinforcement in the element is composed of individual reinforcement units. Single bar or welded fabric mesh as well a set of them is treated as an individual reinforcement unit (Fig. 5). The reinforcement unit is defined by following data:

- the quality of reinforcement steel;
- the diameter;
- the code of typed form;
- the hooks;
- the partial lengths;
- three points for definition of local coordinate system of the reinforcement in the global coordinate system of the element;
- fourth point for the direction of the set of reinforcement; and
- the distance between reinforcement in the set.

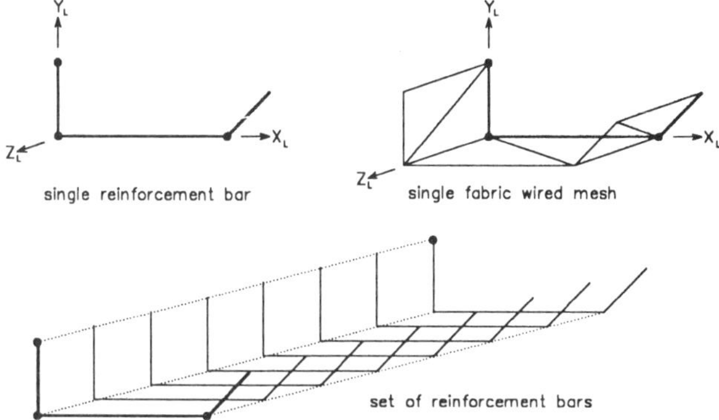

Figure 5 Reinforcement units

<u>Programme complex KONSPP</u>

The data structure of the 3D model is always the same (independent of the element type and its largeness) and so the following treatment is not very exacting and can be made by algorithmic programmes. But, on the other hand, the construction of the 3D model of current example is more exacting. As already explained in the introduction, there are many solutions and the user can very quickly loose the control over them. So he needs a tool to reduce the solution space, but without hindering his creativeness too much. In the system ARCAD 386, the programme complex KONSPP (Fig. 6) is in charge of this.

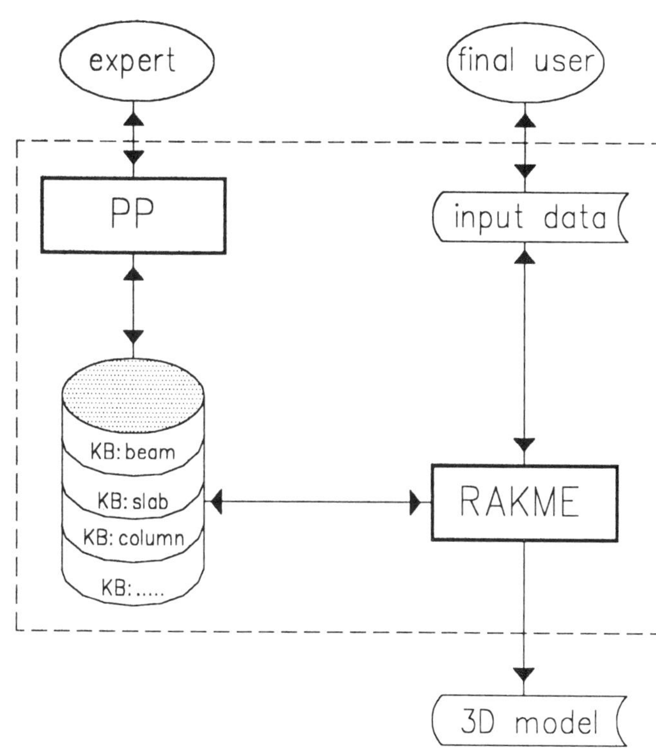

Figure 6 Scheme of programme complex KONSPP

The work with the programme complex should be going on, according to the following way:

- the human expert, knowing the process of the calculation and dimensioning for the chosen type of elements and technology of its fabrication, makes (by the aid of the programme PP) knowledge base for later construction of 3D model of current examples. In the knowledge base all various possibilities of elements for the chosen type must be considered and the base must be written in accordance with valid regulations and standards. In the auxiliary files all possible typed forms, bars' diameter, welded fabric meshes and texts of orders (for easier translation in other languages);
- after testing the knowledge base it is recommended to make a catalogue of knowledge base of each element type for final users. In the catalogue there should be a short instruction and description of: used functions, the topology of the element and the way of reinforcement design (Fig. 7). The production of the catalogue has not been supported by its own programme yet; and
- the next step is the evaluation of current cases. This is done by the final user. The programme RAKME with internal inference engine and knowledge base with auxiliary files make possible an interactive dialogue with final user to put the data in and construct the 3D model of current case. After that the algorithmic programmes are used to draw the reinforcement plans and to make reinforcement schedules.

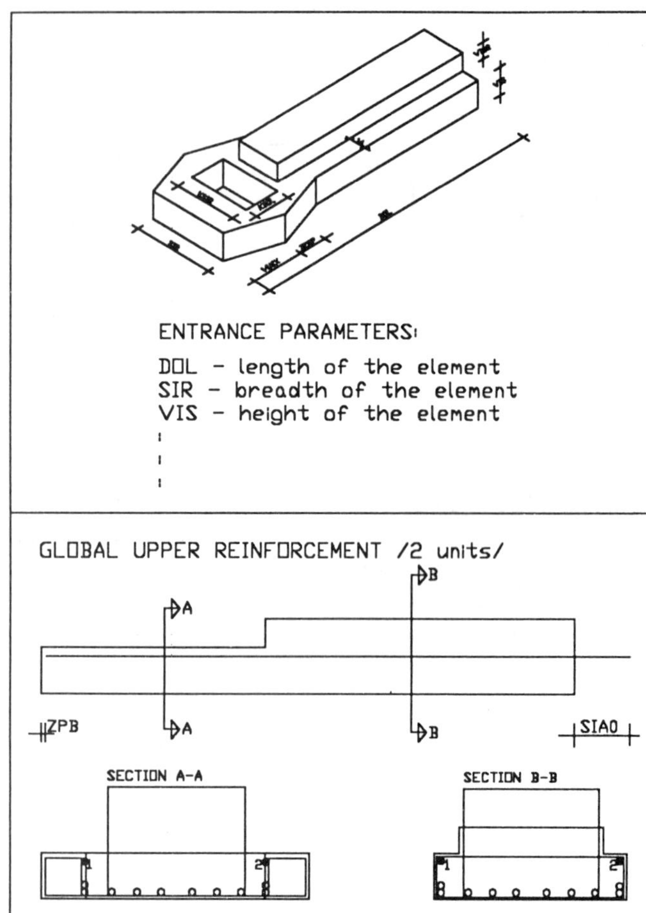

Figure 7 Part of the catalogue

__Knowledge base__

In the knowledge base (each element type requires its own base) the topology of the element and the way of reinforcement design are defined. The knowledge base is divided into the following parts:
- the parameters in form of functions;
- the controls and comments;
- the geometry of the element (points, cross-sections and subelements); and
- the reinforcement units.

the parameters are used locallyt in the programme RAKME for the easier calculation of geometry and reinforcement units. The parameters are composed by the name of the parameter, by the description and by the function determining its value:

name	description

function

Four alphanumeric characters are used for the parameter's name and 50 characters are of maximum length of the description. In the function there can be used real and integer values, FORTRAN-like mathematical operations and other parameters' names. When the parameter has no function (blank line), the parameter is treated as an entrance parameter and the final user must always put in its value during the interactive dialogue. The function can also simulate the rule in the form:

```
IF (cond) THEN (statl) ELSE (stat2)
```

Controls are used for detecting mistakes and irregularities. Comments are used for explanations during the working process. Both are composed by the text and the condition:

text

condition

In the case when the condition is fulfilled, the text is written on the monitor. The difference between the control and the comment is that the programme is interrupted by the control and the user must give a command for continuing. The comment does not interrupt the programme.

In the part where geometry of the element is defined, there are instructions how to compose the cross-sections and subelements. So the topology of the element is fixed. The coordinates of points are defined with the functions, which values are calculated during the working process for each current case separately.

point description

X coordinate of point

Y coordinate of point

Z coordinate of point

cross-section	number of points	points

subelement	initial c.s.	final c.s.

Reinforcement is divided in different groups (main, shear, global, constructive ...). In each group several reinforcement units are defined. The structure of each reinforcement position in the knowledge base is the following:

description of reinforcement unit

typed form	quality	distance	hooks

partial lengths

position defined by coordinates of points

The knowledge base is written or changed by the programme PP. It is also possible to make the knowledge base by one of text editors, but the programme PP has the following advantages:
- hinder mistakes in the structure of knowledge base;
- coordinates of points and reinforcement units can be easily copied, mirrored and moved;
- continuous overview over defined part of knowledge base during its definition;
- knowledge base can be changed from ASCII form into binary and vice versa; and
- knowledge base can be protected by a password.

Inference engine

Inference engine (together with explanation module) is integrated in the programme RAKME. Its task is to find out and define the value of all functions which are needed for the composition of 3D model of current case. At first, the inference engine, using the knowledge base, analyses the functions of parameters and makes the relation net (state space) of parameters (Fig. 8). The defining of the value of parameters starts when the final user selects the reinforcement unit (defines its diameter) and the inference engine must find out all needed parameters for design of the selected reinforcement unit (backward chaining). From the whole relation net of parameters only a part is taken out. In this part depth-first strategy is used, because depth-first strategy enables better dialogue with the final user /HARM85/. During this process the mode of parameters is also defined. For the parameter one of following modes is possible:
- inactive parameter;
- active parameter with the defined value,
- active parameter with the undefined value (the value can be directly calculated); or
- active parameter with the undefined value (the value must be indirectly calculated).

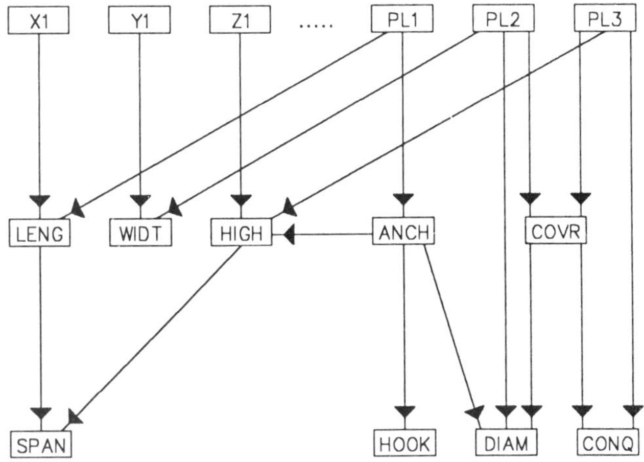

Figure 8 Relation net of the parameters

When inference engine finds out an entrance parameter or an active parameter with the undefined directly calculatable value, the final user's help is required:

```
DIST     /Distance between stirrups /m/
Put-in a new function or value! ----->
(LENG-0.20)/NUMB
```

When the inference engine is caught in the internal trap (Fig. 9) the final user must also put-in the value or a new function of the parameter. It is better to avoid internal trap in the phase of writing the knowledge base.

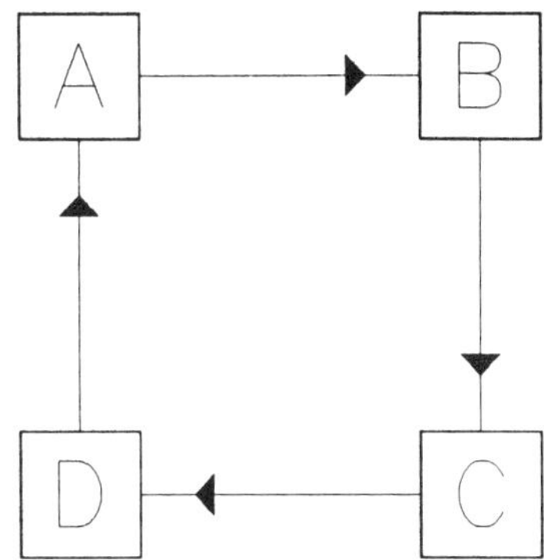

Figure 9 Internal trap

After the defining of the value of the parameter, the whole knowledge base is checked out if there is possible to define values of other parameters (forward chaining).
The whole process of calculating the values of all parameters is repeated (nonmonotonic reasoning), if changing the value or the function of the parameter with defined value.

EXAMPLES FROM THE PRAXIS

The programme system ARCAD 386 is used not only for teaching and training on the university, but is also installed in many companies in Slovenia. It is used on very different structures.

With the same knowledge base is possible to make very different examples. Two examples made by the knowledge base for a prefabricated T beam are in Figure 10. The upper example is very simple and user has to put-in only 10 data. The other example requires 76 inputs.

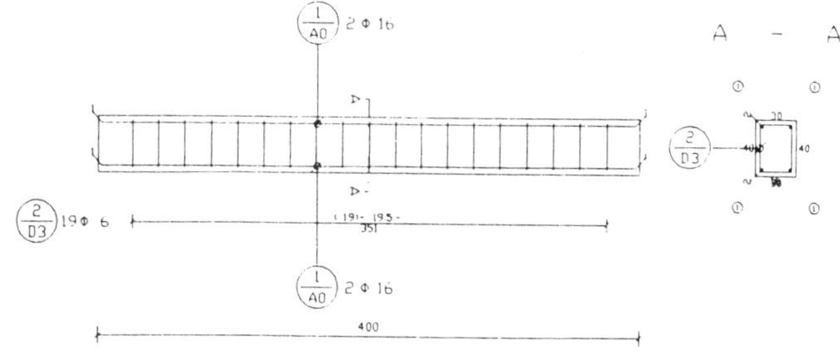

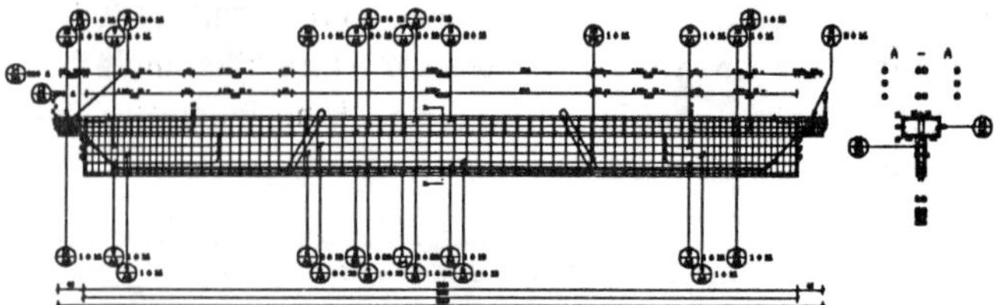

Figure 10 Two examples of T beam

The production of technical documentation of prefabricated elements is very appropriate to be supported by the system ARCAD 386. In the firm INGRAD-Celje a collection of knowledge bases for all elements of their prefabricated system /CERN88/ has been done. To improve the efficiency of the system, it is also possible to combine different elements. In Figure 11 the combination of column and console is shown.

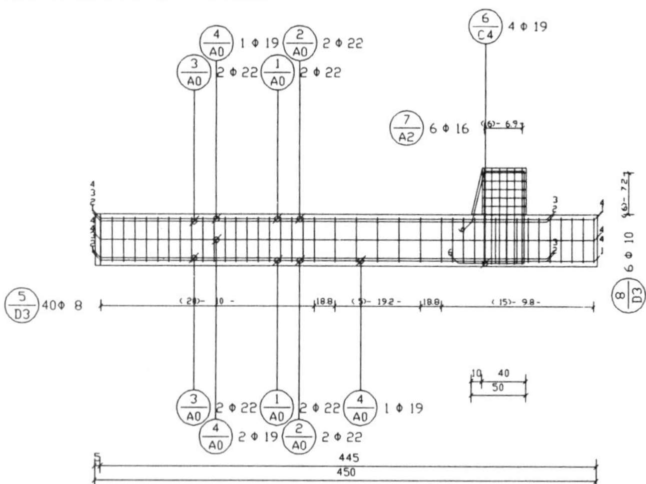

Figure 11 Combination of column and console

The system is not used only for a linear elements, but also walls and slabs can be worked out. In the wall of the large panel system (Fig. 12), there are used reinforcement bars and welded fabric meshes together.

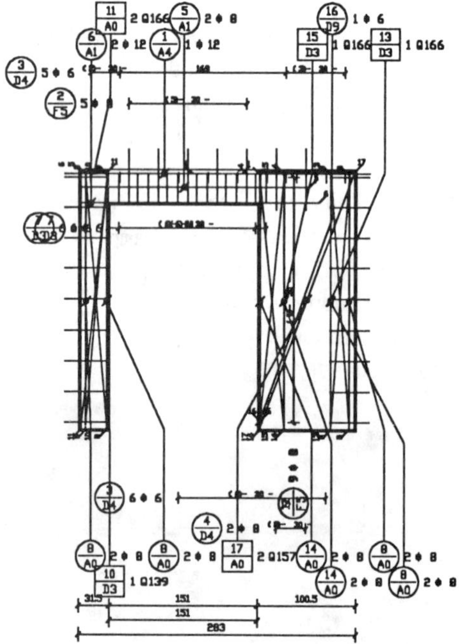

Figure 12 Wall of large panel system

A combination of knowledge based and interactive graphic reinforcement design is very successful in complicated elements. The technical documentation of a highway bridge (Fig. 13) was done in a such way.

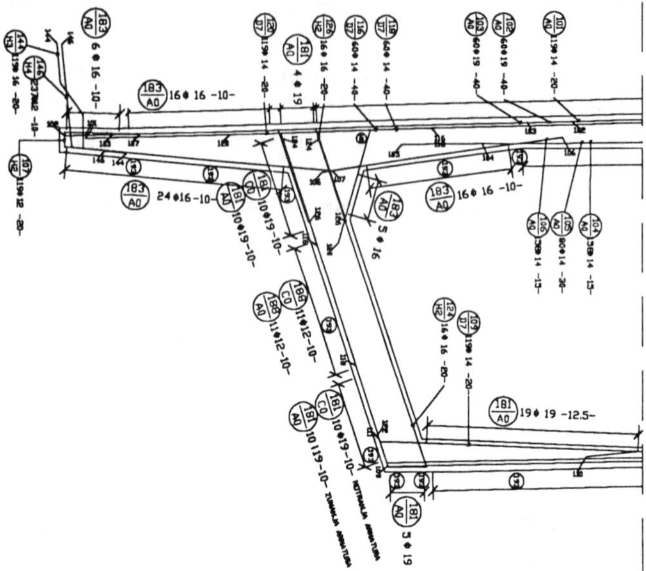

Figure 13 Section of the highway bridge

At the building site it is useful that the reinforcement plans are equipped with details (Fig. 14) and reinforcement schedules in form of sketches (Fig. 15).

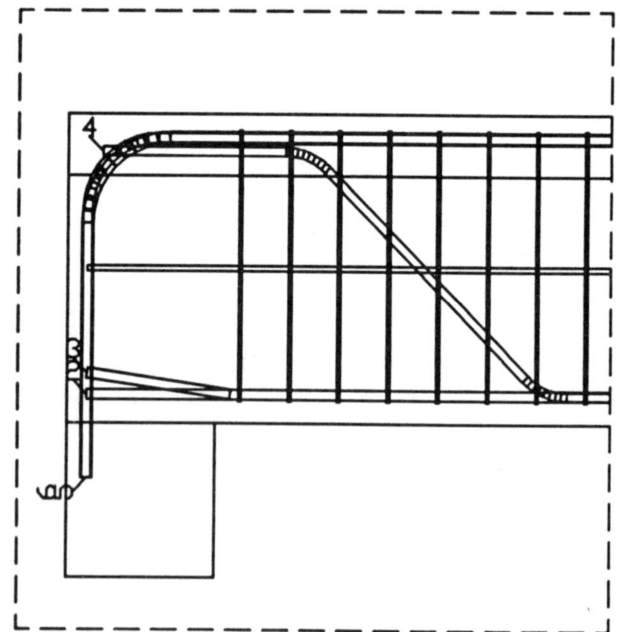

Figure 14 Detailed reinforcement plan

REINFORCING BARS WEIGHT					
SMOOTH RA 240/360		DEFORMED RA 400/500-1		DEFORMED RA 400/500-2	
FI	kg	FI	kg	FI	kg
		22	333.7		
		12	138.6		
16	15.6	16	76.7		
6	76.1				
Total:	91.7	Total:	449.8	Total:	91.7
FI =< 12:	76.1	FI =< 12:	138.6	FI =< 12:	.0
FI > 12:	15.6	FI > 12:	318.4	FI > 12:	.0

FI =< 12:	214.7 kg
FI > 12:	326.0 kg
Total:	540.7 kg

Figure 15 Reinforcement schedule

CONCLUSIONS

The goal of the system ARCAD 386 is not only the theoretical research but the usage in the praxis. Therefore, the development of the system must go on in a way which is acceptable for final users. We would be greatful for every suggestion from praxis and we would try to improve the system, but new version must be compatible with the old one. The inclusion of graphics in the phase of knowledge base creation and specially integration of all needed software in the field of civil engineering design will be our main preoccupation in the future.

ACKNOWLEDGEMENT

This report is based on a part of the research work on the thematic complex COMPUTER IN CIVIL ENGINEERING as a part of RESEARCH PROJECT COMPUTER INTEGRATED MANUFACTURE - CIM AND EARTHQUAKE ENGINEERING supported by the Research Community of Slovenia and Special Research Community of Civil Engineering. We would also like to thank all them who help to accelerate the introduction of the system into praxis.

REFERENCES

COX86 Cox, B.J., Object Oriented Programming, Addison Wesley, Reading, 1986.

CERN88 Cernelc, S., Catalogue of Prefabricated system INGRAD, GIP INGRAD Celje, internal publication, 1988.

DIN DIN 1356, Building and Civil Engineering Drawings, Drawings for Concrete Reinforcement, 1980.

DUHO84 Duhovnik, J., et al, Programme System for the Design of Reinforced-Concrete Structure, Proceedings Computer Aided Analysis and Design of Concrete Structures, Vol.2, 923-936, Pineridge Press, Swansea, 1984.

HARM85 Harmon, P., King, D., Expert Systems, J. Wiley, 1985.

KNIF90 Knific, A., et Al, Programme GA (AutoCAD), IKPIR, Ljubljana, Pub. No.6, 1990.

KROP90 Kroplin, B., et Al, "Ein Konzept fuer die wissensbasierte Konstruktion von Standardbauteilen in Stahlbetonbau", <u>KI-Forschung in Baubereich</u>, Ed. Gauchel, J., Ernst&Sohn, Berlin, 183-195, 1990.

JUS JUS U.a0.017, proposal (in Serbo-Croatian).

WINS84 Winston, P.H., Artificial Intelligence, Addison Wesley, Reading, 1984.

ŽLAJ88 Žlajpah, D., Rule-based Computer Aided Design and Reinforcement Drawing of Prefabricated Elements, Automatika, Vol.29, No.1-2, 213-218, 1988.

BRUTUS: AN EXPERT SYSTEM FOR REPAIR OF CONCRETE STRUCTURES

Aa.N. Blankvoll and G. Horrigmoe
FORUT Technology Ltd, Narvik, Norway

Inspection, diagnostics, maintenance and repair of concrete structures are problems of high complexity which require expertise in several fields. Knowledge-based expert systems (KBES) are now being effectively applied to a variety of engineering problems. It is the authors' opinion that KBES will be a useful tool to solve problems related to durability of concrete. This is the main motivation behind the development of the BRUTUS system. Since BRUTUS is developed as an interactive computer program with a wide variety of supporting elements containing organized and systematized knowledge, the system will also be a valuable tool for training civil engineers in this field.

The BRUTUS system will transfer knowledge and technology from experts to personnel responsible for planning and performing inspection, diagnostics, maintenance or repair of concrete structures. These categories include owners of buildings and structures, consulting engineers and contractors. The use of the BRUTUS system may lead to increased service life and maintenance intervals of concrete structures thereby reducing the annual costs for the structures. Consequently, if a KBES like BRUTUS leads to better planning and performance in the years to come the need for maintenance and repair is so huge that it will represent significant savings for owners of buildings and structures of concrete and the society as a whole.

The present paper points out areas of applications of the BRUTUS system in different phases of management and repair of concrete structures. The basic structure of the complete program is outlined by means four different models. In addition, problem solving strategy and handling of uncertainty in the BRUTUS system are discussed.

INTRODUCTION

Applications of expert systems in civil engineering are growing at a prodigious rate. In the future, expert system technology will be applied to solve a wide variety of tasks in civil engineering. This is clearly indicated by the large number of knowledge based systems that were presented at the IABSE colloquium "Expert Systems in Civil Engineering" in Bergamo, Italy in 1989 (reference 1). These expert systems fall into three categories, which may be described as systems for design and construction, systems for structures in service and systems for codes and standards. The BRUTUS system is an example of an expert system for structures in service.

The objective of the system is to transfer knowledge and technology from experts to personnel responsible for planning and performing inspection, diagnostics, maintenance and repair of concrete structures. These categories include owners of buildings and structures, consulting engineers and contractors. In this way, the BRUTUS system may contribute to increase service life and maintenance intervals of concrete structures and decrease the annual costs for the structures. In Norway, as well as in the rest of the industrialized world, there is an increasing need for maintenance and repair of concrete structures. To meet the needs in Norway alone in the coming decade, the associated costs are estimated to 4 - 10 billion NOK (0.5 - 1,5 USD).

The present paper focuses attention on the prototype of the BRUTUS system. The paper points out possible areas of application for the system and describes the program structure. In addition, problem solving strategy and handling of uncertainty are discussed.

BACKGROUND

Deterioration mechanisms for concrete

There exist a large number of deterioration mechanisms for concrete. Mechanisms of deterioration lead to several visible symptoms, such as cracks in different patterns, spalling of concrete or different kinds of efflorescence. At the same time, a specific symptom can be caused by various deterioration mechanisms. Moreover, different types of concrete structures are exposed to various environments influenced by both climate and pollution. Deterioration mechanisms and durability of concrete structures are dealt with in detail in references 2 - 5. It is possible to use several methods of repair for damaged concrete depending on the factors mentioned above as described in references 6 and 7. Consequently, the different phases in management of concrete structures such as inspection, diagnostics, maintenance and repair, pose problems of high complexity which require expertise in several fields.

The prototype of the BRUTUS system is restricted to concrete buildings and only a limited number of damage mechanisms are included in the prototype. The major emphasis is on corrosion of steel reinforcement and on diagnostics based on information on damage observations. This is one of the most difficult tasks in the development of the system. It is of vital importance that the causes of damage are identified before any repair work is carried out.

Development software

The prototype was developed using the shell Nexpert Object. The Nexpert Object development system is a hybrid rule- and object- based tool for expert system development, see reference 8. Its open AI architecture allows one to integrate it within existing environments rather than being an incompatible, stand-alone technology. Nexpert Object is written in "C". Among its features are integrated forward and backward chaining using the same symmetric rule format, automatic goal generation, pattern-matching, interpretations, dynamic creation of objects, classes, and properties, methods, demons, multiple and user-defined inheritance, and non-monotonic reasoning.

Nexpert Object has a comprehensive graphic interface which allows developers and domain experts to edit rules and objects as well as establish control structures with an overview of the rule and object structures available at all times through a dynamic, graphic browsing mechanism.

DESCRIPTION OF THE BRUTUS SYSTEM

Applications

The BRUTUS system has four different areas of application:

i) As a book of reference.
The system will contain a database with general information on durability, inspection, maintenance and repair of concrete structures.

ii) As a tool for inspection.
BRUTUS will give recommendations, requirements and instructions on how to carry out inspection depending on type of structure, environment and other special conditions for the structure under consideration.

iii) Diagnostics tool.
The system will make use of the information the user gives and then provide diagnosis and present suggestions for maintenance and repair.

iv) As a tool for maintenance and repair.
BRUTUS will use instructions, requirements and specifications for repair products together with the

results from the diagnosis to provide guidelines for products and procedures. In addition, the system will give instructions on how to plan and carry out maintenance and repair.

It is important to have in mind that the system will act as a tool for decision makers in different phases of management and repair of concrete structures. The system will make available knowledge that will help decision makers make better decisions. Consequently, the BRUTUS system may contribute to increase service life and maintenance intervals of concrete structures and decrease the annual costs for the structures.

The relation model defines the relations between the concepts in the object models. The example in Figure 2 is limited to deal with three relations and serves as an illustration only. Far more complex relation models are also used in the prototype.

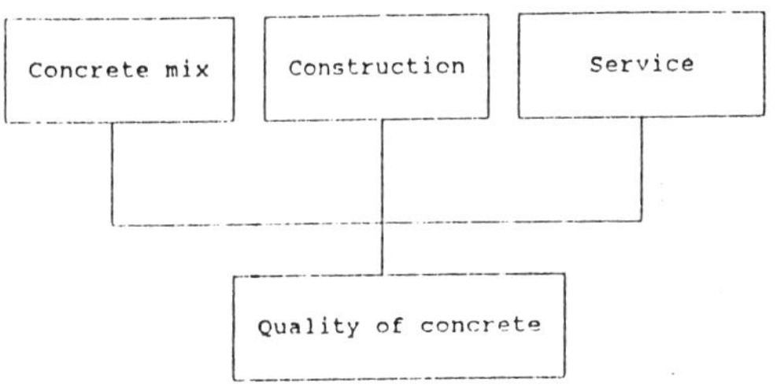

Figure 2 Relation model for concrete quality.

Structure of the system

In the prototype of the BRUTUS system four different models are used to describe the domain knowledge and the use of the system. These models are briefly described in the following. Further details can be found in reference 9.

The object model contains the concepts used in the BRUTUS system. Each concept has a name and may have attributes with specific values. The object model describing deterioration mechanisms for concrete is illustrated in Figure 1. It is to be noted that accident loads such as fire, explosion, earthquake, etc are not included in the BRUTUS system.

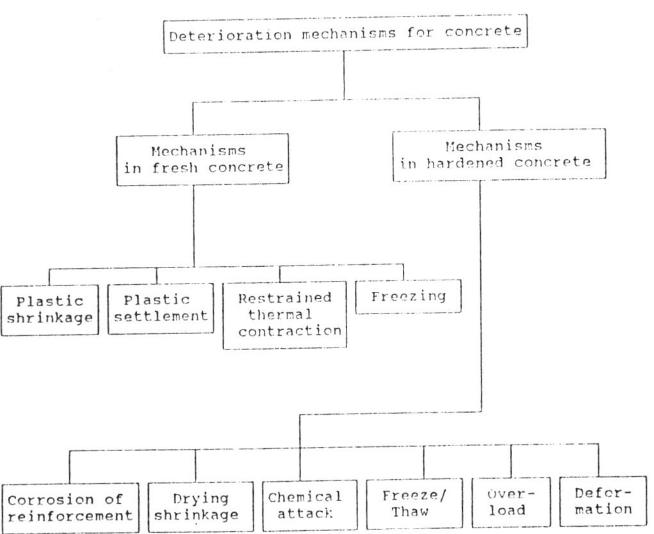

Figure 1 Object model for deterioration mechanisms.

The work process model describes the way concrete experts deal with management problems such as inspection, diagnostics maintenance and repair of concrete structures. The system is to be used in several of the different phases mentioned above. Consequently, it is important that the system gives a reflection of common procedures. The work process model illustrated in Figure 3 is used in the development of the system in order to take into account the order of operations involved in different phases in management of concrete structures. This model have a significant influence on the structure of the BRUTUS system.

The dialogue model describes how the user and the system interact. The dialogue with the system tries to copy the procedures used by concrete experts in analyzing structural damage. The dialogue model used in the prototype which is illustrated in Figure 4, is created by use of the work process model.

Problem solving

The very core of the prototype is the methodical way of handling input data for making diagnostics and providing guidance for repair. The starting point is the visual symptoms which indicate one or several deterioration mechanisms. The visual symptoms are part of the information used for making diagnostics. During visual inspection an estimate is made of how serious the damage is (the visual degree of damage VD).

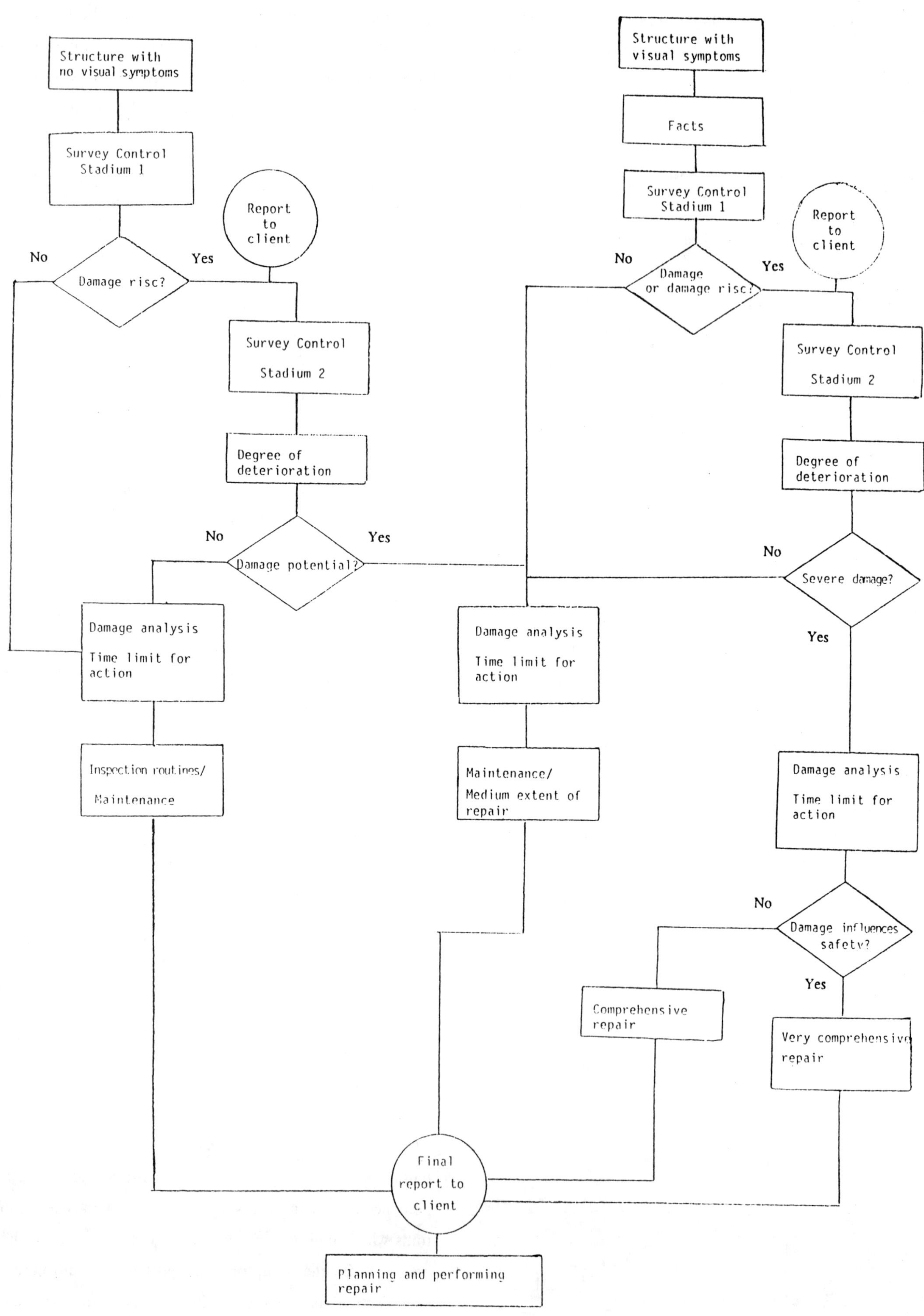

Figure 3 Work process model.

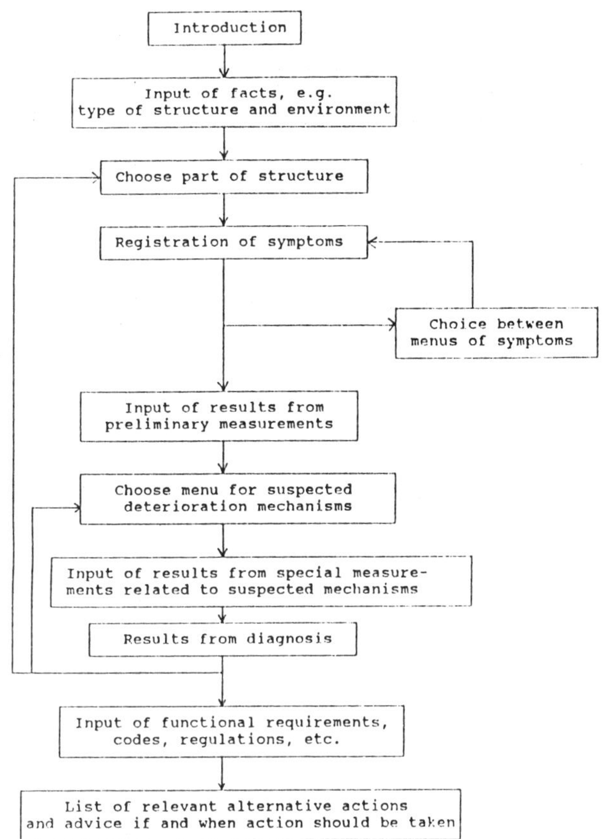

Figure 4 Dialogue model for the BRUTUS prototype.

The next step is to analyze the results from measurements taken in-situ and in the laboratory. These data are used to make diagnostics and for defining the deterioration of concrete, or degree of damage (DD), which is related to the deterioration mechanisms for concrete, see Figure 5.

The visual degree of damage (VD) is combined with the degree of damage (DD) which describes the deterioration of concrete. In addition, functional requirements, codes, regulations, esthetics, economy, etc is taken into account before the system presents a list of relevant alternative actions and advice if and when action should be taken, see Figure 4.

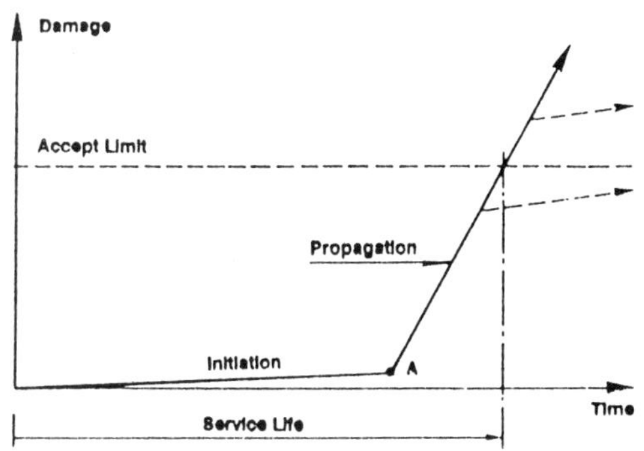

Figure 5 Damage development with age of structures (reference 10).

Uncertainty

Uncertainty is treated in the prototype using certainty factors (CF) as described in reference 11. This method of handling uncertainty seems to be reasonable valid for short inference chains.

Several rules with specific positive or negative weighting factors lead to a hypothesis. The resulting CF for a hypothesis is calculated from the positive or negative weighting factors in the rules. This parameter is a numerical value between zero and one which indicates how confident a hypothesis is. A zero value implies total uncertainty, while CF = 1 is associated with 100 % certainty.

The resulting certainty factor (CF) for a hypothesis is calculated by the following formula,

$$CF(x, y) = x + |y| - x|y| + 0{,}5y(1 - \text{sign}(y)); \quad (1)$$

x = The starting CF value for the first rule ($x \geq 0$).

y = Positive or negative weighting factor.

This equation adds only one positive or negative weighting factor at the time. This means that a new piece of evidence will bring the resulting CF closer to the value one which corresponds to 100 % certainty. Note that the CF value for the hypothesis (the resulting CF) expressed by two rules, is equal zero when the first factor ($x \geq 0$) is zero and the second factor is negative.

CONCLUDING REMARKS

A prototype of the BRUTUS system has been established using the expert system shell Nexpert Object. The objective is to develop a program system which is intended to help civil engineers make the right decisions in condition survey, maintenance and repair of concrete structures.

The prototype demonstrates that it is feasible to handle the problems in the domain, which are of high complexity, in such a way that the system makes diagnostics and gives guidance of alternative actions as well as advice when action should be taken.

ACKNOWLEDGEMENTS

The work reported herein was supported by The Royal Norwegian Council for Scientific and Industrial Research, The Regional Development Fond, The Regional Authorities of Northern Norway and various industry partners and governmental agencies.

Subcontracted work in the BRUTUS project is performed by two groups, one working on concrete technology and the other on software development. The authors wish to recognize the significant contributions of Mr. Ø. Vennesland, Mrs. M. Rønne, Mr. J. Lindland and Mr. B.E. Andersen (concrete technology) and by Mr. R. Fjellheim, Mr. A. Hanssen and Mr. H. Myrland (software development).

REFERENCES

1. IABSE Colloquium, "Expert Systems in Civil Engineering", Bergamo, Italy, October 1989, IABSE, Zürich, Report Volume 58, 1989.

2. Rostam, S. (Editor), "Durability of Concrete Structures", CEB-RILEM International Workshop, Copenhagen, May 1983, Workshop Report, Technical University of Denmark, Copenhagen - Lyngby, 1984.

3. Rostam, S. and Bræstrup, M.W. (Editors), Proceedings of International Symposium on Re-evaluation of concrete structures, Copenhagen, June 1988, Technical University of Denmark, Copenhagen - Lyngby, 1989.

4. Scanlon, J.M. (Editor), "Concrete Durability", Katharine and Bryant Mather International Conference, Atlanta, Georgia, April 1987, American Concrete Institute, Detroit, SP-100, 1987.

5. Malhotra, V.M. (Editor), Second International Conference on Performance of Concrete in Marine Environment, St. Andrews by-the-Sea, Canada, August 1988, American Concrete Institute, Detroit SP-109, 1988.

6. Dhir, R.K. and Green, J.W. (Editors), Proceedings of the International Conference Protection of Concrete, University of Dundee, Scotland, UK, September 1990, E. & F.N. Spon, London, 1990.

7. Allen, R.T.L. and Edwards, S.C. (Editors), "Repair of Concrete Structures", Blackie & Son Limited, Glasgow and London, 1987.

8. Nexpert Object Product Description, Neuron Data, Palo Alto, California, September 1990.

9. Blankvoll Aa. N. and Myrland H., "BRUTUS - Development of prototype" (in Norwegian), FORUT, Narvik, Norway, March 1991.

10. Tuutti, K., "Corrosion of Steel in Concrete", Swedish Cement and Concrete Research Institute, CBI Research Fo 82, Stockholm, 1982.

11. Myrland H., "BRUTUS, Expert System for Concrete Structures - Prototype" (in Norwegian), FUNN-Narvik, Narvik, Norway, February 1991.

9 INNOVATIVE METHODS OF STRUCTURAL DESIGN

HEURISTICS DRIVEN STRATEGIES FOR NEAR-OPTIMAL STRUCTURAL TOPOLOGY DEVELOPMENT

N. Shankar† and P. Hajela‡
†University of Florida, Gainesville, Florida,
†Rensselaer Polytechnic Institute, Troy, New York,
United States of America

The paper describes a quasi-procedural approach for the generation of near-optimal structural topologies. Such an approach is characterized by the use of algorithmic processing to enumerate the design space, and to use this information in a design methodology dominated by heuristic decisions. The paper presents a stagewise decomposition approach, which is thematically similar to dynamic programming, to decompose the given problem into a series of stages (subproblems), and generate a population of trial designs. A genetic algorithm based search is then conducted in a discontinuous space with these designs as seeds, to obtain an optimal design for the problem. The issue of structural stability is addressed by the use of embedded heuristics. A representative structural design problem is used to illustrate the approach.

INTRODUCTION

Optimal structural design for minimum weight is a multistage process, initiated with the definition of a structural topology to transmit applied loads to specified support points. Topology generation forms the initial and perhaps a crucial step in the design process. In most problems, there is no unique topology for a given load-support specification. Once an acceptable topology has been generated, the members of the structure are resized to satisfy structural and dynamic response requirements using formal optimization techniques. The final outcome of this procedure is a minimal weight structure, the characteristics of which are strongly linked to the original topology definition.

Generic CAD systems available today are generally limited to problem solving tasks in the algorithmic or deterministic domain. However, creative aspects in design are often ill structured, and require significant intervention on the part of the designer. Therefore, attempts have been made in recent years to incorporate decision making capabilities in automated design procedures; the topology definition problem described above is naturally amenable to this approach. Such decision making characteristics warrant the use of artificial intelligence methods with unique abstraction capabilities. Artificial intelligence methods, such as search techniques and expert systems, allow us to manage problem decomposition and knowledge representation in a very efficient manner. Furthermore, such expert systems can help in the dissemination of design procedures to a larger user community.

There are two principal drawbacks of such decision support systems in optimal structural design. The first pertains to somewhat limited computational power of such systems which renders them inefficient for meaningful structural design problems. The low computational efficiency is usually attributed to the language selected for the development of such systems. The second stems from its limited interfacing with algorithmic computing. Therefore, most of the expert systems developed for structural design do not extend into the analysis and optimization domains. The present paper explores the development of design systems which realize the advantages of integrated heuristic and algorithmic computations.

A review of the literature stands as a testimony to the abundance of work that has been accomplished in the area of AI based design systems [1-7]. The importance of integrating algorithmic procedures and heuristics in the design process has been examined in the context of structural design [8-10]. Such an integration leads to a 'quasi-procedural' approach, where designs abstracted on the basis of select heuristics are validated through conventional algorithmic procedures. Subsequent sections of this paper describe the architecture of a design system for generating 'near-optimal' structural topologies for a given set of loads and supports in a two-dimensional design domain. The principal components of this design system include a) a formal architecture of the design system b) methods of decomposition which would reduce the given problem to a series of smaller, more tractable subproblems c) quasi-procedural strategies to obtain solutions to the subproblems and d) an approach to integrate the partial solutions into an integrated solution. The focus of the current paper is in the use of decomposition methods inspired by dynamic programming principles; these methods, when combined with a genetic search based optimization approach, provide a powerful design synthesis tool.

PROBLEM DEFINITION AND PRIMITIVES

A representative design problem considered for structural synthesis is shown in Figure 1. The objective is to generate a minimum weight topology, given a set of loads, possible support points, and domains where no structural members are allowed. A heuristics directed load stabilization approach was adopted to transmit each load to the support points with the least resultant structural weight. The design was performed in conjunction with analysis methods to evaluate the relative merits of each topology. These designs were further required to meet prescribed displacement or stiffness requirements, strength requirements, and dynamic considerations as represented by frequency constraints.

Such a problem usually incorporates a set of prescribed primitives, from which the structure is evolved. Table 1 lists the various primitives that were considered for loads, supports, and structural materials. A database of all the primitives with corresponding properties was maintained, and from which relevant information was sought during the design process.

QUASI-PROCEDURAL DESIGN METHODOLOGY

The architecture of the quasi-procedural system used in this work is similar to that described in [8-10]. An overview of this organization is included here for completeness. At the very outset, it is important to recognize three distinct levels at which the automated design process can be organized - the knowledge, function, and program levels. A typical arrangement of the three levels in relation to each other is shown in Figure 2. The figure also portrays typical component modules for the different levels.

A detailed description of the design domain is relegated to the knowledge level. This included all pertinent specifications which define the acceptability of a solution. All necessary and applicable requirements for analysis are also made available at this level. All the heuristics [9] that aid in the synthesis of topology are maintained at this level. This includes the heuristics employed for design decomposition, subproblem integration, and optimal design. Constraint requirements that are to be satisfied for a given topology, are retained at this level. A global database, which stores domain information, taxonomy of topologies, allowable primitives and design requirements, is also provided at this level.

The most significant feature of the function level of organization is a controller which directs the flow of the conceptual design process. The controller is essentially an embodiment of the problem solving strategies that may be invoked for the problem at hand. The design specifications handed down from the knowledge level are attempted to be satisfied at the function level. Although designs can be generated by considering all requirements simultaneously, this methodology is not considered appropriate for the task at hand. Instead, a more natural process of refinement in steps is adopted, wherein the problem is decomposed into smaller, more tractable, and preferably, single goal problems. The underlying principle in such a refinement is that the solution space is more likely to be unique in the presence of a higher degree of specification detail.

At the program level, no problem solving knowledge is explicitly available. However, the implementation of all design steps, on the basis of information received from other levels, is carried out at this level. In addition, tasks of data management and programming procedures are assigned to this level. The database management capabilities of such systems are of particular importance: significant amounts of data are generated and must be managed for a design system to work efficiently. This is even more crucial as large amounts of algorithmically generated numerical data must be stored and post-processed to be used effectively in the iterative process. Such data management facilities are relegated to this level.

DEVELOPMENT OF TOPOLOGY

The framework described in the preceding section was implemented in an automated design system for the generation of near-optimal, two dimensional structural configurations. The problem required the development of both the structural form and member cross-section dimensioning, to carry a prescribed set of concentrated and distributed loads and moments. The least weight structural design was sought with constraints on allowable displacements, fundamental frequency, and structural strength. In previous efforts [8,9], a minimum weight structure was developed by stabilizing each of the given loads, using elements of nominal cross-sectional dimensions. Subsequent to this, problem constraints were considered one at a time, and either additional elements were added or cross sectional dimensions refined to satisfy these requirements.

This approach is not only dependent on the order in which the loads are stabilized and on the sequence of constraints satisfaction, but becomes progressively more cumbersome to implement with increasing number of loads and supports. Hence, a process in which the problem domain was decomposed into a sequence of more tractable subproblems, was proposed [11]. This approach was based on using sensitivities to determine the influence of design changes in one subproblem on the other temporarily decoupled subproblems. Sensitivity information was also used in determining how the designs of each subproblem could be integrated into a unified load carrying structure.

The present approach proposes the use of a stagewise decomposition approach which is thematically similar to dynamic programming. The common component used in this work and the work described in References [8-11], is the heuristic driven approach for stabilization of the applied loads, and this is briefly summarized here for completeness.

KNOWLEDGE BASE FOR TOPOLOGY GENERATION

The integrated heuristic/algorithmic system for topology generation used a forward chaining inference engine CLIPS (C-Language Integrated Production System) [12]. All design heuristics were represented in the form of rules, written in the following rule syntax for CLIPS:

```
(defrule <name> ["comment"]

        [(<first pattern>)
        (<    ..    >)           ;LHS
              ..
              ..
        (<n-th pattern>)

=>

        (<first action>)
        (<    ..    >)           ;RHS
              ..
              ..
        (<m-th action>)])
```

The LHS consists of one or more patterns which form the condition(s) to be satisfied, before a rule can be fired. An implicit 'and' is present in the LHS if more than one pattern is present. The RHS of the rule is a list of one or more action(s) to be performed as a result of the firing of the rule. The rule base also includes a few meta-rules, which are a superset of two or more rules that are generally task specific. In the current application, CLIPS was used as an embedded expert system, where CLIPS was executed from a FORTRAN program, preceded by the loading of the rules. A basic feature of the CLIPS environment is that the rules and data are two separate entities.

In a basic execution cycle, the knowledge base is examined to see if conditions of any rule are satisfied. In case of more than one matched rule, all such rules are activated and pushed on to a stack. The most recently activated rule is assigned the lowest priority. The rules are then activated and their result may in turn, activate other rules. This cycle is recursively applied until no further matched rules are obtained.

In the present stagewise decomposition approach, the heuristics based system was used to generate alternate topologies to transmit loads to the available supports. The various structural elements were chosen so as to stabilize the load support system. The system can also be used to check the stability of an existing topology, and to stabilize unstable structures with the least weight penalty. This aspect was particularly useful in the genetic algorithm based optimal design proposed in subsequent sections.

The approach to stabilize a given load was based on the magnitude, type, and orientation of the load with respect to the supports in some reference axes system. In the present implementation, the controller at the function level initiated the load stabilizing process by reading in as input, the total number of loads, and the magnitude and location of each load in order of increasing distance from the origin of the chosen coordinate system. Corresponding to each of these loads, the supports were classified in terms of distances from the loads.

The synthesis itself was conducted in a sequential manner, with the design heuristics attempting to stabilize one load at a time. If the applied load was a moment, the element necessary to transfer this load was required to allow for rotational degrees of freedom at its node. Hence, a beam element was selected to transfer this load. To stabilize this load and to maintain a minimum weight, the nearest support was examined. If this support was of a clamped type, then a beam element of nominal cross-sectional dimensions was chosen. If this support were a simple support, then the next nearest support was considered. If this support was a clamp, then a beam element was chosen between this support and the load. If not, the load was connected to both the simple supports using two beam elements. A further level of refinement was possible by checking to see if the third nearest support is a clamp, and if so assess the weight penalty of connecting to this support against the use of two beam elements to the two nearest supports. This strategy was also used in the event that at a given point, both applied forces and moments are present.

For the situation where the loads are only applied forces with nonzero x and/or y components, the orientations of the loads stored in the database were used. The angle between the force resultant and the hypothetical element connecting the point of load application and support, was computed. If this angle was less than 20 deg, then the load was considered to have a large axial component, and an axial rod element was preferred.

The other possible scenario was one in which a large transverse component of the load existed in addition to an axial load. The orientation of the load resultant with respect to the next nearest support was considered. In the event that the load could be largely considered axial for such an element, an axial rod element was introduced between the load and this support. In the event that this condition was also violated, the first two supports were considered. If the nearest support was a clamp, then a beam element was chosen between the load and support. If not, then a beam element was chosen between the load and the second support, provided it was a clamp. Otherwise, two beam elements were chosen between the load and the two supports. Multiple loads acting at a point were resolved into the x and y components, and handled as above. Likewise, distributed loads could be present as part of the design problem. In such situations, equivalent end loads and moments were computed by assuming the distributed load to act on a beam segment, and these loads and moments were connected to supports as explained in the foregoing discussion. Once a load was stabilized by the process described above, the point of application of the load was considered as a fixed support, even though physically it is a flexible support.

The stagewise decomposition approach used in this work, and which utilizes the heuristic load stabilization approach, is discussed next.

STAGEWISE DECOMPOSITION (SD)

In the present problem, the application of a stagewise decomposition approach, which is thematically similar to dynamic programming, is largely dependent on the definition of the problem. Dynamic programming is basically a serial optimization process, and hence the identification of a serial structure in the topology optimization problem is of paramount importance. The adaptation of dynamic programming principles in traditional structural optimization problems is somewhat unsatisfactory; such problems, do not often possess the serial structure necessary for an efficient implementation. In the case of structural topology generation problems considered here, the decomposition was viewed from the point of individual load stabilization. The basic approach was based on stabilizing each of the n loads, taken one at a time. The load stabilization task was considered as 'n' distinct subproblems, with each subproblem entrusted with the task of providing all possible configuration that stabilize the load associated with the subproblem. The structure build-up was propagated in this manner from one subproblem to the other, until all the loads were stabilized, simultaneously carrying forward the optimal weight (cost) thus far produced. The final topology at the end of the n-th subproblem was the one that yielded the lowest weight.

In the parlance of dynamic programming [13], the topology generation problem can be represented as follows. Each load was considered as a 'stage' in the design process. All possible supports to which each load could be connected were considered as 'state' variables for that stage. In addition, all previously stabilized loads in the upstream stages were also considered as 'state' variables. The 'cost' function for each stage was the weight of the partial structure for each stage.

Information is transferred from an upstream to a downstream stage as defined by stage function; for this, the optimal cost functions corresponding to each state in the immediate upstream stage must be considered. Figure 3 depicts the flow of information from an upstream stage to a downstream stage. The '*' indicates the optimal value as transmitted from the upstream stage. At each stage in the subsystem, the optimal cost function is given by the equation

$$f_N(S_N) = \min\left(F_N(S_N) + f_{N-1}(S_N)\right) \tag{1}$$

where 'f' is the optimal cost, 'N' is the current stage, and 'F' is the cost (weight) for the current state 'S'. Therefore, the optimal cost function is cumulatively updated during the progress of the analysis of subsequent stages. Ultimately, at the end of the n-th stage, after all the loads have been stabilized, the topology with the least weight is identified. At this point, the analysis is backtracked to identify the states corresponding to each of the upstream stages which contributed to this optimal solution.

The most important advantage in the application of stagewise decomposition to design synthesis is that the less promising partial designs are pruned at each stage in a methodical way. Designs which hold promise are only allowed to propagate to succeeding stages. In the example of Figure 1, the number of stages can be identified as 4, which correspond to the total number of loads to be stabilized. The stabilization proceeds from the load nearest to the origin of the coordinate system and moving away from it. All previously stabilized loads are also considered as possible supports. Hence, the number of states increase with stages in the downstream. For the example problem considered here, more than fifty topologies are possible, but the pruning of less promising paths after the analysis of each subsystem reduces the available paths to a manageable number. This, to a degree, limits the problem associated with a combinatorial explosion in the number of topologies that must be considered. Figure 3 can be used to further demonstrate the pruning of alternatives. The number of stages for the problem is three. The three states in stage 2, combined with two possibilities from stage 1, results in six possible outputs. But since only optimal states are considered for propagation, only three states are identified, one corresponding to each of the three states at stage 2. Similarly,there are two states at stage 3. Correspondingly, for each of the states from stage 2, there are two possible outputs from stage three, with a total of six. However, the actual number of outputs from stage 3 would have been twelve if all the possibilities in stage 2 had been considered.

While this approach for generating 'near-optimal' structural topologies is acceptable, there is some degree of arbitrariness in the process which may detract from its effectiveness. One such factor is the order in which the loads are stabilized. An arbitrary sequence, based on the distance of the load from a specified origin was chosen. Similarly, stage costs are only based on the partial structure weight, itself a function of some arbitrarily specified nominal cross-sectional dimension. Finally, the pruning of possible partial designs does not consider the potential of that design in satisfying system constraints. An optimization strategy that is able to accommodate the discontinuous nature of the present design problem was therefore investigated; this approach is derived from genetic search principles, and was adapted in the present work for the refinement of SD generated designs to produce optimal topologies. The following section provides an introduction to the genetic search approach and describes its implementation in the present problem.

GENETIC SEARCH

These search methods have their philosophical basis in Darwin's theory of survival of the fittest. They belong to a category of stochastic search methods, with an additional strength that randomized search is conducted in those regions of the design space which offer the most significant potential for gain. Analogous to the natural process where a population of a given species adapts to a natural habitat, a population of designs is created and is then allowed to adapt to the design space. The design alternatives representing a population in a given generation are allowed to reproduce and cross among themselves, with bias allocated to the most fit members of the population. Combination of the most favorable characteristics of the mating members of the population results in a progeny population that is more fit than the parent population. If the measure which indicates the fitness of a generation is also the desired goal of a design process, successive generations produce better values of the objective function. The terminology of genetic search and its principal components can be found in [14], and applications of the approach to structural design problems are represented in References [15-17].

There are three basic operations in the genetic search process, namely selection, crossover and mutation, and these are explained in the next section. Representation of the evolution process itself requires completion of three preliminary steps. The first is a definition of a bit string representation scheme to portray possible candidate solutions to the problem. Second, a suitable fitness function to provide for the ranking of the members in the population based on their merit. The last and perhaps most critical is the development of transformations to simulate the evolution process.

Each design variable is represented as a fixed length string of 0's and 1's, typically the components of a binary number. A scaling scheme provides for the conversion of the binary number to decimal and vice-versa. The binary strings for each design variable are placed end-to-end; this constitutes the bit-string representation for one design. Several such designs constitute a population, with a fitness measure for each member of the population. This is a measure of the relative goodness of the design, and is a composite measure of the objective function and the constraint satisfaction.

Genetic Transformations

Once a population of designs becomes available, then selection is the first concept to be introduced. The aim of this process is to select only those candidate designs based on their relative fitnesses, which are better than some others. This is meant to provide bias in the population to contain more fit members and to rid the population of less fit members.

The second component of the search is crossover, and pertains to the exchange of characteristics between select members of the population. Crossover entails selecting a start and end position on a pair of mating strings at random, and exchanging the 0's and 1's between these positions on one string with that from the mating string. This is illustrated as follows:

parent1 = 111000100101
parent2 = 010111001001

child1 = 111011001101
child2 = 010100100001

Mutation is the third transformation operator in the genetic refinement process, and safeguards the premature loss of valuable genetic material during selection and crossover. In the current implementation, this corresponds to selecting few members of the population, determining at random a location on the string, and switching the 0 or 1 at this location.

The above steps are repeated for successive generations of the population until either convergence is achieved on the fitness function, or a predetermined number of evolution cycles have been completed. The approach has shown promise in the current application because of its ability to search in a discontinuous design space where structural members can be either removed or introduced during optimization.

When used in the topology generation problem, this approach is particularly powerful as it accommodates a discrete design space in which structural members are both introduced and removed during the design process. This allows for an exploration of the topology design space until a satisfactory topology is identified within the guidelines of the design requirements. The use of this approach in conjunction with the SD methodology for generating topologies is even more advantageous, as the SD topologies provide good starting designs for the genetic search. The choice of such candidate designs for the initial population has a direct bearing on the number of evaluations necessary for convergence in the genetic search. Powell et. al. [18] report a favorable convergence pattern as a result of good initial seeding of the design space. The use of preferred candidate designs over a randomly initialized starting population, results in fewer function evaluations to convergence. Details pertaining to the actual implementation of genetic search in the topology generation problem, are presented in the following section.

GENERATION OF OPTIMAL TOPOLOGY

Since genetic algorithms work on a representation of the design variables as opposed to the variables themselves, it is fairly easy to incorporate the physical presence or absence of a structural member along with its cross sectional dimensions. If one thinks of a binary string representing the magnitude of a cross sectional area, an additional digit may be added to the string to indicate the presence of a member. The 1's and 0's can represent the presence or absence of a member between a load and a support, respectively. The genetic search approach allows for the optimal design to be obtained in two ways:

a. Both member cross-sectional dimensions and actual definition of optimal topology are considered simultaneously.

b. The optimal topology is first obtained through genetic search methods by using elements with nominal cross-sectional dimensions. This topology is then refined by another application of genetic search, with the cross-sectional dimensions as design variables.

The second approach is presented in this paper. The initial population of designs was generated through the SD method. This population initialization, coupled with the definition of a reliable measure of fitness, are necessary ingredients in a successful implementation of the genetic search.

Population Initialization

Initial population of trial designs were obtained from an application of SD approach described previously. In that approach, only optimal weights corresponding to each state in a stage are propagated to the succeeding stage. Therefore, at the end of the n-th stage, where the last load has been stabilized, the weight of the topology that is optimal until the (n-1)-th stage, is available. This weight can be added to the additional weights which result from all possible ways of stabilizing the last load. This yields an array of topologies with different total weights, and which serve as good initial seeds for the genetic search approach. To increase the number of available topologies, one can simply backtrack to the previous stage, and propogate forward, additional topologies for consideration at the next level.

The approach used in the present work was one where additional partial structures were propagated forward from the very first stage, and all resulting topologies rank ordered according to their weights. A total of 'M' lowest weight topologies were selected from this rank-ordered list, where 'M' is the population size specified for the genetic search, and used as the initial seeds. For the problem reported in this paper, a relatively small population size of twenty was considered.

Fitness Function

Since the artificial evolution processes is dependent on the value of the fitness function, a careful consideration must be given to the formulation of this measure. The fitness function has to take into consideration all factors that establish the relative merits of a given design. However, an option exists of not including some of the more complex design constraints in the fitness function, as they can also be handled at the member resizing level. In the illustrative example presented in this paper, the design constraints were considered at both the topology optimization and the member resizing level.

Most generally, the fitness function in the topology generation problem consists of three components. The first was a measure of the weight of the topology, generated using nominal cross-sectional dimensions. Second, the level of constraint satisfaction for a given topology had to be represented in the fitness function. The third and perhaps the one most critical to this problem, was the necessity to represent the favorability or lack thereof, of the appearance or removal of structural members during redesign. With these requirements in mind, the objective was formulated as follows:

$$z = w + R * \sum_i \langle g_i \rangle^2 + S * \frac{u}{v} \tag{2}$$

where

$$\langle g_i \rangle = \begin{cases} g_i & \text{if } g_i > 0 \\ 0 & \text{otherwise} \end{cases} \tag{3}$$

Here, 'w' is the weight of the structure; 'R' a penalty parameter associated with the violated design constraints; 'g_i' is the i-th design constraint; 'S' is a penalty parameter which penalizes the generation of heavier topologies to a larger degree; 'u' and 'v' are integers used to distinguish between potentially good and poor choices of structural member addition or removal, and their definition will become clear in subsequent discussions. Since z is the objective to be minimized, a fitness measure z* to be maximized, can be written as follows,

$$z^* = z_{max} - z \tag{4}$$

where z_{max} is the maximum value of z in the population.

The representation of structural members which are introduced or removed had to be transformed into a quantitative form, so as to include this process in the fitness function. Each load was stabilized by introducing members to any of the 'n' supports that were available. Also available as additional simple support points, were the previously stabilized loads.

If there are 'm' loads and 'n' supports in the design domain, then for the first load to be stabilized, there are 'n' members that can be introduced. For the second load that is considered, there are 'n+1' selections, as an additional member can be introduced between the load and the previously stabilized load point. Therefore, for the m-th load, the number of possible members that can be introduced is n+m-1. For each load, a binary string of length equal to the number of possible members that can support this load, is specified. Each digit represents a possible connection between a load and a support, with the least length connection represented by the leftmost digit; progressively increasing length members are represented by digits to the right. A '1' indicates that the member between the load and the support in question is actually present, while a '0' denotes its absence. The bit string representation for the entire topology is a concatenation of such strings for all the loads. Therefore, the total bit string length to represent a topology, for a 'm' load and 'n' support domain is given by

$$n + (n+1) + \ldots + (n+m-1) = \left(n + \frac{m-1}{2}\right)$$

The cumulative sum of the position numbers containing a '1' in the bit string is the variable 'u' in Eqn. 2. Likewise, 'v' denotes the total number of digits in the string. As an example, consider the following 10 digit string:

string : 1001001001

Here,

'u' : 1+4+7+10 = 22
'v' : 10

Hence, a higher value of the ratio u/v indicates the presence of longer length members and hence also a heavier topology.

ILLUSTRATIVE EXAMPLE

The example problem considered in this implementation is shown in Figure 3. For this problem, the list of allowed structural members was shortened, and only truss members were considered. The initial population of topologies was derived from the SD approach as described previously. For each of the topologies, nominal cross sectional area members were introduced for analysis. A crossover rate of 0.65 and a mutation rate of 0.009 were assumed in the present simulation, and the designs allowed to evolve over twenty generations. Figures 4-6 depict topologies at the end of first, tenth, and twentieth generations. All topologies in any generated population were checked for stability considerations. This was accomplished by the examination of each topology in the population using the previously developed knowledge-base. The stabilization of each load was checked and member(s) introduced where necessary to ensure stability.

The least weight topology is highlighted in Figure 6. As indicated earlier, this minimum weight topology with nominal values of cross-sectional dimensions was subjected to a second round of optimization. In this cycle, the member cross sectional areas were resized using genetic search. Once again, a crossover rate of 0.65 and a mutation rate of 0.009 was adopted. The objective in this secondary optimization was to minimize the weight of the structure subject to the constraints defined in Eqns. 1-3. The objective function for this problem was defined as follows,

$$z = w + R * \sum_i \langle g_i \rangle^2 \tag{5}$$

where 'w' is the structural weight, 'R' is a penalty parameter such that the second term in the function z is of the same order of magnitude as the weight, and g_i are the normalized design constraints defined as follows:

$$g_i \equiv \left|\frac{\sigma_i}{\sigma_{all}}\right| - 1 \qquad i = 1,2,\ldots nel$$

$$g_i \equiv \frac{d_j}{d_{all}} - 1 \qquad j = 1,2,\ldots m$$

$$g_k \equiv 1 - \frac{\omega_k}{\omega_{all}} \qquad k = 1,2 \tag{6}$$

where nel is the number of structural elements for which stresses must be constrained, and m is the number of load points for which displacements are constrained. Also σ_i and d_j represents the stress in the i-th member and the displacement at the j-th load location, respectively; ω_k is the k-th natural frequency of the final topology; the subscript 'all' indicates allowable values.

Genetic search maximizes a prescribed fitness function. Since our problem is one of function minimization, a transformation of the objective was achieved using eqn. (4). The value z* can serve as the fitness function for the population. The initial population consisted of twenty different topologies, with cross sections of the members in the structure considered as design variables. All the designs in the population were randomly generated. A 9-digit binary number was chosen to represent each individual cross sectional area. A finite element analysis of each of these designs yielded values of nodal displacements, frequencies, and member stresses. These were used in the constraint function evaluations.

Since the population is randomly generated and is very limited in size, some additional steps were necessary to guard against premature convergence. Such a convergence pattern is encountered with small population sizes, particularly if raw values of the fitness function are used to direct the reproduction operator. Here, the more fit designs quickly dominate the population, and the process may converge without adequately searching the design space. The population fitness values should be scaled so that in the event that the average fitness of the population is close to the maximum fitness, the scaling magnifies the difference in relative merits between members. The scaling used in the present implementation was linear, and was defined as follows:

$$f' = c_1 f + c_2 \tag{7}$$

The scaled maximum fitness is given as $k^* f_{avg}$, where k is typically in the range of 1.0 to 2.0. For a chosen value of k = 1.5, the c's in the above equation are obtained as,

$$c_1 = 0.5 f_{avg} / \left(f_{max} - f_{avg}\right)$$

$$c_2 = f_{avg}\left(f_{max} - 1.5\, f_{avg}\right) / \left(f_{max} - f_{avg}\right) \tag{8}$$

where f_{avg} is the average fitness for a given generation.

Scaled fitness values f' were used to detect the most fit members of the population for reproduction.The final design is shown in Figure 7, and the optimized cross sectional areas of members are presented in Table 2.

CLOSING REMARKS

The paper presents a novel technique for the preliminary synthesis of structural systems. The quasi-procedural approach is an efficient process which combines design abstraction and validation capabilities. The genetic search approach proves to be a viable strategy for the structural topology synthesis process. The strengths of this approach are in its ability to account for the discrete nature of the design space. The use of stagewise decomposition provides an efficient approach to both manage the design space, and to provide good initial designs required in the genetic algorithm based optimal design.

REFERENCES

1. Nevill, G. E., Jr., Jackson, L. A., and Clinton, J. H., "Automated Hierarchical Planning for Structural Design" Proceedings of the ASME Computers in Engineering Conference, San Francisco, August, 1988.
2. Shah, J. J., and Pandit, L., "DEZINER - An Expert System for Conceptual Form Design of Structural Parts" Proceedings of the ASME International Computers in Engineering Conference, Vol. 1, Chicago, July, 1986.
3. J.L. Chen and P. Hajela, "A Rule Based Approach to Optimum Design Modeling", Computer Applications in Structural Engineering, (ed. D.R. Jenkins) ASCE, New York, pp. 66-81, 1987.
4. J.J. Shah, "Development of a Knowledge Base for an Expert System for Design of Structural Parts", Proceedings of the ASME International Computers in Engineering Conference, 1985.
5. J. Bennett, L. Creart, R. Englemore and R. Melosh, "SACON: A Knowledge Based Consultant for Structural Analysis", Technical Report STAN-CS-78-699, Stanford University, Stanford, California, 1978.
6. S.C.-Y. Liu, "A Consultative Expert System for Finite Element Modeling of Strip Drawing", presented at XIII North American Research Conference, University of California at Berkeley, May 1985.
7. R.H. Holt and U.V.L. Narayana, "Adding Intelligence to Finite Element Modeling", Expert Systems in Government Symposium (eds. K.N. Karna, K. Parsaye and B.G. Silverman), IEEE, New York, 1986, pp. 326-337.
8. Hajela, P., New Directions for AI Methods in Optimum Design,Proceedings of the NASA/Air Force Symposium on Multidisciplinary Analysis and Optimization, NASA CP-3031, September 1988.
9. Hajela, P., and Shankar, N., "A Coupled Algorithmic and Heuristic Approach for Optimal Structural Topology Generation.", Computers and Structures, 36, 5, 1990.
10. Hajela, P., and Shankar, N., "Recent advances in AI Based Synthesis of Structural Systems.", Vol II, NATO Advanced Study Institute on Optimization and Decision Support Systems, Edinburgh, U.K., 1989.
11. Shankar, N., "Knowledge Engineering Applications in Automated Structural Design.", PhD Dissertation, University of Florida, May 1991.
12. CLIPS Reference Manual, JSC-22948, NASA Johnson Space Center, April 1988.
13. Kirsch, U., Optimum Structural Design-Concepts, Methods and Applications. McGraw-Hill, New York, NY, 1981.
14. Goldberg, D.E., Genetic Algorithms in Search, Optimization, and Machine Learning, Addison Wesley, Reading, 1989.
15. Hajela, P., "Genetic Search - An Approach to the Nonconvex Optimization Problem." AIAA Journal, Vol. 26, No. 7, pp1205-1210, July 1990.
16. Hajela, P., "Genetic algorithms in Automated Structural Synthesis." Vol II, NATO Advanced Study Institute on Optimization and Decision Support Systems, Kluwer Academic, Boston, MA, 1989.
17. P. Hajela and C.Y. Lin, "Genetic Search Strategies in Multicriterion Optimal Design", proceedings of the 32nd AIAA/ASME/ASCE/ASC SDM meeting, Baltimore, MD, April 1991.
18. Powell, D. J., Tong, S. S., and Skolnick, M. M., "EnGENEous-Domain Independent Machine Learning for Design Optimization." Proceedings of the Third International Conference on Genetic Algorithms, Washington DC, 1989.

Table 1. Table of primitives.

Primitives	Attributes
Element	Beam, Truss
Load	Point Loads, Moments, Distributed Loads
Support	Hinge, Clamp
Material	Steel, Aluminum, Aluminum Alloy
Cross Sections	I-section, C-section, Tube, Rod

Table 2. Optimized areas for example in figure 7

Element #	1	2	3	4	5	6	7	8	9
Cross Section Area (in^2)	1.4	2.0	2.8	0.9	2.3	2.2	0.6	0.3	1.0

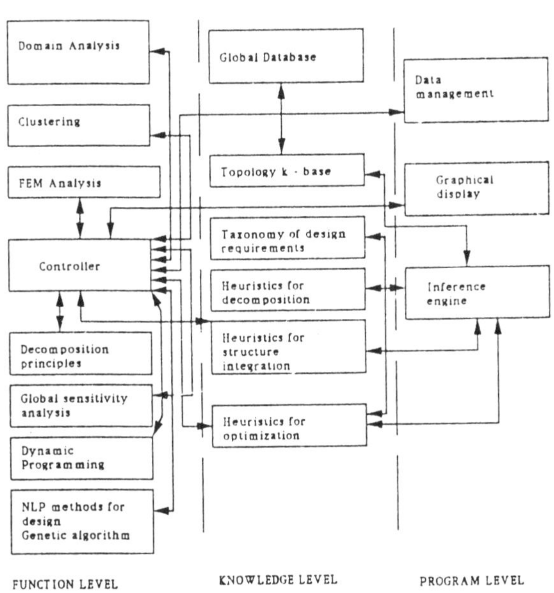

Figure 1. Organization of the three levels.

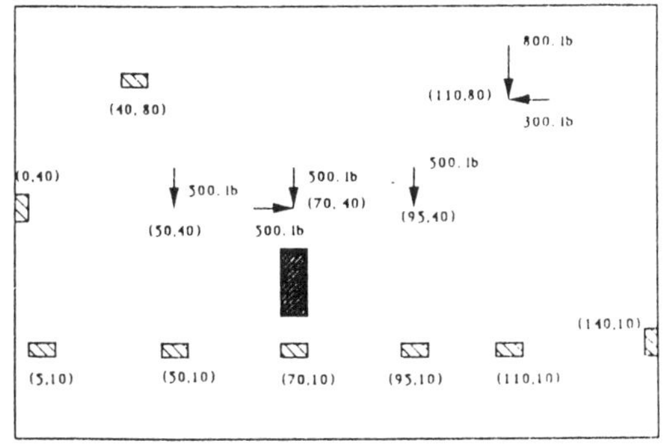

Figure 2. Example design problem.

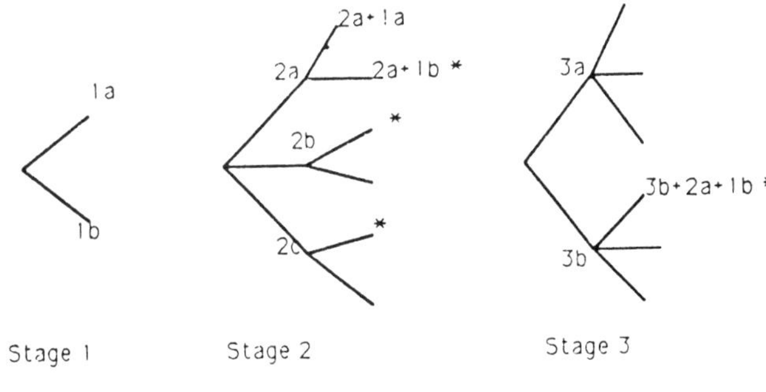

Figure 3. Information flow between stages.

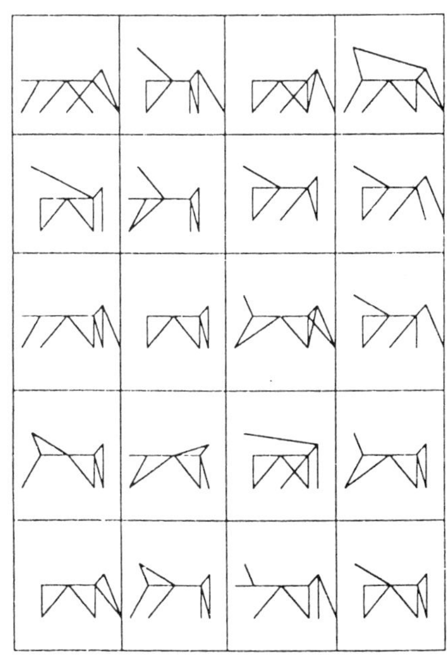

Figure 4. Topologies in generation #1.

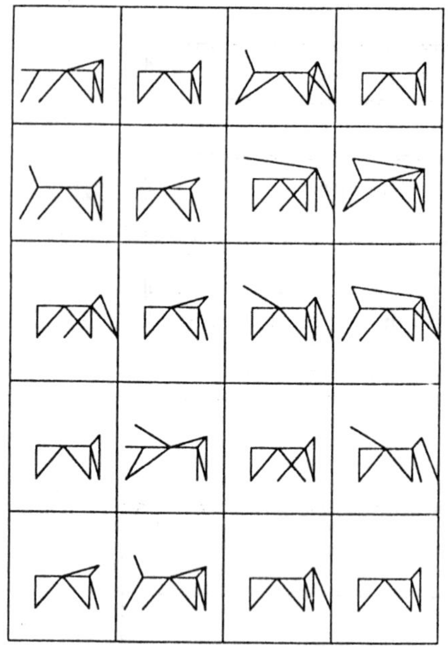

Figure 5. Topologies in generation #10.

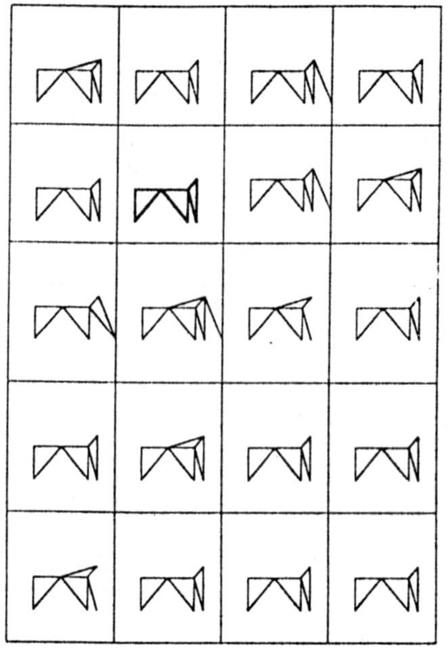

Figure 6. Topologies in generation #20.

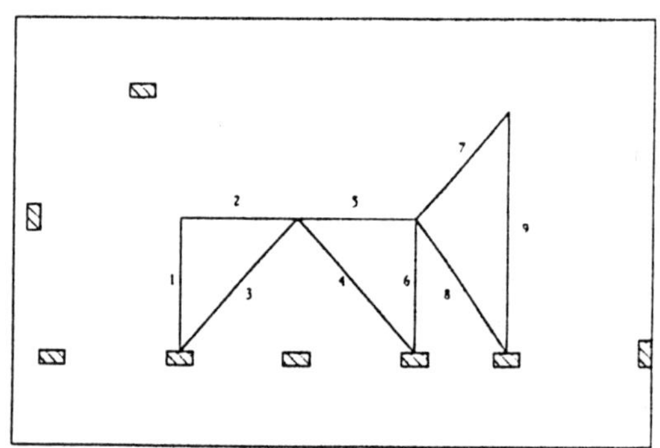

Figure 7. Final topology.

10 REPRESENTATION AND PROCESSING OF CODES OF PRACTICE

A REPRESENTATIONAL SCHEME FOR DESIGN CODE INFORMATION IN AN EXPERT SYSTEMS APPROACH TO BUILDING DESIGN

A. Omari and G.G. Roy
The University of Western Australia,
Nedlands, Australia

The search for new approaches to building design through the use of expert systems invariably stumbles on the need to adopt appropriate representational schemes for the design criteria. Our particular interest lies in the criteria often found embedded in codes of practice, against which the success of a design can be measured. These criteria place a complex set of requirements on the design and the way in which the design process proceeds. In particular, we will use that part of the Building Code of Australia (BCA) concerned with Access and Egress and more specifically part D1, the Provision for Escape as an example.

Our concern is the development of a knowledge representation scheme which is close to the actual code document but still permits formal analysis and inference strategies to be developed and also provides a formal link to the geometric elements of the building specification. This offers the potential of developing an integrated knowledge-based computer-aided architectural design (KBCAAD) approach.

The system in which these schemes are being implemented is PC-based and will attempt a complete integration of geometric design with the formal specification of a code of practice. The focus of this paper is to outline the knowledge representation scheme which provides the necessary foundations for this work. This paper will provide a description of the knowledge representation methods with some preliminary examples of its application to the Provision for Escape requirements from the Building Code of Australia.

INTRODUCTION

Design as a creative endeavour has been practiced since the beginning of the history of our civilization. Mostly, it has been taken for granted, that some particularly gifted people were able to make substantial achievements, and these achievements were then recognised as adding to the utility or welfare of man. Design as a process thus requires two fundamental elements, firstly a creative act followed by an acceptance that the results of the design process are in some way useful.

In modern times we have become rather interested in the process of design, no doubt with the view to either being more successful in attempting design tasks or tackling design problems which have yet to have well defined solutions. The advent of computers has accelerated this interest immensely. The 1960s brought a flurry of activity in the field of design methods. While much interesting work was done we are not so sure today that we are much better off with respect to our knowledge of design processes [Refs. 2, 7]. By and large, design processes have defied synthesis into formal methods.

Rapidly developing computer technologies (both hardware and software) have in the last decade led to new approaches to design problems, though the terminology is often different. We now speak about expert systems, knowledge based systems or the use of AI techniques. In essence these developments have come from the recognition that the creative elements in design may be much harder to grasp than thought earlier, and perhaps may never be formally defined. It is also now accepted that some synergistic man-machine approach might be able to encourage and provide an effective compliment to the skills of both man and machine. In this way it may be possible to make steps forward in improving the quality of certain design tasks.

Design problems for architects and designers of the built environment have been popular areas of interest since the 1960s. The interest in this area is probably due to the complexity of the design problem (ie. there is no real limit to the number of possible solutions) and the fact that mistakes are often made and that it is the users of the built environment (all of us) who must live with the blunders. Building design is thus a particularly rich design problem area.

As a result of these concerns we have many formal controls over the design process, each attempting the avoid catastrophic mistakes. Often these controls are expressed in codes of practice which set a out range of requirements which must be satisfied by a design solution. While codes do not instruct the designer on what the solution to the design problem should be, they do place considerable constraints on the range of possibilities.

The focus of the research, on which this paper is based, is concerned with developing some ideas about how we might improve the design process through an integrated design/knowledge-based system which provides the designer with some powerful tools for ensuring that the evolution of the design can proceed in parallel with a systematic analysis of the appropriate design code. Our efforts have a rather focussed intent at this stage, that is we have limited out interest to a that part of the Building Code of Australia (BCA) [1990] concerned with the means of access and egress, and specifically those clauses dealing with the provision for escape (part D1). This code provides a useful vehicle for our work as it is a good example of the integrated nature of design where there are close relationships between the design geometry and the way in which we must reason about the design to satisfy the rules contained within the code.

In this paper our interest is to present a part of this work, in particular a knowledge representation scheme which is thought to offer some potential for describing the various knowledge types often embodied in design codes and allows an integration with the those parts of the design process concerned with the geometric definition of the design (building). This paper will focus on the knowledge representation scheme, and only cursorily touch on the geometric design issues.

CODES OF PRACTICE

The role of codes of practice in design, and more interestingly methods for formally describing or applying them has been the subject of considerable research work [Refs. 3, 4, 6, 8, 9, 13, 14].

Codes of practice are often intended to represent the legal requirements and obligations of the designer, and as such are usually written in formal legalistic language. Such language forms are intended to provide precise, rather that concise, sets of rules and procedures to be applied in the proposed design. Clearly a designer should have expert knowledge of the relevent codes if a successful design solution is to be efficiently obtained. In reality, however, this is not fully realized and code violations are not always detected until later stages in the design where they might be expensive to resolve. Designers might be expected to have some level of knowledge about the relevent codes, but to assume expert knowledge is not usually realistic. Many designers would therefore accept that obtaining help in validating a design, or partial designs, against the relevent codes would be useful.

Various approaches to the interpretation of design codes have been attempted. A common approach is to attempt to transform code clauses into formal logic, usually first order predicate logic which can then be readily represented within a logic programming environment such as Prolog [Refs. 5, 10, 12, 14].

The transformation process from code clauses into a formal logic representation is not always so easy. For simple (perhaps trivial) examples solutions are quickly developed, but as the complexity increases the difficulty of adding new clauses, or including globally scoped conditions becomes difficult [Ref. 11]. In general the problems of transforming codes of practice into the declarative form of logic programs is a non-trivial problem. It is also generally necessary to utilize the procedural features (side effects!) of particular environments (eg.. expert system shells) to express the full intent of the code provisions [Ref. 13].

The use of knowledge representation schemes developed from the AI area have also been tried. The use of frame concepts allow a natural way of including inheritance, a central element in all object-oriented representations. Frame schemes offer some potential where we need the flexibility to add new slots, or change slot values. Slots can also contain functions which can be triggered to cause required actions to take place. Frame based schemes are thus contenders for areas of the design problem that are strongly hierarchical, such as the basic geometric representation of a building. Related approaches use purer objected-orientated models. While in many ways similar to frame-based schemes a greater emphasis is often attached to the relationships between objects. Object relationships can be mapped onto most codes of practice.

While these representation schemes offer potential, there are problems when such very general schemes are applied to real codes of practice. Common problems arise when trying to represent *components and categories, specific and general conditions, default and overriding conditions* and even *conflicting behaviour* which adds to the complexity of the representation problem [Ref. 6].

A more interesting approach to the problem of representing codes of practice has been given by Hamer [Ref. 6]. His approach has been based on a functional representation proposed earlier by Fenves [Ref. 4]. This has been derived from a concept of classification which is described as the process of selecting an appropriate category for an object. From this basis Hamer has developed a formal scheme of representation, through *Class Language*, which combines the functional representation of Fenves with an object-oriented approach. This provides a basis for describing the necessary hierarchies in codes of practice.

REPRESENTATIONAL SCHEMES AND THEIR IMPACT ON DESIGNERS

Codes of practice are intended to be read directly by designers. While, by necessity, they are often difficult to read this process is still feasible and to some degree successful. The interpretation of the code and its interrelated clauses is a human task. The process of formally representing the code in a way that it can be processed automatically requires a transformation to a language form which is most often foreign to the designer. Logic representation schemes, such as Prolog or most of the expert system shells that use rule-based schemes, are perhaps not too far removed from the domain experts' experience. More complex schemes, such as Hamer's Class Language, make it even more difficult for the expert.

Of course the average user may never be directly involved with the formal representation process, but we must remember that expert code-users will most likely not be computer experts either. The transformation of code to formal language is thus a tricky business and fraught with many pitfalls, especially if the transformation process requires language constructs that are foreign to the expert attempting the task. It is central to our hypothesis that the original code clause, or sub-clause, should be retained as a central element in the representation scheme. In this way the user, designer, is never too far removed from the code document. It is also expected that the code text can be used as an aid in the explanation process, even when the actual inference process is being driven by a more formal rule-based process.

Our approach to the representational problem has been driven by the following requirements:

- To provide a consistent representation scheme for all knowledge relating to a design problem. This includes the building geometry, the applicable codes of practice and the users interaction with the system supporting this knowledge. By consistency we mean that these various knowledge forms appear to be consistent to the user who must interact with various knowledge types during different phases of the design process. Consistency also means that we can transform one type of knowledge into another, and back again, without loss.

- To maintain, as far as possible, the original text of the code clauses at each decision point so that users will always have access to the code document to assist in decision making, evaluation and in the interpretation of explanations.

- To provide facilities for the user to choose the depth and breadth of evaluation at any time that is commensurate with the state of the design development. To be effective the user must be reasonably knowledgeable about the design and the code so that sensible queries can be posed. This is, of course, particularly pertinent when the designer is dealing with a partially completed design.

We are thus concerned with an integrated approach to design and the use of the code as an integral part of the design process. This naturally places some constraints on how we can, or should, represent all the relevent knowledge. We are especially interested in schemes which allow suitable transformations to suit the various phases of the design process.

Design is not a linear process, it is universally recognised as being iterative, perhaps converging (eventually) to a final and complete solution. Our model of design attempts to facilitate this process. Code validation on partial designs is thus an essential component of our approach. The designer should be able to validate code requirements at any level of detail, from a first sketch drawing through to final design.

To achieve these desired qualities we have chosen a frame based knowledge representation scheme, with some special properties.

A building design is formally represented by pieces of knowledge that have strong geometric components. We need to be able to represent objects such as rooms, walls, doors, passageways, items of furniture both in terms of their size, shape and location, and all associated properties (material properties, etc). Objects also relate to each other to reflect the common associations, such as which floor level, which room, which wall, etc, a particular object associates. While this object-orientated scheme is appropriate for actually creating/ modifying the design in a CAD environment, we have found it appropriate to convert the resulting data to a formal frame scheme for analysis within the inference process. The inference engine operates on this data.

We will not dwell further in this paper on the structure of the geometric representation and manipulation scheme, but focus on how this knowledge is integrated into the total scheme. Formally, the geometry is represented by a set of objects (defined as predicates) which are then transformed into frames as required for the inference process. The inference process may, and generally does, modify these frames as further knowledge is gathered during the inference processing. These modified frames are then converted back into the predicate form when the user is ready to resume editing the geometric properties of the design.

KNOWLEDGE STRUCTURES

The complete process is currently being developed into the PlanCAAD system which is represented

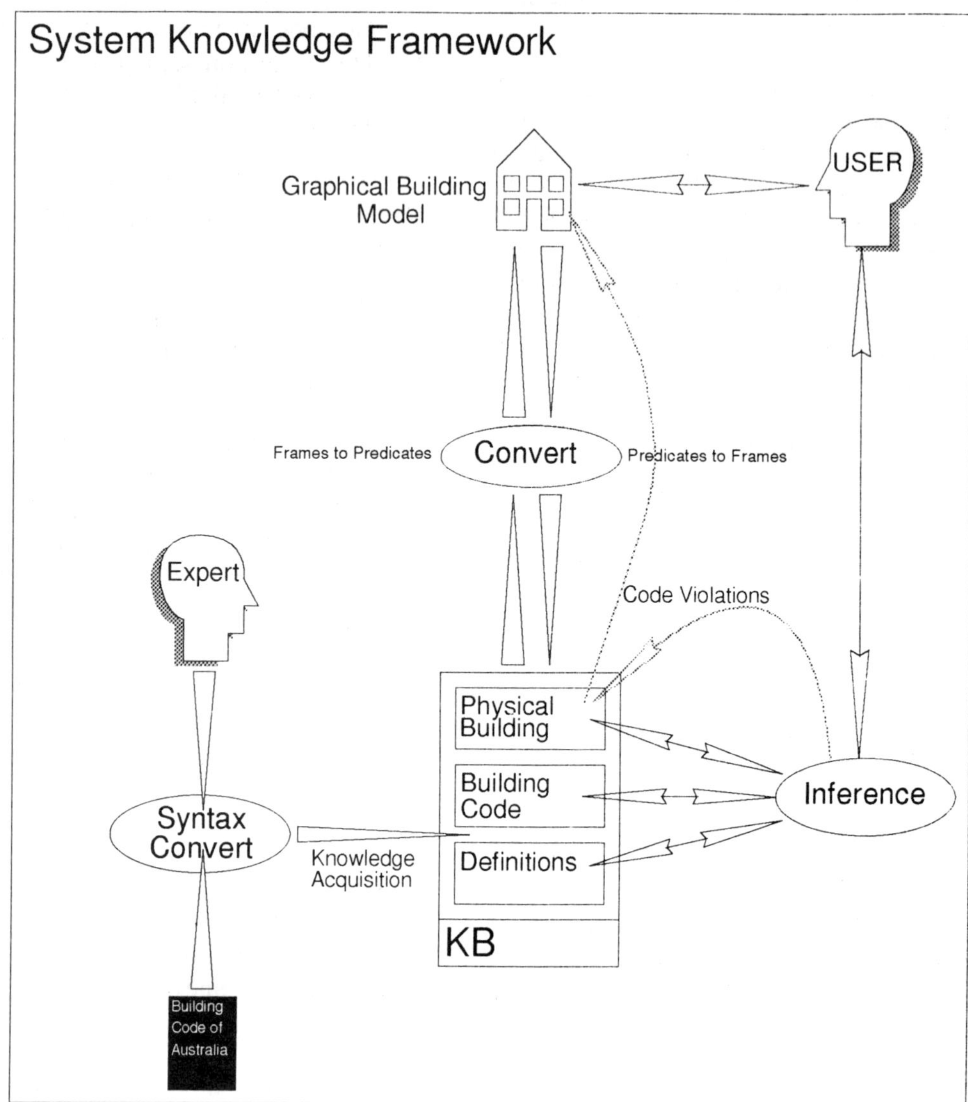

Figure 1 : System Framework

schematically in Figure 1. In this system the Graphical Building Model describes the traditional design representation in geometric terms. It is a $2\frac{1}{2}$D drawing of the design, and in our case it is contained with a PC-based CAAD system which provided all of the basic functions to allow a designer to create and modify the building design.

In overview, the complete system consists of the building model to represent the physical geometry of the design, and a representation of the code and associated information. Taken together this knowledge base can be interrogated by an inference engine which is driven by the user. The user interfaces directly with the graphical building model through a (typical) CAAD interface and to the inference engine through a formally specified dialog language.

The conversion of the code to a form which can be used within the system is a job for the expert. We make no claims that this process is easy or can be automated, but some effort has been spent on designing a formal representation which is compatible with both the geometric definitions and the needs of the user while driving the inference process to validate the design against the code provisions.

Building representation

While most CAAD systems contain only the geometric knowledge (ie. points, lines and perhaps polygons), our model will contain a "complete" object description of every component. By "complete" we mean that it is as complete as the user wishes it to be or the rules of the code require. Any properties that the rule interpreter needs, and the user has not provided, will be requested by the interpreter and subsequently added when the frame system is converted back into predicate form.

At some stage in the design process the user will choose to validate some part of the code against the current state of the design. To initiate this process the building geometry is converted into a frame scheme to facilitate the formal processing by the inference engine. Each frame consists of:

frame_id an identifier which refers to the frame

parent_id the id of the frame from which the current frame inherit properties. If a required frame slot is not found in the current frame it may be inherited from the parent, or the parent of the parent.

properties the properties associated with the object represented by the frame are triples consisting of (property::type::value). The property field is simply an identifier, the type is a valid token identifying the property type and value is a value which is valid for the selected type.

This information is sufficient to describe the complete physical geometry of the design and all associated properties. A typical example might be:

frame_id	DOOR_TYPE_A
parent_id	<NULL>
properties	(width::real::1050)
	(height::real::2100)
	(thickness::real::35)

frame_id	DOOR_103
parent_id	DOOR_TYPE_A
properties	(swing::string::left)
	(colour::string::yellow)
	(x_origin::real::10090.25)
	(y_origin::real::-3054.98)

Code representation

Most codes of practice have a strong rule-based structure, though it is often the complexity of the rules which defeats attempts to accurately represent code clauses and the subtleties of their meaning. We have chosen to accept an underlying rule-based representation, but to wrap it within a strong frame scheme which allows us to attach more information, such as the original clause text and associated (non-code) explanations or interpretations from other sources (eg. legal precedents and expert interpretations).

In addition we have defined a Dialogue Language (𝔻𝕃) which provides a consistent representation for the rules used to represent the code clauses as well as the interaction between the user and the inference engine.

This approach does little in itself to overcome the major difficulties of representing the often subtle and complex code requirements (this is common to all research in this field), but we hope through a sufficiently rich 𝔻𝕃 that the *expert* will be able to satisfactorily express most, if not all, of the code requirements.

The 𝔻𝕃 is the primary tool to manipulate dialogs. Dialogs are frames which contain the code clauses, or sub-clauses. Dialogs are the building blocks of the code representation. In general they match code clauses and are structured as follows:

dialog_id - an identifier which references a dialog

parent_id - an identifier which points to the dialog from which the current dialog was referenced.

code_violation the id of the object to which evaluation error messages are to be attached.

clause the actual text of the clause from the BCA which is embodied within the current dialog.

condition the 𝔻𝕃 interpretation of the conditions which must be satisfied for this dialog to be applicable.

action the 𝔻𝕃 interpretation of the actions to be carried out if the conditions for application of the dialog are met.

comment an explanatory note which describes the reasons for the application of this dialog. This field is used primarily to indicate to the user, why a particular dialog has failed, in simpler terms than can normally be available from the raw code clauses. This text field is attached to the frame identified by the code_violation field to indicate non-conformance of building model elements to the BCA.

dependency A list of property value identifiers which is used during the evaluation of a dialog. This field can be used to indicate the values of the properties used by the dialog. As such they provide a reference by which the user can determine the exact property which is not valid.

Theoretically only the first two slots (*dialog_ID* and *parent_ID*) are essential components of the dialog frame, but more often most slots will be present.

The *comment* slot is used primarily to explain why the dialog might have failed, perhaps in simpler terms than available from the actual clause text. This comment is attached to the object defined by the code_violation slot of the dialog to help explain the reason for non-conformance of the building design to the code clauses.

The evaluation of a dialog condition will result in a true or false result (see latter section Implementation). This may be the consequence of a complex sequence of sub-conditions, of which only one may have caused a failure. In such a case it may be quite difficult for the user to figure out where the real problem lies. Of course we could further divide the dialog into less complex parts, but this would be at the expense of creating an even greater separation between the current rule and the actual clause being represented. To avoid this problem the *dependency* slot can be used to contain a list of properties used by the dialog and provide a reference by which the user can determine which one has failed to satisfy the dialog *condition*.

In general, the failure which actually comes about relates directly to only one element. The code violation thus caused may occur in a number of places, but often relates to a single element.

Definitions

Definitions are central to an understanding of the code and must be represented within the inference system. Definitions are used to give a description of either a frame slot property identifier, or some concept which is included in the clause or comment fields of the dialog. They are used to also impose some limitations on the form or type of data that can be attached to the property or concept, especially for the validation of user-entered data. Definitions are also used to include standard code terms and available globally to the user. Definitions can also be user defined, and used as a part of the code validation process. A definition is also structured as a frame:

definition_id an identifier which refers to the definition. This is normally a string which describes the concept being defined.

value_field one of the following:

1. A <null> value indicating that the definition is only a description of some concept used within either the comment or clause fields of a dialog.

2. A valid token which identifies the type for the definition. Valid types are CHARACTER, INTEGER, REAL, BOOLEAN or STRING.

3. A list of possible values that the definition can have. The list is enclosed within square brackets and each element is delimited by a comma. (e.g. [value1,value2,value3])

description a textual description of the concept being defined. In the case of the BCA clauses, these descriptions are found in Section A - General Provisions.

The knowledge base for the PlanCAAD system is thus composed of the three component parts, the physical model, the code model and the definitions part, each defined using a frame representation. Figure 2 shows the interaction of these three elements of the knowledge base.

The dialog frame thus provides the central focus of our system, and it is through the evaluation of the dialogs (proving them) that the compliance of the design to the code is verified. Dialogs are evaluated by the user initiating a request to evaluate a specified dialog. Remembering that dialogs are essentially mapped onto code clauses, this would mean the user chooses a particular part of the code for evaluation against the current design.

To prove a dialog then, its condition part must be satisfied and all consequential actions proven. The inference engine manages this process through a backward chaining process using a conventional depth-first search strategy. The process continues until the dialog is proven or terminated by the user. Most often processing is terminated when some non-compliance is found or insufficient knowledge is available to continue. Both situations are normal for partial designs.

A complete design could be validated by evaluating the dialog which references the complete code, but this would be unusual, most often the user is interesting in checking either parts of the design or selected parts of the code. Once validation is achieved at the desired level then the designer will move on to complete further stages of the design.

IMPLEMENTATION

These concepts about the representation of the code, the geometry of the design and other data required for the validation process have been built into the PlanCAAD system. This is a completely integrated system for creating schematic designs by drawing them and evaluating the conditions of the appropriate code of practice. We have chosen a part of the Building Code of Australia (BCA), that part concerned with access and egress provisions for escape (part D1), for our current implementation. This part of the BCA is typical of most codes of practice and is sufficiently complex to demonstrate our approach, both from a formal representation of the code and the integration into a CAAD environment.

At this point it is relevant to mention the different approaches to code evaluation which can be implemented. The two approaches which we have identified are, firstly to express the code in a manner by which it is the clauses which determine when and how an element is evaluated. Optionally, the building elements themselves can determine the course of evaluation of the clauses which are expressed. Using the latter approach, the code is organized in a manner that each element or type of element is identified and then the rules which may or may not apply to this are then evaluated. The latter approach is taken by Hamer and is consistent with the object oriented paradigm [Ref.6].

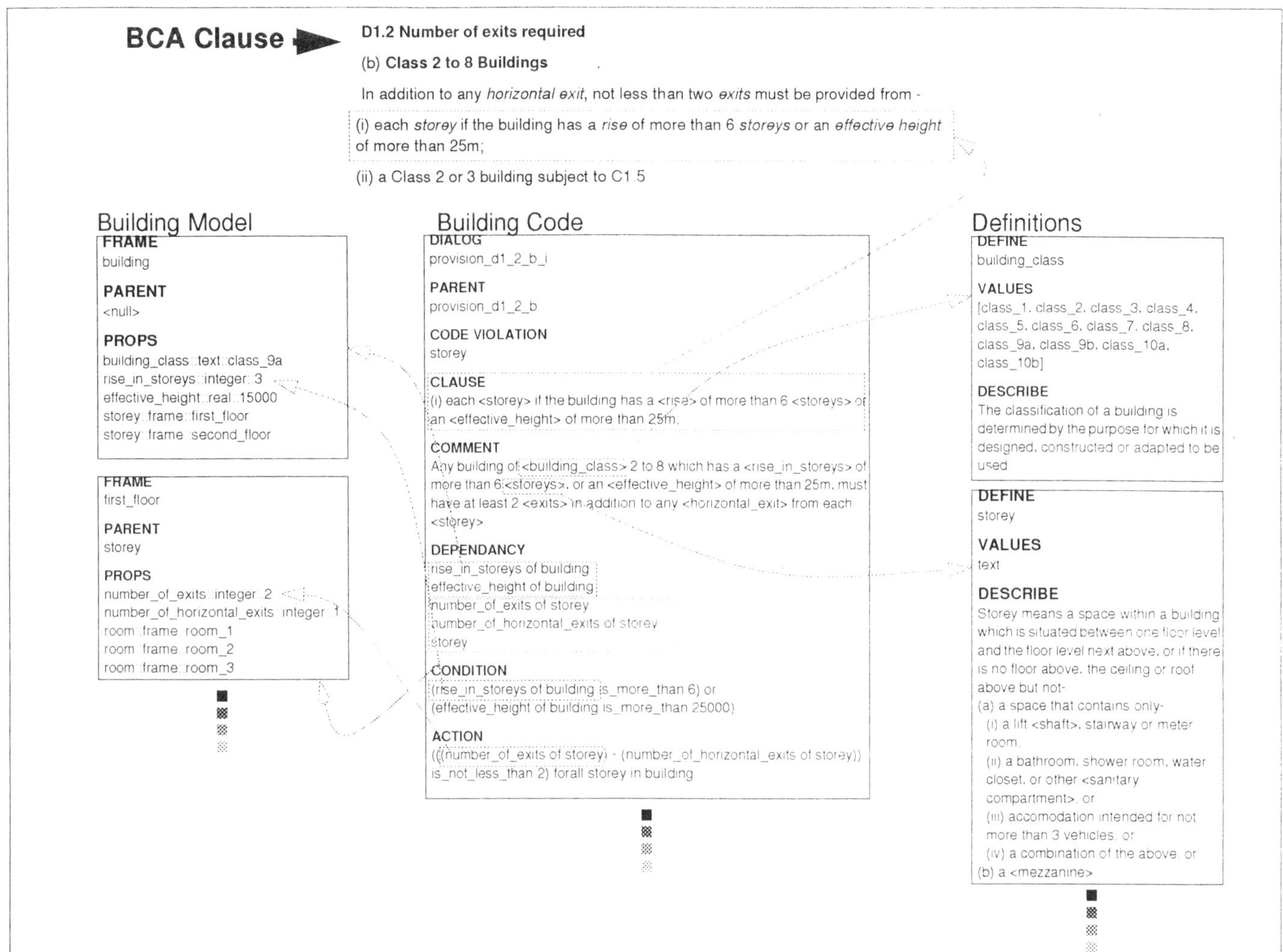

Figure 2 : Relationships between knowledge base elements

Our approach follows the first path, dialogs are examined to determine their applicability and then any elements which may need be evaluated are examined. The main advantage that we see from the approach we have taken is the ability to maintain the consistency of the code as a document, the structure of the dialogs and their execution follows very closely to the code clauses of the BCA. This has advantages in the future for modifying the existing building code rule base, the ability to edit a rule base is of fundamental importance when dealing with codes of any description yet even more so for the BCA which is still relatively new and subject to change.

We will not discuss the drafting system in this paper, but focus just on the dialogs and their manipulation. The dialog editor provide the means of constructing and editing the dialogs, primarily used by the expert constructing the knowledge base. Figure 3 shows the clause used in figure 2 (ie. D1.2.b.i) as it would appear to the user on the editing screen. As shown in figure 2, the BCA italicises any concepts defined elsewhere in the code, in the dialog editor however, the syntax has been changed so that definitions available to the user are shown between "<....>". Also note that concepts are identified by the use of names which contain no white space, hence the BCA concept 'building class' becomes the DL definition '<building_class>'. The syntax used by DL is very much defined by the fact that the language uses similar constructs to interpreted Prolog, but with extra predicates and functions added to allow the expression of code provisions. Further discussion of the DL syntax occurs later in this paper.

Figure 4 shows an example editing session for the definitions. In this case it is showing one of the basic definitions in the BCA, the building classes, which will be used throughout the code.

Figure 5 shows an example editing session for a frame representing a part of the design geometry. While these frames can be edited directly, they are mostly changed throughout the interactive drafting module which allows a direct manipulation of the geometry and the hierarchical structure of the design. The ability to edit the frame definition of a building allows the user to carry out "what-if" type queries of the knowledge base, if the dialog evaluation fails at some point, the user may wish to experiment with different values of an object property to find an acceptable compromise.

At some point in the design process any dialog may be evaluated. This is akin to making a proposition that the dialog, or at least the code clause it includes, is to be proven true. The operation of the system provides several basic modes of operation are available:

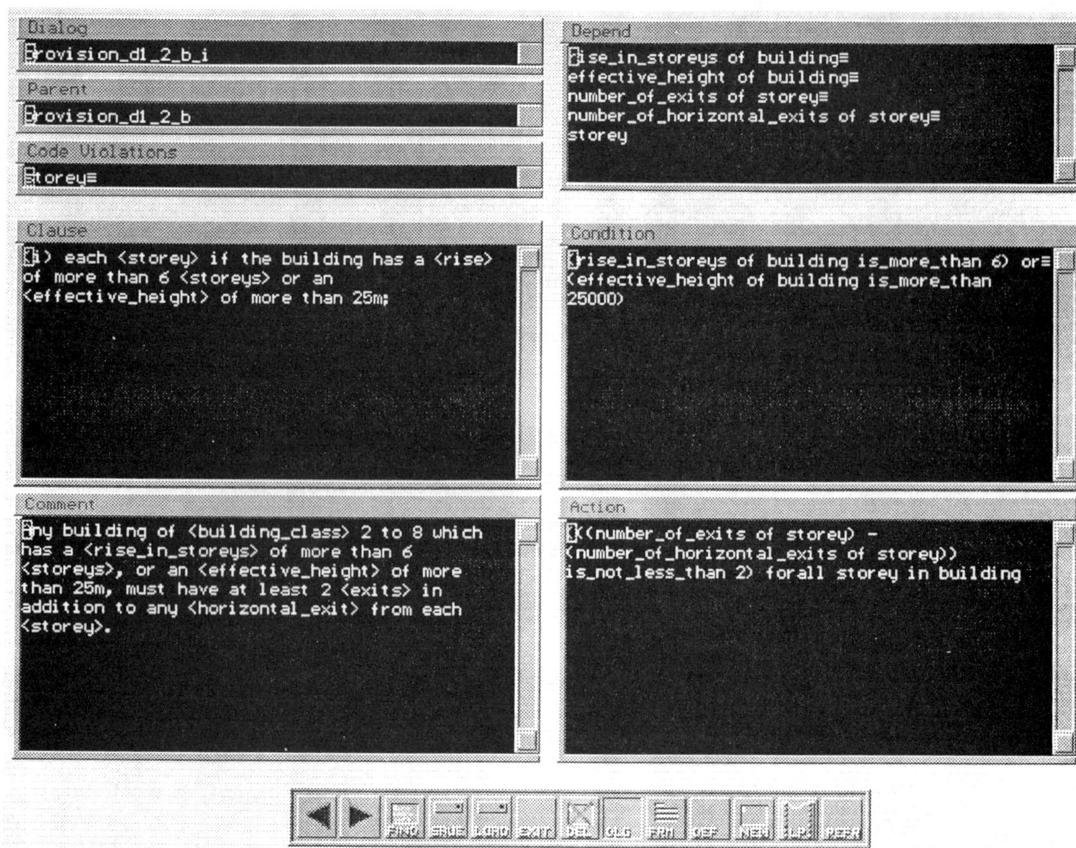

Figure 3 : The dialog editor

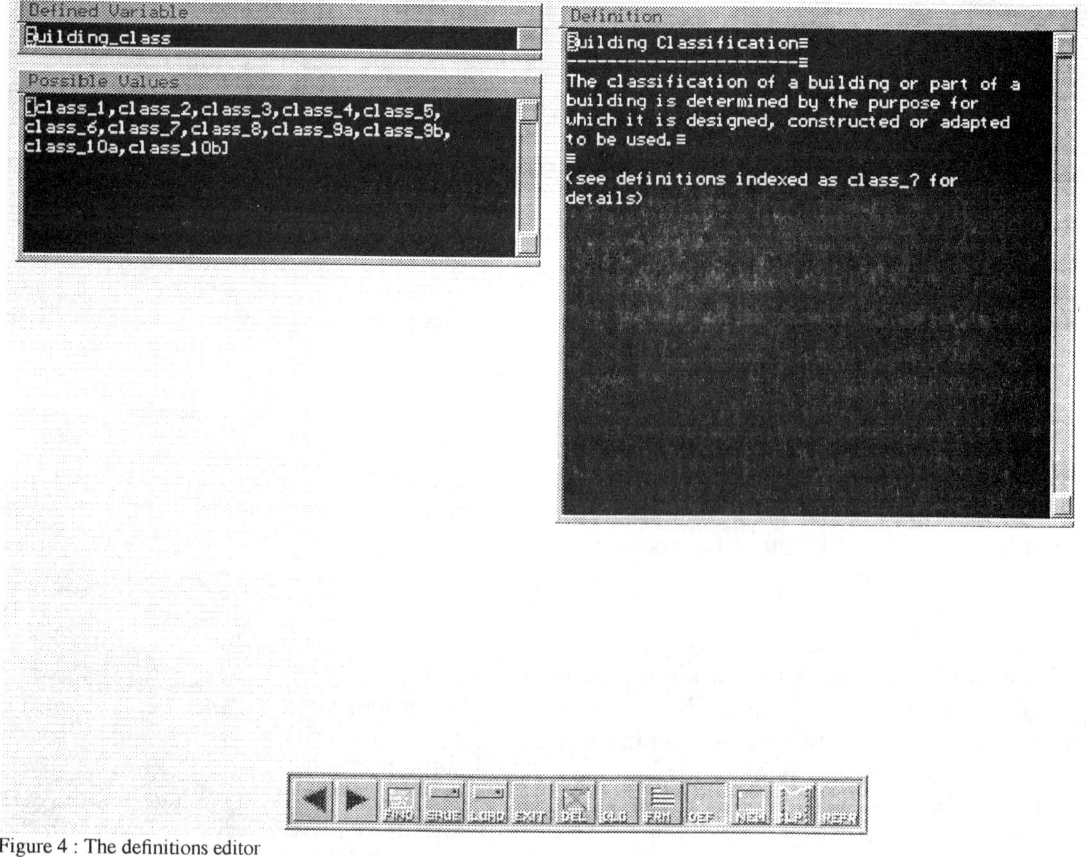

Figure 4 : The definitions editor

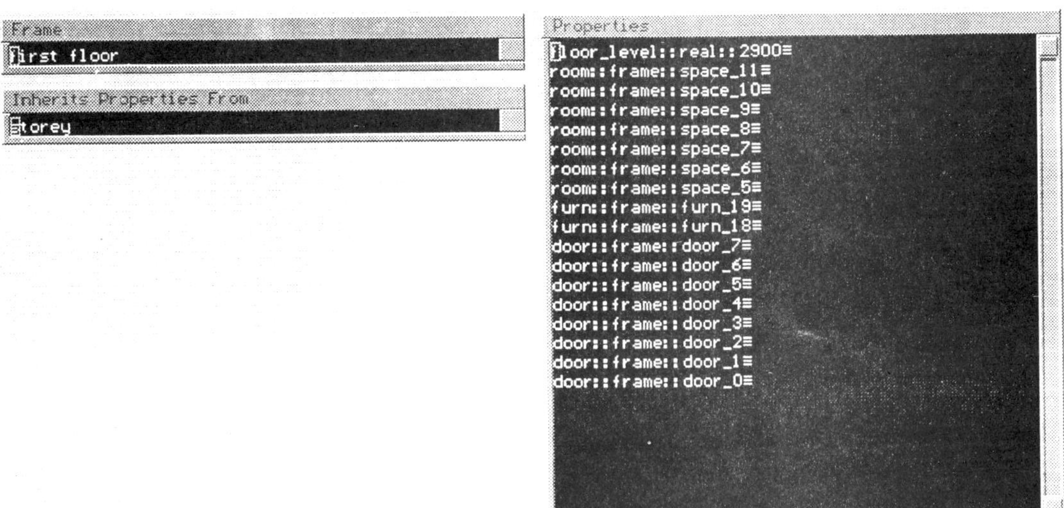

Figure 5 : The frame editor

Evaluation Mode In this mode general functions such as editing of dialogs, frames and definitions can take place. Dialog commands and functions in the DL can be specified and executed. From this mode dialogues can be formally evaluated using the *evaluate (<dialog id>)* function call.

Query Mode During evaluation, if the system finds that a frame slot value which is required to evaluate a condition is not yet defined, *query* mode will be entered. The user is then able to enter a valid value, or otherwise discontinue with the evaluation until more design work is undertaken.

Fail Mode If a dialog fails during evaluation, the system enters *fail* mode. This mode provides opportunities to discover why failure has occurred.

The process of dialog evaluation is implemented as follows:

if <condition> **is** *FALSE* **then** *SUCCEED*

else if <condition> **is** *TRUE* **and** <action> **is** *TRUE*
then *SUCCEED*

else if <condition> **is** *TRUE* **and** <action> **is** *FALSE*
then *FAIL*

If the dialog is not applicable then the *<condition>* part is false and the dialog is made to succeed so that the interpreter can move on to the next condition to be examined. If the *<condition>* part is true, then it does apply and for this dialog to succeed the *<action>* part must also be true. A dialog fails when the action part is false.

When a dialog fails, the frame referred to by the code violation slot has a new slot added of the form:

code_violation::<timestamp>::<text from dialog comment>

Which ensures that a record is maintained, for later reference, of the code violations. In a later design revision cycle, the user is thus able to quickly refer to all those violations which need some attention if the design is to be suitably adjusted and improved. Once the frames have been re-converted into predicate form and the user re-enters the graphics CAAD environment, the option to display code violations is available. The violation is highlighted to bring it to the immediate attention of the user who may then choose to modify the design (geometry or properties) accordingly.

It is within Fail and Query modes that the user interacts with the system to determine problems with the design. Within these modes the user has available a small set of commands:

clause — this command displays the actual clause from the current dialog. If the clause contained within this dialog is really a sub-clause then the clause components from the parent dialog are prefixed to provide the complete clause text, down to this sub-clause.

comment — displays the explanatory comment as provided in the dialog.

define <definition id> displays the definition available for a code concept or frame slot description

how — displays the list of dialogs executed (successfully) so far

quit — exits *fail* mode and continues evaluation (only valid in fail mode, *query* mode terminates only when valid data is provided)

rule — displays the *condition* and *action* fields for the current dialog

where — displays the position of the current dialog within the rest of the dialogs, ie a trace of the progress so far.

?<variable id> displays the value of a selected variable or property value.

depend — displays the values of those slots and global variables defined in the dependency slot of the current dialog.

help — displays help screen.

An example execution of the inference engine is shown in the Appendix, where the use of these commands can be seen. In this printout, user input has been highlighted in **bold**.

THE DIALOG LANGUAGE

The effectiveness of our representation scheme arises from the expressiveness of the *DL*.. While it is inappropriate to explain the *DL* in detail here some general introduction is justified. A code of practice is represented by a set of, one or more, dialogs. Each dialog has a set of possible components (as outlined earlier) which specify its properties.

Within the condition and action slots, the expression of the clause is made using a rule based structure. Rules are expressed as a list of terms. Terms are related to each other using conventional rules of logic with *and* and *or* operations. Terms are strongly typed and can be one of *char, integer, real, string, text* and *list*. The distinction between string and text types, allows text type to incorporate formatting characters (eg. eoln). List types allow lists of terms, of various types, to be defined.

Terms may be formed from arithmetic (eg. + - * / div mod) or relational (eg. < > <= >= <> =) expressions. Relational expressions also include membership and valid range constructs. These relational expressions also have equivalent textual forms, such as *greater_than, less_than, has_value*, etc. which are often used for a clearer description of the required relations. Slot values are accessed using the *of* operator (*<slot> of <frame_id>*), these constructs are then valid terms upon which to apply arithmetic, relational or other operations.

In addition it has proved necessary to include some special expressions to facilitate certain operations characteristic of the scope of certain code clauses. Operators which apply a scope over all frames (*forall <frame id>*) or over all slots in a selected frame (*forall <slot id> in <frame id>*) have been included in the language. Also included are list membership (*<expression> is_a_member_of <list>*) and range operators (*<expression> is_in_range <expression> to <expression>*).

The *DL* thus provides the complete mechanism for the specification and manipulation of the complete knowledge base. We have endeavoured to make the expressions and operations available within *DL*as descriptive as possible in an effort to allow the some equivalence to be made between the actual code text and the *DL* representation.

CONCLUSIONS

The proposals made in this paper have been an attempt to show that through a formal specification of the types of knowledge required it might be possible to develop operational CAAD systems which integrate both the geometric design and the code of practice validation phases.

We know from the reports from many researchers that the formal representation of codes of practice is not always an easy task. In particular we know that the use of simple (and perhaps single) knowledge constructs leads to problems. Taken alone, frame or rule representations are usually inadequate. We need more powerful tools, perhaps embodying elements from all these constructs.

The value of an expert system tool to assist designers also depends on its acceptance and credibility by the users. Our approach has been to explicitly encode the actual code clauses together with the more formal (and more abstract representation) so that direct reference can be made to the code at the time when the user is most pressed to decide on the meaning, or significance, of the information to be provided or the conflicts identified.

REFERENCES

1. Australian Uniform Building Regulations Council (AUBRCC) Building Code of Australia 1990 Australia 1990

2. Broadbent G., Design in Architecture, Wiley, London, 1973

3. Dym, C L, Henchey, R P, Delis, E A and Gonick S, "A Knowledge Based System for Automated Architectural Code Checking", Computer Aided Design volume 20 number 3, 1988

4. Fenves SJ and Garrett JH, Introduction to SASE, National Bureau of Standards, Department of Commerce, Washington DC 20234, 1986

5. Gero JS, Coyne RD, "Logic programming as a means of representing semantics in design languages", Environment and Planning B : Planning and Design, volume 12, pp 351-369, 1985

6. Hamer J , Expert Systems for Codes of Practice, Phd Thesis, University of Auckland, 1990

7. Jones J. C., Design Methods, Wiley-Interscience, London, 1970

8. Mitusch P., "Expert system for the Norwegian building regulations", Journal of CIB : Building Research and Practice, number 4, 1988

9. Oey KH, Passchier E, "Complying with practice codes : achieving a fire-safe design", Journal of CIB : Building Research and Practice, number 1, 1988

10. Rosenman, M A and Gero, J S, "Design Codes as Expert Systems" in Computer Aided Design volume 18 number 9, 1985

11. Sergot M.J., Sadri F., Kowalski R. A., Kriwaczek F., Hammond P. and Cory H. J., " The British Nationality Act as a Logic Program", Comm. ACM 29/5 May, 1986

12. Sharpe, R, Marksjö , B S and Thomson J V A. "Knowledge Representation Language for Engineering Design Codes" in Environment and Planning B Planning and Design volume 14, 1987

13. Sharpe R., Marksjö B. and Ho F., "Wind Loading and Building Code Expert Systems", Proc. Australasian Conference on Expert Systems in Engineering, Architecture and Construction, 6-8 Dec, University of Sydney, 1989

14. Stahl, F I, Wright, R N, Fenves, S J, and Harris, J R. "Expressing Design Standards for Computer Aided Building Design" in Computer Aided Design volume 15 number 6 November, 1983

APPENDIX - A sample inference session

```
Architectural Inference Engine (V0.51)
Arshad Omari
School of Architecture
University of Western Australia
>dialog(egress)
(1 Solution)
>evaluate(a_test)
**** DIALOG [a_test] ****
Building Code of Australia
Access and Egress Provisions
Part D1 : Provision for Escape
Frame File Name : building
**** DIALOG [provision_d1_1] ****
**** DIALOG [provision_d1_1_a] ****

Value of <building_class> of <building> ? : define building_class

Building Classification
The classification of a building or part of a building is determined by
the purpose for which it is designed, constructed or adapted to be used.
(see definitions indexed as class_? for details)

Possible Values :
[class_1,class_2,class_3,class_4,class_5,class_6,class_7,class_8,class_9a,
class_9b,class_10a,class_10b]

Value of <building_class> of <building> ? : define class_9a

Class 9 Buildings :
A building of public nature -
(a) Class 9a - a <health_care_building>;
(b) Class 9b - an <assembly_building>, and

Class 9a includes a pathology laboratory in a<health_care_building> and
Class 9b includes a trade workshop, laboratory or the like in a primary or
secondary <school>, but excludes any other part of these buildings that
are of another Class.

Value of <building_class> of <building> ? : class_9a
**** DIALOG [provision_d1_1_b] ****
**** DIALOG [provision_d1_2] ****
**** DIALOG [provision_d1_2_a] ****
Value of <number_of_exits> of <ground floor> ? : clause

CODE CLAUSE RELATING TO provision d1 2 a :
D1.2 Number of Exits Required
(a) All Buildings
Every building must have at least 1 <exit> from each <storey>.

Value of <number_of_exits> of <ground floor> ? : define
number_of_exits

Exit
The definition of <exit> is -
(a) Any, or any combination of the following if
they provide egress to a road or <open_space>:
        (i)  An internal or external stairway
        (ii) A ramp complying with Section D.
        (iii)A <fire_isolated_passageway>.
        (iv) A doorway opening to a road or <open_space>
(b) A <horizontal_exit> or a <fire_isolated_passageway> leading to a
<horizontal_exit>

Possible Values :
integer

Value of <number_of_exits> of <ground floor> ? : 2
Value of <number_of_exits> of <first floor> ? : wrongtype
ILLEGAL TYPE
Valid Type is integer

Value of <number_of_exits> of <first floor> ? : 2
**** DIALOG [provision_d1_2_b] ****
**** DIALOG [provision_d1_2_c] ****
**** DIALOG [provision_d1_2_c_a] ****


Value of <number_of_horizontal_exits> of <ground floor> ? : define
horizontal_exit

Horizontal Exit
A <required> doorway between 2 parts of a buildng separated from each
other by a <fire_wall> with an FRL as <required> by Specification C1.1
```

Value of <number_of_horizontal_exits> of <ground floor> ? : **1**
**** DIALOG [provision_d1_2_c_a] ****
Value of <number_of_horizontal_exits> of <first floor> ? : **1**
Value of <egress_level> of <building> ? : **define egress_level**

Level of Egress
The floor level of the <storey> which provides egress to a road or <open_space>.

Possible Values :
real

Value of <egress_level> of <building> ? : **define open_space**

Open Space
means a space on an allotment, or a roof or similar part of a building complying with D2.12, open to the sky and connected directly with a public road.

Value of <egress_level> of <building> ? : depend

DIALOG <provision d1 2 c a> DEPENDANCIES :
number_of_exits of storey = 2
number_of_horizontal_exits of storey = 1
storey = first floor
egress_level of building = <undefined>
floor_level of storey = 2900

Value of <egress_level> of <building> ? : **0**
**** DIALOG [provision_d1_2_d] ****
**** DIALOG [provision_d1_2_d_i] ****
Value of <rise_in_storeys> of <building> ? : **2**
Value of <effective_height> of <building> ? : **define effective_height**

Effective Height
Means the height to the floor of the topmost <storey> (excluding the topmost <storey> if it contains only heating, ventilating, lift or pther equipment, water tanks or similar service units) from the floor of the lowest <storey> providing direct access to a road or <open_space>.

Possible Values :
real

Value of <effective_height> of <building> ? : **6100**
**** DIALOG [provision_d1_2_d_ii] ****
Value of <is_patient_care_area> of <space_11> ? : **define is_patient_care_area**

Patient Care Area
means a part of a <health_care_building> usedfor the treatment, care, accomodation,recreation, dining and transit of patients.

Possible Values :
boolean

Value of <is_patient_care_area> of <space_11> ? : **no**
Value of <is_patient_care_area> of <space_10> ? : **no**
Value of <is_patient_care_area> of <space_9> ? : **yes**

(fail)**depend**
DIALOG <provision d1 2 d ii> DEPENDANCIES :
is_patient_care_area of room = yes
room = space_9
storey = first floor
number_of_exits of storey = 2
number_of_horizontal_exits of storey = 1

(fail)**clause**
CODE CLAUSE RELATING TO provision d1 2 d ii :
D1.2 Number of Exits Required
(d) Class 9 buildings
In addition to any <horizontal exit>, not less than 2 <exits> must be provided from -
(ii) any <storey> which includes a <patient care area> in a Class 9a building;

(fail)**comment**

COMMENT FROM <provision d1 2 d ii>
Any <storey> which includes a <patient_care_area> in a building of <class_9a>, must have not less than 2 <exits> in addition to any <horizontal_exits>.

(fail)**rule**

RULE <provision_d1_2_d_ii> :
CONDITION
building_class of building is class_9a
ACTION
((((number_of_exits of storey) - (number_of_horizontal_exits of storey)) is_not_less_than 2) or
((is_patient_care_area of room is no) forall room in storey)) forall storey in building

(fail)**depend**

DIALOG <provision d1 2 d ii> DEPENDANCIES :
is_patient_care_area of room = yes
room = space_9
storey = first floor
number_of_exits of storey = 2
number_of_horizontal_exits of storey = 1

(fail)**comment**

COMMENT FROM <provision_d1_2_d_ii>
Any <storey> which includes a <patient_care_area> in a building of <class_9a>, must have not less than 2 <exits> in addition to any <horizontal_exits>.

(fail)**?number_of_horizontal_exits of storey**
number_of_horizontal_exits of storey = 1

(fail)**quit**
Value of <is_patient_care_area> of <space_8> ? : **no**
Value of <is_patient_care_area> of <space_7> ? : **no**
Value of <is_patient_care_area> of <space_6> ? : **no**
Value of <is_patient_care_area> of <space_5> ? : **no**
**** DIALOG [provision_d1_2_d_iii] ****
**** DIALOG [provision_d1_2_d_iv] ****
Value of <building_usage> of <building> ? : **define building_usage**

Building Usage
The primary purpose for a which a building is to be used.

Possible Values :
[gallery,museum,cafe,church,dining_room,dormitory,early_childhood_centre,factory,garage,gymnasium,hostel,hotel,motel,guest_house,stadium,kiosk,kitchen,laboratory,library,hospital,school,shop,showroom,open_spectator_stand,theatre,hall,transport_terminal,workshop]

Value of <building_usage> of <building> ? : **hospital**
**** DIALOG [provision_d1_2_d_v] ****
Value of <number_of_occupants> of <ground floor> ? : **define number_of_occupants**

Number of Occupants
means the number of persons accomodated in a <storey>, room, or <mezzanine> as calculated under table D1.13.

Possible Values :
integer

Value of <number_of_occupants> of <ground floor> ? : **clause**

CODE CLAUSE RELATING TO provision_d1_2_d_v :
D1.2 Number of Exits Required
(d) Class 9 buildings
In addition to any <horizontal exit>, not less than 2 <exits> must be provided from -
(v) any <storey> or <mezzanine> that accomodates more than 50 persons, calculated under D1.13.

Value of <number_of_occupants> of <ground floor> ? : **100**
Value of <number_of_occupants> of <first floor> ? : **100**

(fail)**comment**

COMMENT FROM <provision d1 2 d v>
Each <storey> or <mezzanine> with more than 50 occupants (table D1.13) must have at least 2 <exits> in addition to any <horizontal_exit>.

(fail)**depend**
DIALOG <provision_d1_2_d_v> DEPENDANCIES :
number_of_occupants of storey = 100
number_of_horizontal_exits of storey = 1
number_of_exits of storey = 2
storey = first floor

(fail)**where**
DIALOG POSITION :
--->provision_d1_2_d_v
--->provision_d1_2_d
--->provision_d1_2
--->part_D_bca

(fail)**how**

DIALOG EXECUTION TRACE :
--->provision_d1_2_d_iv.....not applicable
--->provision_d1_2_d_iii.....not applicable
--->provision_d1_2_d_ii.....dialog passed
--->provision_d1_2_d_i.....not applicable
--->provision_d1_2_c.....dialog passed
--->provision_d1_2_c_a.....not applicable
--->provision_d1_2_c_a.....not applicable
--->provision_d1_2_b.....not applicable
--->provision_d1_2_a.....dialog passed
--->provision_d1_1.....dialog passed
--->provision_d1_1_b.....not applicable
--->provision_d1_1_a.....not applicable

(fail)**rule**

```
RULE <provision_d1_2_d_v> :
CONDITION
ACTION
((number_of_occupants of storey is_at_most 50) or
((number_of_exits of storey) - (number_of_horizontal_exits of storey)
is_not_less_than 2)) forall storey in building

(fail)quit
**** DIALOG [provision_d1_2_e] ****
MESSAGE :
Part C2.7 of the CODE should be checked here
**** DIALOG [provision_d1_2_f] ****
(1 Solution)
>
>
>QUIT
```

APPROACHES TO FORMAL PROCESSING FOR DESIGN CODES IN STRUCTURAL ENGINEERING INTELLIGENT CAD

Zhong Wanxie, Qiu Chunhang, Qin Xiaolin and Liu Xiaojian
Research Insitute of Engineering Mechanics,
Dalian University of Technology, Dalian, PR China

Some knowledge in design codes can not be directly applied in CAD system. They must be described in formal language, and then used in CAD system for engineering design. In this paper some approaches to the formal processing of codes for reinforced concrete structural design are studied. For the purpose of applying design knowledge expressed in natural language and/or diagram form to CAD system, the concept of formalization description is proposed and used for knowledge processing. For design knowledge about calculation and inference sequence implied between provisions in design codes, a code-based problem-solving control structure is established to explicitly describe this part of knowledge. Using the methods presented in this paper, we have successfully applied design knowledge with different forms in design codes to the intelligent CAD system-ICAD, and provided the design type knowledge-based system with a new way to expand and refine its knowledge base.

INTRODUCTION

Design codes, as legal documents for design, are bases of building structural design. They are basic sources of knowledge base of CAD system. But some knowledge in design codes is prescribed in natural language and/or diagram form, and is ill-structured, therefore it can not be directly applied in CAD system. It must be formalized in order to satisfy the requirements of computer system which is suitable for symbolic treatment.

This paper presents the formal processing of design knowledge about wind load coefficient of structural shape and problem-solving control process of reinforcement design in ICAD [Ref. 4], a reinforced concrete space frame structural intelligent CAD system.

WIND LOAD COEFFICIENT OF STRUCTURAL SHAPE

It is prescribed in the code [Ref. 3] that the wind load exerting at a point on the surface of a structure, W, is determined by the formula:

$$W = \mu_s \mu_z \beta_z W_0 \qquad (1)$$

where μ_s = wind load coefficient of structural shape; μ_z = wind load coefficient of height; β_z = wind load coefficient of vibration; and W_0 = basic wind load.

The value of μ_s varies with the shape and dimension of a structure, and is prescribed in diagram form in the design code. It must be formalized for developing a CAD system.

Mathematical model

We define the plane geometrical contour of a building structure as a digraph represented by

$$G = (V,E,E') \qquad (2)$$

where V and E are the sets of vertices and edges of the contour, respectively; and E' is the set of virtual edges which are introduced in order to construct a convex polygon from the plane geometrical contour of the structure.

As shown in Fig. 1, the two endpoints of a virtual edge (in dotted line) are called virtual vertices. Regions with virtual edges are called virtual regions (excluding the outside region or exterior region). The digraph G is oriented in such a way that the area bounded by the contour is always on the right side of a person as he moves along the contour in the arrow direction. The virtual edges are oriented in such a way that the two directed paths between the two v-vertices of a virtual edge can not form a circuit, where a v-vertex means that it is a end of a virtual edge.

In Fig. 1, $V = \{v(i) | i \in [1,12]\}$; $E = \{e(i,j) | i \in [1,12], j = i \bmod 12 + 1\}$; $E' = \{e'(2,4), e'(5,8), e'(10,12)\}$.

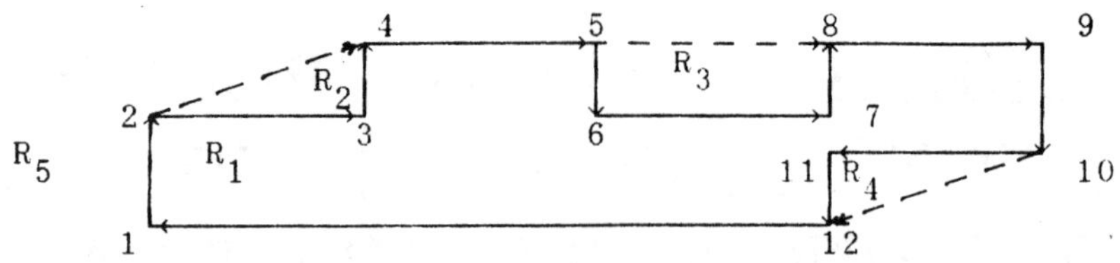

Fig. 1 An example of the digraphs.

Table 1 The dependence of wind Load Coefficient μ_s to Structural shape (For orthogonal structures)

DS	ES	OTHER CONDITIONS	μ_s
1	1		0.8
	0	There is only one virtual region in the digraph G and ES'=1	0.7
		Otherwise	0.6
	0.7		0.3
	-0.7		-0.6
	-1		-0.5

DS	ES'	OTHER CONDITIONS	μ_s
0	1		0.9
	0.7	There is only one virtual region in the digraph G	0.8
		Otherwise	0.6
	-0.7		-0.5

For structures with simple rectangular plane geometrical contours μ_s is determined according to the given formula in the code [Ref. 3].

Area R_1, bounded by the contour, is called real region. R_2, R_3, and R_4 are virtual regions. R_5 is the outside region or exterior region. According to the definition above, we can describe G with the attributions {ES,DS,ES',X,Y}, which can be described as follows:

- The indicator of edge direction, denoted as ES, is defined as the direction sine of the edge in the cartesian coordinate system in which the building structure is located.

- The region indicator of edge, denoted as DS, is defined as:

$$DS = \begin{cases} -1 & \text{if the edge is a virtual edge (i.e. for } e' \in E') \\ 0 & \text{if the edge is a public edge (i.e. the edge of both the real region and a virtual region)} \\ 1 & \text{otherwise} \end{cases} \quad (3)$$

For instance in Fig. 1, DS(e'(2,4)) = -1, DS(e(2,3)) = 0, DS(e(4,5)) = 1.

- The direction indicator of the public edges of a virtual region, denoted as ES', is defined as the direction sine of the virtual edge of the virtual region.

- The coordinates of the vertices in structural coordinate system, X, Y.

Representation of μ_s

So far, the value of μ_s at any point on the surface of a structure can be uniquely determined by means of the corresponding digraph G=(V,E,E'), for there exists a mapping between {ES,DS,ES',X,Y} and μ_s. Using the digraph model established above, the prescription about μ_s in the design code can be described in a revised form of decision table [Ref. 4] so as to be used in ICAD. Table 1 illustrates a part of the formal description result of μ_s in the form of the revised decision table.

Formalization of μ_s

When we apply the knowledge about μ_s determined in the form of table 1 to ICAD, we must transform the given design condition into a suitable form to match the knowledge in the knowledge base of ICAD, because the knowledge about μ_s is already transformed from diagram form into the form of the decision table. This part of design knowledge is applicable if and only if wind is along the direction of positive X axis of the structural coordinate system OXY. If the direction of wind is different from the positive X axis, a coordinate transformation for ES and ES' must be carried out in order to use table 1. Suppose that the angle rotating anticlockwise from the +X direction to the wind direction is β, then we have a new coordinate system OX'Y', shown in Fig. 2. For edge e , whose angle are α and α' in OXY and OX'Y' respectively, we have:

$$\alpha = \alpha' + \beta \quad \text{or} \quad \alpha' = \alpha - \beta \qquad (4)$$

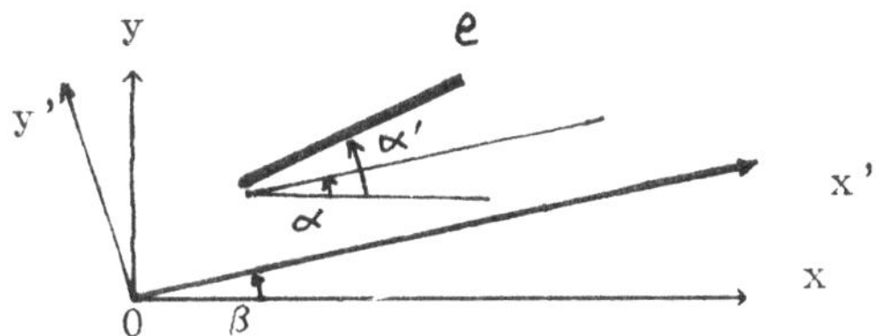

Fig. 2 The coordinate transformation.

Thus, for a edge e with ES=sin(α) in OXY, its ES is changed to ES'=sin(α-β) in OX'Y'. This means that the μ_s of edge e should be determined according to ES' in the new wind direction.

For a building structural design, another proposition to use table 1 is to determine the plane geometrical contour of the structure. Let a side joint of the structure be the initial vertex of searching for the contour. For instance, we can start to search from the joint with coordinates x=0 and y=0. Let **i**, **j**, **k** be the unit vectors in cartesian coordinate system OXYZ. Then an arbitrary directed edge **e** in digraph G in the OXY plane can be expressed in vectorial form:

$$\mathbf{e} = a\mathbf{i} + b\mathbf{j} \qquad (5)$$

where a and b are constants. The vectorial product of two vectors $\mathbf{e}_1$ and $\mathbf{e}_2$ is given by

$$\mathbf{e}_1 \times \mathbf{e}_2 = \begin{vmatrix} \mathbf{i} & \mathbf{j} & \mathbf{k} \\ a_1 & b_1 & 0 \\ a_2 & b_2 & 0 \end{vmatrix} = (a_1 \times b_2 - a_2 \times b_1)\,\mathbf{k} \qquad (6)$$

Thus, based on the relations between each edge and the real region in digraph G=(V,E,E'), the search direction for determining the geometrical contour of a building structure can be described as follows:

Suppose that e_0 is a directed edge incident into vertex v_0 and $\mathbf{e}_r$ (r=1,2,3, for orthogonal frame structures, shown in Fig. 3) are outward. Their vectorial products are $(a_0 \times b_r - a_r \times b_0)\mathbf{k}$ (r=1,2,3). The edge incident out of v_0 corresponding to maximum$(a_0 \times b_r - a_r \times b_0)$ is selected as the search direction for determining the geometrical contour of the structure at vertex v_0.

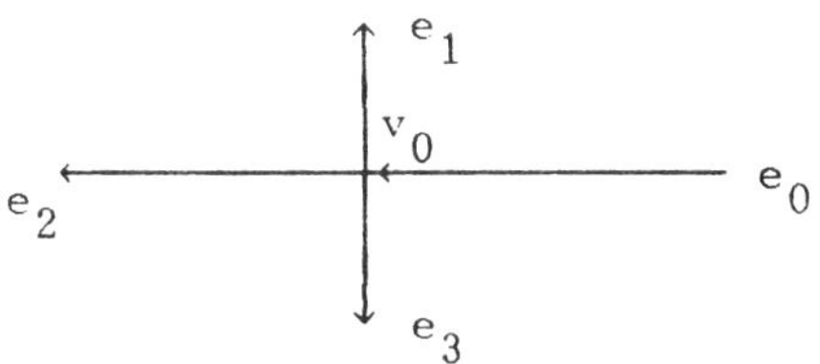

Fig. 3 The search direction at vertex v_0.

CONTROL STRUCTURE OF PROBLEM-SOLVING FOR REINFORCEMENT DESIGN

A design process can be subdivided into a number of weakly interrelated sub-processes. For a low level sub-process, it only concerns with a limited feasible region of solution, so that the problem of combinatorial explosion may be greatly reduced. For this purpose, we can divide design knowledge given in design codes [Ref. 1,2] into a number of sub-sets of design rules, and represent them in rule network diagrams (RN thereafter). Each sub-set of design rule is corresponded with an RN, which controls the low level sub-process. For example, Fig. 4 is a simple RN, where [i] denotes the *i*th rule in the rule set; and the directed edge represents the interrelation of rules, where the arrowhead stands for the dependence of rules; and μ_{ij} is the correlative coefficient, i.e. the evaluation function of relative intensity between the *i*th rule and the *j*th rule in the rule set.

The values of μ_{ij} can be supplied by means of the experience of the user, who can adjust these coefficients interactively or by learning examples through practical engineering design. For an arbitrary low level sub-process, it is easy to determine an initial rule r_0 as the root of a search tree in the solving process. In running the sub-process, ICAD retrieves the rules related to r_0 in the corresponding RN, and selects and executes the rules according to the order of importance represented by the correlative coefficient μ_{ij}'s.

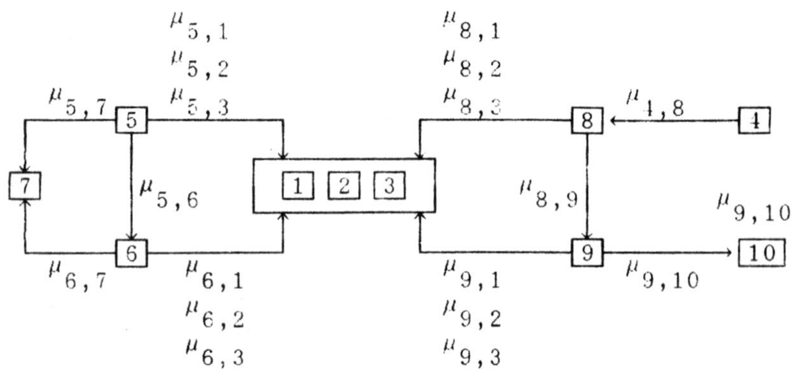

Fig. 4 AN example of rule network diagrams.

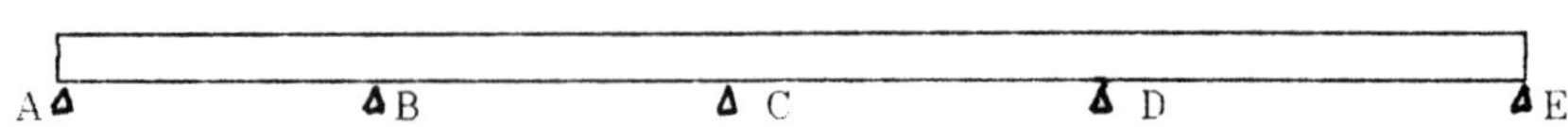

Fig. 5 A continuous beam.

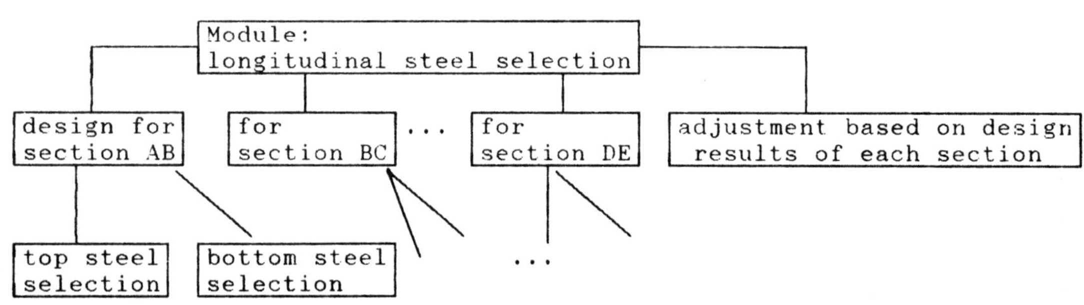

Fig. 6 The decomposition of the design module for the selection of longitudinal steels

As an example of applications of RN, let us discuss the solving of a continuous beam reinforcement design in a frame structure.

Consider a continuous beam shown in Fig.5 with given area of reinforcement steel cross section, the goal of the program module is to determine the number, diameter and length of the longitudinal steels in the beam. Let us decompose the design problem into sub-processes, shown in Fig. 6.

Let B, CD represent the width of the beam cross section and aseismatic reguirement grade respectively; D, N are the diameter and the number of steels in a row respectively; N1, N2 denote the numbers of steels spaced in one and two rows respectively. In Fig. 6, the lowest level sub-process must satisfy the following rules of constraint [Ref. 1,2]:

1. $Dmax = \phi_1(B)$, $Nmax = \phi_2(B)$;
2. $Dmin = \phi_3(B)$, $Nmin = \phi_4(B,CD)$;
3. The number of steels should be as few as possible;
4. $NT \geq \phi_5(CD)$, NT is the number of continuing steels across all spans of the beam;
5. $N1 \in [Nmin,Nmax]$ and $D1 \in [min,Dmax]$ for bottom steels;
6. $N2 \in [2,N1]$ and the determination of N2 is influenced by N1 for bottom steels;
7. All top steels and all bottom steels should be as uniform as possible, respectively;
8. $N1 \in [Nmin,Nmax]$ and $D1 \in [Dmin,Dmax]$ for top steels;
9. $N2 \in [2,N1]$ and the determination of N2 is influenced by N1 for top steels;
10. For top steels cut off at the middle of a span, if they come up to certain lengths, they should be continuously across the span (or the two adjoining supports of the span).

where ϕ_i $(i=1,2,3,...)$ denote some constraints given in the design codes. The interrelations of the constraints listed above can be represented in an RN, such as in Fig. 4.

Representation and management of RN

From the viewpoint of graph theory, an RN is actually a weighted digraph $G=(V,E)$ (without self-loops and parallel edges),

order no. of vertex: N
pointer to the rule corresponding the vertex: PTN
pointer to next vertex: PTR

Fig.7 The nodal data of the one-dimensional linked data structure.

where V is the set of vertices (numbers of rules) and E is the set of edges (relations) between the pair of vertices in the RN. Let us construct a matrix:

$$A_G = [\ a_{ij}\ ]_{N \times N} \tag{7}$$

where N is the total number of vertices in the RN. The elements a_{ij} of A_G are defined as

$$a_{ij} = \begin{cases} \mu_{ij} & \text{if } \langle v_i, v_j \rangle \in E \\ 0 & \text{otherwise} \end{cases} \tag{8}$$

where μ_{ij} is the correlative coefficient between the *i*th rule and the *j*th rule in the RN. Obviously the matrix A_G is highly sparse, so it can be represented with the data structure of N one-dimensional linkers. The *i*th linker contains all the adjacent vertices of the *i*th vertex in the RN, and the adjacent vertices are arranged in the order of μ_{ij}'s. The nodal data of the one-dimensional linked data structure corresponding to each element of A_G is shown in Fig. 7.

Thus, the manipulation of RN, such as deletion,insertion, and retrieval, is completely equivalent to the manipulation of the corresponding one-dimensional linked data structure.

Since the recursions of rules in an RN would possibly make ICAD fall into "deadlock", so each of the RNs should be checked in advance to erase the possibility of the recursions. Provided that a digraph G=(V,E) is constructed from an RN, and its adjacency matrix is defined as

$$A = [\ a_{ij}\ ]_{N \times N} \tag{9}$$

where elements a_{ij} are given by

$$a_{ij} = \begin{cases} 1 & \text{if } \langle V_i, V_j \rangle \in E \\ 0 & \text{if } \langle V_i, V_j \rangle \notin E \end{cases} \tag{10}$$

Construct matrix B_k from the matrix A:

$$B_k = [\ b_{ij}\ ] = A + A^2 + \ldots + A^k = \sum_{m=1}^{k} A^m \tag{11}$$

and its elements b_{ij} are given by

$$b_{ij} = \sum_{m=1}^{k} a_{ij}^{(m)} \quad \text{for } i,j = 1,2,\ldots,N \tag{12}$$

The value of the element b_{ij} means the number of the directed paths of length less than or equal to k from vertex v_i to vertex v_j. For a digraph G = (V,E) with total number of vertices $|V|$=N, if there exists no path of length less than or equal to N from v_i to v_j, it is impossible to reach v_j starting from v_i.

Based on the above discussion, we can construct a corresponding reachability matrix for an RN:

$$P = [\ p_{ij}\ ]_{N \times N} \tag{13}$$

with its elements given by

$$p_{ij} = \begin{cases} 1 & \text{if } b_{ij} \neq 0 \\ 0 & \text{if } b_{ij} = 0 \end{cases} \tag{14}$$

where b_{ij} are computed from formula (12) and let k=N.

From the main diagonal elements p_{ii} (i = 1,2,...,N) of the reachability matrix P, it can be detected whether there exist recurrence calls of rules in an RN. If there are any $p_{ii} \neq 0$ (i=1,2,...,N), the RN should be revised.

The calculation of the reachability matrix P can be performed by introducing logical operators $\vee$, $\wedge$ or adopting Warshall algorithm [Ref. 6,7].

SUMMARY AND CONCLUSIONS

By means of formal processing, design knowledge prescribed in diagram form and/or natural language in design codes is transformed into symbolic system, which has been utilized conveniently in ICAD. Practically this paper has presented the methods and procedures of formal processing of design knowledge about wind load coefficient of structural shape and control process of problem-solving for reinforcement design. Finally, although the paper discusses the methods and procedures based on especial design knowledge, the fundamental concepts presented are applicable to any alike design knowledge.

ACKNOWLEDGEMENTS

This research is sponsored by the National Natural Science Foundation. The authors also want to thank Prof. Chen Gengdong for his helpful suggestions during the course of this work.

REFERENCES

1. Concrete Structural Design Code (GBJ 10-89), Chinese Building Industry Press. 1989, Beijing, China.

2. Building Aseismatic Design Code (GBJ 11-89), Chinese Building Industry Press. 1989, Beijing, China.

3. Building Structural Loading Code (GBJ 9-87), Chinese Plan Press. 1989, Beijing, China.

4. Zhong Wanxie and Liu Junxian, "ICAD: An Intelligent CAD system for Reinforced Concrete Frame Structures", Proceedings CCAAE 1987, Guangdong, China, Dec. 1987, pp. 48-51.

5. R.B. Hurley, Decision Tables in Software Engineering, Van Norstrand Reinhold Company, 1983.

6. Wai-Kai Chen, Applied Graph Theory-Graphs and Electrical Networks, North-Honand Publishing Company, 1976.

7. R. Skvarcius, W.B.Robinson, Discrete Mathematics with Computer Science Applications, The Benjamin/Cummings Publishing Company, inc.

STANDARDS PROCESSING IN AN INTEGRATED ENGINEERING SYSTEM

T.S. Sakthivel and V Kalyanaraman
Structural Engineering Laboratory, Department of Civil Engineering, Indian Institute of Technology, Madras, India

This paper discusses a standards processing architecture in integrated systems to handle different standards at different stages of the engineering process, using the blackboard problem solving model. The standards processing model is based on the semantic mapping between the standard independent and standard dependent knowledge sources, represented separately. The semantic mapping allows direct interaction of independently developed knowledge sources for efficient reasoning. The architecture provides a knowledge based backtracking facility to handle design failures. The proposed model forms a part of a project on the development of an integrated KBES for engineering of steel structures (covering planning, analysis, detailed design and construction management activities). The standards processing concepts are illustrated with the help of a loading standards processing and a design standards processing examples, developed for engineering of steel industrial structures.

INTRODUCTION

Standards are the legal documents for planning, design, detailing and construction of structures and the standards are developed over the years with the help of professional experience contributed by many experts in the area. The efficient representation and the use of standards in computer programs for the design and construction of buildings is very essential. But the various codes of practice and specifications are subject to modifications depending upon the findings of continuing research and development.

For the implemention of standard provisions in computer programs, three approaches can be identified, namely the hard coding, the decision table and the knowledge based approaches [Ref.1]. The approach involving hard coding standards information in to the application programs suffers seriously from early obsolescence, since changes in standards are difficult to make. Decision table approach provides a systematic organizational model to represent and process design specifications, wherein the decision tables are arranged as a network of tables representing the standards organization[Ref.4]. The standards specifications represented as decision tables are translated in to program codes by a preprocessor which are then linked to a generic processor for design checking purposes. This approach also suffers from the well known limitations of the procedural methodology. The third is the knowledge based approach for the implementation of standards. In this case, the provisions of the standards are represented in a knowledge base in the form of production rules or other forms of knowledge representation. Some compilers are also capable of generating production rules from decision table representation of design standards. The standards knowledge represented in such a separate knowledge base can be used by an application program at the runtime. In this form, the provisions of the standard are represented explicitly in an easy to understand syntax, usually in the same order as in the standard. This visible form makes modification of the standard provisions in the application programs easy, yet they can be used efficiently.

An ongoing project at IIT, Madras involving the development of an integrated KBES for engineering of steel structures (covering planning, analysis, detailed design, detailing and construction management activities) uses the blackboard problem solving model. In the usual blackboard problem solving model, multiple knowledge sources working cooperatively produce solution incrementally and opportunistically. The interaction between the knowledge sources is not direct, but takes place only through the blackboard, which acts as the global solution space. This problem solving model has been adopted since it allows modular, independent and incremental development of the software in an experimental way with out suffering from large scale modification in midstream.

This paper describes the standards processor developed to handle the processing of different standards at different stages of the engineering process in the blackboard model. The approach described here is based on semantic mapping between standard dependent knowledge sources and standard independent knowledge sources in the blackboard environment. The approach facilitates the active design process as well as the process of standards specifications checking with the knowledge based backtracking facility, implemented using dependency network. The standard processing model is illustrated with two applications, involving Loading Standards and Design Standards for General Steel Construction.

PRIOR RESEARCH IN KNOWLEDGE BASED STANDARDS PROCESSING

Review of literature on the first two approaches of standards processing has been well covered in [Refs. 1,3]. This section only briefly reviews the methodology used in knowledge based standards processing.

SPECON (Specification Consultant) [Ref.11] is a knowledge based system where standards specifications were represented as rules . The system helps the designer in checking the structural steel elements for conformance with the AISC steel design specifications. The knowledge base is divided into two levels. The first level consists of production rules which identify the applicable constraints. The second level is made up of the specific design constraints. The second level constraints are called from first level rules whenever applicable preconditions are satisfied. SPECON was implemented in OPS5 and LISP.

SICAD (Standards Interfaces in CAD) [Ref.9] also uses a knowledge based approach to standards processing. Standards are represented in the common decision table form and SICAD uses knowledge base and inference mechanism to use the standards. The knowledge base contains classifier trees to relate engineering terminology to provisions of a standard, Information Network which represents the provisions of a standard in a network of decision tables and mappings to relate data items in a design database. The design system with which SICAD is interfaced, invokes SICAD to perform compliance checking by identifying and checking applicable provisions. Checking uses a goal driven strategy to determine all data items required to evaluate a provision. Checking automatically accesses this data in the program's database through mappings stored in the knowledge base.

SPEX (Standards Processing Expert) [Ref.3,5] is a knowledge based standard independent structural component design system. The basic approach involves classifying each standard requirement in to the structural behaviour to which it applies thus forming a behaviour limitation classification system. The design focus is identified based on the classification system which is translated in to a set of requirements from the governing standard to be satisfied. SPEX is composed of several functional modules including Design focus module to generate design focus hypothesis using a set of expert rules for choosing the proper focus of the design process, Requirement retrieval module to generate list of requirements to be satisfied, Constraint set generations module to generate a set of constraints from the requirements that are to be satisfied during design, Constraint set satisfaction module to determine the optimal component within the solution space defined by the constraint set, Conformance verification module to check the resulting component design for all applicable standard requirements apart from the design requirements. SPEX was implemented as a blackboard system in which the system controller using a set of control rules invokes appropriate knowledge modules based on the current state of the design. The current state of design is represented by messages on the message blackboard and design information on the design information blackboard.

SPIKE (Standards Processing In a Knowledge - based Expert system Environment) [Ref. 10] represents a standard as a data base of facts which can be processed by an expert system. Facts are classified in to provisional and organizational facts. Provisional fact represents a datum used in standards and its associated decision table entry. Organizational fact represents the precedence relationships among the datums of a standard. The standards processor is implemented as a set of generic production rules that manipulates standards represented as facts. Since, there is no fixed precedence among provisional facts, they can be manipulated by standard independent rules. But, precedence exists among organizational facts and must be handled in a standard dependent manner. However, all the data of a standard were represented explicitly in a well defined information network, the traversal of which were handled by a set of standard independent production rules. SPIKE was implemented in OPS5.

Feijo et.al [Ref. 2] presented an alternate model for the representation of design codes based on first order logic and their incorporation in to design automation systems. They also suggested a formal link with hypertext systems, in which data is stored in a network of nodes connected by links.

DESIRABLE FEATURES OF A STANDARDS PROCESSOR IN INTEGRATED SYSTEMS

This section provides a general discussion on the issues to be addressed in computer aided standards processing in an integrated environment.

- Standards processing in integrated systems is not confined to design standards alone. Engineering process involves checking of the artifact being developed conforming to the specified standards at various abstraction levels starting from the very general planning specifications to detailed design specifications. Loading, material, equipment, construction and various other standards are involved in evolving an engineering artifact and a generic standards processor should be able to process different standards at different stages of the engineering process in a uniform way.

- Frequently, for the same engineering activity (such as design) different alternate standards are used, depending upon the use of the structural system being designed, importance of the structure and the geographical location of the structure. Standards processor should be able to perform this alternate standards processing depending upon the context of the problem.

- Conformance checking of any standards requires a thorough knowledge about both the applicable standards as well as the way it is utilized. It is recognized that a considerable amount of domain knowledge is involved in conformance checking that is not a part of the specified standards. There is much "micro" interactions between the standards knowledge and standard independent domain knowledge [Ref.9]. Standards processor should have "separate" standard dependent and standard independent knowledge sources and still be able to handle the "micro" interactions involved between them efficiently.

- Standard independent domain knowledge may involve different bodies of expertise and hence different syntax and semantics for the information involved in the interaction with standard dependent knowledge. In addition, different knowledge representation strategies may be required for efficient representation and inferencing. If constrained to communicate with each other and the standard in a common language or representation, the expressiveness and transparency of the knowledge is lost. Standards processor should be able to maintain different bodies of knowledge as originally produced by the human domain experts and standards knowledge in separate knowledge sources. This makes it easier for any domain expert to criticize or revise a knowledge source.

- Frequently, the same standards may be involved in various engineering process or activities. For example, steel design standard may be used both in the detailed design of structural members as well as at the detailing of structural connections. The standards processor should facilitate the use of appropriate representation strategy for the knowledge in the standards and yet allow more than one type of binding between the standard and the knowledge sources that use them.

- Interactions not only exist between standard dependent and standard independent knowledge but also between different standards. In standards, frequently references are made to related standards to obtain relevant information. For example, in the impact allowances for crane girders for various classes of cranes in the loading standards, cross reference has to be made to the code of practice for cranes and hoists, where the information about different classes of cranes are available. Hence direct interaction between standards having different syntax and semantics should be possible.

- In design standards processing, many iterations may be necessary before evolving an acceptable design conforming to all design specifications. For example, steel components design may involve iteration over different available sections before arriving at a suitable section. Most of the systems perform this following a linear search, by checking the available sections one after another, where sections are arranged in a data base in an increasing order. In this way, systems ensure that section with lowest weight has been picked up. This time consuming process may be acceptable in stand alone systems, but in integrated systems, wherein hundreds of components may have to be designed, this process may retard the system performance. More over, expert designers do not perform component design by the linear search. Depending upon the failure performance of the assumed section and the context information about the capacity and other requirements, they use their domain expertise to "jump" between intermediate sections to arrive at acceptable designs with in one or two iterations.

- Standards processor should be able to analyze and recover from design failures based on the domain specific failure handling knowledge. For example, in the design of steel column in a truss bent, if the failure of the assumed section occurs because the deflection limit at column-truss junction is exceeded, it may be wiser to select a section based on higher moment of inertia than based on area. Also, while undoing the decisions made, it should undo only the dependent decisions.

- In integrated systems, hundreds of structural components may have to be designed. In practice, designers essentially know which component they are designing. In computing systems, this specific instance information will be kept implicit and the processor will be made to "loop" over the number of components to be designed. As long as information sought is confined to the specific instance which is being designed, no special problem arises. But problems arise when information is sought about other instances to carry out the current specific instance design. For example, while girders are being designed, the information about the columns to which the girders are connected are required in order to ensure that proper connection can be made between the girder and the column. Another kind of switching of contexts exists within a component design. For example, in the design of battens in a builtup column, both the intermediate and end tie battens need to be designed. Only a few requirements differ, such as the depth, while most of the specifications such as design forces are the same for the design of both the battens. This problem was also observed by Lopez et.al.,[Ref. 9]

Most of the standards processing systems described earlier do not address all the issues discussed above. In the next section the standards processing architecture including the representational model that would provide necessary interaction between knowledge sources, by taking in to account most of the issues discussed above, is presented.

STANDARDS PROCESSING MODEL

The overall architecture of the system includes a blackboard (partitioned into solution, control as well as postulation blackboard panels), knowledge sources and a control unit. In general, each knowledge source contains the following segments: criterion for scheduling, inferencing strategy to be used and the knowledge. The knowledge segment contains the domain knowledge represented in the form of production rules, which may call external functions or relational data base for some values. The solution blackboard provides a structured representation of network of objetcs in the form of frames. In addition, failure handling rules present in the knowledge source are used to represent the domain advice needed when failures are encountered. This problem solving architecture has been implemented in 'C' language, running on a Sun workstation. Since all the details of the architecture are not important for the present discussion, this article omits them.

In most blackboard systems, knowledge sources do not interact directly. This has been maintained to help modularity and opportunistic scheduling capability of knowledge sources. However, sometimes it is desirable to have interaction between the knowledge sources directly when their interaction is well defined in advance, frequent and direct in order to reduce the control and scheduling overheads in the above standard blackboard implementation. In our **Integrated Engineering System (IES)** although the knowledge sources mostly interact through the blackboard, facilities for direct interaction between the knowledge sources are also provided in situations where it is required.

The direct interaction has been made possible through a Coordinating Knowledge Source (CKS) having a communication segment to provide the necessary direct interface between the different Interacting Knowledge Sources (IKS) as desired. The CKS includes specific declarative knowledge about the design information in different IKSs having the same semantics. These variables are referred to as Mapped Variables. For example, in the case of design standards processing, the knowledge in the design standards and the standard independent component design knowledge are represented as independent and interacting knowledge sources (IKS), and are interfaced through the communication segment of the CKS. Standards processor uses this declarative knowledge during the reasoning process.

The representation of Mapped variables as declarative knowledge in the CKS provides more flexibility for independent development of different IKSs. IES allows knowledge to be represented in each knowledge source in its own design language and no constraint is placed on the syntax and semantics used by different knowledge sources. Solution blackboard provides a structured global data representation, shared among different knowledge sources. The matching of variables having the same semantics in the different knowledge sources and in the blackboard is by using the same syntax for the variables as declared in the blackboard or by mapping them in CKS declarative knowledge if the syntax of the variable in the IKS is different from that in the blackboard. However, matching of the variables in the different IKSs and CKS having the same semantics can be brought about only by the explicit mapping through declarative knowledge in CKS even if their syntax in different IKSs and CKS are the same. This is done to avoid inadvertant mapping of variables in the KS developed independently.

The IES has a three level memory model organization. A local working memory within each knowledge source serving as a scratch pad, which is lost once the next knowledge source (or a set of CKS and IKSs) are scheduled, a working memory associated with CKS for all the mapped variables which all the IKSs in the group can access and a global memory in the blackboard. In the next section the three level memory model utility is discussed.

To invoke the kind of reasoning process to be described in the following sections, the IKSs and CKS should be declared as INTERACTING and COORDINATING, respectively in KS_TYPE slot in the knowledge sources. If the KS_TYPE is

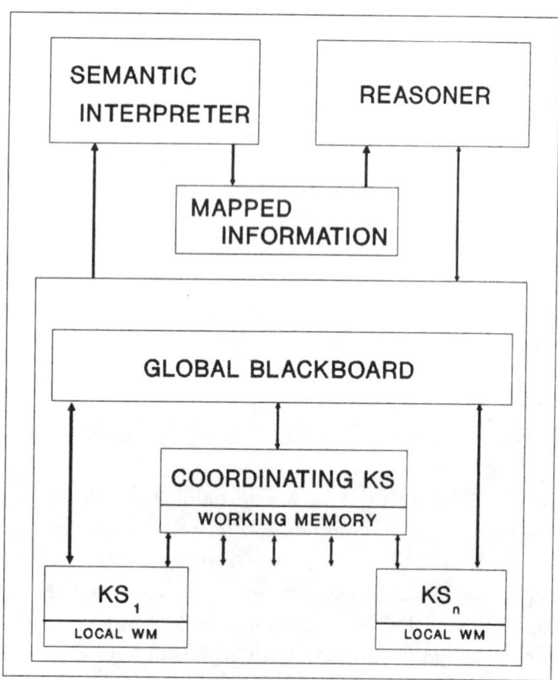

Fig. 1 STANDARDS PROCESSOR ARCHITECTURE

COORDINATING, then the IKSs involved in the interaction should be specified in the CKS. Any number of CKS can be present in the IES and any number of IKSs can be made to interact with a CKS. Also, there is no constraint in an IKS cooperating with more than one CKS. This standards processor architecture is as shown in Fig.1.

Semantic Interpreter

The IES has a knowledge base compiler for converting the knowledge sources represented in to an efficient internal representation. Knowledge base compilation produces the internal representation in local data format for each knowledge source with links to the global solution blackboard data format, wherever required. Local data format changes from one knowledge source to the other, but global solution blackboard data format is unique for all the knowledge sources.

The semantic interpreter is provided as a component in the knowledge base compiler which helps in the compilation of the interacting knowledge sources. The IKSs are compiled in the usual manner (Fig.2), to create compiled information about the presence of variables in the antecedent and consequent portions of the rules, whether the respective variable can encounter any constraint violation or design failure and the failure handling rules related to these variables.

Variable	c3
Memory_type	Local
Var_type	float
Consequent In	R_i
Antecedent In	R_j, R_k
Demon Variable	Yes
Mapped Variable	Yes

FIG. 2 COMPILED INFORMATION IN AN IKS

While compiling a CKS, the semantic interpreter using the variable binding declared in the CKS and the appearance of these variables in the antecedent and consequent portions of the rules in the CKS and IKSs creates a Mapped information which is similar to the compiled information of the other knowledge sources.

A Mapped variable, if not available in the global blackboard can be only deduced in one of the IKS or CKS unless it has to be obtained from the user. The Mapped information includes a record about the deduction of each Mapped variable indicating the knowledge source, where the deduction is made. Furthermore, the mapping also records the information about the presence of global blackboard variables in the different IKSs as well as the KS where it is deduced. Without this information the reasoner has no way of knowing such information. It can be seen that in the case of Mapped variables, the IKSs rather than the rules in the IKSs is mapped while compiling. In Mapped information, the IKSs are essentially treated as large size rules. Fig.2 shows the compiled information in the IKS and in Fig. 3, a sketch of the mapped information in CKS is shown.

As mentioned earlier, compiled information about the variables in each IKS is kept in local data format only. Seman-

VARIABLE BINDING AND VIOLATION STATUS						
CKS	KS1	KS2	KS3	DEDUCED KS	DEMON VARIABLE	DEMON KS
a	a1	a2	a3	KS1	NO	-
b	-	b2	-	CKS	NO	-
C.d	-	c2	c3	KS2	YES	KS3
D.e	d1	-	D.e	KS3	NO	-
-	e1	e2	-	KS1	YES	KS2

C.d -> <Object>.<Attribute> (Global Blackboard Variable)
a,a1 ... (Local Variables)

FIG.3 MAPPED INFORMATION IN CKS

tic interpreter creates necessary links between the mapped variables represented in local data format in each IKS with the Mapped information generated in CKS. Each knowledge source can have a local working memory at the time of its invocation for local computations in addition to the global blackboard panels for storing and retrieving global data. In order to bring in effective coordination and cooperation between different design participants, IES facilitates the access to the local working memory with CKS, by all the IKSs. Hence, IES essentially provides a three level memory model with a local working memory for each KS to be used as a scratch pad for local computations, a global blackboard accessible to all knowledge sources and a local working memory associated with CKS serving as a common pool of information which all the IKSs in the group can use.

The IES does not constrain the knowledge representation of any IKS in any unique design language as described earlier. Any IKS can be represented in its own design language or can use the global data representation or a combination of the two. Hence, the same design information with the same semantics might have been represented in local language in some IKSs and in global data representation in some other IKSs. While compiling, if the semantic interpreter identifies any design information with the same semantics, represented in global data structure in any of the IKS or CKS, it creates links between the design information in each IKS and CKS with the global data . If any design information is represented in their respective local design languages in all IKSs and the semantics of that design information can be matched with any global data representation only by mapping that design information in CKS to be in the global data representation; otherwise such design information will be lost when inferencing of CKS is over. The local working memory of any KS is deleted as soon as the inferencing of that KS is over.

Reasoning Process

The criterion part in a CKS for scheduling it along with the IKSs in the group by the IES scheduler represents the collective capability and requirements of all the coordinating knowledge sources. On the basis of this criterion and the priority rating of the CKS at any state of information on the blackboard, the scheduler of the IES schedules a CKS along with the associated IKSs, declared in the CKS. The inferencing strategy to be used for the inferencing process of a KS is specified in the respective KS. The inferencing process of the CKS is started with the suggested inference strategy in CKS to reach the goal state. During the inferencing process, if any design information is represented in the local design language and it is not a Mapped variable, then IES looks for that information from the local working memory of that KS. If it is a Mapped variable and in the CKS mapping has a link to the solution blackboard, then IES looks for that information from the solution blackboard. If it is a Mapped variable having no link to the solution blackboard, then IES looks for the information from the local working memory of CKS. However, if the design information is directly represented in the syntax of the global data representation, then IES looks for that information directly from the solution blackboard.

Whenever the IES looks for some design information during inferencing, it accesses appropriate memory as described above. If the information sought is not deduced yet and if that information can be deduced within the current KS itself, then IES continues its normal inferencing process by making it as a subgoal. If the sought information can not be deduced in any of the IKSs or CKS and the information has not been deduced yet, then IES assumes that information as the input parameter and asks the user to supply the value. If that information is to be deduced from some other knowledge source, the IES temporarily suspends the execution of the current KS. Then the IES invokes the KS where the information can be deduced, with that information as the current goal for that KS. In the process, IES maintains the local working memories of all IKSs involved in the collective solution process till the common objective is achieved.

Each KS is suspended when it has satisfied the goal with which it is invoked. Otherwise it may look for some design information from some other KS. In this process, in the CKS as well as in all the IKSs, independent goal stacks and local working memories are created and maintained. The IES also maintains a KS stack to record the problem solving trace. Whenever a KS satisfies the goal with which it is invoked, it is removed from the KS stack maintained by IES. Only when a KS is suspended in the middle to get some information from other KS, the suspended KS will be placed on the top of the KS stack with the current goal of the suspended KS. Whenever a KS completes its inferencing task, the KS on the top of the KS stack is invoked next with the recorded goal information in the KS stack as the current goal. This process is continued until the goal of the CKS is reached or no further inferencing is possible.

Reasoner is capable of inferencing both in the goal driven and data driven fashion. Whenever an IKS or CKS is scheduled in data driven fashion, reasoner identifies the information available at all the three levels of the memory model described and creates rule list for firing. Different IKSs and CKS in a set can be scheduled with different inferencing strategies, which should be specified in the respective knowledge source itself. In default, reasoner resorts to goal driven inference.

While the inferencing process is in progress, the IES maintains a dependency network of the decisions made by different IKSs and CKS. If a design failure occurs, IES performs dependency directed backtracking with necessary consistency maintenance procedures and continues the inferencing process with the changed value of parameters provided by failure handling rules in the knowledge base.

EXAMPLES

This section provides two examples through the implementation of Loading standards (IS 875 - 1989) and Design standards (IS 800 - 1984) in the above described standards processing architecture. While Loading standards has been implemented with one CKS and three IKSs, Design standard has been implemented with three CKS with different IKSs for different component designs, which demonstrates multiple bindings between standards and the knowledge sources that use them. These applications form a part of an integrated system which includes planning, preliminary design, analysis, detailed design, detailing and construction management activities for engineering of single storey steel industrial structures.

Implementation Of Loading Standard(IS 875 – 1989)

The Loading Standard has been implemented as three interacting knowledge sources Dead Load standards IKS(DIKS), Live Load standards IKS (LIKS) and Wind Load standards IKS (WIKS) with Industrial Loading Evaluation Coordinating Knowledge Source (ILCKS), which contains the knowledge about the method of evaluation of loads for the single storey industrial buildings for which loading calculations are to be carried out and the knowledge about plans and objectives of the load calculation task. The declarative knowledge about the mapping variables in the ILCKS is shown in the appendix.

For the purpose of illustration, it is assumed that the planning and preliminary design phases are over and the system has already arrived at A_type truss as the main roofing system for the industrial building to be designed. ILCKS will be scheduled by the scheduler in the IES for the loading calculation event based on priority at this stage. ILCKS decides about the load cases for which loading calculations have to be made, namely dead load, live load and wind load cases in the industrial structures. ILCKS contains the information required to complete the calculations for the specified load cases. For example, ILCKS knows that in order to complete the wind load calculation, it should evaluate all the three cases involving wind blowing parallel to the ridge, wind blowing perpendicular to the ridge with internal suction and with internal pressure, depending upon the enclosure permeability. This forms standard independent knowledge about the load calculation task. But on the other hand, for ILCKS to perform these calculations, it requires design wind pressure, external and internal wind pressure coefficients, which forms a part of the loading standard. Once the wind pressure and coefficients are known, the calculation of nodal loads for the geometry chosen forms a part of standard independent domain knowledge and hence is represented in ILCKS. Partial network of parameters involved are shown in Fig.4 with references to where the parameters are deduced given in the parentheses.

As shown in the network in Fig.4, reasoner continues its reasoning process with ILCKS in a goal driven fashion until it

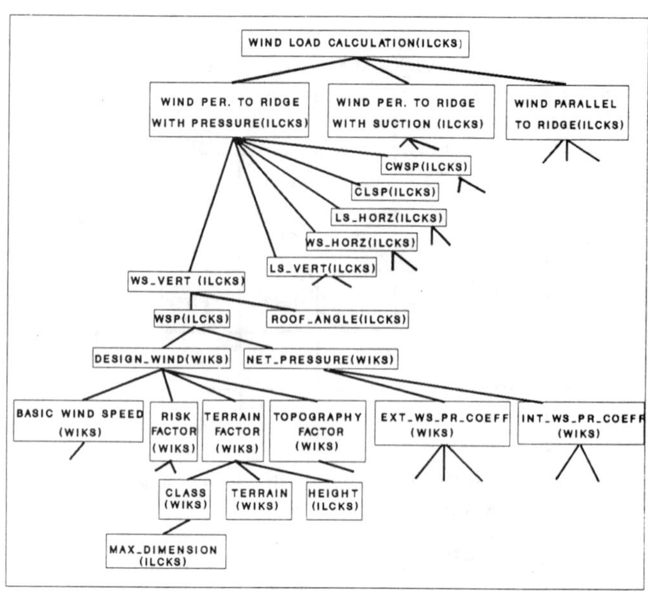

FIG. 4 PARTIAL PARAMETER NETWORK FOR WIND LOAD CALCULATIONS

finds that in order to calculate windward side nodal load, it requires design wind pressure which is not deduced in ILCKS and can be inferred only from WIKS. Hence, reasoner puts ILCKS in doze mode and invokes WIKS with the design wind pressure as its current goal. For the design wind pressure to be calculated, basic wind speed, risk factor, terrain factor and topography factor should be determined. Basic wind speed is determined based on the location of the building specified by the user and the risk factor is calculated within the WIKS. In order to determine the terrain factor, max_dimension of the building is required. Maximum dimension calculation is not a part of the codal knowledge. The knowledge to calculate max_dimension is represented in the ILCKS and hence, the WIKS is put in doze mode and the ILCKS is invoked with max_dimension as its current goal. After the determination of the max_dimension in ILCKS, WIKS gets rescheduled and inference with the calculation of design wind pressure as the subgoal pursued. This process is continued with the reasoner looking for information wherever it is available with the help of Mapped information to achieve the main goal in the ILCKS with which the reasoning process started. Appendix gives snap shots of the goal stacks and local working memory of ILCKS and WIKS at two different stages of the inferencing process.

Implementation Of Design Standard(IS 800 – 1984)

EDSEL [Ref. 6], a stand alone expert system for the design of steel structural elements, has been reimplemented in the blackboard system for the design phase of the integrated engineering process using this standards processing model. Three sets of CKS and IKSs are implemented based on the structural components to be designed. Fig. 5 shows separate implementation of the design standards as an IKS (DSIKS), which interacts with the three CKS (Axial Member design (ACKS), Beam design (BCKS) and Beam Column design (BCCKS)). ACKS interacts with tension member design KS (TMKS), compression member design KS (CMKS) in addition to Design standards. BCKS interacts with beam design KS (BDKS) in addition to the Design standards. BCCKS interacts with BDKS and CMKS in addition to Design standards. In all the three cases, no change in Design standards KS is involved and only relevant mapping knowledge has

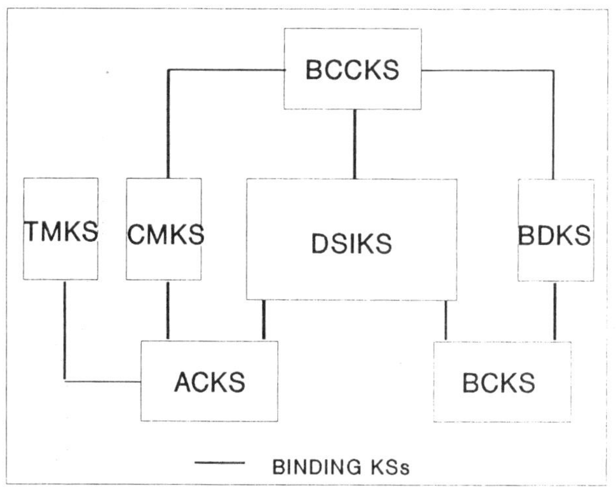

FIG. 5 DESIGN STANDARD IMPLEMENTATION

been established in respective CKS. This application shows that any number of interacting knowledge sources can be combined with a CKS and that the same IKS can be used with different binding with different CKSs.

Based on the component to be designed, the respective CKS gets invoked. CKS stores general objectives and initialization procedures. For example, ACKS contains procedures for retrieving member instance information about length, axial forces acting on it and the member properties based on preliminary design. Based on the axial forces acting on the member for different load combinations, ACKS decides whether design has to be carried out for both tension and compression forces or for tension or compression alone. TMKS and CMKS contain standard independent knowledge about tension and compression member designs such as what constitutes a tension or compression design, selection of shapes, failure handling knowledge etc.,. For example, in the case of compression member in a truss chord, if the design satisfies slenderness ratio and have the required capacity then compression member design is over and this knowledge is represented in CMKS. But CMKS doesn't contain knowledge about what is the requirement for slenderness ratio because it is a part of the Design standards. Similarly, permissible stress in compression multiplied by the area of the cross section gives the capacity of the section and forms a part of CMKS, since it is standard independent. But the knowledge for the calculation of permissible stress in compression lies in the Design standards.

In the load calculations described above, failure is not a possible event and the process involves only the sharing of relevant information . But, in the case of Design standards, design failures are possible and many iterations with trial sections may be involved. In order to recover from failures, standard independent, domain specific failure handling knowledge is provided in TMKS and CMKS. This knowledge includes how to choose next trial section, switching from single to built up section if no single section is available and also the different contexts under which design failure in tension and compression designs may occur. Whenever design failure occurs, reasoner looks for this kind of failure handling knowledge to recover from failures. For example, when the slenderness ratio limit is exceeded, as inferred in the DSIKS, it causes the reasoner to look for knowledge to handle this type of failure which is available in CMKS and hence invokes it to make corrections in design. If no failure handling knowledge is available within the knowledge base, user is called upon to help to overcome the failure.

SUMMARY

The paper has discussed the details of a standards processor implemented in a blackboard system used for the integrated engineering of steel structures. The main features of the standards processor are the following :

- Representation of standards as independent, separate knowledge source, facilitating easy extension and modification.
- The architecture supports the direct interaction between standards knowledge source and standard independent knowledge sources for efficient inference.
- Standards processor handles the design failures using domain specific backtracking knowledge and the dependency network generated during the inferencing process.
- The present architecture allows multiple binding between the standards knowledge source and the knowledge sources that use them.
- The proposed model has been able to process different standards at different stages of the engineering process.
- The underlying knowledge representation allows the different bodies of knowledge to be maintained as originally produced by the domain experts with out constraining the syntax and semantics of the design language involved.
- In addition to the standards processing application, the architecture provides an excellent test bed to study the interactions between multiple design participants in the cooperative problem solving.

The concepts involved were illustrated using loading standards processing and design standards processing as examples.

ACKNOWLEDGEMENT

This work on standards processing forms a part of a project on integrated expert systems for engineering of steel structures, sponsored by a research grant from the Department of Science and Technology, Govt. of India. Authors would like to thank Mr. Narain Hegde, Computer science dept., IIT, Madras for providing initial version of a rule based engine over which the current blackboard system has been implemented.

REFERENCES

1) Cronembold, J.R., and Law, K.H., "Automated processing of design standard", ASCE Journal of Computing in Civil Engineering, Vol.2, 1988, pp. 255-273.

2) Feijo, B., Dowling P.J., and Smith, D.L., "Incorporation of steel design codes in to design automation systems",

Expert Systems in Civil Engineering, IABSE Colloquium, Bergamo, 1989.

3) Fenves, S.J., and Garrett, J.H., "Knowledge based standards processing ", Artificial Intelligence, Vol. 1, 1986, pp.3 - 14.

4) Fenves, S.J., "Representation of the computer-aided design process by a network of decision tables", computers and structures, Vol.3, 1973, pp 1099-1107.

5) Garret, J.H., and Fenves, S.J., "Knowledge based standard - independent member design ", ASCE Journal of Structural Engineering, Vol.115, 1989, pp. 1396-1411.

6) Ghosh, D.K., "EDSEL: An Expert System for Design of Steel Structural Elements", M.S thesis, Dept. of Civil Engineering, IIT, Madras, 1990.

7) IS 800 - 1984. Indian standard code of practice for general construction in steel, Bureau of Indian Standards, New Delhi.

8) IS 875 - 1987. Indian standard code of practice for design loads (other than earthquake) for buildings and structures, Bureau of Indian Standards, New Delhi.

9) Lopez, L.A., Elam, S., and Reed, K., "Software concept for checking engineering designs for conformance with codes and standards", Engineering with Computers, Vol. 5, 1989, pp.63-78.

10) Rasdorf, W.J., and Wang, T.E., "Generic design standards processing in an expert system environment", ASCE Journal of Computing in Civil Engineering, Vol.2, 1988, pp. 68-87.

11) Sriram, D., Maher, M.L., and Fenves, S.J., "Knowledge - based expert systems in structural design", Computers and structures, Vol.20, 1985, pp. 1-9.

APPENDIX

The following code shows a part of the coordinating segment in the ILCKS.

```
KNOWLEDGE SOURCE : ILCKS ;

KS_TYPE : COORDINATING;

INTERACTIVE KS : DIKS,LIKS,WIKS;

      ILCKS                            DIKS    LIKS      WIKS

ext_ws_pressure_coeff                   -        -     ext_pr_per_ridge_w
ext_ls_pressure_coeff                   -        -     ext_pr_per_ridge_l
ext_pressure_coeff_par                  -        -     ext_pr_p_ridge
int_wind_pressure_coeff                 -        -     int_pr_coefficient
column_ws_pr_coeff_per                  -        -     ext_ws_wall_pr_coeff
column_ls_pr_coeff_per                  -        -     ext_ls_wall_pr_coeff
column_pr_coeff_par                     -        -     ext_wall_pr_coeff_par
Industrial_Building.                    -        -     design_wind
       design_wind_pressure
Main_frame.height                       -        -     height
Industrial_Building.cladding clad    cladding cladding
roof_live_load                          -     roof_udl   -
max_building_dimension                  -        -     max_dimension
roof_angle                              -      angle   roof_angle
Industrial_Building.              sht_type   -         -
    sheeting_type
```

The following code shows the goal stack and ILCKS working memory when WIKS is invoked to infer about wind pressure.

```
KNOWLEDGE SOURCE : ILCKS

Goal Stack Contents FROM TOP TO BOTTOM are as follows

VARIABLE                                             ICID    RULE

140)wsp                                               F       29
71)x1                                                 F       -1
51)wsp_vert                                           T       34
52)lsp_vert                                           F       -1
6)wsp_horz                                            F       -1
7)lsp_horz                                            F       -1
112)clsp                                              F       -1
```

142)cwsp	F	-1
39)wl_cal perpendicular to ridge with int_pressure	T	45
118)wl_cal perpendicular to ridge with int_suction	F	-1
32)wl_cal parallel to ridge	F	-1
135)wind load calculations	T	5
134)live load calculations	F	-1
164)load cases	F	-1
137)load calculations	T	3

THE WORKING MEMORY CONTENTS (ILCKS)

VARIABLE NAME	OPTR	VARTYPE	RULE	VALUE
column_ecc	=	float	23	0.125000
total_clad_ecc	=	float	21	0.250000
purlin_wt_per_m	=	float	51	92.000000
roof_angle	=	float	11	18.434948
column_depth	=	float	50	0.250000
sheeting_wt	=	float	51	170.000000
girts_wt	=	float	14	2760.000000
cladding_wt	=	float	15	6120.000000
clad_miscellaneous_wt	=	float	18	144.000000
side_clad_wt	=	float	17	9024.000000
purlin_wt	=	float	8	92.000000
miscellaneous_wt	=	float	8	50.000000
no_of_purlins	=	float	9	12.000000
dead load calculations	IS	string	26	'over'
no_of_girts	=	float	20	5.000000
no_of_panels	=	float	10	5.000000
dl_per_node	=	float	7	1581.165000
load_on_column	=	float	25	9247.800000
ecc_from_face_of_column	=	float	19	0.125000
column_weight	=	float	50	223.799988
tot_dl_per_sq_m	=	float	6	402.000012
clad_moment	=	float	24	1891.711792
hypot	=	float	11	6.555412

The following code shows the goal stack and working memory of WIKS when WIKS goes in to doze mode and invokes ILCKS to infer about maximum dimension.

KNOWLEDGE SOURCE : WIKS

Goal Stack Content FROM TOP TO BOTTOM are as follows

VARIABLE	ICID	RULE
109)class	F	10
0)terrain_category	F	-1
154)height	F	-1
75)k2	T	13
25)basic_wind_speed	F	-1
142)design_wind_speed	T	1
79)design_wind	T	2

THE WORKING MEMORY CONTENTS (WIKS)

VARIABLE NAME	OPTR	VARTYPE	RULE	VALUE
design_life	=	float	0	50.000000
basic_wind_speed	=	float	0	33.000000
k1	=	float	3	0.993697
k3	=	float	16	1.000000
a	=	float	4	83.199997
topography	IS	string	14	'Plain'
b	=	float	4	9.200000
upwind_slope	=	float	0	2.000000

A LISP REPRESENTATION FOR DESIGN STANDARDS

A.I. Hansen, B. Kumar and B.H.V. Topping
Department of Civil Engineering,
Heriot-Watt University, Riccarton, Edinburgh, Scotland

Abstract

Earlier research [1, 2] proposed the representation of the provisions of design codes as facts. These facts may be manipulated by standard-independent generic rules. This concept of standards processing is derived from PROLOG where a data-base consisting of facts may be manipulated by generic rules. This paper is a description of an implementation of this factual representaion of standards in LISP. Section 4.2 of BS5950 [3] for the use of steel in buildings has been used for illustration of the concept.

1 Representing Standards in LISP

One of the main objectives of the research described in this paper was to find a way to represent clauses from a code of practice using LISP [4, 5] in a similar way to that used in STAPRO [1]. STAPRO is a prototype knowledge-based system for processing codes of practice for structural engineering design. STAPRO was written in PROLOG [6]. The object of this project was to illustrate that standard clauses may be efficiently represented as facts in a nested LISP structure. Section 4.2 of BS 5950:Part1:1985 "Structural use of Steelwork in Buildings" [3] is used as an illustrative example. This will be explained further in section 4.

2 The Representational Model of a Standard

Considerable research has been undertaken into standards representation and processing since the late sixties. This research has resulted in a model which uses four basic components to represent the contents of a design standard [7]:

- Data items representing every variable which is found in a standard.
- Decision tables representing the logic used to determine the values of data items.
- Information networks representing the precedence relationship among the data items.
- Organizational system representing organization (arrangement and scope) of the standard.

These four components are discussed further in the next four sub-sections.

2.1 Data Items

Data items may be either;

- a "boolean", which evaluates to TRUE or FALSE, SATISFIED or VIOLATED.
- a numeric item.
- a multi-valued item.
- a single-valued item.

Data items represent the lowest level in the structure and the relationship between the data items is finally used in evaluating each clause in the standard. The clause may then be evaluated to satisfied or violated. A data item may be either basic or derived. This classification of data items will be discussed in section 2.1.2

2.1.1 Decision Tables

A decision table represents the orderly presentation of the reasoning process leading to assignment of a value to a data item (see figure 1). Decision tables have a tabular form and each condition represents a boolean expression which may be evaluated to true or false only. Thus decision tables may be prepared to give an exhaustive set of rules for evaluating the data item. Decision trees may be automatically generated from a decision table as shown in figure 1. LISP provides a function called cond which can be directly used to implement a decision table. cond has the following syntax:

```
(cond (<test 1> <action 1>)
      (<test 2> <action 2>)
            :
      (<test n> <action n>))
```

This represents: IF <test 1> is not false then evaluate <action 1> and return the value of <action 1>. IF <test 1> is false (NIL) then jump to the second statement and so on. In LISP figure 2 would be represented as:

```
(cond (<condition 1> <action 1>)
      (<condition 2> <action 2>)
      (t <action 3>))
```

	Rule 1	Rule 2	Rule 3
Condition 1	T	F	T
Condition 2	F	T	F
Condition 3	F	F	T
Action 1	X		
Action 2		X	
Action 3			X

Figure 1
Decision Table for a Data Item

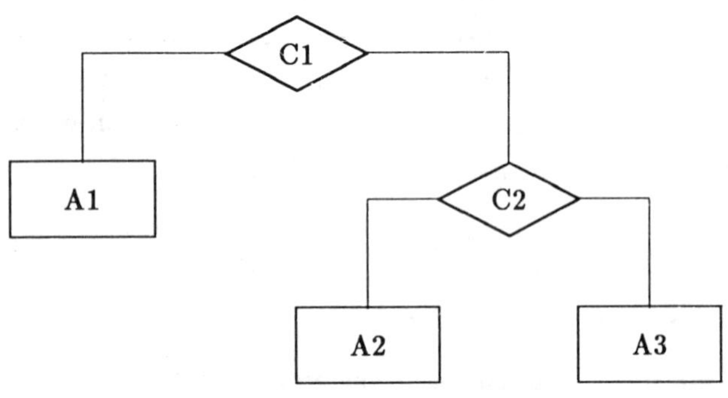

Figure 2
Decision Tree for a Data Item

2.1.2 Information Networks

An information network is used to represent the precedence relationship between data items. As shown in figure 3, an information network consists of nodes, each representing a

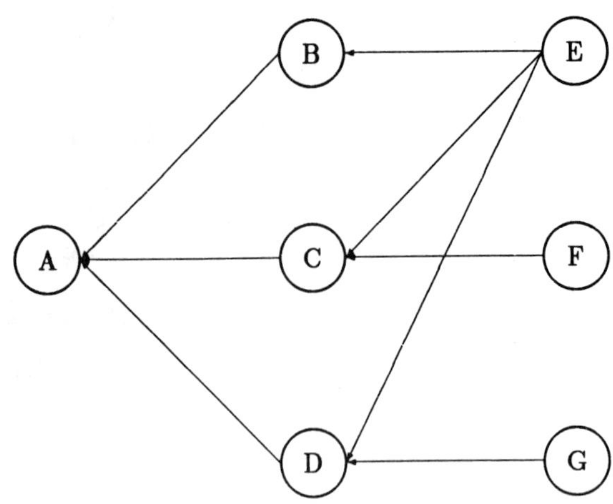

Figure 3
An Information Network

data item, and lines representing the precedence between the data items. Data item E, F, and G are basic data items, i.e. they have no ingredients. Data items A, B, C and D are all derived data items and contain ingredients to evaluate them. For example, D has the ingredient list (E G). The information network plays an important role in order effectively to extract the information necessary for each step in the process. The system will only search for the information contained in the ingredient list and will in this way work its way through the information network and finally determine a basic data item.

Derivation of the data item ϵ from table 7 of BS 5950 [3] will be used to demonstrate this concept. As given in table 7: $\epsilon = \sqrt{\frac{275}{p_y}}$. Clearly ϵ depends on data items p_y. ϵ (epsylon) is represented in the list "derived_data" as:

```
(epsylon(ingredient_list (py))
        (expression (sqrt (/ 275 py))))
```

2.2 Organization System

An organizational system is required in order to identify the provisions to be tested in the design process. This is fairly straight forward for section 4.2 of BS5950 [3] since there are few possible paths to choose from.

3 The Standards processing approach adopted

The natural and intuitive approach used to represent the clauses of design standards is production rules. However, the approach based on representation of clauses as facts was found to offer certain advantages [1] compared to the production-rule approach. Drawbacks observed in the production rule approach were that:

- rules were found to require more computer memory than facts;
- all the information in a standard did not need to be in rule form (e.g. definitions, tables), in which case the rules were found not to be feasible representations; and
- the generic standard-independent nature of the processor was found to be difficult to accomplish because all the information was hard-coded in the rules; whereas facts could be treated as data upon which generic rules could easily operate.

Casting standards as facts was proposed in [1, 2]. For the reasons given in [1], the more direct approach described in [1] is adopted as a basis for the approach chosen in this work.

4 Representation of standard clauses as facts in LISP

As in STAPRO [1] the representation of clauses in LISP was undertaken using facts. The information contained in a standard may be divided into three categories of facts:

- Provisional facts
- Organizational facts
- General facts

4.1 Provisional Facts

Provisional facts contain the information provided in a clause, and consist of the following two parts:

- applicability criteria; and
- performance criteria

The applicability criteria represent the conditions which have to be met before the performance criterion is evaluated. For example, clause 4.2.3 of BS 5950 [3] which pertains to shear indicates that: before the performance criteria of $F_v \leq P_v$, where the shear capacity $P_v = 0.6P_yA_v$ is tested, the applicability criteria stating that depth to thickness ratio of the web, $d/t \leq 63\epsilon$ must be satisfied.

The representation of clauses as facts in a nested LISP structure has the following syntax:

```
(clause-no. (ingredient list)
            (applicability criteria)
            (performance criteria))
```

For example, the representation of clause 4.2.3 of BS 5950 [3] is shown in figure 4. When the inference mechanism is applied to the clause, the criteria in the clause can only evaluate to T (true) or NIL (false).

4.2 Organizational Facts

Organizational facts are required to guide the inference mechanism in deciding how the precedence relationship between the different provisions of a standard are related.

Before a clause is found to be applicable certain constraints must be satisfied. This may be summarized as:

$$< subject > \text{ must satisfy } < constraints >$$

The subject is the clause and the constraints are those that have to be satisfied before the inference mechanism evaluates the clause. The organizational system representing section 4.2 of BS 5950 [3] is expressed in a nested LISP structure as shown in figure 5. As demonstrated in figure 5, clauses 4.2.1 to 4.2.4 BS 5950 [3] will always return T (true) when the constraint is evaluated. These clauses will therefore always be considered by the inference mechanism. Clause [4.2.5] will return T for $F_v \leq 0.6P_v$, i.e. the clause will only be fired by the inference mechanism if this constraint is satisfied.

4.3 General Facts

General facts represent the general information regarding

```
(4.2.3 (ingredient_list ((FV PV))
                         (FV PY AV LD LT EPSYLON))
       (applicability_criteria (<= (/ LD LT) (* 63 EPSYLON)))
       (performance_criteria (<= FV (use 'PV (* 0.6 PY AV)))))
```

Figure 4
Representation of clause 4.2.3 BS 5950 [3]

the data items contained in the standard. These facts support and provide the base for evaluation of the provisional and organizational facts in the inference mechanism.

All data items found in section 4.2 BS 5950 [3] are listed in a list called "notation" together with an explanation

```
(setq org_facts
     '((4.2.1 (ingredient_list ())
              (constraint t))
       (4.2.2 (ingredient_list ())
              (constraint t))
       (4.2.3 (ingredient_list ())
              (constraint t))
       (4.2.4 (ingredient_list ())
              (constraint t))
       (4.2.5 (ingredient_list (FV PV))
              (constraint (<= FV (* 0.6 PV))))
       (4.2.6 (ingredient_list (FV PV))
              (constraint (> FV (* 0.6 PV))))))
```

Figure 5
Representation of Organizational Facts

of the data item and a tag which is used to differentiate between basic and derived data items. For example steel strength, p_y, appears in the notation list as:

```
(py d "Design strength of steel")
```

The tag "d" tells that this is a derived data item. Derived data items contain an ingredient list describing on which data items they depend. The syntax of a derived data item is represented as a general fact:

```
(data item (ingredient list)
           (expression))
```

Using the same example for p_y, the associated representation in the list derived data is shown in figure 6. Here, the ingredient list contains the data items `lt` (web thickness), `tf` (flange thickness) and `steel-grade` (steel grade as decribed in clause 3.1.1 of BS 5950 [3]). The functions `cond` and `case` are used to tabulate the different alternatives for p_y.

5 Comparisions with STAPRO

As in STAPRO [1] the representation of clauses is as facts. An example of how clause 5.3.5 of BS 5950 [3] would be represented in LISP is demonstrated in this section. The performance criteria of clause 5.3.5 is given by:

$$L_m \leq \frac{38 r_y}{\left(\frac{f_c}{130} + \left(\frac{p_y}{275}\right)^2 \left(\frac{x}{36}\right)^2\right)^{1/2}}$$

where f_c is the average compressive stress due to axial load (in N/mm^2)
p_y is the design strength (in N/mm^2)
r_y is the design strength (in mm)
x is the torsional index

The representation of Clause 5.3.5 of BS 5950 [3] in the system is given in figure 7. The representation of the performance criteria of clause 5.3.5 of BS 5950 [3], from STAPRO [1] given in the PROLOG-based Edinburgh Blackboard

```
(py (ingredient_list (lt tf steel_grade))
    (expression (case steel_grade
                 ((43) (cond ((and (<= lt 16) (<= tf 16) t) '275)
                             ((and (<= lt 40) (<= tf 40) t) '265)
                             ((and (<= lt 100) (<= tf 100) t) '245)
                             (t (format t "&%Not standard steel"))))
                 ((50) (cond ((and (<= lt 16) (<= tf 16) t) '355)
                             ((and (<= lt 63) (<= tf 63) t) '340)
                             ((and (<= lt 100) (<= tf 100) t) '325)
                             (t (format t "&%Not standard steel"))))
                 ((55) (cond ((and (<= lt 16) (<= tf 16) t) '450)
                             ((and (<= lt 25) (<= tf 25) t) '430)
                             ((and (<= lt 40) (<= tf 40) t) '415)
                             (t (format t "&%Not standard steel"))))
                 (t (format t "&%Steel grade A does not comply to
                             BS-4360" steel_grade)))))
```

Figure 6
Representation of p_y

```
(5.3.5 (ingredient_list ((lm lmcalc))
                        (fc py x ry lm d))
       (applicability_criteria
        (my-y-or-n-p "
         Torsional restraint should be provided at a plastic hinge
         location if practicable (except as in 5.5.5.3).  Where
         impracticable the restraint should be provided within a
         distance of D/2 of the plastic hinge location.
         Is this satisfied?")
       (performance_criteria
        (<= lm (use 'lmcalc (/ (* 38 ry)
                               (sqrt (+ (/ fc 130) (* (/ py 275)
                                                      (/ py 275)
                                                      (/ x 36)
                                                      (/ x 36)
                                                      ))))))))
```

Figure 7
Representation of clause 5.3.5

Shell [8], is below:

```
[perf('5.3.5'),(lm,[ry,fc,py,x],((38*ry)/
((fc/130)+((py/275)^2)*((x/36)^2)),'=<'),true]
```

The above facts concern the data item `lm` whose ingredients are `[ry,fc,py,x]`. The expression plus the operator (`=<`) for evaluating `lm` are also given. In the LISP representation the ingredients list also contains `lm`. When the system runs in *conformance mode*, this data item is basic and must therefore be supplied by the user in order to test `lm` against `lmcalc` which is the constraint value derived from the expression. In *determinaton mode*, when the system is asked to determine the value of `lm`, the system uses the first element of the short list `(lm lmcalc)` to zero the value of `lm`. `lmcalc` is then calculated and `lm` is set to be `<=` to the value of `lmcalc`.

In the PROLOG version, the operator was given at the end of the list whereas in the LISP representation the operator will always appear first in the list. Using this known fact, the operator may be extracted from the list using simple list manipulation for use in output to describe the limits of the data item. Given the lists are built up in a systematic manner, LISP's flexibility makes it possible for the inference mechanism to extract any data item or variable from a list. The flexibility offered by LISP makes it possible to build up an inference mechanism which may perform complex knowledge extraction from a relatively transparent and simple knowledge base.

6 Problems with Standards Representation

The major difficulty when representating standards, regardless of the model adopted, is the fact that standards are not generally written with computer aided processing in mind, but more as a designers' guide. Therefore, from a "computer" point of view, they have shortcomings, such as incompleteness and unsystematic cross-references. This missing information is meant to be covered by the engineer's theoretical knowledge and experience. One example from section 4.2 is clause 4.2.4 , "Elastic shear stress", which state that:

> *In sections where the web varies in thickness or have holes significantly larger than those reqired for fasteners, the shear stress should be calculated from first principles assuming elastic behaviour. In such cases the maximum shear stress, f_v should not exceed $0.7p_y$.*

A representation of this clause in LISP is given in figure 8. Clearly this is a problem which requires the theoretical knowledge and experience of the engineer, and therefore creates a problem in a computer aided system for standards processing. The approach adopted for such problems is to make the engineer aware of the problem and make him take a decision based on theoretical knowledge and experience. A knowledge base could of course theoretically be built to decide if the web has "holes significantly larger than those

```
(4.2.4 (ingredient_list ()
                        (py))
       (applicability_criteria
        (text
         "&%?:  Does the web vary in thickness or does it have
         holes significantly%larger than those required
         for fasteners?")
        (cond ((r-my-y-or-n-p "") t)
              (t (my-y-or-n-p
                  "&%Then the shear stress should be calculated from
                  first principles%assuming elastic behaviour.  In such
                  cases the maximum%shear stress (LFV) should not exceed
                  0.7py.%Is this fact considered?"))))
       (performance_criteria t))
```

Figure 8
Representation of clause 4.2.4 BS 5950 [3]

required for fasteners", but this will only increase the complexity of a problem which is trivial for the human brain, but very difficult to automate.

7 Summary

This paper has demonstated that an exisiting model for the representation of standards as facts which was previously implemented in PROLOG may be implemented easily in LISP.

Acknowledgement

The research described in this paper was supported by the Science and Engineering Research Council grant reference: GR/F/07385.

References

[1] Topping, B.H.V., Kumar, B. "Knowledge representation and processing for structural engineering design codes", Eng. Appli. of AI, Vol. 2, 214-227, 1989.

[2] Rasdorf, W.J., Wang, T.E., "Generic Design Standards Processing in an Expert System Environment", Journal of Computing in Civil Engineering, Vol.2, No.1, January, 68-87, 1988.

[3] "BS 5950 Structural use of steelwork in buildings: Part 1. Code of practice for design in simple and continious construction: Hot rolled sections: 1985", British Standards Institution, London, 1985.

[4] Steele, G.L., "Common Lisp, The Language", Digital Press, 1984.

[5] "Sun 2.1 Common Lisp User's Guide", Sun Microsystems Europe, Camberley, Surrey, England, 1987,

[6] Bratko, I., "Prolog Programming for Artificial Intelligence", Addison-Wesley, 1987.

[7] Fenves, S.J., "Knowledge Based Standards Processing", Artificial Intelligence in Engineering, Vol.1, No.1, 3-13, 1986.

[8] Chan, N., Johnson, K., "Edinburgh Blackboard Shell User's Manual", Artificial Intelligence Applications Institute, Edinburgh University, 1987.

11 APPLICATION OF NEURAL NETWORKS TO STRUCTURAL ENGINEERING

ART NETWORKS IN AUTOMATED CONCEPTUAL DESIGN OF STRUCTURAL SYSTEMS

P. Hajela†, B. Fu† and L. Berke‡
†Rensselaer Polytechnic Institute, Troy, New York,
‡NASA Lewis Research Center, Cleveland, Ohio,
United States of America

The present paper examines applications of adaptive resonance theory (ART) based neural networks in structural design. ART networks are vector classifiers, and their proposed use in the present work is to classify given conceptual design situations into one of several categories for which solutions are known. Once such a category is established, the actual generation of design is relegated to a procedural process associated with the category. In the present work, this approach is applied to the design of beam and frame structures for minimum weight and for maximum load carrying capacity. For such problems, the significant issues are shown to reside in selection of problem features used in the model abstraction, and in the choice of network parameters during training and classification.

INTRODUCTION

There has been significant research activity in the general area of neurobiological computing. The field had its early beginnings in the 1950's, in the guise of the Hebbian learning rule [1] which derived from a model in which biological associative memory was determined by synaptic connections between nerve cells. This idea was advanced by Rosenblatt in 1958 in the form of the perceptron model [2] which viewed biological learning as a dynamic sensory process. The entire effort received a setback in the late 1960's as the result of a critical assessment of the approach in a paper by Minsky [3], asserting the limitations of the perceptron approach. In 1972, independent efforts of Kohonen [4] and Anderson [5] produced the linear associator model for learning. This work was important in that it removed the earlier objections of Minsky, and gave a fresh start to research in neural networks.

Neurobiological methods of computation are distinctly different from the traditional numerical computing and the more recently emergent strategies of symbolic processing and manipulation. The basic computational processor in such neural networks is an artificial neuron which is analogous to its biological counterpart. These basic processing elements can interact in various ways depending upon the nature of the interconnections. Processing elements may have feed-forward characteristics only, or may have feedback loops. Likewise, networks may be fully connected in that each neuron is linked to every other neuron in the network, or the processing elements may be sparsely connected. While the feed-forward or feedback architecture has an impact on whether a network can be trained, the property of connectivity is intimately related to the parallelism in the system.

Applications of the neural computing paradigm in automated design have been recently proposed [6,7]. In these efforts, the basic approach has been one where neural networks were used to provide a map between some input and output quantities of interest. Typical of these are maps between the sizing variables in a structural optimization problem and the response constraints of interest, or a map between design requirements and the corresponding optimal design. The first provides a rapid reanalysis capability which is computationally advantageous in an iterative optimization sequence, while the second may be viewed as an approach wherein an estimate of the optimal design may be made, given a set of design requirements. Both strategies require a set of input-output pairs spanning the domain of interest with which to train the neural networks. Such training is generally referred to as supervised training.

The present paper deals with the application of neural networks in the conceptual design of structural systems. The process of conceptual design is a difficult one to quantify. In the present application, conceptual design describes the selection of a structural model capable of transmitting applied loads to specified support points, so as to meet some design requirements. This geometry definition is not necessarily unique, and falls into the category of topology assignment problems. Human designers, largely as a result of vast experience with similar problems, are able to sort through the given information, extracting only those vital features which must be considered in the design task. The understanding of human memory for purposes of representing in an artificial neural network, has been the subject of extensive research [8-10]. Not only must the memory allow for recognition of previously encountered patterns but must also accommodate and store new input without destroying the existing contents. This stability-plasticity requirement has contributed to the proposition of adaptive resonance theory, or ART, networks.

ART networks belong to the general category of self-organizing or unsupervised-learning systems. A description of the network architecture, including its key elements, is provided in subsequent sections of the paper. The adaptation of such systems in the conceptual design of structures is illustrated with simple representative design problems including beams, trusses, and frame structures. The influence of network parameters and the domain representation scheme, on the overall network performance, is also evaluated.

ART NEURAL NETWORKS

Two forms of ART networks have been studied, and can be essentially distinguished on the basis of the form of input information they can accept (binary or continuous), and on the approach used to process this information. The present paper describes an implementation where the network can only process binary inputs. As stated earlier, ART networks attempt to model the process by which humans learn and recognize invariant properties of a given problem domain. A special characteristic of such networks is the plasticity that allows the system to learn new concepts, and at the same time retain a stability that prevents destruction of previously learned information. ART networks accommodate these requirements through interactions between different subsystems, designed to process previously encountered and unfamiliar events, respectively.

A schematic representation of an ART network architecture is shown in Figure 1. These networks are vector classifiers in that they can classify an input vector as belonging to one of the categories of stored patterns. As shown in the figure, the interacting subsystems in such a network are a set of two gain controls, an attentional subsystem consisting of two layers of neurons referred to as the comparison and the recognition layers, respectively, and a reset control, also termed as the orienting subsystem. The attentional subsystem is designed to process familiar events. By itself, however, it is unable to process new patterns and at the same time retain stability of existing patterns. In an isolated mode, it is either rigid to new patterns or exhibits extreme instability with regard to stored patterns. This shortcoming is addressed by the orienting subsystem or the reset control. The function of each component subsystem is summarized in the following sections.

Attentional Subsystem

This subsystem consists of two layers of artificial neurons - the comparison and the recognition layer. In the present paper, these layers will be referred to as F_1 and F_2, respectively. As shown in Figure 1, the F_1 layer receives an input pattern, which activates the neurons in the F_1 layer. The output signals from the F_1 neurons then stimulate the neurons in the F_2 layer through the bottom-up filtering. A lateral inhibition mechanism, in which the output of each F_2 layer neuron is fed back to itself with a positive weight (enforcement), and is presented to all other neurons in the layer with a negative weight (inhibition), ensures that only one F_2 neuron fires in this "winner-take-all" competition. The output of this active neuron then activates, through a top-down filtering, a template pattern for comparison with the input pattern. If the similarity between them is greater than a preassigned vigilance level, the input pattern is automatically classified under the same category as the flag of the fired F_2 neuron. Otherwise, a reset wave is generated from the orienting subsystem.

Orienting Subsystem

The orienting subsystem resets the attentional subsystem by disabling the fired neuron in the F_2 layer. The next dominant neuron in the F_2 layer fires as a result, and the process of similarity assessment is repeated until a match is located. If no similar category can be found, as would be true for an unfamiliar pattern, a new category is established. The orienting subsystem exists primarily to process unfamiliar patterns.

Gain Controls

The gain controls G_1 and G_2 are vital for the process of training and classification. They regulate the whole system by a "two-thirds rule" as described by Carpenter and Grossberg [11]. This rule states that an F_1 layer neuron will fire only if at least two of the three input signals that it receives prompt its activation.

IMPLEMENTATION OF ART NETWORKS

A C-language implementation of the ART network was used in the present work. The system described in the preceding sections must be initialized to start the network training process. This entails setting G_1 and G_2 to unity, the status of the orienting subsystem to zero, and selecting a value for the vigilance parameter between 0 and 1. It also requires the prescription of initial values of the bottom-up long term memory (LTM) trace b_{ij}, and top-down LTM trace t_{ji} between the i-th neuron in the F_1 layer and the j-th neuron in the F_2 layer (see Figs. 2a and 2b). As described in [11], b_{ij}'s were set to some low values as,

$$b_{ij} = \frac{L}{L-1+m} \tag{1}$$

where $L > 1$, and m is the number of neurons in the F_1 layer as determined by the dimensionality of the input pattern. All t_{ji}'s were initialized to unity.

$$t_{ij} = 1 \tag{2}$$

As shown in Figure 2a, with the gain G_1 set to unity, the "two-thirds rule" dictates that the activity of the F1 layer neurons denoted by the short term memory (STM) activity vector $X = (x_1, x_2, ..., x_m)$ be exactly equal to the input pattern $I = (I_1, I_2, ...I_m)$.

$$X = I \tag{3}$$

This activity vector is processed through an activation function to produce an output signal of the F_1 neurons as $S = (h(x_1), h(x_2), ..., h(x_m))$. This signal is processed through the bottom-up filtering process to generate an input signal pattern $T = (T_1, T_2, ..., T_n)$ for the F_2 layer, where n is the number of the neurons in the F_2 layer, and

$$T_j = \sum_i h(x_i) b_{ij} \tag{4}$$

The STM activity vector for the F_2 layer $Y = (y_1, y_2, ..., y_n)$ is assumed to be equal to T. Clearly, the most active neuron j* is the one corresponding to the largest T_j, and is the only one to be fired in the F_2 layer.

In Figure 2b, the output signal $U = (f(y_1), f(y_2), ..., f(y_n))$ from the F_2 layer generates a template pattern $V = (V_1, V_2, ..., V_m)$ through the top-down filtering, where

$$V_i = \sum_j f(y_j) t_{ij} \tag{5}$$

Since the lateral inhibition mechanism in the F_2 layer allows only one neuron to fire, the strength of the output signals from the F_2 layer is defined as

$$f(y_j) = \begin{cases} 1 & \text{if } T_j = \max \{T_k, k \in J\}, \\ 0 & \text{otherwise,} \end{cases} \tag{6}$$

where J is the set of all neurons in the F_2 layer that have not been reset in the current training cycle. At the beginning of each cycle, J includes all the n neurons. From Equation (5),

$$V = \left(t_{j^*1}, t_{j^*2}, t_{j^*m}\right) \tag{7}$$

Once the F_2 layer neurons have been activated, the gain G_1 is set to zero.

In the F_1 layer, the template pattern V is compared with the original input pattern I, to produce a new STM activity vector X* as follows.

$$X^* = I \cap V \tag{8}$$

The similarity between the patterns is measured in terms of the ratio $|X^*| / |I|$. A more meaningful measure of similarity would be one which would take into consideration a component by component comparison of vectors X* and I. Clearly, a ratio of 1 indicates a perfect match and a ratio of 0 denotes a complete mismatch. Frequently, a trained network may be presented with an input pattern that is a little different from previously encountered patterns. This may result in the excitation of an F_2 layer neuron, which, when fed back through the top-down filtering elicits an STM activity vector X* in the F_1 layer with a similarity lower than the specified vigilance level. When this happens, the orienting subsystem changes its status to unity, and disables the firing neuron in the F_2 layer. This would allow another recognition layer neuron to fire, in turn producing a new X* for which the similarity must be assessed. This process is continued until all stored patterns are examined. If no trained neuron meets the vigilance test, an uncommitted neuron is assigned the new pattern. It is worthwhile to note [11] that this search can be made more efficient if implemented on parallel processors.

The classification of input patterns is contingent upon the development of a network memory. This memory is embodied in the network weights, or LTM traces, which must be trained to ensure that similar input patterns result in the firing of the same F_2 layer neuron. This process of training is unsupervised in that there are no target patterns to emulate. Rather, the memory is built up as a result of successive reinforcement with similar patterns. The learning process can be categorized as belonging to either slow or fast training. In the former, the input patterns are presented so briefly that the weights, or LTM traces, do not reach their converged asymptotic values. In fast training, however, the network weights do actually attain their converged values. The mathematical details of these competitive learning models are presented in [11], and their essence summarized here for completeness.

The differential equations describing the dynamics of learning can be written as,

$$\frac{d}{dt} b_{ij^*} = K\left[\left(1 - b_{ij^*}\right)L - b_{ij^*}\left(|X^*| - 1\right)\right] \quad \text{if } I_i = 1, \tag{9}$$

$$\frac{d}{dt} t_{j^*i} = -t_{j^*i} + 1 \qquad \text{if } I_i = 1, \tag{10}$$

where K = O (1). For the fast training assumed in the present work, these LTM traces are considered to have attained their converged steady state values and are defined as follows,

$$b_{ij^*} = \begin{cases} \dfrac{L}{L - 1 + |X^*|} & \text{if } I_i = 1, \\ 0 & \text{otherwise,} \end{cases} \tag{11}$$

$$t_{j^*i} = \begin{cases} 1 & \text{if } I_i = 1, \\ 0 & \text{otherwise,} \end{cases} \tag{12}$$

which indicates that a steady state weight vector B_{j^*} (with components b_{ij^*}) is obtained as the error between this vector and the F_1 STM activity vector X^* is reduced. Typically, a steepest descent direction is chosen for the reduction of this error.

A Computer Implementation of ART Networks

A stepwise description of the present ART network implementation may be described as follows:

Step 1: Initialization

With a value of L = m/(m-1), the initial values of the bottom-up and the top-down LTM traces are prescribed according to Equations (1) and (2). A value for the vigilance level ρ is prescribed, and the gain controls and orienting subsystems are initialized.

Step 2: The input pattern I is presented to the network.

Step 3: I -> X -> S -> T

The bottom-up filtering is completed as described by Equations (3) and (4). In the present work, a simple threshold function was used as follows:

$$h(x_i) = \begin{cases} 1 & \text{if } x_i = 1, \\ 0 & \text{otherwise.} \end{cases} \tag{13}$$

Step 4: Y -> U -> V

With only one firing neuron in the F_2 layer as determined by the lateral inhibition (Eqn. (6)), the top-down filtering generates a template pattern as defined by Eqn. (7).

Step 5: Vigilance Test

Since the present work was restricted to binary inputs, an estimate of similarity between V and I is obtained as a ratio of the number of 1's in the X* and I as follows.

$$|X^*| = |I \cap V| = \sum_i I_i t_{j^*i} \tag{14}$$

$$|I| = \sum_i I_i \tag{15}$$

A more refined approach can be used which not only looks for the number of 1's, but also weights their relative locations in the vector.

If

$$\frac{|X^*|}{|I|} > \rho \tag{16}$$

an acceptable match is obtained. We transfer to Step 6. If this criterion is not met, the orienting subsystem disables the j*-th F_2 layer neuron, and the search domain is modified as follows:

$$J = \left(1, 2, \ldots, j^* - 1, j^* + 1, \ldots, n\right) \tag{17}$$

The process resumes with step 4 unless either a pattern similarity is established, or the pattern is confirmed to be unfamiliar and a new category is created.

Step 6: Adapt the LTM traces

The modifications of the bottom-up and the top-down LTM traces are made according to Equations (10) and (11).

Step 7: Go to Step 2

APPLICATIONS IN CONCEPTUAL DESIGN

The ART network described in preceding sections was used in the conceptual design of structural systems, including beams, trusses, and frame assemblies. There are two distinct design processes that were considered in the present work. The first encompasses a class of structures where the structural layout was generally known and the load and support conditions were allowed to vary. The second class of problems is one where the load and support points were assumed to be given, and the object of the design was to generate a near optimal structural topology. In either case, significant design experience with similar problems was assumed to be available, and dependent upon the identifiable critical features of the problem, a distinct procedural design process associated with the problem could be formulated. The problem may therefore be best understood as a use of ART networks to provide a memory capacity or a knowledge base for design, from which information can be recovered upon presentation of relevant features. The approach, therefore, draws upon the notion that human memory operates according to associative principles.

It is quite obvious that such an approach would be most meaningful if the problem could be described in terms of distinctive and discrete features, including an unambiguous statement of the associations between these attributes. If such a detailed description were indeed available, the problem would be amenable to solution methods available in rule-based systems.

Neural networks, however, attempt to build such associations by varying degrees of connectivity between the neurons, and by adjusting the strength of these connections. For the present discussion, we will assume that the critical problem features, for the class of structural design problems considered, are distinct and discrete. As different combinations of feature values are presented to a network, it either creates a distinct category, or classifies the feature set as belonging to an existing category. In a structural design problem, distinct features may correspond to types of load or support conditions. Once a combination of such features is presented to a network and categorized, an associated procedural processes may be invoked to generate the appropriate design.

As described in earlier sections, the ART network implemented in the present work can only accept binary input. The representation of critical features in a form compatible with ART networks, and the influence of network parameters on the ability of the network to train and classify, are discussed next in the context of representative structural design problems.

Beams of Uniform Strength

The first problem considered in the present work is the uniform strength sizing of an end-supported beam of length L, and a rectangular cross section as shown in Figure 3.

For an allowable bending stress of σ_{al}, the beam section dimensions can be obtained as follows:

$$h = \sqrt{\frac{6M}{b\sigma_{al}}} \quad \text{if b remains constant.} \tag{18}$$

$$b = \sqrt{\frac{6M}{h^2\sigma_{al}}} \quad \text{if h remains constant.} \tag{19}$$

In the above expressions, the bending moment M is directly related to the support conditions and the applied loads. The type of loads and supports can thus be considered as critical features of the problem. Combination of these features result in a specific moment distribution, and hence a distinct sizing of the beam section for uniform strength. In the present work, the problem features were represented by a 4-digit binary string; the relationship between the binary numbers and the physical features is summarized in Table 1. A total of 16 distinct combinations can be obtained in this representation. These combinations and their associated procedural processes are shown in Table 2.

For the comparison layer in the attentional subsystem, this problem requires a total of four nodes (one node per digit of the input pattern). The recognition layer in this subsystem should typically have as many nodes as the number of distinct patterns presented for categorization. The first step in the present work was intended to exercise the ART implementation, and to study the influence of the vigilance parameter on the network's ability to learn and classify the given patterns. Six different training strategies were considered and these are summarized as follows:

Strategy 1:

All 16 input patterns were presented to the network, and the vigilance parameter systematically varied to determine its influence. In general, with a higher vigilance level (i.e. finer level of distinction between patterns), more training cycles were required before attaining an equilibrium. During the course of training, a given input pattern may change the category to which it belongs. After equilibrium is attained, however, the category does not change further. As may be expected, a lower level of vigilance resulted in the creation of fewer number of categories. As a result of the choice of similarity definition between patterns (see Equation (16)), a continuous variation in the vigilance level does not result in a continuous variation in categories. Lastly, since ART networks treat all coded digits equally, no physical significance can be attached to clustering several training patterns in any single category.

Strategy 2:

In this strategy, the first 12 input patterns were first presented to the network, and training of the 16 available output nodes continued to an equilibrium state. Following this, the remaining 4 patterns were presented to the network. The general conclusions are the same as in strategy 1. In addition, for the same vigilance level, the number of categories obtained in either training strategy remained the same. However, the clustering order of these strategies did change.

Strategy 3:

In this case, after the first 12 input patterns had been classified by the system, one more pattern was introduced. If a higher vigilance level was enforced, additional cycles of training were required to attain equilibrium. This also resulted in a change in the clustering order. For lower vigilance levels, the network clustering order remained unchanged. The new pattern was either classified into one of the existing clusters or a new cluster was created.

Strategy 4:

The number of output nodes was fixed at 12 for this training strategy, restricting the total number of output clusters to 12. The network was trained with the first 12 input patterns. When a 13th pattern was presented to the network, it had to be forced into one of the existing clusters. This may be considered an approach where the network is asked to identify the closest category in which pattern 13 belongs. For a vigilance parameter greater than 2/3, a general reclustering of all patterns was found. For lower vigilance levels, conclusions similar to strategy 3 may be drawn.

Strategy 5:

Here, as in strategy 4, the number of output nodes was restricted to 12, and the network trained with 12 input patterns and a high vigilance level. When the 13th input pattern was presented with a lower value of the vigilance parameter, the clustering achieved was similar to one that was obtained by maintaining a higher vigilance level during the presentation of the 13th pattern.

Strategy 6:

This is similar to strategy 5, with the exception of reversal in vigilance levels between the initial training with 12 patterns, and subsequent clustering of the 13th pattern. The conclusion is quite different from that in the Strategy 5, and somewhat similar to that of Strategy 4.

Results of these strategies are summarized in Tables 3-8.

To further explore the extent to which the vigilance level influences clustering of input patterns, consider the simply supported beams shown in Figure 4.

The loads and supports correspond to the input pattern [0 0 0 0] of the previous example. This beam was further divided into ten segments as shown in the figure. The presence of a 0 in any segment indicates that there is no load, while a 1 indicates a concentrated load at the center of the segment. An ART network was established with 10 neurons each in the comparison and the recognition layers of the attentional subsystem. Four patterns shown in the figure were presented to the network. With a vigilance level of 0.9 (fairly high), the network clustered these patterns into four distinct clusters as follows:

Pattern 1 ⇒ Category 9

Pattern 2 ⇒ Category 4

Pattern 3 ⇒ Category 8

Pattern 4 ⇒ Category 7

With the vigilance set at 0.8, the patterns were placed in only three categories.

Pattern 1 ⇒ Category 9

Pattern 2 ⇒ Category 4

Pattern 3 ⇒ Category 8

Pattern 4 ⇒ Category 8

Clearly, with a higher vigilance level, minor differences in input patterns are treated as critical parameters, and therefore clustered into different categories. For lower vigilance levels, such minor differences are treated as noise. For the four patterns considered in this example, the distribution of cross-sectional depth 'h' for a uniform strength sizing is shown in Figure 5.

Here, each concentrated load was taken to be 1,000N, b = 10cm, $\sigma_{al} = 140 \times 10^6 N/m^2$. Clearly, patterns 1 and 2 result in very different material distribution, while the loading in patterns 3 and 4 produces only minor differences. In order for these minor differences to be recognized, higher vigilance levels would have to be enforced.

Truss and Frame Structures

ART networks can be used to classify patterns at various stages of the design process. This was shown in the preceding examples where, at the first level of feature detection only the type of loads and supports was considered. Once a pattern was appropriately clustered, more details about the features (such as load distribution) could be presented to identify the appropriate procedural process for design. In the present example, the topology generation of truss and frame structures is considered as a problem at the first level of feature abstraction. The problem statement may be summarized as follows:

> "Given a set of external loads, their locations and orientations, and a set of supports and their locations, obtain a near optimal topology to transmit the loads to the available supports".

At the first stage of the design process, all assumptions and other characteristic problem features that are to be used in making design decisions should be identified. Alternative specification of these assumptions and features, each related to a different procedural design process, may be used in providing the input training sets for an ART system. As an example, consider a problem where a structural topology must be generated from cable, bar, and beam elements. One set of assumptions for this problem may be summarized as follows:

1. The only requirement of the structure for the given system of loads and supports is that it be equilibrium.
2. Weight of elements used in the design may be ordered as cable < bar < beam.
3. Element type is of greater significance in "optimizing" topology than element length. As an example, the topology of Figure 6a is preferred to that shown in Figure 6b.
4. Element weight is proportional to element length alone.
5. Only simple supports are provided to stabilize the load systems.
6. The problem is two-dimensional.

The procedural process for these set of assumptions and problem features can be stated as follows:

Given

Supports: $s(x_0, y_0), s(x_1, y_1), ..., s(x_m, y_m)$,
Loads: $f(x_0, y_0), f(x_1, y_1), ..., f(x_n, y_n)$,

develop a structural configuration that meets the requirements for equilibrium.

Process:

for $f(x_k, y_k)$, find two nearest supports $s(x_i, y_i)$ and $s(x_j, y_j)$, where the distance ik $\leq$ distance jk,

```
IF the applied load k is a concentrated load, THEN
   decompose force along directions ik and jk;
   IF any of these load components is tensile, THEN
     use a cable element;
   IF one of the load components is compressive,THEN
     IF the support to which load is directed is simple, THEN
        use a bar element;
     ELSE IF the corresponding support is clamped, THEN
             use a beam element; eliminate the element used
             for tensile component; a new flexible support is
             created at the point of load application (xk,yk);
     ENDIF
   ELSE IF both components are compressive,THEN
     IF one of the supports is of a clamped type, THEN
        use a beam to connect to the nearest support;
        a new flexible support is created at the point of load
        application (xk,yk);
     ELSE IF both supports are simple, THEN
             use two bar elements; a new simple support
             is created at the point of load application
             (xk, yk);
     ENDIF
   ENDIF
ELSE IF the applied load k is a moment,THEN
   IF both supports i and j are simple, THEN
      use a frame structure ikj to connect ik and jk;
   ELSE IF one of the supports is clamped, THEN
           for the nearest clamped support, use a beam
           element to connect k and the corresponding
           support;
   ENDIF
   a new flexible support is created at the point of load
   application (xk,yk);
ENDIF
repeat with f(xk+1, yk+1) for all n loads.
```

Another possible variation in problem features may be the requirement of structural system/component stability as the primary consideration in place of system equilibrium. An additional set of rules would have to be formulated if both equilibrium and stability of the final topology were to be considered. For purposes of ART implementation, one approach of coding the assumptions and problem features is as follows:

Assumption 1.

only equilibrium	⇒ 1
stability also	⇒ 0

Assumption 2.

element weight order cable<bar<beam	⇒ 1
otherwise	⇒ 0

Assumption 3.

element type is primary consideration	⇒ 1
element length is primary consideration	⇒ 0

Assumption 4.

weight is proportional to length	⇒ 1
otherwise	⇒ 0

Assumption 5.

only simple supports available ⇒ 1

both simple and fixed supports available ⇒ 0

Assumption 6.

two-dimensional ⇒ 1

three-dimensional ⇒ 0

Input patterns for ART classification can be created from the coding scheme described above. Some illustrative patterns are as follows:

(1 1 1 1 1 1), (1 1 1 1 1 0), (1 1 1 1 0 1), (0 1 1 1 0 1).

For the case where only equilibrium requirements are considered, the entire inference process above can be represented by the following ART input patterns and their associated processes.

(0 0 0 * *) => frame ikj; k becomes a new flexible support;
(0 1 * * *) => beam ik; k becomes a new flexible support;
(0 0 1 * *) => beam jk; k becomes a new flexible support
(1 * * 0 0) => cable ik; cable jk
(1 0 0 1 0) => bar ik; cable jk
(1 0 0 0 1) => cable ik; bar jk
(1 0 0 1 1) => bar ik; bar jk; k becomes new simple support
(1 1 * 1 0) => beam ik; k becomes a new simple support
(1 1 0 0 1) => cable ik; bar jk;
(1 1 1 0 1) => beam ik; k becomes a new simple support
(1 1 * 1 1) => beam ik; k becomes a new simple support
(1 0 1 1 0) => beam jk; k becomes a new simple support
(1 0 1 0 1) => beam jk; k becomes a new simple support
(1 0 1 1 1) => beam jk; k becomes a new simple support

In the above patterns, the first digit denotes the type of load, with 0 indicating a moment and 1 representing a point force; the second and third digits represent support types, with a 0 indicative of a simple support and 1 denoting a clamped support; the fourth and fifth digits represent types of load components along directions of ki and kj, with 0 and 1 denoting tensile and compressive components, respectively. The '*' denotes a wild card specification, and that either a 0 or 1 can be input with no effect on the result.

The process described above was implemented for the problem shown in Figure 7. Two different combinations of support types were assumed to demonstrate the influence of this selection on the final structural design, and the results are shown in Figures 8a and 8b. The first case considered was one in which each of the supports A, B, and C were simple supports. For the applied load X, A and B are the two nearest supports. Resolving the applied load in direction XA and XB, cable and truss elements were proposed to carry the tensile and compressive loads, respectively. Likewise for the applied moment Y, B and C are the nearest supports. Due to the inability of the hinge to carry applied moments, a frame BYC was proposed by the system. Following a similar line of reasoning, cable element ZY and bar element ZC were introduced to complete the structure shown in Figure 8a. If the supports B and C were assumed to be of a clamped type, two beam elements XB and YB were introduced to carry loads X and moment Y. Node Y is assumed to be an elastic support to which another beam ZY was added to carry load Z. ZYB thus becomes a frame structure. The final structure is shown in Figure 8b.

The above representation can be extended to include decisions concerning element selection on the basis of load and support locations and magnitudes of load. Both of these features are available as real-valued numerical data, and can be converted to binary form for use in the present ART implementation. Alternatively, ART networks capable of accepting real-valued input patterns can also be studied. In either case, it is evident that a decomposition of the problem into a larger number of problem features creates more categories, and also allows for a lesser degree of mismatch when categorizing a new combination of problem features that does not belong to one of the existing categories.

CLOSING REMARKS

The usefulness of ART based neural networks in problems of conceptual design of structural systems has been examined in the present paper. The approach is based upon a premise that for this class of design problems, it is possible to identify certain critical features that mandate the classification of the problem into distinct categories. ART networks may be trained to recognize combinations of problem features as belonging to a distinct category, where each such category has a unique procedural process associated with it, and which may be invoked to obtain the appropriate solution. An overview of ART network architecture and operation has also been included for completeness. These networks belong to the general category of neural networks that can be trained without supervision. The influence of network parameters on the ability to learn, has also been examined. Representative problems in structural design have been included to demonstrate the implementation of the proposed approach. It has also been shown that the traditional IF_THEN_ELSE format of representing design decisions can be more generally transferred into an ART network architecture. The use of continuous, real-valued input patterns in ART are currently under study, and results from these studies will be reported in forthcoming publications.

ACKNOWLEDGEMENTS

This work was supported under research grant NAG 3-1196 from the NASA Lewis Research Center.

REFERENCES

1. Hebb, D.O. (1949), *The Organization of Behavior*. New York: Wiley.
2. Rosenblatt, F. (1958). The Perceptron: A Probabilistic Model for Information Storage and Organization in the Brain. *Psychological Review*. **65**, 386-408.
3. Minsky, M. and Papert, S. (1969). *Perceptrons*. Cambridge, MA: MIT Press.
4. Kohonen, T. (1972). Correlation Matrix Memories. *IEEE Transactions on Computers*. **C - 21**, 353-359.
5. Anderson, J.A. (1972). A Simple Neural Network Generating an Interactive Memory. *Mathematical Biosciences*. **14**, 197-220.
6. Hajela, P. and Berke, L. (1990), Neurobiological Computational Models in Structural Analysis and Design, 31st AIAA/ASME/ASCE/AHS/ASC SDM Conference, Long Beach, California.
7. Hajela, P. and Berke, L. (1991), "Neural Network Based Decomposition in Optimal Structural Synthesis," International Conference on Computational Structures Technology, Edinburgh, U.K.
8. Grossberg, S. (ed.) (1987). *The Adaptive Brain, Vol. I and II*. Amsterdam, North-Holland: Elsevier.
9. Grossberg, S. (ed.) (1988). *Neural Networks and Neural Intelligence*. Cambridge, MA: MIT Press.
10. Carpenter, G.A. and Grossberg, S. (1988). The ART of Adaptive Pattern Recognition by a Self-Organizing Neural Network. *Computer*, **21**, 77-88.
11. Carpenter, G.A. and Grossberg, S. (1987). A Massively Parallel Architecture for a Self-Organizing Neural Pattern Recognition Machine. *Computer Vision, Graphics, and Image Processing*, **37**, 54-115.

Table 1. Binary representation of problem features

A	0	The left end is a simple support.
	1	The left end is a clamped support.
B	0	The right end is a simple support.
	1	The right end is a clamped support.
C	0	The load is a force.
	1	The load is a moment.
D	0	The load is concentratedly applied.
	1	The load is uniformly distributed.

Table 2. ART Codes and related procedure processes

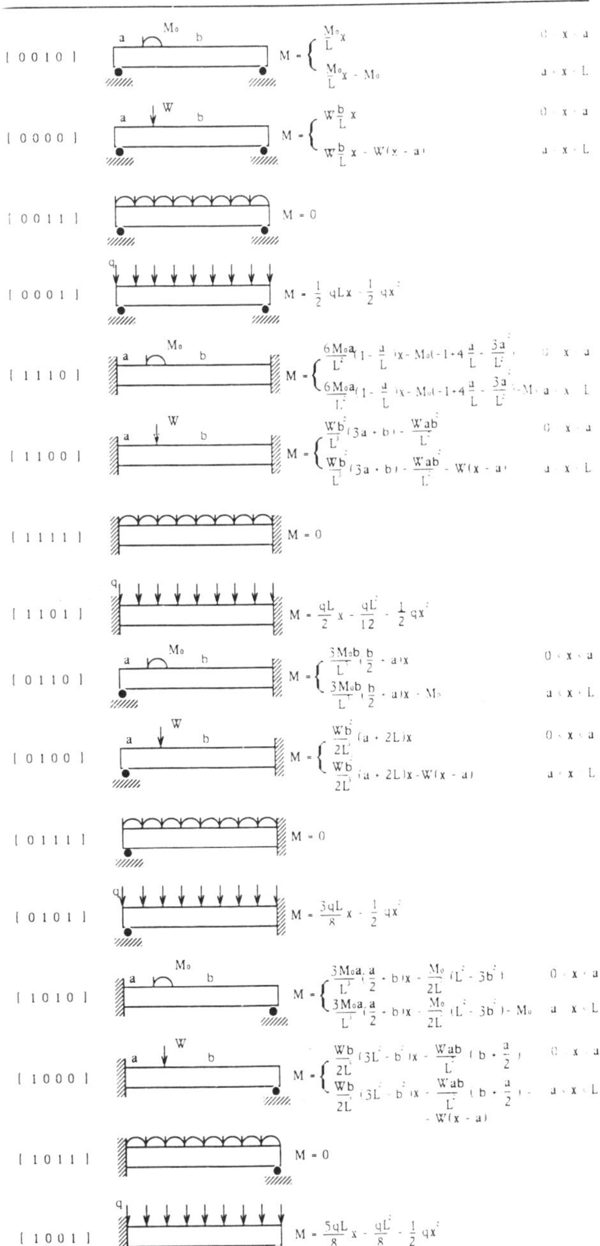

[0 0 1 0]	$M = \frac{M_0}{L}x$	$0 < x < a$
	$M = \frac{M_0}{L}x - M_0$	$a < x < L$
[0 0 0 0]	$M = W\frac{b}{L}x$	$0 < x < a$
	$M = W\frac{b}{L}x - W(x-a)$	$a < x < L$
[0 0 1 1]	$M = 0$	
[0 0 0 1]	$M = \frac{1}{2}qLx - \frac{1}{2}qx^2$	
[1 1 1 0]	$M = \frac{6M_0a}{L^2}(1-\frac{a}{L})x - M_0(-1+4\frac{a}{L}-\frac{3a^2}{L^2})$	$0 < x < a$
	$M = \frac{6M_0a}{L^2}(1-\frac{a}{L})x - M_0(-1+4\frac{a}{L}-\frac{3a^2}{L^2}) - M_0$	$a < x < L$
[1 1 0 0]	$M = \frac{Wb^2}{L^3}(3a+b) - \frac{Wab^2}{L^2}$	$0 < x < a$
	$M = \frac{Wb^2}{L^3}(3a+b) - \frac{Wab^2}{L^2} - W(x-a)$	$a < x < L$
[1 1 1 1]	$M = 0$	
[1 1 0 1]	$M = \frac{qL}{2}x - \frac{qL^2}{12} - \frac{1}{2}qx^2$	
[0 1 1 0]	$M = \frac{3M_0b}{L^3}(\frac{b}{2}+a)x$	$0 < x < a$
	$M = \frac{3M_0b}{L^3}(\frac{b}{2}+a)x - M_0$	$a < x < L$
[0 1 0 0]	$M = \frac{Wb^2}{2L^3}(a+2L)x$	$0 < x < a$
	$M = \frac{Wb^2}{2L^3}(a+2L)x - W(x-a)$	$a < x < L$
[0 1 1 1]	$M = 0$	
[0 1 0 1]	$M = \frac{3qL}{8}x - \frac{1}{2}qx^2$	
[1 0 1 0]	$M = \frac{3M_0a}{L^3}(\frac{a}{2}+b)x - \frac{M_0}{2L}(L^2-3b^2)$	$0 < x < a$
	$M = \frac{3M_0a}{L^3}(\frac{a}{2}+b)x - \frac{M_0}{2L}(L^2-3b^2) - M_0$	$a < x < L$
[1 0 0 0]	$M = \frac{Wb}{2L^3}(3L^2-b^2)x - \frac{Wab}{L^2}(b+\frac{a}{2})$	$0 < x < a$
	$M = \frac{Wb}{2L^3}(3L^2-b^2)x - \frac{Wab}{L^2}(b+\frac{a}{2}) - W(x-a)$	$a < x < L$
[1 0 1 1]	$M = 0$	
[1 0 0 1]	$M = \frac{5qL}{8}x - \frac{qL^2}{8} - \frac{1}{2}qx^2$	

Table 3. Strategy 1 (n = 16)

Vigilance Level	Cycles of Training	Number of Categories
0.7	4	16
0.6	3	12
0.5	3	10
0.3	2	6

Table 4. Strategy 2 (n = 16)

Vigilance Level	Cycles of Training		Number of Categories	
	for the first 12 patterns	for the remaining 4 patterns	for the first 12 patterns	for the remaining 4 patterns
0.7	4	3	12	16
0.6	3	3	9	12
0.5	3	2	8	10
0.3	3	2	5	6

Table 5. Strategy 3 (n = 16)

Vigilance Level	Cycles of Training		Number of Categories	
	for the first 12 patterns	for the 13-th pattern	for the first 12 patterns	for the 13-th pattern
0.7	4	2	12	13
0.6	3	1	9	10
0.5	3	1	8	8
0.3	3	1	5	5

Table 6. Strategy 4 (n = 12)

Vigilance Level	Cycles of Training		Number of Categories	
	for the first 12 patterns	for the 13-th pattern	for the first 12 patterns	for the 13-th pattern
0.7	4	2	12	12
0.6	3	1	9	10
0.5	3	1	7	7
0.3	3	1	6	6

Table 7. Strategy 5 (n = 12, initial vigilance level = 0.7

Vigilance Level	Cycles of Training	Number of Categories
0.6	2	12
0.5	2	12
0.3	2	12

Table 8. Strategy 6 (n = 12, initial vigilance level = 0.3)

Vigilance Level	Cycles of Training	Number of Categories
0.7	3	12
0.6	2	9
0.5	2	6

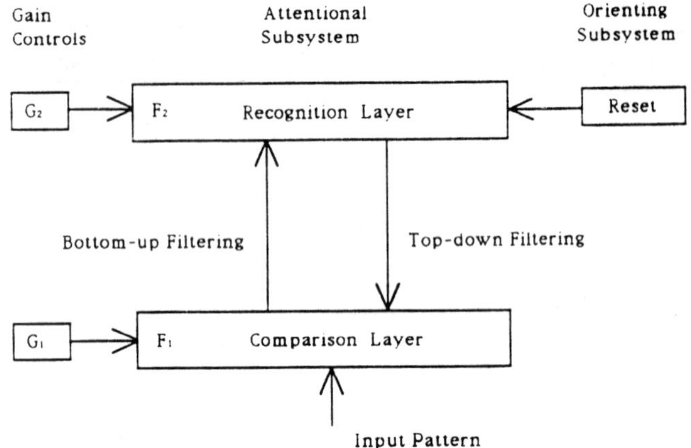

Figure 1. An ART network architecture

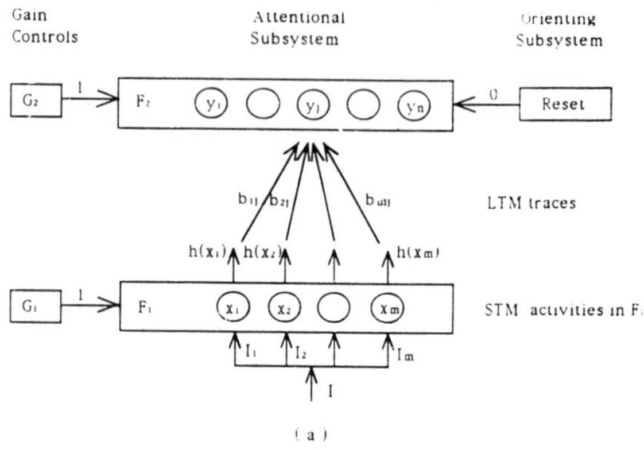

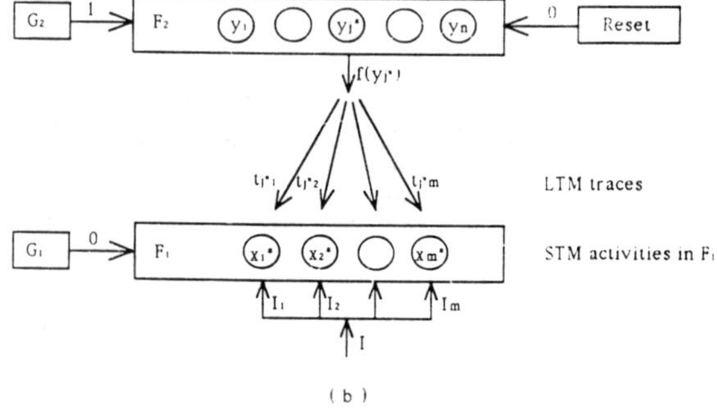

Figure 2. An implementation of ART networks

Figure 3. A beam with rectangular cross section

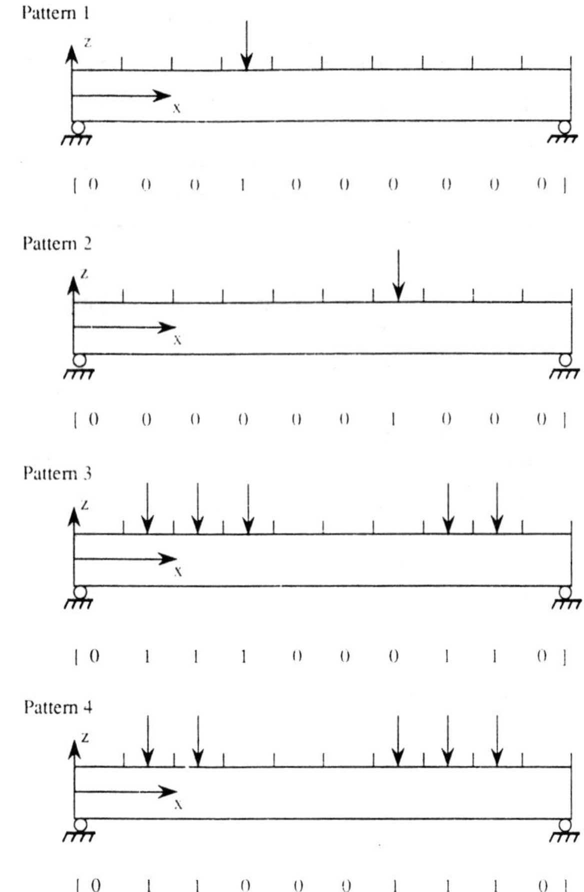

Figure 4. Simply supported beams and their corresponding ART codes

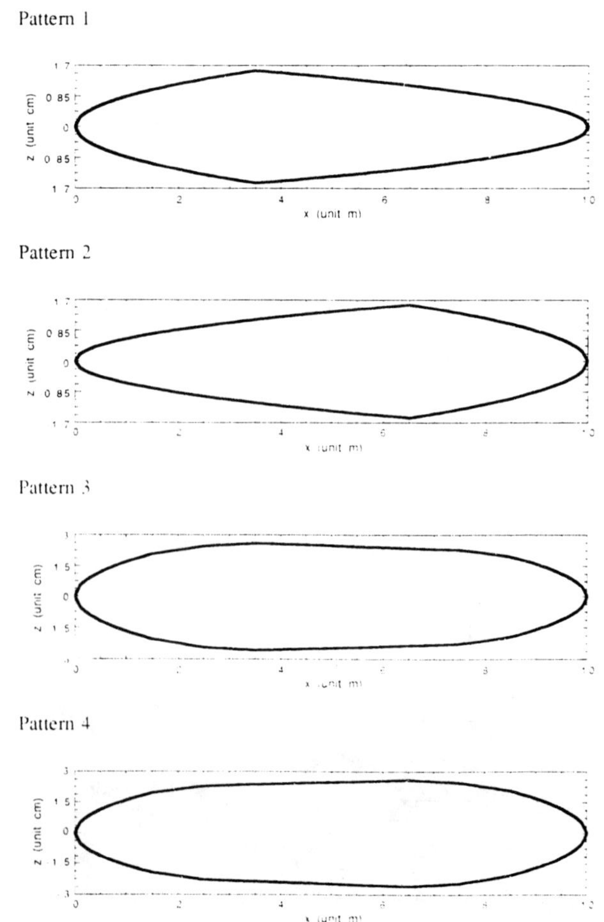

Figure 5. Optimal distribution of cross-sectional depth for uniform strength

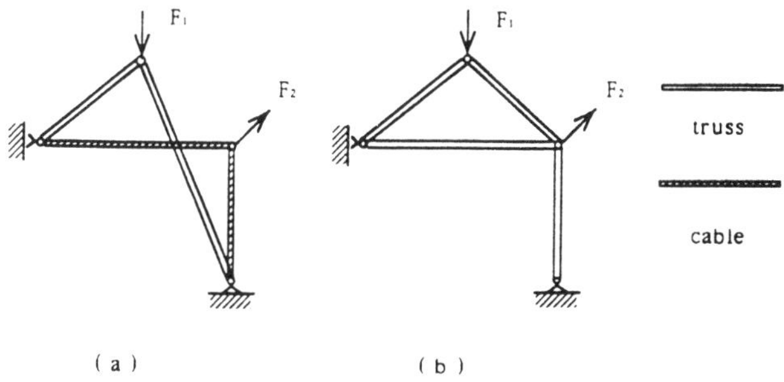

Figure 6. Assumption 3

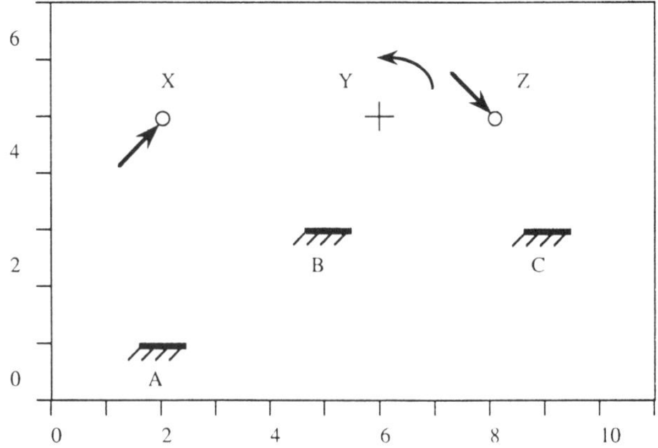

Figure 7. A problem domain with concentrated loads and moments

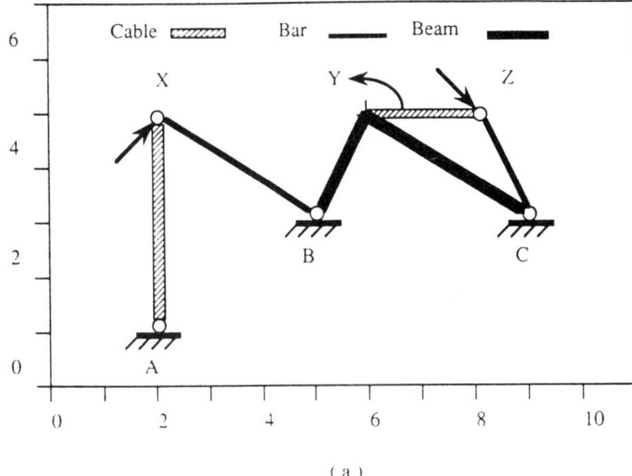

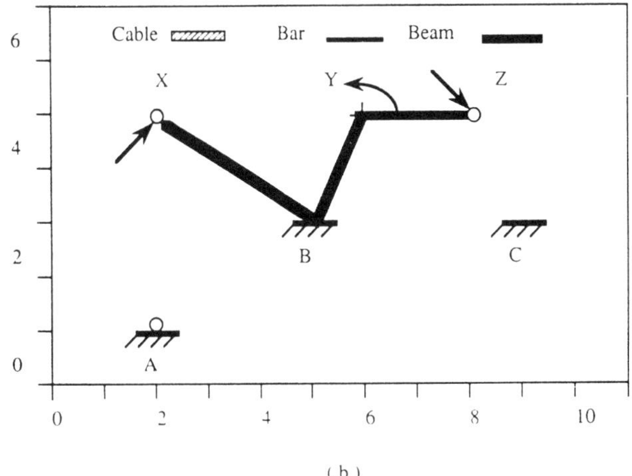

Figure 8. Solution to the problem shows in Figure 7
(a) With only simple supports
(b) With both simple and clamped supports

A NEURAL NETWORK SYSTEM FOR AESTHETIC DESIGN OF DAM STRUCTURES

H. Furuta, K. Sugiura, T. Tonegawa and E. Watanabe
Department of Civil Engineering, Kyoto University, Japan

In this study, paying attention to the steps of making master plan and evaluating dam configuration and view for the aesthetic design of dam structures, an attempt is made to develop a decision-making supporting system aided by neural network system and computer graphics. The present system was built on a 32-bit microcomputer whose main memory is 640KB and extended to 2MB RAM and Math coprocessor; therefore, anyone can use this system without difficulty. Using the neural network computation, it is possible to choose an adequate configuration from several alternatives without difficulty.

INTRODUCTION

Recently, the aesthetic design of structures has been focused because of the increasing needs to consider the view of the structures as one of important design conditions. In the past, perspective and photomontage techniques have been utilized to image the completion of structures in real state. However, this methods have needed a lot of efforts and are time-consuming. In addition, the view from other eye-points can be evalutated by repeating the same process. On the contrary, due to the remarkable developments of computer technology and data processing, the image of structures can be constructed easily on display, and the view from several different locations can be quickly created, so that this computer graphics(abbreviated CG) can be suitable and applied for the aesthetic design of structures at present time.

In this study, the applicability of CG for the aesthetic design of structural configurations and details is investigated. The gravity type of dam structure is taken as an example. First of all, computer simulation of the view of structures is implemented on a microcomputer, supplying the basic information of dam stuructures. CG can easily provide us with the many view of structures by changing detail configutrations[Ref. 1,2]. The knowledge base is constructed based on the answers of questions for experienced designers[Ref. 3]. The system can select the proper configuration among several alternatives by referring the knowledge base through neural network computation. It is the purpose to construct a system which is very easy to use on a microcomputer.

GENERAL DESCRIPTION OF THE SYSTEM AND VIEW SIMULATION

The system is constructed on a 32-bit microcomputer whose main memory is 640KB and extended to 2MB by ram board and Math coprocessor for faster processing, coupling with image scanner to take in the background behind the structures at construction site and video printer to print out the composed view for the view prospects. Therefore, the system can be purchased in low cost and anyone can use the system without difficulty.

In order to accomplish the high quality of view simulation on a microcomputer, Personal LINKS (Links Corp.)[Ref. 4] is utilized, which consists of Full-Color Frame Buffer, Rendering Tools (TRACY, RISCAN), Animation Tool, and two dimensional Paint Tool. This software can simulate the view by rendering the input data describing coordinates of objects, lights, and others. The input data can be categorized into five types: (1)cluster data describing relative position of objects; (2)shape data describing the shape and orientation of objects; (3)attribute data describing color and brightness of objects; (4)staff data describing the view point and light; and (5)environmental data to back up the operation of software. All of the input data should be editted in advance to simulate the view of structures.

The input data are prepared for the model structure as follows.

type : gravity dam
structural details : full overflow

Important is to make alternatives of detail configuration of dam structures. For each element such as surface conditon, shape of top structure, several alternatives are prepared and the over-all view of the structure can be composed by selecting designer's preference or by the learning process of neural network. As needed to store data set in the mass-strage previously , it is a difficult task in the microcomputer with limited memory.

The viewpoint and light source are assumed and edited as many as possible. In additon, brightness can be varied by using optional parameters of ATTRIBUTE EDITOR. The surface geometry of landscape can be assembled by patches. Based on the input coordinates of patches according to geological map, each patch is smoothened by B-spline. In additon, in order to create the natural view of mountains, Bump-mapping method is utilized.This can describe the surface roughness by the brightness of surface based on the light reflection. (Refer Fig. 1)

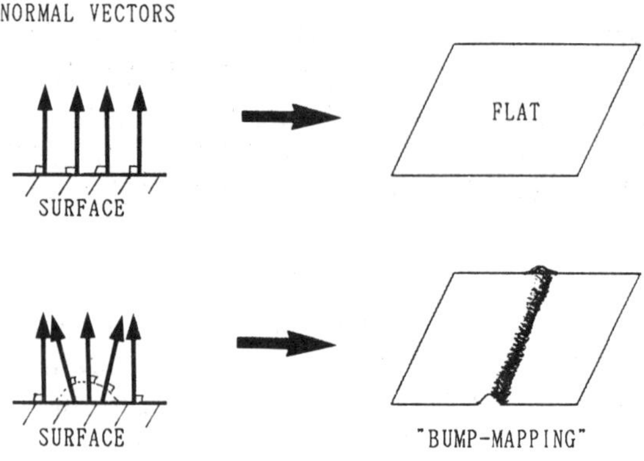

Fig.1 Bump-mapping.

In order to assess the applicability of the current software, an example will be given. First of all, the two-dimentional view of existing dam is recreated on display by using image scanner to digitalize and modifying the overview of dam structures. The several alternatives of surface configuration of dam are drawn up as shown in Photos. 1 and 2. To frame this portrait took about a couple of hours. In general, the process to modify the shade and overlap of each objects needs experiences. Therefore, this method is recommended to be used in creation of image, not to make a good view.

Photo.1 Existing gravity dam.

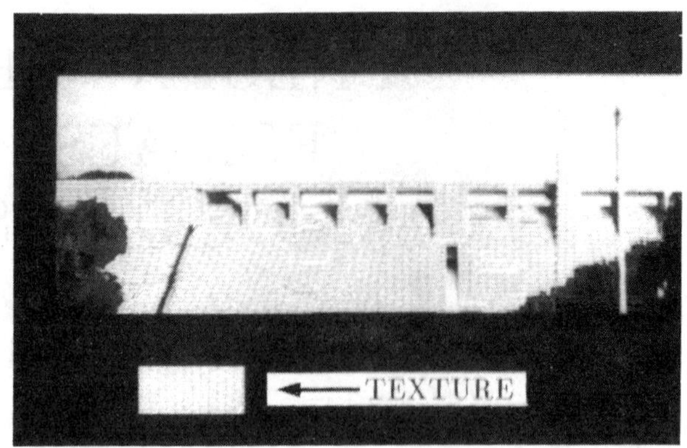

Photo.2 Two dimensional modification

Secondly, the configuration of dam structures is input according to the drawing made in design. Then, the color and brightness of dam structures can be editted interactively by using the ATTRIBUTE EDITOR. Taking the background by scanning the photograph (shown in Photo. 3) of expected dam site, the composed view of dam and background is constructed on display. After the minor modification on the superimposed view, the final simulated view is shown in Photo. 4. About 25 minutes was taken to construct a new view. This flow is shown in Fig. 2.

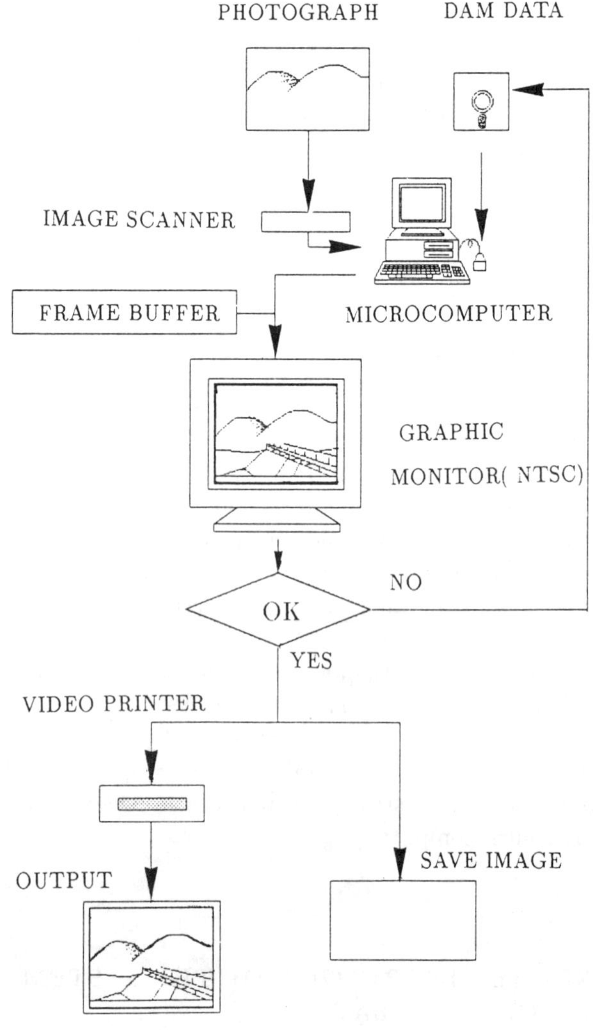

Fig.2 Flow of view simulation.

Photo.3 Background.

Photo.4 The compotion of background and CG image of dam structure.

Thirdly, in addition to the data describing the structures, configuration of construction site is also the input. The data are created by digitalizing the three-dimensional position of lattice on topographical map. This is the most advanced and preferable because the visual point can be set arbitrarily and different view can be easily constructed as shown in Photo. 5. The reality of view is of course superiority in taking the background from photos, but it needs the same number of photographs as the number of visual points.

Photo.5 Example of rendering.

The applicability of CG in the field of civil engineering, depends on the compatibility of data transform with CAD(Computer Aided Design). Few software compatible to CAD software is available, and then different software have been used in the past. It is obvious that the view with reality made by animation utility is not required and that it is difficult because of time limitation in design. It is desired that the application of CG depends on the required quality of construction project concerned with aesthetic assessment of structures.

APPLICATION OF NEURAL NETWORK

The neural network is developed by mathematically modelling the neuron system. Recently, numerous research works are being carried out in the field of the analysis of sentence structure, voice recognition, portrait process and so on. The purpose of such network is to create the system which can make the pattern recognition and judgement automatically based on the knowledge base. For example, at first; instructional cause and effect signals are used to construct the knowledge base, which are voices, characters and figures, and arbitrary signal may be identified by checking or learning the causality of input data. The authors have applied this system to the selection of repairing method of existing damaged structures[Ref. 5], and the efficiency to use neural network has been shown.

In order to carry out a neural network computing on microcomputer coupled with CG, RHINE(CRC Corp.)[Ref. 6] is utilized. A brief outline of the neural network used in RHINE is as follows;

Learning Method	: back-propagation
Parameters for learning	: synapses weight, threshold values
Transfer function	: sigmoid function

$$S_{out} = \frac{1}{1 + exp(-\frac{x-\theta}{T})} \tag{1}$$

where, x, S_{out}, θ and T are summation of input data, output data, threshold values and temperature respectively. An example is given concerned with color recognition at first. The color will be composed by mixing the three primary colors, namely, Red(R), Green(G) and Blue(B). Personal LINKS can change the intensity of each color by 256 steps, then 16.7 million kinds of color can be simulated. In neural network computing, n of effect factors, A_j $(i = 1, 2, ...n)$ and m of cause factors, B_j $(i = 1, 2, ...m)$ are given and if only A_l and B_k are 1, they are expressed as follows;

$$(A_1, A_2, .., A_l, .., A_n) = (0, 0, .., 1, .., 0) \tag{2}$$

$$(B_1, B_2, .., B_k, .., B_m) = (0, 0, .., 1, .., 0) \tag{3}$$

The causality of 8 effect factors(Red, Green, Blue, Black, White, Sky Blue, Yellow, Violet) and 3 cause factors are evaluated. If Violet is chosen as follows;

$$(R, G, Blu, Bla, W, SB, Y, V) \doteq (0, 0, 0, 0, 0, 0, 0, 1) \qquad (4)$$

The computer uses the combination of

$$(R, G, B) = (1, 0, 1) \qquad (5)$$

A simple back-propagation network architecture for solving the X-OR problem.[Ref. 7] The inputs will be 0's or 1's in either of the input nodes, and we want the output to be 1 if the input pattern is (0,1) or (1,0). In additon to adapting the connection weights, the thresholds for activating both the hidden and output nodes as shown Fig. 3. can also undergo adaptation, again using the back-propagation learning rule to find optional values.

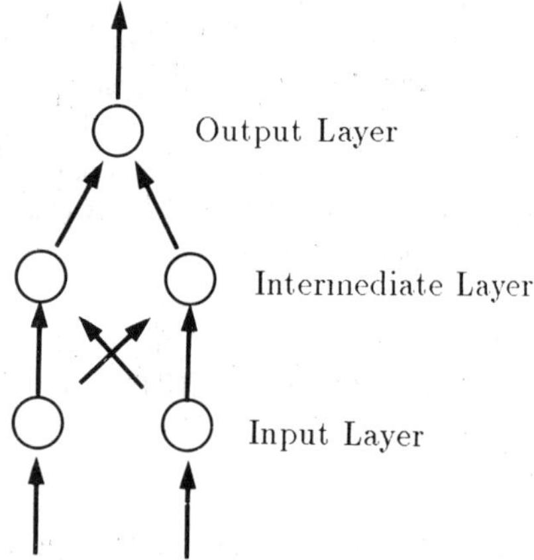

Fig.3 A back-propagation network to solve the "X-OR" problem.

The applicability of RHINE is examined by the color tests. Input data of causality is shown in Table. 1. The eight kinds of colors have been learning by RHINE, and it took almost 5 minutes. By using Personal LINKS, several colors similar to given 8 kinds are created and checked by RHINE, and the results for recognitions of colors are given in Table. 2. There exists a little difference of recognition from exact answers. However, it makes no difference just by eye-checking, so that it can be applied without any practical problem(Refer Photo. 6).

Table.1 Input data of causality.

EFFECT FACTORS								CAUSE FACTORS		
R	G	Blu	W	Bla	SB	V	Y	R	G	B
1	0	0	0	0	0	0	0	1	0	0
0	1	0	0	0	0	0	0	0	1	0
0	0	1	0	0	0	0	0	0	0	1
0	0	0	1	0	0	0	0	1	1	1
0	0	0	0	1	0	0	0	0	0	0
0	0	0	0	0	1	0	0	0	1	1
0	0	0	0	0	0	1	0	1	0	1
0	0	0	0	0	0	0	1	1	1	0

Table.2 Results for recognitions of colors.

	R	G	B
R	0.96	0.08	0.08
G	0.06	0.96	0.08
Blu	0.04	0.09	0.93
W	0.93	0.98	0.92
Bla	0.08	0.05	0.06
Y	0.99	0.96	0.08
V	0.92	0.05	0.92
SB	0.02	0.92	0.93

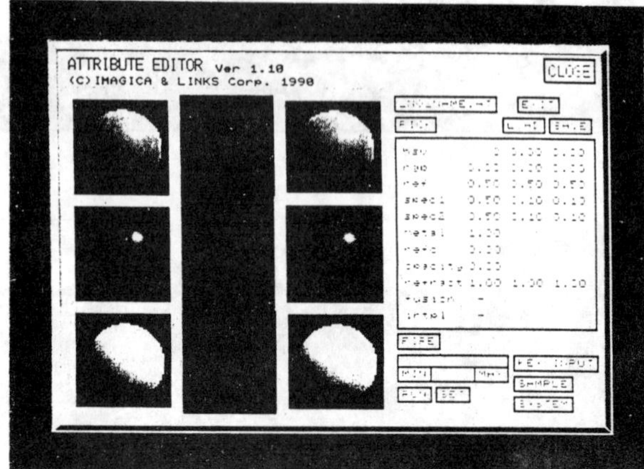

Photo.6 Comparison of exact and simulated color.

Several factors affecting the view have been chosen from reference[Ref. 3]. In the aesthetic design of dam structures, the harmony of structures to the environment is most important. Because of developing the supporting system of design, the following factors have been considered:

(1) Surface of Dam Body
(2) shape of gate piers
(3) shape of intakes
(4) design of handrails
(5) location and shape of lights
(6) shape of crown

For these facotrs, the 11 questions shown in Fig. 4 was

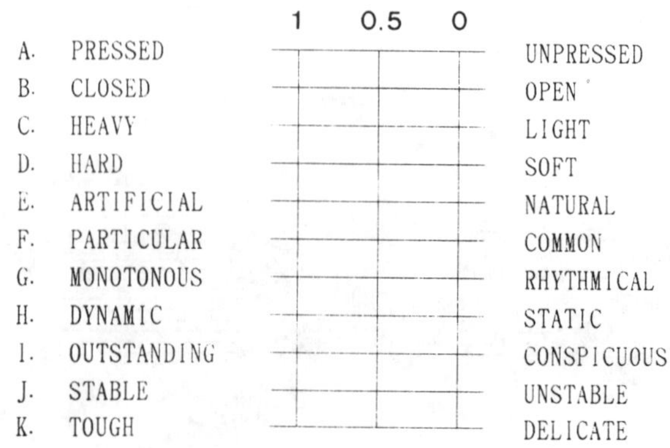

Fig.4 Several factors affecting the views.

distributed and asked to answer by 3 kinds of impression. For example, the 4 kinds of surface of dam body are simulated as shown in Photo. 7 and Table. 3; and then experienced designers have been impressed as "common" and "monotonous". The following data can be implemented:

$$(A, B, C, D, E, F, G, H, I, J, K) = (*, *, *, *, *, 0, 1, *, *, *, *) \qquad (6)$$

$$(a, b, c, d) = (1, 0, 0, 0) \qquad (7)$$

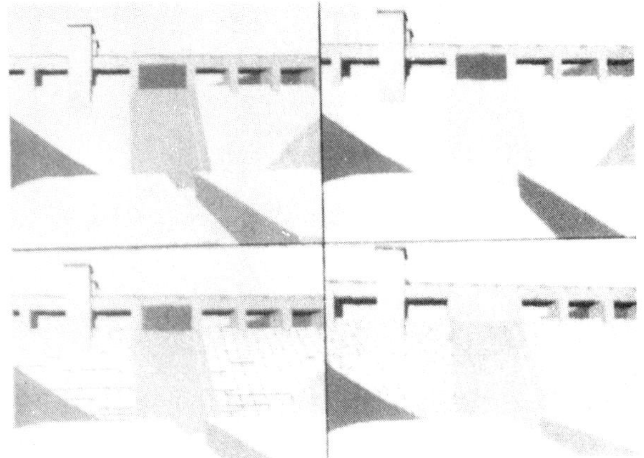

Photo.7 4 kinds of surface of dam body

Table.3 4 kinds of surface of dam body

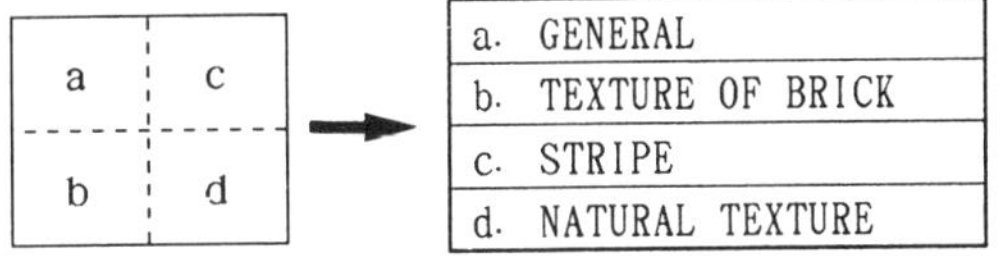

a. GENERAL
b. TEXTURE OF BRICK
c. STRIPE
d. NATURAL TEXTURE

in which, $* = 0.5$. Based on the answers from 10 designers, the knowledge base has been constructed. Aforementioned 6 items are considered and alternatives of structural details are constructed. By the interactive operation, where the user selects items and alternatives can be shown on display quickly, and finally the system select the desired view by using the neural network. For example, the factors such as "solid", "artificial", "monotonous" and "unattractive" are chosen;

$$(A, B, C, D, E, F, G, H, I, J, K) = (*, *, *, 1, 1, *, 1, *, 0, *, *) \qquad (8)$$

The result is obtained as follows:

$$(a, b, c, d) = (0.29, 0, 0.92, 0) \qquad (9)$$

This means that the "stripe " of surface is preferred. In the same manner, other items can be considered. Photos. 8, 9 and 10 shows the views from four different viewpoints simultaneously by separating the field of display into four blocks. Fig. 5 shows the flow of this system.

Photo.8 Example of rendering.

Photo.9 Example of rendering.

Photo.10 Example of rendering.

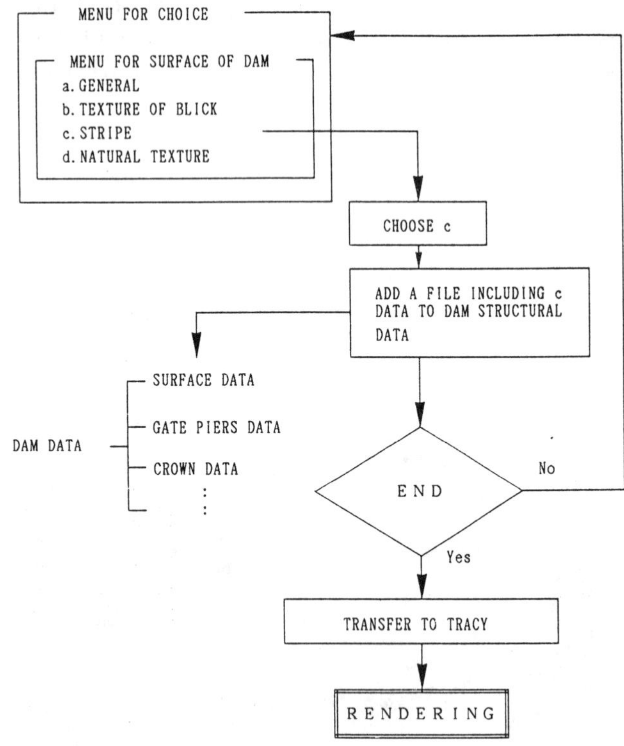

Fig.5 Flow of system.

CONCLUSIONS

The view simulation can be carried out with reality even by using the relatively cheap computer system(costs less than 20 thousands US$ for hardware and software). By reffering the knowledge base constructed based on the answers for questions distrubuted to experienced designers, the system can choose the proper configuration of structures according to designer's preference. In addition, the visualization of detail configuration can be made easily by using CG. The current study has considerd only the shapes of structures, so that it can be improved to the general-purpose system incorporating others factors such as colors, materials and environments.

REFERENCES

1. CENTER FOR TECHNOLOGY AND RESEARCH OF LANDSCAPE, "The AESTHETIC DESIGN OF DAM STRUCTURES," 1991.(in Japanese)

2. Abe, T., "Aesthetic Design of Miharu Dam," Dam Technology, No.51, 1990, pp.64-72. (in Japanese)

3. Furuta H., Ohtani, H., Nakabayashi, M. and Shiraishi, N., "Application of Neural Network to Aesthetic Design of Bridge Structures, Journal of Structural Engineering, JSCC, Vol.37A, 1991, pp.669-675.(in Japanese)

4. Links Corp., "Reference Manual of Personal LINKS," 1986.

5. Furuta, H., Ohtani, H. and Shiraishi, N., "Application of Neural Network to the Selection of Repairing Method," Proc. of Symposium on Numerical Computation, JSCC, Vol.14, 1990, pp.319-344. (in Japanse)

6. CRC Corp., "User's Manual of RHINE," 1986.

7. Maren. A, Harston. C, Pap. R, "Handbook of Neural Computing Applications," 1990. pp.60-63.

FUZZY REASONING WITH ENGINEERING APPLICATIONS USING NEURAL NETWORKS

Liu Xihui, Sun Baocheng and Feng Wenyuan
China Academy of Electronics and Information Technology, Beijing, PR China

This paper addresses itself to the expert system building technology by employing fuzzy reasoning and artificial neural network. An expert system building environment developed recently by authors named Fuzzy Neural Network Knowledge System f-NNKS is presented herein and its potential for engineering applications is discussed as well. The fuzziness, ranking itself amongst various kinds of uncertainties, of human knowledge is widely recognised in recent years for developing powerful expert systems with engineering applications.Though some methods have been proposed, the ability of present-day knowledge systems for handling such a kind uncertainty is still very weak. To make use of the artificial neural network for fuzzy reasoning is a hopeful,and new way.In order to utilize the neural network effectively, a fuzzy reasoning framework is formulated. The architecture of f-NNKS is described. The preliminary result of applying f-NNKS to earthquake induced damage forcasting is outlined as well.

INTRODUCTION

The management of various kinds of uncertainties in building expert system has attracted more and more attention of knowledge engineers for making effective use of artificial intelligence technology in engineering domains. Especially, the fuzziness, which takes place in the situation that the division of involved objects or concepts is vague and without crisp boundaries between categories of them, is widely recognised in the process for developing expert systems with engineering applications. The methodology of dealing with fuzziness is based on the theory of fuzzy sets and systems, which was first advanced by professor L.A.Zadeh some twenty years ago.Since then a number of fuzzy reasoning methods have been established for implementation of inference by making use of expertise of human experts' knowledge which always including fuzzy experience. The effectiveness of these methods was discussed by some researchers and theshortcoming of some of them was revealed during putting them to practical problems, since the problem is so complicated that there are many difficult points in knowledge acquisition,knowledge representation and knowledge utilization.In order to overcome the obstacles in fuzzy reasoning such as smoothenization of the membership functions after every reasoning step which often brings the users to a useless result from inference. So it is an urgent task for knowledge engineers to find a new way to deal with the fuzzy uncertainty more effectively.

The artificial neural network has been showing its powerful potential for developing a new generation of computer, and some excellent acheivements have been got in recent years.Preliminary results of applications to pattern recognition, information processing have been encouraging, but examples applied to knowledge systems, especially that incorporating fuzzy reasoning with neural computing is still lacking.Though almost every knowledge engineer extends the recognition of the comprehensiveness of fuzziness of human knowledge, the complicatedness of it makes the progress of combination of both into an effective system being much slower than in other fields. After serious investigation of the features of neural network,authors are comfirmed that there is not only a potential use of it, but the neural network itself is a much powerful tool for implementation of fuzzy reasoning if some appropriate ways to coordinate them is found.

This paper presents a knowledge system newly developed the Fuzzy Neural Network Knowledge System f-NNKS, which acts as a developing environment of intelligent systems such as expert systems, decision making support systems, carrying out fuzzy reasoning by employing artificial neural network. In order to get more reasonable results, a new framework of fuzzy reasoning, which is much suitable for neural computing is constructed. The system consists of some basic elements, each of them contains several neural networks organized by an input transformation unit and an output processing unit. The knowledge base of the system is divided into two parts-- a guiding knowledge base, which is a production system involving only a few rules, and several basic knowledge bases, each of which is in correspondence with a neural net work and represents a rule set with same antecedents and consequence but having different fuzzy values. As a result, the knowledge representation is getting much simpler. Since the neural net may be trained by samples, and the internally distributed parameters of the nets after training represent the knowledge of that sample set, so the system f-NNKS has an advantage over others of self-learning. The membership function of factors in the rule are adopted as trapezoid L-R fuzzy numbers, the problem of smoothenization, as encountered by many suggested fuzzy reasoning methods, is overcome.

A simple example,yet meaningful enough,the expert system for earthquake induced damage to various kinds of buildings is presented herewith for the purpose to examine its effectiveness.The architicture of f-NNKS,the framework of fuzzy reasoning proposed by authors and the application example are outlined in following paragraphes.

FRAMEWORK OF FUZZY REASONING

A new framework for fuzzy reasoning is suggested in this paper in order to make it more efficient to incorporate the neural network with fuzzy reasoning.The fuzzy reasoning may be viewed as following:

1. one dimension and one stage
The rule is written as

IF a is A THEN c is C (1)

The propositions in both antecedent and consequence are fuzzy sets, it may be rewritten as

$$A \longrightarrow C \qquad (2)$$

The inference from new known fact A' to get the new conclusion is

$$\begin{array}{l} A \longrightarrow C \\ A' \\ \hline \quad C' \end{array}$$

There are two important issues involved for carrying out the reasoning, which are implication A ---> C and composition operator.

2. multiple dimensions and one stage
The rule of multiple dimensions and one stage is in the manner:

IF a1 is A1, a2 is A2,..., am is Am THEN C (3)

where A1,A2,..Am and C are all fuzzy sets.Then the reasoning is

$$\begin{array}{l} A1,A2,\ldots,Am \longrightarrow C \\ A1',A2',\ldots,Am' \\ \hline \quad C' \end{array} \qquad (4)$$

3. multiple dimensions and multiple stages
This is the most common type of fuzzy reasoning, the rule sets are written as

$$\begin{array}{l} A1(1),A2(1),\ldots,Am(1) \longrightarrow C(1) \\ A1(2),A2(2),\ldots,Am(2) \longrightarrow C(2) \\ \ldots\ldots\ldots\ldots \\ A1(n),A2(n),\ldots,Am(n) \longrightarrow C(n) \\ A1',A2',\ldots,Am' \\ \hline \quad C' \end{array} \qquad (5)$$

Many suggestions to treat the multiple dimensions and multiple stages reasoning have been made. There are mainly two kinds of treatment methods as following:

$$\begin{array}{l} C'=[(A1',A2'\ldots,Am')\circ(A1(1),A2(1),\ldots,Am(1))\longrightarrow C(1)] \cup \\ \quad [(A1',A2'\ldots,Am')\circ(A1(2),A2(2),\ldots,Am(2))\longrightarrow C(2)] \cup \\ \quad \cup \ldots\ldots\ldots\ldots \cup \\ \quad [(A1',A2'\ldots,Am')\circ(A1(n),A2(n),\ldots,Am(n))\longrightarrow C(n)] \end{array} \qquad (6)$$

or

$$\begin{array}{l} C'=[(A1',A2'\ldots,Am')\circ(A1(1),A2(1),\ldots,Am(1))\longrightarrow C(1)] \cap \\ \quad [(A1',A2'\ldots,Am')\circ(A1(2),A2(2),\ldots,Am(2))\longrightarrow C(2)] \cap \\ \quad \cap \ldots\ldots\ldots\ldots \cap \\ \quad [(A1',A2'\ldots,Am')\circ(A1(n),A2(n),\ldots,Am(n))\longrightarrow C(n)] \end{array} \qquad (7)$$

The composition may be treated by two ways, that is

$$\begin{array}{l} C(i)'=(A1',A2,\ldots,Am')\circ(A1(i),A2(i),\ldots,Am(i)\longrightarrow C(i)) \\ \quad =[\,A1'\circ(A1(i)\longrightarrow C(i))] \cup [\,A2'\circ(A2(i)\longrightarrow C(i))] \cup \\ \quad \cup \ldots\ldots\ldots\ldots \cup \\ \quad \cup [\,Am'\circ(Am(i)\longrightarrow C(i))] \end{array} \qquad (8)$$

or

$$\begin{array}{l} C(i)'=(A1',A2,\ldots,Am')\circ(A1(i),A2(i),\ldots,Am(i)\longrightarrow C(i)) \\ \quad =[\,A1'\circ(A1(i)\longrightarrow C(i))] \cap [\,A2'\circ(A2(i)\longrightarrow C(i))] \\ \quad \cap \ldots\ldots\ldots\ldots \cap \\ \quad \cap [\,Am'\circ(Am(i)\longrightarrow C(i))] \end{array} \qquad (9)$$

Both two methods may be viewed as according to a common principle to treat the complicated problem, that is the principle "seperately first, then combining". That means to treat every rule seperately and to treat each proposition in the antecedents of every rule seperately first and then to combine the results got by this way. The key points are the selection of a suitable implication operator and the method to carry out the composition. Obviously, the criterion to formulate the procedure is that the conclusions must fit the way that human being inferences by making use of his knowledge. Some judgements have been made about the methods proposed and only a few are considered to be satisfactory.

In order to fully utilize the advantage of artificial neural network, a new framework of fuzzy reasoning is suggested in this paper. The fuzzy reasoning is considered as a mapping:

$$\begin{array}{l} N\!: F(U1)*F(U2)*\ldots*F(Um)\longrightarrow F(V) \\ \qquad (A1,A2,\ldots,Am) \mapsto C \end{array} \qquad (10)$$

where $Aj \in F(Uj)$, j=1,2,...,m and $C \in F(V)$. U1,U2,...,Um, V are universe of discourse of a1,a2,...,am and c, and Aj, C are fuzzy sets on Uj and V.

The mapping N is considered consisting several mappings:

$$\begin{array}{l} N1\!: F(U1)\longrightarrow F(V) \\ N2\!: F(U2)\longrightarrow F(V) \\ N3\!: F(U3)\longrightarrow F(V) \\ \ldots\ldots \\ Nm\!: F(Um)\longrightarrow F(V) \end{array} \qquad (11)$$

and

$$N0\!: \underbrace{F(V)\times F(V)\times\ldots\ldots\times F(V)}_{m}\longrightarrow F(V) \qquad (12)$$

The mapping Ni is induced by samples (Ai(1), C(1)), (Ai(2), C(2)),...(Ai(n),C(n)). Hence the inference from the new facts A1',A2'...Am' may get the conclusion C'(1),C'(2),...,C'(m).Then the mapping N0 synthesize all these conclusions into the final one C'. That is

$$C'=N0(C'(1),C'(2),\ldots\ldots,C'(m)) \qquad (13)$$

Perhaps we can say that the method proposed here is also along the way "seperately first, combining then". But as it will be shown this framework has more advantages over some others.

IMPLEMENTATION OF FUZZY REASONING BY EMPLOYING NEURAL NETS

The artificial neural network has given very strong impact to AI technology. Though its application to pattern recognition is successful, to fuzzy reasoning is very rare. Some examples have been published, which commonly utlised the feature that the output of a neuron is like a membership function since the activation function of the neuron is a sigmoid curve.In this paper, the mapping characteristics of neural nets are fully considered and used for the purpose of implementation of fuzzy reasoning.

The system f-NNKS consists of several basic elements(BE), each one of these contains an input transformation unit(ITU), a synthesis device(SD), an output processing unit(OPU),and a number of basic networks(BN). The number of basic nets is equal to m, that is the number of propositions in the antecedent of the fired rule during reasoning.

1) the fuzzy set of A(i), i=1,2,...,m and C in any rule is represented by L-R fuzzy numbers, of which membership functions are adopted as trapezoid. As a result, every menbership function can be represented by four characteristic points a1,a2,a3,a4 (or c1,c2, c3,c4) as shown in Fig 1.

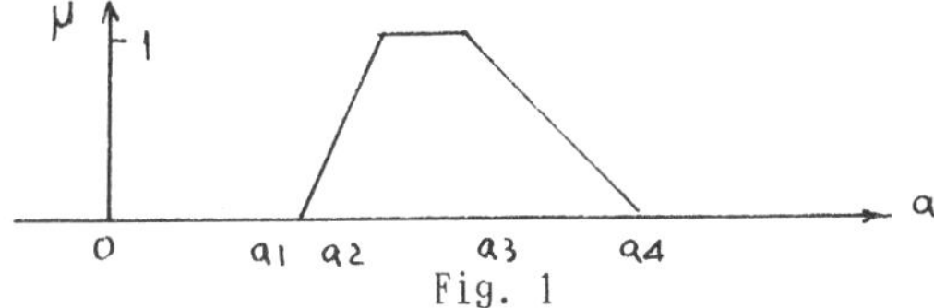

Fig. 1

2) input transformation unit(ITU)---the function of ITU is to transform the input A1', A2', ... , Am' into sets of a1(i)',a2(i)',a3(i)',a4(i)', i=1,2,...,m, and arrange those as the input of each basic network.

3) basic network(BN) --- the basic network consists of a couple of feedforword layered model of neural networks, Fig.2, each one of which conducts the mapping of one side of the L-R fuzzy numbers, that is a1', a2' to c1', c2'(or a3', a4' to c3', c4').So the output of all the basic nets are c1'(i),c2'(i),c3'(i), c4'(i), i=1,2,...,m. These sets of cj'(i), j=1,2,3,4, i=1,2 ...,m, constitute the fuzzy sets C(i)' of conclusion generated by mapping N1,N2,...,Nm as shown in Eq.(11). The weights wij between neurons i and j and threshold values θj of neurons in hidden and output layers generated after training by sample sets

$$\begin{matrix} [I(1) & O(1)] \\ [I(2) & O(2)] \\ \cdots & \cdots \\ [I(n) & O(n)] \end{matrix} \qquad (14)$$

where I(i) stands for A1(i),A2(i),....,Am(i), and O(i) for C(i), represent the knowledge reflected by that sample set.

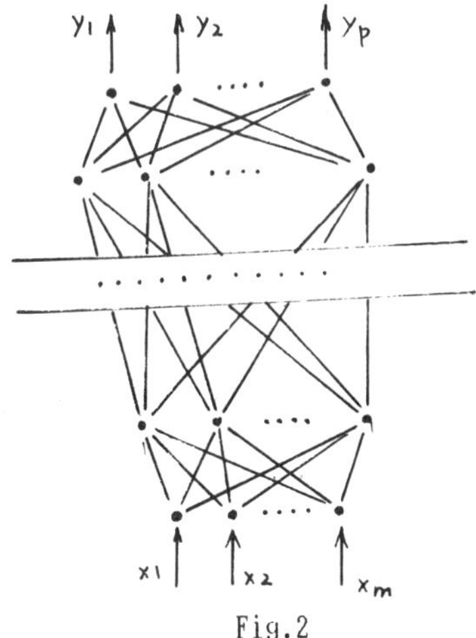

Fig.2

4) synthesis device(SD) --- it carries out the mapping from C(i)', i=1,2,...,m to C' as shown in Eq.(12) and (13). The method of synthesis will be given in detail later.

5) output processing unit(OPU) --- it can translate the results obtained by the synthesis device into natural language if the linquistic terms are used to input the new facts and there is a requirement to make decision by utilising linquistic terms.

The kernel of the basic element(BE) is the feature of BNs which conduct the mapping of Eq(11) and that of SD to combine the results C(i)' to C'.Whether the system f-NNKS will be succesful or not is determined by these two important points. The methods will be discussed here thoroughly.

The basic nets are feedforward layered models with only one hidden layer of neuron elements. The activation function (output function) of neurons in hidden and output layers is taken as a sigmoid curve, Fig.3. The relation of the input(supposed as x1,x2) and the output (supposed as y1,y2, for convenience's sake, only the input of output layer neurons g1,g2 are discussed) may be shown in Fig.4.

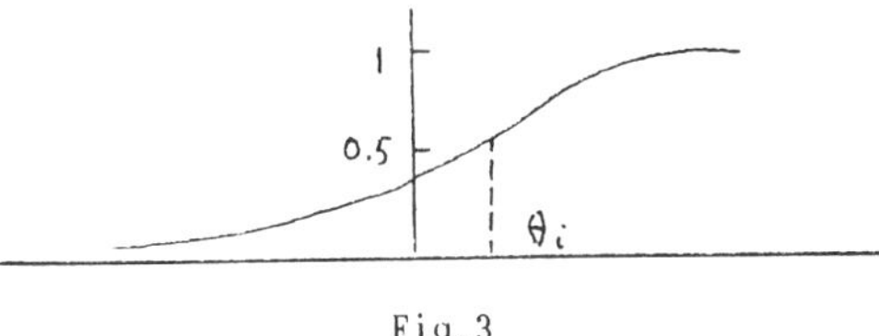

Fig.3

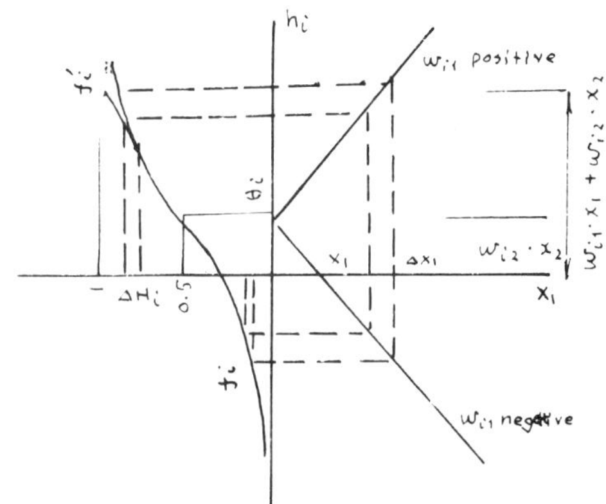

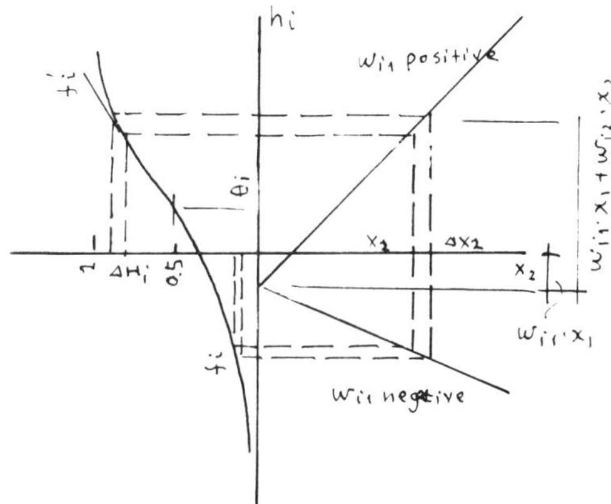

Fig.4

The input g1, g2 of output layer elements transmitted from hidden layer can be written as follows:

$$\Delta g1(t)=\sum_{i=1}^{K} w1i'[wi1*fi'|_{x_1(t-1)}\Delta x1(t)+w2i*fi'|_{x_2(t-1)}\Delta x2(t)] \quad (15)$$

$$\Delta g2(t)=\sum_{i=1}^{K} w2i'[wi1*fi'|_{x_1(t-1)}\Delta x1(t)+w2i*fi'|_{x_2(t-1)}\Delta x2(t)]$$

where $\Delta x1(t)$, $\Delta x2(t)$ are the increments of input x1 and x2 from the sample (t-1) to t, the similar meaning are $\Delta g1(t)$, $\Delta g2(t)$. fi' indicates the tangent of activitation function curves of hidden element i. What the Eq. (15) tells is that the basic net(BN) possesses the ability of interpolation, that is the input x1', x2' in Fig.5 will bring about the output y1' and y2', if the number of samples is large enough. As a result, the following features are worth pointing out:

1) the membership functions will not be flattened after every inference step.

2) the input is fuzzier, the output is fuzzier too. This is very satisfactory for an effective model of fuzzy reasoning.

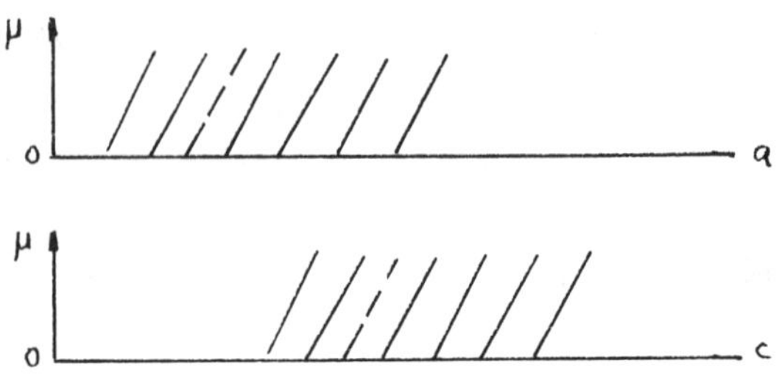

Fig.5

3) the output interpolates at the similar place of input locating in training sample set. This feature constitutes the essential of the mapping, as well as the fuzzy reasoning implemented by f-NNKS.

The procedure of fuzzy reasoning accomplished by BE may described as follows. The facts A1', A2',...., Am' that are put in by the user or transmitted from other BEs are processed first by ITU forming the input of BNs. What the BNs do is to bring about C'(1), C'(2),...., C'(m). Then SD completes the task to generate C' from C'(i),i=1,2,...,m. From the standpoint of view of knowledge representation, the BE keeps the knowledge of a rule set shown in Eq.5, so it causes the knowledge representation much simpler. Fig.6 showsthe construction of BE.

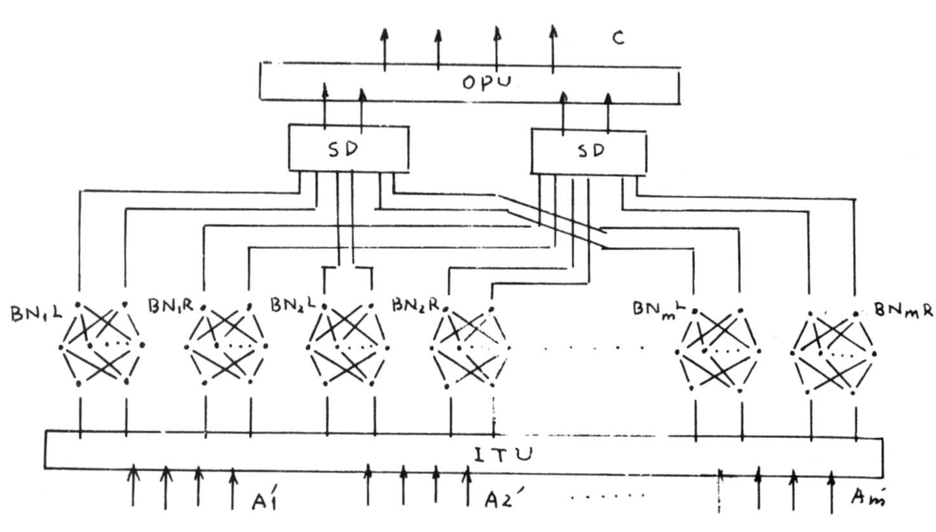

Fig.6

ARCHITECTURE OF f-NNKS

To illustrate the architecture of f-NNKS, suppose the reasoning tree is like that as shown in Fig.7. The knowledge base is devided into two parts: a guiding knowledge base which contains only the rules as following:

IF A1,A2,...,Am THEN C
IF C,D,G THEN M
IF F1,F2,..,Ft THEN J2 (16)
IF E,H THEN J2
IF J1,J2,...,Js THEN L
IF L,K THEN M

and six basic knowledge bases, each of them corresponding to a BN and a rule set read as (for instance, rule 1 in Eq.16).

IF A1(1),A2(1),...,Am(1) THEN C(1)
IF A1(2),A2(2),...,Am(2) THEN C(2) (17)
....
IF A1(n),A2(n),...,Am(n) THEN C(n)

So the architecture of f-NNKS may be depicted in Fig 7. The adoption of twofold structure of f-NNKS manifests wonferful advantage that a production system embedded in it contains only a few rules which guide the search process automatically, and several BNs are embodied to carry out the fuzzy reasoning by employing neural nets.

f-NNKS is developed into twin systems, one is for training another is for reasoning.It has a friendly interface to show every step of reasoning, and prompt users to respond if there is a requirement to ask the input. It is important to point out that the users' input during the process of inference are A1,A2,...,Am,D,G, E,H, (or F1,F2,..,Ft), J1,J3,...,Js,and K. But the sample for training must involve C,J2,L,M.

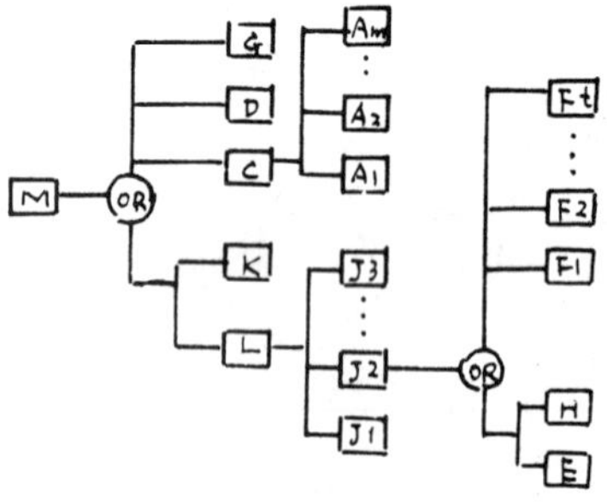

Fig.7

APPLICATION TO EARTHQUAKE ENGINEERING

An example to apply f-NNKS to practical problem is expressed herewith. The example is rather simple, yet meaningful enough to examine the efficiency of the system. The topic chosen for demonstration is forcasting earthquake induced damage to various kinds of buildings. This problem is of great importance for earthquake-prone areas. The preparedness for expected earthquake attack in coming time span is based on the prediction of potential disaster caused by possible strong ground motion.The plan for disaster mitigation, especially that of urban region, requires the information about damage to various kinds of buildings. The estimation of damage situation can only be made by carefully using experts' knowledge, which always possesses ambiguous and vague expressions of the relationship between the causes and consequences. For example, there may exist rules like:

If the site soil is solid soil
and the distance from active fault is rather far
then the site condition is considered being very good

if the site soil is rather weak
and the distance from active fault is rather far
then the site condition is considered being in general case

So the knowledge engineering places an important role in relevant fields of anti-earthquake studies.

The causes of damage to specific building may be classified into two groups. One is attributed to the seismic environment, another to the resistance capacity of building itself. The knowledge of Building Damage Prediction BDP-1 is depicted in Fig.8. The left branch is with regard to the resistance capacity of specified building, while the right being in relation with the seismic environment.

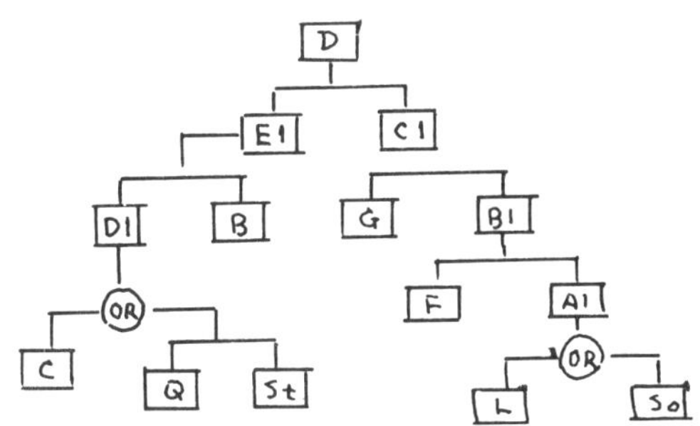

Fig.8

The guiding knowledge base embraces rules:

rule 1. IF So THEN A1
rule 2. IF L THEN A1
rule 3. IF A1, F THEN B1
rule 4. IF B1,G THEN C1 (18)
rule 5. IF Q,St THEN D1
rule 6. IF C THEN D1
rule 7. IF D1,B THEN E1
rule 8. IF C1, E1 THEN D

The symbols in the rule set stand for following meaning respectively:

So -- soil type of building site
L -- liquifaction potential
F -- distance apart from active faults
G -- expected ground motion
St--strength of load-bearing structure of the building
Q -- construction quality
C -- lateral force resistance capacity of main structure
B -- type of building
D -- damage to the building

There are still several factors in the reasoning tree, such as A1,B1,C1,D1,E1. All these have not explicit meaning, they only stand for the results induced from lower level knowledge. Hence the problem for training emerges, since the sample sets must include all these factors other than those with explicit meaning. So the knowledge acquisition requires great care.

Obviously, there are five basic elements (BEs) corresponding to rule 3,4,5,7,8.The collection of samples for training is a painstaking work. Generally speaking, experts' experience can only be expressed by linguistic terms. Fuzzy sets for descriptions of these linguistic terms should be carefully chose. For instance, rule 3 embodies two factors in the antecedent A1,F. The site condition A1 may be considered as very bad, bad, rather bad, not so good, general, rather good, good, very good, very very good, and so on.The distance apart from active fault may be grouped into categories as very very far, very far, far, not so far, medium, rather near by, near, very near, and so on. The grades of concerned factors perhaps are not equal to each other. Some special treatments are needed. For instance,some intermediate classes may be added in. The rule 3 may be shown as in Fig. 9.

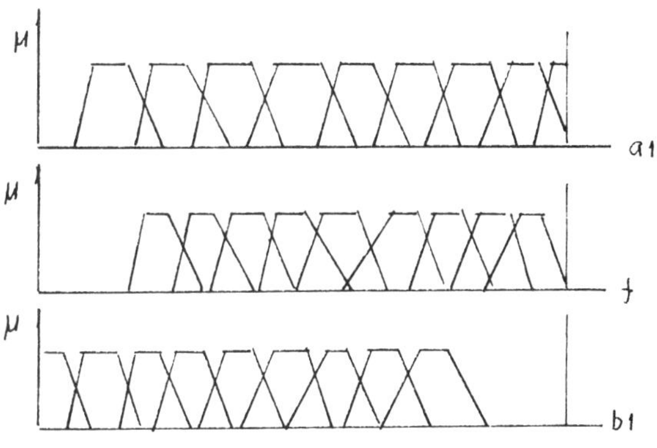

Fig.9

The problem of combination explosion is serious for every knowledge engineer.In the case of Fig.9, the rule number may be as high as 10x10x10. How to extract most useful knowledge from expertise of domain experts is tough. The implicit principle is the coverage of knowledge base represented by rather less rules over those which are not included in the knowledge base is as good as expected. To achieve the goal, the fuzzy propositions are better than crisp ones. Since the former covers more possible situation by employing fuzzy sets than the later. But the problem does still require serious consideration.In the system f-NNKS a method named Increment Representation is proposed,it arranges all factors in an increasing orderin combining sample pairs. So deep investigation of the influence of each factor in antecedent upon the concequence is badly needed.

The man-machine interface included two kinds. First, the system shows the question like that:

Do you know the information about the distance of the building site from active fault ?
1. know, and can put it in as fuzzy sets.
2. unknow.

If the user responds by choosing 1, then the rule is fired and the interface will be changed to the neural nets. It first prompt the user to put in the parameter filename of wij, and Oj, and the filename of results obtained by lower level nets if there exist.During every step the system will show the user to what place within the resoning tree, it enable the users to avoid possible wrong actions. If the choice of the answer to the question is 2,the match of the rule fails. The interface is friendly to the users.

CONCLUSION

The artificial neural networks give a strong impetus to the development of AI technology. To make use of it for dealing with fuzzy uncertainty is only a test. The preliminary result demonstrates the vitality.The philosophy of developed system f-NNKS is to utilize the nonlinear mapping of neural nets. Why can't the fuzzy reasoning be viewed as nonlinear mapping ? The focus is how we match these two kinds of mapping in an appropriate way and organize a system in a right manner. For this purpose, the framework of fuzzy reasoning formulated here and the architecture of BE and BN constructed here can match them so well that more reasonable results having been got by the demonstration. The system f-NNKS is only a first step, there is a long way to go. The possible extension is to incorporate knowledge representation methods other than production rules in the system and make it more powerful.

ACKNOWLEDGE

The work to establish the system f-NNKS is supported by National Natural Science Foundation of China. Professor Liu Huixian organizes the research work of project " The Intelligent Decision-making Support Systems in Construction", the work presented in this paper is a part of that project.

REFERENCES

1, R.P.Lippmann,An Introduction to Computing with Neural Nets, IEEE ASSP Magazine, 1987

2, G.E.Hinton, Connectionist Learning Procedures, Artificial Intelligence, vol.40, 1989

3, Liu Xihui et al., Approximate reasoning in Earthquake Engineering, in Approximate Reasoning in Expert System,Edited by M. Gupta, et al., North-Holland,1987

4, Liu Xihui et al., Generalised Knowledge Base for Uncertainty Treatment, Collected Works of Intelligent Decision making Support Systems in Construction, Publishing House of Electronic Industry, 1990

5, Tatsumi Furuya, et al., NFS: Neural Fuzzy Inference System,Proceedings of International Workshop on Fuzzy Application, Japan, 1988

6, M.Mizumoto, Fuzzy Control under Various Fuzzy Reasoning Methods,Information Sciences, vol.45, no.2,1988

7, H.Takagi, et al., Neural Network Designed on Approximate Reasoning Architecture and its Application to the Pattern Recognition, Proceedings of ICFLNN, Japan, 1990

8, H.Takagi,Fusion Technology of Fuzzy Theory and Neural Networks -- Survey and Future Directions, Proceedings of ICFLNN, Japan, 1990

9, Liu Xihui et al.The Fuzzy Mathematic Model for Earthquake-induced Damage Prediction, Journal of Building Structure, vol.5, NO.1, 1984

10,Liu Xihui et al., Power Thinking Mode with Engineering Applications,Proceeding of International Conference on Expert Sustems and Engineering Application, Wuhan, China,1989

NEURAL NETWORKS IN STRUCTURAL PRELIMINARY DESIGN

Ming Gan† and Xila Liu‡
†Beijing Architecture Design Institute,
‡Tsing Hua University, Beijing, PR China

This paper presents an expert system for preliminary design of spatial gridwork structures based on the neural networks, which consists of the prediction subsystem, evaluation subsystem, and the control subsystem. This system may select a feasible scheme from possible schemes. Coordinate structural engineers with the engineers worked in other fields, and treat different ideas among designers with the neural nets. In present system, B-P algorithm is modified to speed up the convergence in training procedure. Finally, some practical examples are introduced. The results obtained by the present system are satified.

INTRODUCTION

Computer, as a design tool, has been used in civil engineering for over 30 years and it is very popular in the numerical calculation and the graphics. However, it still is a new topic to use computer to develop expert system in preliminary design. Obviously, the thinking mode in this stage can not be expressed only the numerical vaule, single path, and the series calculation either. Therefore, the knowledge based CAD systems are needed. But the first generation of expert systems still has some problems. The logic thinking mode is used, but another mode , thinking in image, which is often used in preliminary design is neglected. Combining the both thinking modes to perform the preliminary design is the object of the present paper.

In China, structural design processes are divided into four stages, i. e, the conceptural design, preliminary design, numerical calculation, and construction design. The conceptural design only provides the spatial layout decisions affecting the structural system, which are usually made by architects. In this area, AI have not been used. The preliminary design begins at the conceptural design end. In this stage, the designers first performs a schematic design, producing one or more structural configurations that may provide the intended function. This stage is considered as the most creative phase of the structural design. In this case, both thinking in terms of the images and some calculation are necessary. Finally, some key parameters will be given. It should be mentioned that, high accuracies are unnecessary in preliminary design stage. It is very often to provide feedback to structural configurations or originally assumed key parameters to satify the architetual decisions. In the construction design, however, almost each step has been computerized. As mentioned above, the important point is how to formally represent the knowledge, which includes both the logic mode and the image mode.

Artificial neural networks, as a new AI tool, is receving a lot of attention. The past six years or so have been a great amount of work being done in this area. Neural networks are good tools which are suit for solving certain types of engineering problems that required a mapping from a large-distributed set of input features to a large - distributed set of output features. In the present paper, a prototype for the preliminary design of space grid structures (SPRED-1) is introduced. In this prototype, each neural networks is treated as a block which has an independent function. Such as various konwledge representation modulus, simulation modulus (instead of the finite element method), and optimization modulus. Combining these blocks, the system can imitate human being thinking mode for preliminary design stages.

In the first part of the present paper, the prototype (SPRED-1) is introduced in more detail, followed by the modification of the B-P algorithm. Finally, some engineering examples are given.

SPRED-1 SYSTEM

The present system (SPRED-1) is an expert system for preliminary design of the space gridwork. As mentioned previously, the designers have to interact with and coordinate with architects, ventilation engineers, environment engineers, and engineers who work in other fields, when they perform preliminary design. Discrepancies, discussing ,feedback, and redisign usually can not be avoided. In this case, time is very important. The view from the knowledge engineering, both the thinking in logic and thinking in image are involved. Therefore, more friendly interface with user, more quickly response from the system, and efficent knowledge representation moduli are very important for preliminary design.

The system architecture is shown in Fig. 1. Each subsystem of the prototype is given as follows.

The Prediction Subsystem

Using the information obtained from the architect, the present subsystem predicts a set of parameter. In this case, a three layer preceptron is used. The function of the present subsystem is a simulation modulus. The present subsystem is used for simulation to obtain design parameters in preliminary design of space gridwork.

The input nodes in present subsystem are (a) plane dimension, (b) structural depth, (c) grid-line(so-called the frequency number), (d) the number of the different member type, (e) support condition and (f) applied force for the new structure. The output nodes can be presented such as (a) the total amount of steel for the new project, (b) the maxmium internal forces, (c) the maxmium deflection, (d) the maxmium reaction, and (e) the maxmium size of the joint. From the input to the output,it is obvious, that the present subsystem is a typical calculation simulation modulus. After training procedure, the outputs from the present subsystem are of enough accuracy for the preliminary design. Many kinds of the space grid structures, such as grid dome, grid deck ect. , can all be considered in the present subsystem。 For a particular space structure, a related database, which includes almost forty previous designed projects, are used for training procedure. From the view of the knowledge engineering point, this is also called the example-based reasoning[Ref. 2]. Usually, it is very often that designers would like to prepare a number of input sets, which all seems reasonable for the new project, and check the outputs provideed by the prediction subsystem.

The Evaluation Subsystem

The present subsystem consists of the two neural network blocks. Each of them has a three layer preceptron and their independ function. For the first block, the input nodes are the inputs and outputs of the prediction subsystem, which means eleven terms in total. The output data in this block have five terms, each of them are divided into five levels and shown the credit of the consponding output from the prediction subsystem. Credits are expressed by the define linguistic variables, such as "excellent", "good", "fair", "bad", and "very bad ". After training from a database, in which a lot of the evaluation marks are collected from a group of experienced chief engineers, the whights in the block are determined.

In second block, the overall evalution based on the outputs of the first block is given. It means that, the inputs of the block are the evaluation values obtained from the outputs of the previous block, and output is a overall evaluation value only. During the training procedure of the two blocks, the B-P algorithm is also used.

It should be mentioned that, during investigation, some engineers may give a very unusually credit, which makes the training procedure convengence very slow. After a certain number of oscillations, the special data can be deleted automatically [Ref. 1]. It means that some contradictory samples will be filtered.

The Control Subsystem

The present subsystem has several function, which check, filter and modify a number of suggest parameter sets (in this case, each set has six parameters). The data obtained from the present subsystem are treated as the final data or feedback to the prediction subsystem. There are three blocks in the present subsystem, which make this subsystem has intelligent control function, each block will be described as follows.

The filter block in the subsystem is used for checking the suggested sets of five basic parameters from the evaluation subsystem. There is a database which stores many rules collected from the design code and domain experts. The threshold of the overall evaluation is given.

In general, if the overall credit is worse than "fair", the suggested set will be deleted. As example, for a suggested set A, if each term is worse than the corresponding term in other set B, then A is defined as an inferior solution.

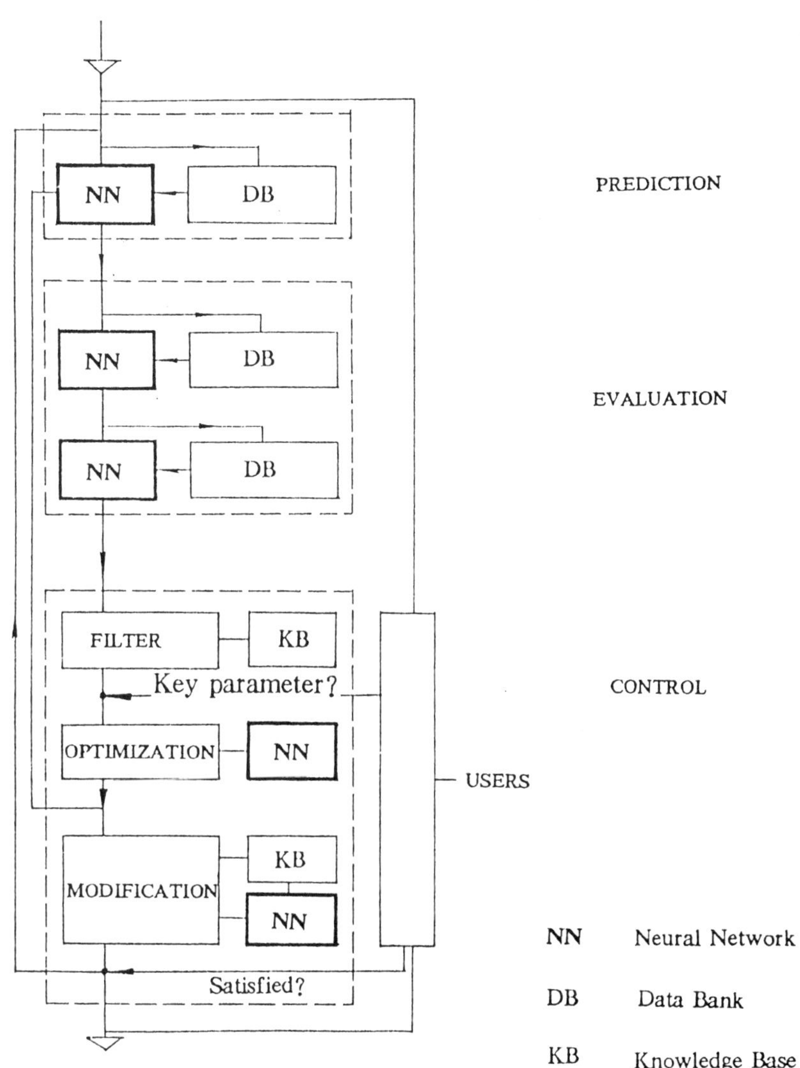

Figure 1 System architecture (SPRED—1)

After comparison, inferior solutions are deleted. Finally, a set of the solution, or several so-called satisfied sets of eleven basic parameters are remained and sent to the optimization block.

In the optimization block of the present subsystem, one or two parameters can be chosen as a objective(s) by an interface contacted with the designers. Usually, the total amount of steel is the most important in China. In this block, another preceptron, in which there are three layers, is used for optimization. The output(s) of this neural net is (are) the objective(s) which is (are) chosen by the designer, and inputs are the same as the inputs of the prediction subsystem. After training procedure from the basic parameters which are stored in special database used for supporting decision, the weights of the neural net will be determined. Following Hopefield idea [Ref. 3], a new function is defined as

$$E(q) = \sum (g - g')^2 + \varphi(\sum ln(cq)) + \varphi_1(\sum q - F_{tot}) \quad (1)$$

where q is the same as the inputs of the prediction subsystem borrowed from thermodynamics, g is (are) the ideal value(s) of the objective(s), g' is (are) outputs(s)from the present subsystem, φ is the entropy constrain function, φ_1 is an additional function, F_{tot} is the possible range of q, and c is a constant show as

$$c = \frac{N}{e^{\sum(1/q)}} \quad (2)$$

where N is a constant.

It should be explained that, sometimes the constrains of the optimization in the preliminary design are very soft, which make algorithm ill-pose. Eq. (1) has to includes the three terms. The second term of Eq. (1) is the entropy constrain function. Using Eq. (2), the entropy will be well-distributed. When the minimum E is reached the optimal q can be obtained, and transfered to the prediction subsystem again to obtained the optimal outputs.

The third block in present subsystem is the modification block. After optimization, only one set of eleven basic parameters are remained and will be transfered to the modification block. In the block, all design parameters are checked by a database which stored many rules obtained from the human being experts. According to the comments from the rule base, if some parameters should be changed in given ranges, one more three layer preceptron is needed for the sensitivity analysis. In this preceptron, the output is the parameter which should be modified and the inputs are ten rest parameters. After the sensitivity analysis, several rules can be obtained for modification, such as "if the maxmium force of the ball joint is larger than 78 ton, then the structural depth should increase". Finally, the most satisfied feasible design parameters are obtained.

If it is necessary, the block can interact with and coordinate the chosen parameters with parallel design activities by an interface. If something should be changed, the changed parameters can be fed back to the prediction subsystem.

MODIFICATION OF THE BACKPROPAGATION ALGORITHM

As mentioned previously, the neural net is a parallel model. Because of the lack of parallel computers, mostly, neural nets are performed on conventional serial computers. Therefore, the back propagation algorithm may be prone to the poor convergence problem in some cases, and often lead to the local minima of errors. In the present system, two methods are used to speed up convergence of the B-P algorithm, which are introduced as follows.

(1) Use Daridom-Fletcher-Powel (DFP) method in stead of the steepest descent method in backpropagation algorithm. The steepest descent method is a very simple and robust method, but its convergent speed in training procedure is very slow , especially when the results approach to the optimal point. This method use the first order derivative to determine the descent direction. Using conventional computers which supply serial computing envivonment, more CPU time for convergence is needed. However, the optimal strategy of DFP method use the gradient information from both this step and the last step. It builds a direction matrix H, which is very close to the invert Hesse matrix. The direction matrix is adopted to find the descent direction and to calculate the descent scale of the next step. Therefore, this method is called the second order convergent speed. Table 1 is a comparison between the DFP method and the steepest descent method. The testing Function is taken as a conditional failure rate function of structures (sometimes called the " bathtub curve"). For example, $y = 100(x_2 - x_1^2) - (1 - x_1)^2$, where x_1 and x_2 are variable.

When the present system is applied to the primary design of space grid structures, a feasible schme set can be formed through accurately computing behaviour parameters of various grid structures by the prediction subsystem, all the schemes in the set are put into evaluaton subsystem and the inferior schemes are deleted from the set. Afterwards a set of better feasible schemes is displayed on computer terminal. Through the feasible scheme set, a scheme of projects can be reached by adjusting design parameters with the control subsystem. Now a primary design of a Golden Monkey Exhibltion Hall in Beijing Zoological Garden is used as an example to illustrate the present system. The geometrical configuration of this poroject is a semi-spherical single-layer grid dome with 28 m diameter(Fig. 2). The knowledge inference based on rules gives out the variation of space grid height from 5 m to 14 m and the frequency number from 7 to 13. In this feasible scheme set, the intervals of space grid height is 5 cm, and the intervals of the frequency number is 1. This feasible scheme set includes 1267 feasible schemes. At first, all those scheme are computing by prediction sub-system, evaluated by evaluation sub-system. In the process of deleting inferior schemes, the constrained condition are: the displacement less than 3 cm, elastic stability factor greater than 20, and consume of steel material less than 1.5 times of the minimum value among the feasible schemes. After that, 804 scheme are deleted from the set and 403 scheme are remained. Following explained process and continuing to carry out inferior deleting, then 104 schemes are purged and 359 schemes remained. At this time, the terminal screen displays 24 schemes with credit more than 4.5. After discussion with the responsible engineers of different disciplines, one scheme is selected as implemental scheme. The present system has been used for primary design of 2 single-layer grid domes and 7 space grid decks. The comparison between preliminary design with and without using the present system is listed on Table 3 and Table 4.

Table 3 The Comparison of Space Grid Deck

Project Name	Structural Plan (m * m)	Height of Structure (m)	Grid Line	Maximal Internal Force (ton)	Total Amount of Steel (ton)	Maximal Displacement (mm)
Madasgascar Sport Center (gymnassium)	60×60	3.71/3	13×13/14×14	50/62	184/202	70/110
No. 1 Workshop of Water Equipment Factory of Shun Yi County	72×15	2.4/3	18×5/18×5	9.684/8	11/12.3	16/12
No. 2 Workshop of Water Equipment Factory of Shun Yi County	72×21	2.1/3	18×7/18×7	23.23/17	17.2/19.5	15/11
Lecture Hall of High Procuratorate	24×17.6	1.5/2.21	8×6/8×5	23.36/14.8	7/7.8	54/43
Lecture Hall of Metallurgical Industry	24×24	2.667/2.667	8×8/8×8	11.5/11.5	12.3/12.3	21/21
Qui Huang Dao Project	24×24	1.5/2.1	8×8/8×8	14.6/11.2	8/8.9	15/8

NOTE: above/ = scheme value given by the present expert system.
below/ = scheme value given by the traditional method.

(2) Classify and carefully select the samples sets which are used in the neural net training procedue. It is obvious, that the classification of the sample sets before training procedure can increase the convergent speed of the B-P algorithm, especially when the modification DFP method has been adopted. For example, as metioned above, if a grid work is to be designed, a related database, which collects almost 40 previons designed grid decks, has to be prepared for the training procedure to determine the weights of this neural net. It means that only those structures which are similar to the new deck, can be taken as the samples used for training procedure.

On the other side, the possible ranges of each parameter are taken for avoiding the local optimal point in backpropagation algorithm. The orthogonal tables are used for selecting the previous designed samples.

After training procedure, another previous designed project, which is a space grid structure similar to the pvevious designed project, collected in the database, is used to check the network ability. It is found that the selected samples, which are chosen by orthogonal tables, has much higher efficiency than that from randomly set samples (Table 2).

APPLICATION EXAMPLES

In the real procedure on the space grid design, especially for the single-layer grid structures, the structural deformation is very large, therefore, its geometrical non-linearity should be considered. Furthermore, since this kind of structures has so-called defect-sensitivity, the post-buckling characteristics have to be taken into account. In the incremental method, when the effect of geometrical non-linearity is considered, the computing will be time consuming. When the post-buckling characteristics of the structure is considered, the pre-buckling point must be passed in calculation. Thus, the structural stiffness matrix. becomes singular. Therefore, the displacement control method must be used to complete this calculation. In this case, designer needs to have rather profound theoretic background and compution experience. Eventhough, the Compution sometimes still fails. In this case, it is impossible to select the best scheme from all the computed feasible schemes. Even in a simple case, such as a double-layer space grid deck, the repeat calculation needs a lot of data preparation and adjustment works. So, when a tradition method is used for the space grid design, designer only uses such a scheme which decided in the primary design and all design factors of the scheme do not violate requirements of Building Codes.

Table 1 Comparison between the DFP method and the steepest descebnt method

	The Max. Error(%)	Iteration Times	*CPU(hrs)
The Steepest Descent Method	21	1100	4
DFP Method	18	137	0.128

* It is perfomed on Vax 780

Table 2 Comparison of the efficiencies between different sample selections

	error(%)	Iteration Times	Number of Samples	*CPU (hrs)
Random Selection	43	1405	21	3
Selection by Orthognal Tables	15	148	8	0.4

* it is perfomed on Vax 780

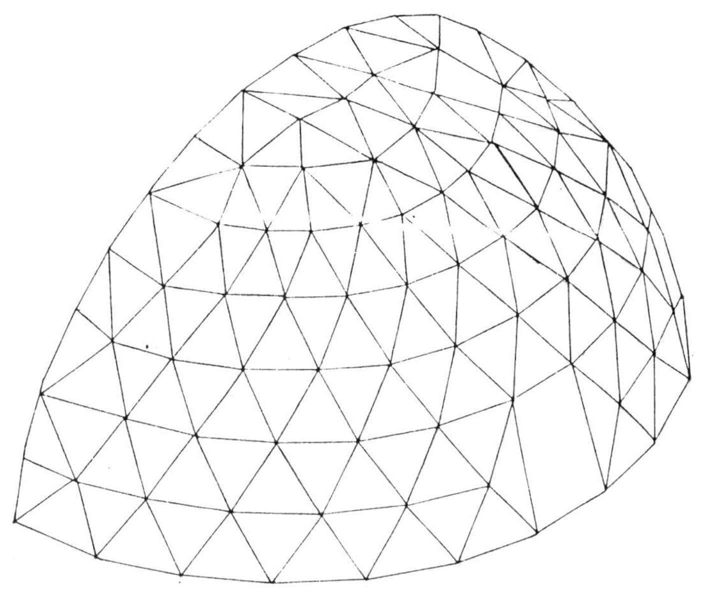

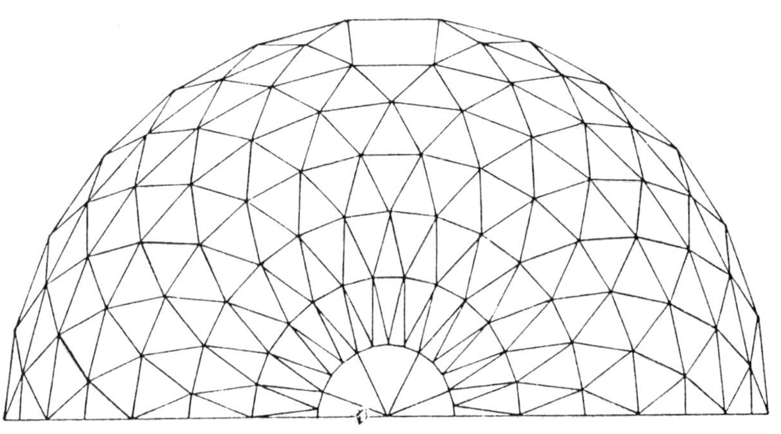

Figure 2 Geodesic dome in Beijing Zoological Garden

Table 4 Comparison of Single-layer Grid Domes

Project Name	Maximal Internal Force (ton)	Maximal Reaction (ton)	Maximal Displacement (cm)	Total Amont of steel (ton)	Structure Dimension (m * m)	Camber (m)	Frequency Number
Golden Monkey Exhibition Hall	4. 2/5. 4	4/6. 7	1. 2/1. 6	6/10	D=24m	14/14	13/8
the Palace of Monkey King	6. 3/7	5/6	5. 45/1. 7	5. 4/11	D=28	6. 1/7	6/9

NOTE: above/ = scheme value given by the present expert system.
below/ = scheme value given by the traditional method.

Table 5 The Comparison of Space Grid Decks

Project Name	Maximal Internal Forcekl (ton)	Maximal Displacement (cm)	Reaction Force (ton)	Total Amount of Steel (ton)	Diameter of Sphere (mm)
Madasgascar Sport Center (gymnassium)	62/65	7/7. 6	36/38. 9	184/176	450/450
No. 1 Workshop of Water Equipment Factory of Shun Yi County	9. 684/11. 2	1. 6/2	6/7. 15	11/12. 3	250/250
No. 2 Workshop of Water Equipment Factory of Shun Yi County	23. 23/25. 4	1. 5/2. 1	9. 1/9. 65	17. 2/17. 7	250/250
Lecture Hall of High Frocuratorate	23. 36/26	1. 5/1. 7	11/12. 4	7/8. 1	250/250
Lecture Hall of Metallurgical Industry	12. 30/14. 32	2/2. 8	7/8. 1	11. 50/14. 50	110/120
Qui Huang Dao Project	14. 6/17	1. 5/2	15. 5/19	8/8. 5	250/250

NOTE: above / = by Finite Element Method
below / = by Neural Net Work

Table 6 Comparison for Single-layer Grid Dome

Project Name	Maximal Internal Force (ton)	λ (Linear-elastic)	Maximal Displacement (cm)	Total Amount of Steel (ton)	Maximal Diameter of Sphere (mm)	Maximal Reaction (ton)
Golden Monkey Exhibition Hall	4.2/5.1	24/20	1.2/1.37	6/7.6	250/200	4/5.3
the Palace of Monkey King	6.3/6.5	26/21.5	1.45/1.6	5.4/6.2	180/200	5/6

NOTE: above / = by Finite Element Method
below / = by Neural Net Work

The comparisons of predicting results between values resulted from the finite element method and the neural network are listed on Table 5 and Table 6.

From above tables, it can be seen that: (1) The total amount of steel material can be reduced considerably, especially for the single-layer grid dome structure, i. e. 40% to 60% of steel material can be saved in this case. This may be due to lack of design experience by traditional method in present time. Stronger steel elements are often used by designer to avoid repeated computation because of the complicated calculation and the lack of information. (2) The relative differences between behavioal parameters predicted by this system and values computed by finite-element methos is from 10% to 20%. It has been proved that the less 30% differences can not influence the optimal decision.

CONCLUSION

1. The preliminary design expert system of space grid structure composed of prediction, evaluation and control sub-systems integrates the inference processes of logical thinking and of image thinking together, which well simulates the thinking procedure of human experts.
2. Using the present improved learning algorithm, the convergent process of B-P algorithm is speeded up.
3. From practical results, the present system has significant advantages to save material and to reduce the design period.

REFERENCES

1. Shi, C. G., and Liu, X. L. (1990). Learning strategy of neural network and its application to earthquake engineering. Proceedings of "Research of the Application of AI Aided Decision-making Systsms in Civil Engineering", Electronic Industry Publisher, Beijing, PP. 160-165 (in Chinese)
2. Wang, J., and Tai, R. W. (1990). An approach to build knowledge system with artificial neural net. Proceedings of "Neural Networks and Its Application", vol. 1, Institute of Automation, Acadmic Sinica, PP. 22-27, (in Chinese).
3. Zeidenberg, M. (1990). Neural networks in artificial intelligence, ELLIS HORWOOD.

12 BIBLIOGRAPHY

ARTIFICIAL INTELLIGENCE AND STRUCTURAL ENGINEERING: A BIBLIOGRAPHY

B.H.V. Topping, M. Jadid and B. Kumar
Department of Civil Engineering,
Heriot-Watt University, Riccarton, Edinburgh, Scotland

The following bibliography is concerned with the application of Artificial Intelligence Techniques to Structural Engineering. The bibliography has been divided into the following sections:

1.0 BOOKS: General

Bramer, M.A., (1984), (editor), "Research and Development in Expert Systems", Proc. of the Fourth Technical Conference of the British Computer Society Specialist Group on Expert Systems, University of Warwick, December, Cambridge University Press, Cambridge.

Frost, R.A., (1986), "Introduction to Knowledge Base System", Collins, London.

Hayes-Roth, F., Waterman, D.A., Lenat, D.B., (editors), (1983), "Building Expert Systems", Addison-Wesley Publishing, Reading, Massachusetts.

Nilsson, N.J., (1980), "Principles of Artificial Intelligence", Springer-Verlag, Berlin.

Rich, E., (1983), "Artificial Intelligence, McGraw Hill Inc., London.

Winston, P.H., (1984), "Artificial Intelligence", second edition, Addison-Wesley, Reading, Massachusetts.

2.0 BOOKS: Structural Engineering

-, (1989), "Proceedings: USACERL/ASCE First Joint Conference on Expert Systems June 29-30, 1988", USACERL Conference Proceedings P-89/10, US Army Corps of Enginees, Construction Engineering Research Laboratory, Champaign, IL, USA.

Adeli, H., (1988), (editor), "Expert Systems in Construction and Structural Engineering", Chapman and Hall, London.

Adeli, H., (1988), (editor), "Microcomputer Knowledge-Based Expert Systems in Civil Engineering", American Society of Civil Engineers, New York.

Adeli, H., Balasubramanyam, K.V., (1988), "Expert Systems for Structural Design: A New Generation", Prentice Hall, New Jersey.

Coyne, R.D., (1989), "Logic Models of Design", Pitman, London.

Dym, C.L., (1985), "Applications of Knowledge-Based Systems to Engineering Analysis and Design", American Society of Mechanical Engineers, New York.

Gero, J.S., (1985), (editor), "Knowledge Engineering in Computer-Aided Design", Elsevier Science Publishers B.V.(North Holland).

Gero, J.S., (1988), (editor), "Artificial Intelligence in Engineering: Design", Computational Mechanics Publications, Southampton.

Gero, J.S., (1988), (editor), "Artificial Intelligence in Engineering: Robotics and Processes", Computational Mechanics Publications, Southampton.

Gero, J.S., (1988), (editor), "Artificial Intelligence in Engineering: Diagnosis and Learning", Computational Mechanics Publications, Southampton.

Gero, J.S., (1989), (editor), "Artificial Intelligence in Design", Computational Mechanics Publications, Southampton.

Gero, J.S., (1989), (editor), "WAID: Workshop on research directions for artificial intelligence in design, Stanford University, March, 1989", The Department of Architectural and Design Science, University of Sydney, Australia.

Gero, J.S., (1990), (editor), "Applications of Artificial Intelligence in Engineering V, Vol 1: Design", Computational Mechanics Publications, Southampton.

Gero, J.S., Stanton, R., (1988), (editors), "Artificial Intelligence Developments and Applications", edited selection of papers to the Australian Joint Artificial Intelligence Conference, Sydney, November, 1987, North-Holland, Amsterdam.

Kostem, C.N., Maher, M.L., (1986), "Expert Systems in Civil Engineering", American Society of Civil Engineers, New York.

Maher, M.L., (1987), (editor), "Expert Systems for Civil Engineers: Technology and Application", American Society of Civil Engineers, New York.

Mohan, S., Maher, M.L., (1989), (editors), "Expert Systems for Civil Engineers: Education", American Society of Civil Engineers, New York.

O'Connor, M.J., Garza, de la, J.M., (1987), (editors), "Proceedings: Workshop on Expert Systems for Construction Scheduling" USA-CERL Conference Proceedings P-87/13, US Army Corps of Engineers, Construction Engineering Research Laboratory, Champaign, IL, USA.

Rzevski, G., (1990), (editor), "Applications of Artificial Intelligence in Engineering V, Vol 2: Manufacture and Planning", Computational Mechanics Publications, Southampton.

Rzevski, G., Adey, R.A., (1991), (editors), "Applications of Artificial Intelligence in Engineering VI", Computational Mechanics Publications, Southampton.

Sriram, D., (1987), "Knowledge-Based Approaches for Structural Design", Computational Mechanics Publications, Southampton.

Siram, D., Adey, R., (1986), (editors), "Applications of Artificial Intelligence in Engineering Problems: Proceedings of the 1st Int. Conf., Southampton University, April 1986", Computational Mechanics Publications, Southampton.

Siram, D., Adey, R.A., (1987), (editors), "Knowledge Based Expert Systems in Engineering: Planning and Design", Computational Mechanics Publications, Southampton.

Siram, D., Adey, R.A., (1987), (editors), "Knowledge Based Expert Systems in Engineering: Tools and Techniques", Computational Mechanics Publications, Southampton.

Siram, D., Adey, R.A., (1987), (editors), "Knowledge Based Expert Systems in Engineering: Classification, Education and Control", Computational Mechanics Publications, Southampton.

Topping, B.H.V. (1987), (editor), "The Application of Artificial Intelligence Techniques to Civil and Structural Engineering", Civil-Comp Press, Edinburgh.

Topping, B.H.V. (1989), (editor), "Artificial Intelligence Techniques and Applications for Civil and Structural Engineers", Civil-Comp Press, Edinburgh.

Winstanley, G., (1991), "Artificial Intelligence for Engineering", John Wiley, Chichester, England.

3.0 JOURNALS

Artificial Intelligence in Engineering, (vol.1, 1986, quarterly), Computational Mechanics Publications, Southampton.

AIEDAM (Artificial Intelligence in Engineering Design and Manufacture), (vol.1, 1987, quarterly), Academic Press.

Computer and Structures, (vol. 31, 1989), Pergamon Press, Oxford.

Computing Systems in Engineering, (vol. 1, 1990), Pergamon Press, Oxford.

Civil Engineering Systems, (vol. 6, 1989, quarterly), E. & F.N. Spon, London.

Engineering Applications of Artificial Intelligence, (Vol. 1, 1988, quarterly), Pineridge Press, Swansea, U.K.

Knowledge Based Systems, (vol. 2, 1989, quarterly), Butterworths, Guildford, Surrey.

Metastates, The Newsletter of Creative Thinking, vol. 1, 1990, published by C. Santamarina, Polytechnic University, 333 Jay Street, Brooklyn, NY 11201, USA.

4.0 BIBLIOGRAPHIES IN ENGINEERING

Bento, J.P., Feijo, B., Dowling, P.J., (1987), "An Anotated Bibliography on Artificial Intelligence Applications for Structural Engineering", Expert Systems Laboratory Internal Report TR 2/87, Deptartment of Civil Engineering, Imperial College of Science, Technology and Medicine, London, UK, 1987, 1-24.

Chung, P., Chan, N., Filby, I., (1988), "A Bibliography of AI in Engineering", in "Workshop on AI Tools", Artificial Intelligence Applications Institute, University of Edinburgh, 28 pages.

Duffy, A., (1987), "Bibliography-Artificial Intelligence in Design", Artificial Intelligence in Engineering, v.2, n.3, 173-179.

Garett, J.H., Jr., (1989), "Expert Systems and AI Texts and References", in Mohan, S., Maher, M.L., (1989), (editors), "Expert Systems for Civil Engineers: Education", American Society of Civil Engineers, New York.

Gero, J.S., (1985), "Bibliography of Books on Artificial Intelligence with particular reference to expert systems and Knowledge Engineering", Computer Aided Design, v.17, n.9, 463-464.

Sriram, D., (1989), "Knowledge-Based Expert Systems in Civil Engineering: An Annotated Bibliography" in Mohan, S., Maher, M.L., (1989), (editors), "Expert Systems for Civil Engineers: Education", American Society of Civil Engineers, New York.

Topping, B.H.V., Kumar, B., (1989), "The Application of Artificial Intelligence to Civil and Structural Engineering: A Bibliography", in Topping, B.H.V., (Editor), (1989), "Artificial Intelligence Techniques and Applications for Civil and Structural Engineers", Civil-Comp Press, Edinburgh, 285-303.

5.0 GENERAL REPORTS

Allwood, R.J., Stewart, D.J., Hinde, C., Negus, B., (1985), "Report on Expert System Shells Evaluation for Construction Industry Applications", Department of Civil Engineering, Loughborough University of Technology.

Kim, S.S., Maher, M.L., Levitt, R.E., Rooney, M.F., Siller, T.J., Richie, S.G., (1986), "Survey of the State-of-the-Art Expert/Knowledge Based Systems in Civil Engineering", USA-CERL Special Repart P-87/01, US Army Corps of Engineers, Construction Engineering Research Laboratory.

Lansdown, J., (1982), "Expert Systems: Their Impact on the Construction Industry", RIBA Conference Fund, London.

Pavelin, C.J., Wilson., (1987), "Knowledge Acquisition for Engineering Applications", Rutherford Appleton Laboratory, Science and Engineering Research Council, Chilton, Didcot, RAL-87-055.

Sriram, D., Maher, M.L., Bielak, J., Fenves, S.J., (1982), "Expert Systems for Civil Engineering - A Survey", R-82-137, Department of Civil Enginering, Carnegie Mellon University, Pittburgh, U.S.A.

Wager, D.M., (1984), "Expert Systems and the Construction Industry", Construction Industry Computing Association, Cambridge, England.

6.0 SOME GENERAL PAPERS

Alty, J.L., (1985), "Use of Expert Systems", Computer-Aided Engineering Journal, 2-9.

Alvey, P.L., Myres, C.D. and Greaves, M.F., (1984), "An Analysis of the Problems of Augmenting a Small Expert System", in Bramer, M.A., (1984), (editor), "Research and Development in Expert Systems", Proc. of the Fourth Technical Conference of the British Computer Society Specialist Group on Expert Systems, University of Warwick, December, Cambridge University Press, Cambridge, 61-72.

Bramer, M.A., (1984), "Expert Systems: Vision and Reality" in Bramer, M.A., (1984), (editor), "Research and Development in Expert Systems", Proc. of the Fourth Technical Conference of the British Computer Society Specialist Group on Expert Systems, University of Warwick, December, Cambridge University Press, Cambridge, 1-12.

Feigenbaum, E.A., (1979), "Themes and Case Studies of Knowledge Engineering", in Michie, D., (1979), (editor), Expert Systems in Micro-Electronic Age, Edinburgh University Press, 3-25.

Gammack, J.G., Young, R.M., (1984), "Psychological Techniques for Elicting Knowledge", in Bramer, M.A., (1984), (editor), "Research and Development in Expert Systems", Proc. of the Fourth Technical Conference of the British Computer Society Specialist Group on Expert Systems, University of Warwick, December, Cambridge University Press, Cambridge, 105-112.

Hertz, D.B., (1990), "The Knowledge Engineering Basis of Expert Systems", Expert Systems with Applications, v.1, n.1, 79-84.

Sloman, A., (1984), "Why we need many Knowledge Reperesentation Formalisms", in Bramer, M.A., (1984), (editor), "Research and Development in Expert Systems", Proc. of the Fourth Technical Conference of the British Computer Society Specialist Group on Expert Systems, University of Warwick, December, Cambridge University Press, Cambridge, 163-183.

Stefik, M., (1981), "Planning with Constraints" (MOLGEN: Part 1)", Artificial Intelligence, North-Holland Publishing Company, Amsterdam, Vol. 16, pp 111-139.

Wigan, M.R., (1986), "Engineering Tools for Building Knowledge-Based Systems on Microsystems", Microcomputers in Civil Engineering, v.1, 52-68.

7.0 GENERAL PAPERS: Structural Engineering

Corby, O., (1986), "Blackboard Architectures in Computer Aided Engineering", Artificial Intelligence in Engineering", v.1, n.2, 95-98.

Dowling, P. J., (1990), "Engineering Structures Towards 2000", The Structural Engineer, 6-8.

Dym, C.L., (1985), "Expert Systems: New Approaches to Computer-aided Engineering", Engineering with Computers, v.1, 9-25.

Fenves, S.J., (1986), "What is an Expert System" in Kostem, C.N., Maher, M.L., (1986), "Expert Systems in Civil Engineering", American Society of Civil Engineers, New York.

Garrett, Jr., J.H., (1990), "Knowledge-Based Expert Systems: Past, Present and Future", Proceedings of the Int. Association of Bridge and Structural Engineers, (IABSE Surveys S-45/90), August 1990, 21-40.

Howard, H.C., Levitt, R.E., Paulson, B.C., Pohl, J.G., Tatum, C.B., (1989), "Computer Integration: Reducing Fragmentation in AEC Industry", Journal of Computing in Civil Engineering, ASCE, v.3, n.1, 18-32.

Hutton, G., (1990), "Information Technology for Structural Engineering", The Structural Engineer, 36-38.

Naeim, F., DiJulio,Jr. R.M., Martin,Jr., J.A., (1985), "Potential Applications of Artificial Intelligence in Structural Engineering", Proceedins of the 2nd International Conf. on Computing in Civil Engineering, June 1985, Hanazhou, China, 146-156.

Rasdorf, W.J., Parks, L.M., (1986), "Expert Systems and Engineeering Design Knowledge", Proceedings of the Ninth Conference on Electronic Computation, ASCE, Birmingham, Al, 28-42.

Taffs, D., (1984), "Expert or Knowledge-Based Systems: Consequences for the industry?", The Arup Journal, London, v.19, n.4, 11-13.

8.0 EXPERT SYSTEM SHELLS, LOGIC PROGRAMMING AND INTERFACES

Adeli, H., (1988), "Artificial Intelligence and Expert Systems", in Adeli, H., (1988), (editor), "Expert Systems in Construction and Structural Engineering", Chapman and Hall, London.

Adeli, H., (1988), "AI Techniques and the Development of Expert Systems", in Adeli, H., (1988), (editor), "Expert Systems in Construction and Structural Engineering", Chapman and Hall, London.

Adeli, H., (1988), "AI Languages and Programming Environments", in Adeli, H., (1988), (editor), "Expert Systems in Construction and Structural Engineering", Chapman and Hall, London.

Adeli, H., (1988), "Expert Systems Shells", in Adeli, H., (1988), (editor), "Expert Systems in Construction and Structural Engineering", Chapman and Hall, London.

Adeli, H., (1989), "Personal Consultant Plus", Microcomputers in Civil Engineering, v.4, n.3, 243-244.

Adeli, H., (1989), "Personal Consultant Easy (Release 2.0)", Microcomputers in Civil Engineering, v.4, n.2, 163-165.

Aiken, M.W., Sheng, O., (1990), "NEXPERT OBJECT", Expert Systems: The International Journal of Knowledge Engineering, v.7, n.1, 54-57.

Alim, S., Munro, J., (1987), "PROLOG-based expert systems in civil engineering", Proc. Instn. Civ. Engrs, Part.2, v.83, 1-14, Discussion, 185-187.

Allen, R.H., Boarnet, M.G., Culbert, C.J., Savely, R.T., (1987), "Using Hybrid Expert System Approaches in Engineering Applications", Engineering with Computers, v.2, 95-110.

Allwood, R.J., Stewart, D.J., Trimble, E.G., (1987), "Some Experiences from Evaluating Expert System Shell Programs and Some Potential Applications", in Topping, B.H.V. (1987), (editor), "The Application of Artificial Intelligence Techniques to Civil and Structural Engineering", Civil-Comp Press, Edinburgh.

Allwood, R.J., Stewart, D.J., Trimble, E.G., (1985), "Some Experiences from Evaluating Expert System Shell Programs and Some Potential Applications", Proceedings 2nd int. conf. on Civil and Structural Engineering Computing, 1985, Civil-Comp Press, Edinburgh, v.2, 415-420.

Aylett, R., Beck, H., Chung, P.W.H., Fraser, J., and Kingston, J.K.C., (1991), IKBS Development Environments in Software Engineer's Reference Book (ed. McDermid, J.), Butterworth-Heinemann Ltd. ,

A-Razzak, M., Hassan, R., Pettipher, R., (1984), "EX-TRAN; A Fortran-Based Software Package for Building Expert Systems" in Bramer, M.A., (1984), (editor), "Research and Development in Expert Systems: Proceedings of the 4th Technical Conf. of the British Computer Society Specialist Group on Expert Systems, December 18-20, 1984, University of Warwick, 23-30.

Berry, D.C., Hart, A.E., (1990), "Evaluating Expert Systems", Expert Systems: The International Journal of Knowledge Engineering, v.7, n.4, 199-208.

Bodkin, T., Graham, I., (1989), "Case Studies of Expert Systems Development Using Microcomputer Software Packages", Expert Systems: The International Journal of Knowledge Engineering, v.6, n.1, 12-16.

Boley, H., (1990), "Expert System Shells: Very-High-Level Languages for Artificial Intelligence", Expert Systems: The International Journal of Knowledge Engineering
v.7, n.1, 2-8.

Cestnlk, B., Kononenko, I., Bratko, I., (1987), "ASSISTANT 86: A Knowledge-Elicitation Tool for Sophisticated Users", in Bratko, I., Lavrac, N, (1987), (editors), "Progress in Machine Learning - Proceedings of EWSL 87: 2nd European Working Session on Learning", Bled, Yugoslavia, May 1987, Sigma Press, Wilmslow, 31-45.

Chung, P.W.H., Aylett, R., Bental, R.I., and Lydiard, T., (1990), "Overview of Artificial Intelligence Tools in Knowledge-Based Systems for industrial Control", (Eds. McGhee, J., Grimble, M.J., and Mowforth, P.), IEE Control Engineering Series 44, Peter Peregrimus Ltd.

Citrenbaum, R., Geissman, J.R., Schultz, R., (1987), "Selecting a Shell", AI Expert, September, 1987, v.2, n.9, 30-39.

Derfler, Jr. F.J., (1985), "Expert-Ease Makes Its Own Rules", PC MAGAZINE, Software Review, April 1985, v.16, 119-124.

Don, Baker,C.-L., (1990), "NeuroShell", PC AI, v.4, n.2, 48-52.

Drake, K.C., Hess, P., (1990), "Abduction", PC AI, v.4, n.5, 58-61.

Dym, C.L., (1987), "Implementation Issues in the Building of Expert Systems", in Maher, M.L., (1987), (editor), "Expert Systems for Civil Engineers: Technology and Application", American Society of Civil Engineers, New York.

Forsyth, R., (1989), "XpertRule Advanced", Expert Systems: The International Journal of Knowledge Engineering, v.6, n.4, 270-273.

Forsyth, R., (1988), "Leonardo", Expert Systems: The International Journal of Knowledge Engineering, v.5, n.2, 160-164.

Forsyth, R.S., (1981), "BEAGLE - A Darwinian Approach to Pattern Recognition", Kybernetes, v.10, 159-166.

Freedman, R., (1987), "27-Product Wrap-Up: Evaluating Shells", AI Expert, September, 69-74.

Gatton, T.H., Lawrence, D.J., (1988), "Multi-Level User Domain Knowledge in Expert System Development", in "Proceedings: USACERL/ASCE First Joint Conference on Expert Systems June 29-30, 1988", USACERL Conference Proceedings P-89/10, US Army Corps of Enginees, Construction Engineering Research Laboratory, Champaign, IL, USA, 115-119.

Gevarter, W.B.,(1987),"The Nature and Evaluation of Commercial Expert System Building Tools", Computer, May, 24-41.

Gilmore, J.F., Pulaski, K.A.,(1985), "A Survey of Expert System Tools", Proceedings of the 2nd Conf. on Artificial Intelligence Applications: The Engineering of Knowledge-Based Systems, IEEE, December 1985, Miami Beach, Fl., USA, 498-502.

Gowri, K., Bedard, C., Fazio, P., (1988), "Evaluation of Knowledge-Basd System Development Tools for Building Engineering Design", Proceedings of the Third International Conference on Computing in Civil Engineering, Vancover, Canada, 430-438.

Harmon, P., (1986), "Expert Systems-Building Tools",Expert Systems Strategies, v.2, n.8, 17-22.

Harris, R.A., Cohn. L.F., (1989), "A Control Structure for Micro-Based Design Expert Systems using the Shell EXSYS", in "Proceedings: USACERL/ASCE First Joint Conference on Expert Systems June 29-30, 1988", USACERL Conference Proceedings P-89/10, US Army Corps of Enginees, Construction Engineering Research Laboratory, Champaign, IL, USA, 50-55.

Hassan, T., A-Razzak, M., Michie, D., Pettipher, R.,(1985), "EX-TRAN 7: A Different Approach for an Expert System Generator",Proceedings of the 5th int. Workshop on Expert Systems and their Applications, August 1985, Avignon, France, 153-168.

Kostem, C.N., (1986), "Attributes and Characteristics of Expert Systems", in Kostem, C.N., Maher, M.L., (1986), "Expert Systems in Civil Engineering", American Society of Civil Engineers, New York.

Kumar, B., Chung, P.W.H., Topping, B.H.V., (1987), "Approaches to Fortran-Prolog Interfacing for an Expert System Environment", in Topping, B.H.V. (1987), (editor), "The Application of Artificial Intelligence Techniques to Civil and Structural Engineering", Civil-Comp Press, Edinburgh.

Lakmazaheri, S., Rasdorf, W.J., (1989), "The Analysis and Partial Synthesis of Truss Structures via Theorem Proving", Engineering with Computers, to appear.

Lange, R., Chien, S., (1989), "Exception Hierarchy Based Knowledge Representations for Expert Systems", in "Proceedings: USACERL/ASCE First Joint Conference on Expert Systems June 29-30, 1988", USACERL Conference Proceedings P-89/10, US Army Corps of Enginees, Construction Engineering Research Laboratory, Champaign, IL, USA, 148-150.

Leung, K.S., Felix-Wong, W.S., Lam, W., (1989), "Applications of a Novel Fuzzy Expert System Shell", Expert Systems: The International Journal of Knowledge Engineering,
v.6, n.1, 2-10.

Li, T., Fang, L., (1986), "CODAR: general purpose design tool for rule-based engineering expert systems", Artificial Intelligence in Engineering, v.2, n.2, 76-81.

Ludvigsen, P.J., Grenney, W.J., Dyreson, D., Ferrara, J.M., (1986), "Expert System Tools for Civil Engineering Applications", in Kostem, C.N., Maher, M.L., (1986), "Expert Systems in Civil Engineering", American Society of Civil Engineers, New York.

Mackerle, J., (1989), "A Review of Expert Systems Development Tools", Engineering Computations, v.6, n.1, 2-17.

Maher, M.L., (1986), "Problem Solving Using Expert System Techniques", in Kostem, C.N., Maher, M.L., (1986), "Expert Systems in Civil Engineering", American Society of Civil Engineers, New York.

Maher, M.L, Allen, R., (1987), "Expert System Components", in Maher, M.L., (1987), (editor), "Expert Systems for Civil Engineers: Technology and Application", American Society of Civil Engineers, New York.

Malasri, S., (1988), "PC Plus and Lotus 123", in Adeli, H., (1988), (editor), "Microcomputer Knowledge-Based Expert Systems in Civil Engineering", American Society of Civil Engineers, New York.

Michie, D., Muggleton, S., Riese, C., Zubrick, S., (1984), "RuleMaster: A Second-
Generation Knowledge Engineering Facility", Radian Technical Report MI-R-623: To be presented at the First Conf. on Artificial Intelligence Applications Sponsored by IEEE Computer Society in Cooperation with AAAI, December 5-7, 1984, Denver, Co., USA, 1-7.

Michie, D., Muggleton, S., Rieses, C., Zubrick, S., (1984), "RULEMASTER: A Second - Generation Knowledge - Engineering Facility", Artificial Intelligence Applications, Proceedings of the 1st conf. IEEE Computer Society, December 1984, Denver, Co. USA,
591-597.

Mullarkey, P.W., (1987), "Lanaguages and Tools for Building Expert Systems", in Maher, M.L., (1987), (editor), "Expert Systems for Civil Engineers: Technology and Application", American Society of Civil Engineers, New York.

Nguyen, R.P., (1989), "Learning Expert Systems by Rule-Based VP-EXPERT", COED: Computer in Educational Division of ASEE, v.9, n.4, 11-16.

Nolen, T., (1991), "The Intelligent Applications of AI Tools: Integration of Tools can Simplify Classical Programming",", PC AI v.5, n.1, 24-27.

Oki, A., Lloyd-Smith, D., (1989), "Meta-Knowledge Reasoning in Civil Engineering Expert Systems", in Topping, B.H.V., (Editor), (1989), "Artificial Intelligence Techniques and Applications for Civil and Structural Engineers", Civil-Comp Press, Edinburgh, 33-36.

Ortolano, L., Perman, C.D., (1987), "Software for Expert Systems Development", Journal of Computing in Civil Engineering, ASCE, v.1, n.4, 225-240.

Pedersen, K., (1988), "Connecting Expert Systems and Conventional Environments", AI Expert, May 1988, v.3, n.5, 26-35.

Richer, M. H., (1986), "An Evaluation of Expert Systems Development Tools", Expert Systems: The International Journal of Knowledge Engineering, v.3, n.3, 166-183.

Rowley, D.T., (1990), "PC/BEAGLE", Expert Systems: The International Journal of Knowledge Engineering, v.7, n.1, 58-60.

Rowlinson, S. (1987), "Expert System Development - Problems in Practice", in Topping, B.H.V. (1987), (editor), "The Application of Artificial Intelligence Techniques to Civil and Structural Engineering", Civil-Comp Press, Edinburgh.

Slagle, J.R., Wick, M.R., (1988), "A Method for Evaluating Candidate Expert System Applications", AI Magazine, Winter, 44-53.

Surya (1984), "Expert-Ease", Personal Computer World v.7, n.6, 208-213.

Thompson, B., Thompson, W., (1986), "Finding Rules in Data", Byte, v.11, n.11, 149-158.

Zhang, R., Gu, M., Lui, X., (1989), "A Generalised Knowledge Base", in Topping, B.H.V., (Editor), (1989), "Artificial Intelligence Techniques and Applications for Civil and Structural Engineers", Civil-Comp Press, Edinburgh, 49-56.

Van Koppen, J., (1988), "A Survey of Expert System Dvelopment Tools", in Pham, D.T., (1988), (editor), "Artificial Intelligence in Industry: Expert Systems in Engineering", IFS Publications, Springer-Verlag, Berlin, 43-57.

Van Koppen, J., Philips, C., (1986), "A Survey of Expert System Development Tools", Proceedings of the 2nd int. Expert Systems Conf., September 1986, London 157-173.

Vedder, R.G., (1989), "PC-Based Expert System Shells: Some Desirable and Less Desirable Characteristics", Expert Systems: The International Journal of Knowledge Engineering, v.6, n.1, 28-42.

Wigan, M.,R., (1986), "Engineering Tools for Building Knowledge-Based Systems on Microsystems", Microcomputers in Civil Engineering, v.1, n.1, 52-68.

Williams, C., (1986), "Expert Systems, Knowledge Engineering, and AI Tools-An Overview, IEEE Expert, 66-70.

9.0 UNCERTAINTY AND FUZZY LOGIC SYSTEMS

Blockley, D.I., (1987), "Uncertainty Analysis in Expert Systems", Civil Engineering Systems, v.4., 3-6.

Blockley, D.I., Henderson, J.R., (1988), "Knowledge Base for risk and cost benefit analysis of limestone mines in the West Midlands", Proc. Instn. Civ. Engrs., Part 1, v.84, 539-564.

Chameau, J.-L., Santamarina, J.C., (1987), "Membership Functions I: Comparison Methods of Measurement", International Journal of Approximate Reasoning, v.1, 287-301.

Chang, T.-C., Ibbs, C.W., Crandall, K.C., (1989), "A Fuzzy Logic System for Expert Systems", Journal of Artificial Intelligence for Engineering Design, Analysis and Manufaccturing, v.2, n.2.

Futura, H., Shiraishi, N., Fu, K.-S., Yao, J.T.P., (1985), "Evaluation of Structural Serviceability Based on Fuzzy Reasoning", in Deyi, F., Xihui, L., (1985), "Fuzzy Mathematics and Earthquake Reasearches", Proceedings of International Symposium on Fuzzy Mathematics in Earthquake Researches", Seismological Press, Beijing, China, 218-234.

Norris, D., Pilsworth, B.W., Baldwin, J.F., (1987), "Medical Diagnosis from Patient Records-A Method Using Fuzzy Discrimination and Connectivity Analysis", Fuzzy Sets Systems, v.23, 73-87.

Ross, T.J., Hadipriono, F.C., (1987), "Fuzzy Production Rules for Damage Assessment Expert System", Proceedings of the International Symposium on Fuzzy Systems and Knowledge Engineering, Guangdong, China,

Santamarina, J.C., Chameau, J.L., (1989), "Limitations in Decision Making and System Performance", Journal of Performance of Constructed Facilities, v.3, n.2, 78-86.

Santamarina, J.C., Chameau, J.-L., (1987), "Membership Functions II: Trends in Fuzziness and Implications", International Journal of Approximate Reasoning, v.1, 303-317.

Shiraishi, N., Futura, H., (1985), "Assessment of Structural Durability with Fuzzy Sets", Proceedings of the Workshop on Civil Engineering Applications of Fuzzy Sets, 193-216.

10.0 NEURAL NETWORKS

Coyne, R.D., Postmus, A.G., (1990), "Spatial Applications of Neural Networks in Computer-Aided Design", Artificial Intelligence in Engineering, v.5, n.1, 9-22.

Garrett,Jr., J.H., Case, M.P., Westervelt, J., Hall, J.W., Yerramareddy, S., Herman, A., Sun, R., Ranjithan, S., (1991), "Engineering Applications of Neural Networks", Journal of Intelligent Manufacturing, Special Issue on Neural Networks, 1-20.

Ghaboussi, J., J.H. Garrett, Jr., X. Wu., (1990), "Material Modelling with Neural Networks", presented at NUMETA 90 - The International Conference on Numerical Methods in Engineering: Theory and Applications, Swansea, U.K., January 7-11.

Ghaboussi, J., Garrett,Jr., J.H., Wu, X., (1991), "Knowledge-Based Modeling of Materials Behavior with Neural Networks", Journal of Engineering Mechanics, v. 117, 1-31.

Goldberg, D.E, Kuo,C.H., (1987), "Genetic Algorithms in Pipeline Optimization", Journal of Computing in Civil Engineering, ASCE,v.1,n.2,128-141.

Goldberg, D.E, Samtani, M.P., (1986), "Engineering Optimization Via Genetic Algorithm", in Will, K.M., (1986), (editor), "Electronic Computation: Proceedings of the 9th conf. on Electronic Computation", ASCE, New York, N.Y., USA,471-482.

Goldberg, D.E., Thomas, A.L., (1986), "Genetic Algorithms: A Bibliograph (1962-1986)", TCGA Report No. 86001, May 1986, The Clearinghouse for Genetic Algorithms, Department of Engineering Mechanics, The University of Alabama, Tuscaloosa, Al., USA, 1-17.

Hajela, P., Berke, L., (1990), "Neurobiological Computational Methods in Structural Analysis and Design", 31st AIAA SDM Conf. April 1990, Long Beach, Ca., USA, 345-355.

Johnson, R.C., (1990), "A CASE for Neural Networks: Automating the Learning Mechanisms of the Brain", PC AI, v.4, n.5, 47-48.

Kilmasauskas, C.C., (1991), "Applying Neural Networks: Part1: An Overiew of the Series", PC AI v.5, n.1, 30-33.

Kottai, R.M., Bahill, A.T., (1989), "Expert Systems Made with Neural Networks", The Int. Journal of Neural Networks, v.1, n.4, 211-226.

Maren, A.J., (1990), "Neural Network Structures: Form Follows Function", in Maren, A.,J., Harston, C.T., Pap, R.M., (1990), (editors), "Handbook of Neural Computing Applications", Academic Press, Inc, Harcourt Brace Jovanovich, Publishers, San Diego, USA, 45-57.

McAulay, A.D., (1988), "Optical Neural Network for Engineering Design", NAECON, 1988, Dayton, Oh., USA, 1302-1306.

Rehak, D.R., Thewalt, C.R., Doo, L.B., (1989), "Neural Network Approaches in Structural Mechanics Computations", in Nelson,Jr. J.K., (1989), (editor), "Computer Utilization in Structural Engineering: Proceedings of the Sessions Related to Computer Utilization at Structures Congress'89, May 1-5, 1989, San Francisco, Ca., ASEC, N.Y, USA, 168-176.

Wasserman, P.D., Schwartz, T., (1987), "Neural Networks, Part 1: What are They and Why is Everybody so Interested in Them Now?" IEEE Expert, IEEE Computer Society, Winter 1987, 10-12.

Wasserman, P.D., Schwartz, T., (1988), "Neural Networks, Part 2: What are They and Why is Everybody so Interested in Them Now?" IEEE Expert, IEEE Computer Society, Spring 1988, 10-15.

11.0 MACHINE LEARNING

Adeli, H., (1990), "Explanation-Based Machine Learning in Engineering Design", Engineering Application of Artificial Intelligence, v.3, n.2, 127-137.

Adeli, H., Yeh, C., (1990), "Explanation-based machine learning in engineering design", v.3, n.2, 127-137.

Andreae, P.M., (1984), "Constraint Limited Generalization: Acquiring Procedures from Examples", Proceedings of Artificial Intelligence National Conf., AAAI , August 1984, Austin, Texas, USA, 6-10.

Arbab, B., Michie, D., (1985), "Generating Rules from Examples", Proceedings of the 9th Int. Joint Conf. Artificial Intelligence, August 1985, Los Angeles, USA, 631-633.

Arciszewski, T., (1989), "Inductive Learning in Engineering", in Proceedings: USACERL/ASCE First Joint Conference on Expert Systems, June 29-30, 1988", USACERL Conference Proceedings P-89/10, US Army Corps of Engineers, Construction Engineering Research Labboratory, Ill, USA, 7-11.

Arciszewski, T., (1989), "Teaching Expert Systems Techniques at Wayne State University", in Mohan, S., Maher, M.L., (1989), (editors), "Expert Systems for Civil Engineers: Education", Published by ASCE, New York, N.Y., USA, 15-22.

Arciszewski, T., Ziarko, W., (1988), "Verification of Morphological Table Based on Probabilistic Rough sets Approach", Int. Journal of Particulate Science and Technology, v.6, 193-205.

Arciszewski, T., (1989), "Inductive Learning in Civil Engineering", Expert Systems in Civil Engineering: Proceedings of the 1989 Int. Association of Bridge and Structural Engineers (IABSE), 1989, Italy, 1-8.

Arciszewski, T., (1989), "Inductive Learning in Engineering", in "Proceedings: USACERL/ASCE First Joint Conference on Expert Systems June 29-30, 1988", USACERL Conference Proceedings P-89/10, US Army Corps of Enginees, Construction Engineering Research Laboratory, Champaign, IL, USA, 7-11.

Arciszewski, T., Mustafa, M., (1989), "Inductive Learning: The User's Perspective", in Forsyth, R., (1989), (editor), "Machine Learning: Principles and Techniques", Chapman and Hall Computing, London, n.3, 39-61.

Arciszewski, T., Mustafa, M., Ziarko, W., (1987), "A Methodology of Design Knowledge Acquisition for Use in Learning Expert Systems", Int. Journal of Man-Machine Studies, v.27, 23-32.

Arciszewski, T., Ziarko, W., (1987), "Adaptive Expert System for Preliminary Engineering Design", Revue Internationale D.E. CFAO et D'intographie, January 1987, v.2, n.1, 27-39.

Attarwala, F.T., Basden, A., (1985), "A Methodology for Constructing Expert Systems", R & D Management, v.15, n.2, 141-149.

Bareiss, E.R, Porter, B.W., Wier, C.C., (1987), "Protos: An Exemplar-Based Learning Apprentice", Proceedings of the 4th Int. Workshop on Machine Learning, June 22-25, 1987, University of California, Irvine, Ca., USA, 12-23.

Bloomfield, B.P., (1986), "Capturing Expertise by Rule Induction", Knowledge Engineering Review, v.1, n.4, 30-36.

Carbonell, J.G., (1983), "Learning by Analogy: Formulating and Generalizing Plans from Past Experience", in Michalski, R.S., Carbonell, J.G. and Mitchell, T., (1983), (editors), "Machine Learning: An Artificial Intelligence Approach Part I", n.3, 137-161.

Carbonell, J.G., Michalski, R.S., Mitchell, T.M., (1983), "Machine Learning: A Historical and Methodological Analysis", The AI Magazine, v.4, n.3, 69-79.

Carbonell, J.G., Michalski, R.S., Mitchell, T.M., (1983), "An Overview of Machine Learning", in Michalski, R.S., Carbonell, J.G. and Mitchell, T., (1983), (editors), "Machine Learning: An Artificial Intelligence Approach Part I", v.I, 3-23.

Clark, P., (1989), "Knowledge Representation in Machine Learning", in Kodratoff, Y., Hutchinson, A., (1989), (editors), "Machine and Human Learning: Advances in European Research", Michael Horwood Limited, England, 35-49.

Clark, P., Niblett, T., (1987), "Induction in Noisy Domains", in Bratko, I., Lavrac, N, (1987), (editors), "Progress in Machine Learning – Proceedings of EWSL 87: 2nd European Working Session on Learning", Bled, Yugoslavia, May 1987, Sigma Press, Wilmslow, 11–30.

DeJong, G., (1981), "Generalizations Based on Explorations ", Proceedings of the 7th IJCAI, August 1981, Vancouver, B.C. Canada, 67–69.

Dietterich, T.G., (1986), "Learning at the Knowledge Level", Machine Learning, v.1, 287–316.

Dietterich, T.G., Michalski, R.S., (1983), "A Comparative Review of Selected Methods for Learning from Examples", in Michalski, R.S., Carbonell, J.G. and Mitchell, T., (1983), (editors), "Machine Learning: An Artificial Intelligence Approach Part I", n.2, 41–81.

Dietterich, T.G., Michalski, R.S., (1981), "Inductive Learning of Structural Descriptions: Evaluation Criteria and Comparative Review of Selected Methods", Artificial Intelligence, v.16, 257–294.

Flann, N.S., Dietterich, T.G., (1986), "Selecting Appropriate Representations for Learning from Examples", Proceedings of the 5th National Conf. on Artificial Intelligence, August 1986, Philadelphia, Pa., USA, 460–466.

Fisher, D., Langley, P., (1985), "Approaches to Conceptual Clustering", Proceedings of the 9th Int. Joint Conf. on Artificial Intelligence, Los Angeles, Ca., USA, 691–697.

Fisher, D.H., (1987), "Conceptual Clustering, Learning from Examples, and Inference", Proceedings of the 4th Int. Workshop on Machine Learning, June 22–25, 1987, University of California, Irvine, Ca., USA, 38–49.

Fisher, D.H., (1987), "Knowledge Acquisition Via Incremental Conceptual Clustering", Machine Learning, v.2, 139–172.

Forsyth, R., (1989), "The Logic of Induction", in Forsyth, R., (1989), (editor), "Machine Learning: Principles and Techniques", Chapman and Hall Computing, London, 3–22.

Forsyth, R., (1984), "Machine Learning Systems", Aslib Proceedings, May 1984, v.36, n.5, 219–227.

Forsyth, R., (1989), "Machine Induction as a Form of Knowledge Acquisition in Knowledge Engineering", in Forsyth, R., (1989), (editor), "Machine Learning: Principles and Techniques", Chapman and Hall Computing, London, 23–38.

Forsyth, R., (1989), "The Evolution of Intelligence", in Forsyth, R., (1989), (editor), "Machine Learning: Principles and Techniques", Chapman and Hall Computing, London, 65–82.

Forsyth, R., Rada, R., (1986), "Towards the Learning Machine" in Forsyth, R., Rada, R., (1986), (editors), "Machine Learning: Applications in Expert Systems and Information retrieval", 96–103.

Forsyth, R., Rada, R., (1986), "Introduction to Machine Learning", in Forsyth, R., Rada, R., (1986), (editors), "Machine Learning: Applications in Expert Systems and Information retrieval", 14–23.

Forsyth, R., Rada, R., (1986), "A Theme for Machine Learning in Information Retrieval", in Forsyth, R., Rada, R., (1986), (editors), "Machine Learning: Applications in Expert Systems and Information retrieval", 128–142.

Forsyth, R., Rada, R., (1986), "Introduction" in Forsyth, R., Rada, R., (1986), (editors), "Machine Learning: Applications in Expert Systems and Information retrieval", 107–111.

Forsyth, R., Rada, R., (1986), "Learning Structural Descriptions", in Forsyth, R., Rada, R., (1986), (editors),"Machine Learning: Applications in Expert Systems and Information retrieval", Ellis Horwood Limited, England, 54–77.

Gallant, S.I., (1985), "Automatic Generation of Expert Systems from Examples", Proceedings of the 2nd Conf. on Artificial Intelligence Applications, December 1985, Miami Beach, Fl., USA, 313–319.

Gams, M., Lavrac, N., (1987), "Review of Five Empirical Learning Systems Within a Proposed Schemata", in Bratko, I., Lavrac, N, (1987), (editors), "Progress in Machine Learning – Proceedings of EWSL 87: 2nd European Working Session on Learning", Bled, Yugoslavia, May 1987, Sigma Press, Wilmslow, 46–66.

Grossberg, S.,(1987),"Competitive Learning: From Interactive Activation to Adaptive Resonance", Cognitive Science, v.11, 23–63.

Hajo, P., Arciszewski, T., Ziarko, W., Aktan, H., (1988), "Inductive Shallow Approach for Generation of Engineering Models",Proceedings of the 9th European Meeting on Cybernetics and Systems Research, April 1988, Vienna, Austria, 933–940.

Jackson, A.H., (1988), "Machine Learning", Expert Systems: The International Journal of Knowledge Engineering, v.5, n.2, 132–150.

Kedar–Cabelli, S., (1985), "PURPOSE–DIRECTED ANALOGY: A Summary of Current Research", Proceedings of the 3rd Int., Workshop on Machine Learning, June 1985, Skyton, PA., USA, 80–83.

Kibler, D., Aha, D.W., (1987), "Learning Representative Examples of Concepts: An Initial Case Study", Proceedings of the 4th Int. Workshop on Machine Learning, June 22–25, 1987, University of California, Irvine, Ca., USA, 24–30.

Kolokouris, A.T., (1986), "Machine Learning", Byte, v.11, n.11, 225–231.

Langley, P., Carbonell, J.G., (1984), "Approaches to Machine Learning", Journal of the American Society for Information Science, v.35, n.5, 306–316.

Lenat, D.B., (1983), "The Role of Heuristics in Learning By Discovery: Three Case Studies", in Michalski, R.S., Carbonell, J.G. and Mitchell, T., (1983), (editors), "Machine Learning: An Artificial Intelligence Approach Part I", n.4, 243–306.

Lu, S.C.–Y., Chen, K., (1987), "A Machine Learning Approach to the Automatic Synthesis of Mechanistic Knowledge for Engineering Decision–Making", AIEDAM v.1, n.2, 109–118.

Mackenzie, C.A., Gero, J.S., (1987), "Learning Design Rules from Decisions and Performances", Artificial Intelligence in Engineering", v.2, n.1, 2–10.

McDermott, J., (1979), "Learning to Use Analogies", Proceedings of the 6th IJCAI, August 1979, Tokoyo, Japan, 568-576.

Michalski, R.S., (1980), "Pattern Recognition as Rule-Guided Inductive Inference", IEEE Transactions on Pattern Analysis and Machine Intelligence, v.PAMI-2, n.4, 349-361.

Michalski, R.S., (1980), "Knowledge Acquisition Through Conceptual Clustering: A Theoretical Framework and an Algorithm for Partitioning Data into Conjunctive Concepts", International Journal of Policy and Analysis Information Systems, v.4, n.3, 219-244.

Michalski, R.S., (1983), "A Theory and Methodology of Inductive Learning", Artificial Intelligence, v.20, 111-161.

Michalski, R.S., (1986), "Understanding the Nature of Learning: Issues and Research Directions", in Michalski, R.S., Carbonell, J.G. and Mitchell, T., (1986), (editors), "Machine Learning: An Artificial Intelligence Approach Volume II", v.II, 3-25.

Michalski, R.S., (1987), "Learning Strategies and Automated Knowledge Acquisition: An Overview", in Bolc, L., (1987) "Computational Models of Learning", Springer-Verlag, 1-19.

Mozetic, I., (1987), "Learning of Qualitative Models", in Bratko, I., Lavrac, N, (1987), (editors), "Progress in Machine Learning - Proceedings of EWSL 87: 2nd European Working Session on Learning", Bled, Yugoslavia, May 1987, Sigma Press, Wilmslow, 201-217.

Pazzani, M.J., (1987), "Explanation-Based Learning for Knowledge-Based Systems", Int. Journal of Man-Machine Studies, v.26, 413-433.

Quinlan, J.R., (1987), "Decision Trees as Probabilistic Classifiers", Proceedings of the 4th Int. Workshop on Machine Learning, June 22-25, 1987, University of California, Irvine, Ca., USA 31-37.

Quinlan, J.R., (1979), "Discovering Rules by Induction from Large Collection of Examples", in Michie, D., (1979), (editor), "Expert Systems in Micro-Electronic Age", Edinburgh University Press, Edinburgh, UK, 168-201.

Quinlan, J.R., (1987), "Generating Production Rules from Decision Trees", Proceedings of the 10th Int. Joint Conf. on Artificial Intelligence, August 1987, Milan, Italy, 304-307.

Quinlan, J.R., (1988), "Inductive, Knowledge and Expert Systems", in Gero, J.S., Stanton, R., (1988), (editors), "Artificial Intelligence Developments and Applications", Elsevier Science Publishers B.V.(North-Holland), Amsterdam, 253-271.

Quinlan, J.R., (19??) "Inductive Knowledge Acquisition: A Case Study", in Quinlan, J.R., (19??), (editor), "Application of Expert Systems: Based on the Proceedings of the 2nd Australian Conf." 157-173.

Quinlan, J.R., (1986), "Induction of Decision Trees", Machine Learning, v.1, 81-106.

Reich, Y., Fenves, S.J., (1989), "The Potential of Machine Learning Techniques for Expert Systems", AIEDAM v.3, n.3, 175-193.

Rich, E., (1983), "Learning, Chapter 11", in Rich, E., (1983), (author), "Artificial Intelligence", McGraw-Hill Book Co., Singapore, 361-388.

Simon, H.A., (1983), "Why Should Machines Learn?", in Michalski, R.S., Carbonell, J.G. and Mitchell, T., (1983), (editors), "Machine Learning: An Artificial Intelligence Approach Part I", v.I, 25-37.

Sleeman, D., Langley, P., (1982), "Learning from Solution Paths: An Approach to the Credit Assignment Problem", The AI MAGAZINE, v.3, n.2, 48-52.

Smith, R.G., Mitchell, T.M., Chestek, R.A., Buchanan, B.G., (1977), "A Model for Learning Systems", Proceedings of the 5th IJCAI-Int. Joint conf. on Artificial Intelligence, 338-343.

Steels, L., Van De Velde, W., (1988), "Learning in Second Generation Expert Systems", in Kowalik, J.S., (1988), (editor), "Knowledge Based Problem Solving", Prentice Hall, New Jersey, USA, 270-295.

Stirling, D., Buntine, W., (1988), "Process Routings in a Steel Mill: A Challenging Induction Problem", in Gero, J.S., Stanton, R., (1988), (editors), "Artificial Intelligence Developments and Applications", Elsevier Science Publishers B.V.(North-Holland), Amsterdam, 301-313.

Stone, J.R., (1989), "Expert System Learning from Structural Failures", PhD Thesis, Department of Civil Engineering, Universiyt of Bristol, England.

Stone, J.R., Blockley, D., (1989), "Finding Patterns in Structural Failures by Machine Learning", Expert Systems in Civil Engineering: Proceedings of the 1989 Int. Association of Bridge and Structural Engineers(IABSE), 1989, Italy, v.58, 47-56.

Stone, J.R., Blockley, D., (1989), "Structural Reliability Through Machine Learning from Case Histories", Proceedings 5th Int. conf. on Structural Safety and Reliability (ICOSSAR 89), ASCE, August 1989, San Francisco, Ca., USA, 1755-1761.

Stone, J.R., Blockley, D.I., Pilsworth, B.W., (1989), "Towards Machine Learning from Case Histories", Civil Engineering Systems, v.6, n.3, 129-135.

Thathachar, M.A.L., (1985), "Learning Systems: Stochastic Automata Models", Def. Sci. Journal, v.35, n.3, 361-366.

Van Horn, M., (1987), "RuleMaster", Byte, January 1987, 341-342.

Van Someren, M., (1989), "Using Attribute Dependencies for Rule Learning", in Morik, K., (1989), (editor), "Knowledge Representation and Organization in Machine Learning" "Lecture Notes in Artificial Intellignce, 347, edited by Siekmann, J., Springer-Verlag, Berlin, 192-210.

Whitehall, B.L., Lu, S.C.-Y., Stepp, R.E., (1990), "CAQ: A Machine Learning Tool for Engineering", Artificial Intelligence in Engineering, v.5, n.4, 189-198.

Williams, G.J., (1987), "Some Experiments in Decision Tree Induction", The Australian Computer Journal, v.19, n.2, 84-91.

Williams, G.J., (1988), "Combining Decision Trees: Initial Results from MIL Algorithm", in Gero, J.S., Stanton, R., (1988), (editors), "Artificial Intelligence Developments and Applications", Elsevier Science Publishers B.V.(North-Holland), Amsterdam, 273-289.

Winston, P.H., (1975), "Learning Structural Descriptions from Examples, Chapter 5", in Winston, P.H., (1975), (editor), " The Psychology of Computer Vision", McGraw-Hill Book Company, New York, N.Y., USA, 157-209.

Winston, P.H., (1982), "Learning New Principles from Precedents and Exercises", Artificial Intelligence v.19, 321-350.

Wong, S.K.M., Ziarko, W., Ye, R.L., (1986), "Comparision of Rough-Set and Statistical Methods in Inductive Learning", Int. Journal of Man-Machine Studies, v.24, 53-72.

11.1 KNOWLEDGE ACQUISITION AND ELICITATION

Chung, P.W.H., Kumar, B., (1987), "Knowledge Elicitation Methods: A Case Study in Structural Design", in Topping, B.H.V. (1987), (editor), "The Application of Artificial Intelligence Techniques to Civil and Structural Engineering", Civil-Comp Press, Edinburgh.

Cohn, L.F., Harris, R.A., Bowlby, W., (1988), "Knowledge Acquisition for Domain Experts", Journal of Computing in Civil Engineering, v.2, n.2, 107-120.

Cullen, J., Bryman, A., (1988), "The Knowledge Acquisition Bottleneck: Time for Reassessment?", Expert Systems: The International Journal of Knowledge Engineering, v.5, n.3, 216-225.

Gatton, T.M., Ferguson, B., Johnson, P., Lawrence, D.J., Kearney, F.W., (1989), "Interactive Domain Model (IDM): A System for Automated Knowledge Acquisition", in "Proceedings: USACERL/ASCE First Joint Conference on Expert Systems June 29-30, 1988", USACERL Conference Proceedings P-89/10, US Army Corps of Enginees, Construction Engineering Research Laboratory, Champaign, IL, USA, 37-44.

Harandi, M.T., Lange, R., Buchner, B., Kearney, F., (1989), "Distributed Knowledge Acquisition using Abstarct Domain Models", in "Proceedings: USACERL/ASCE First Joint Conference on Expert Systems June 29-30, 1988", USACERL Conference Proceedings P-89/10, US Army Corps of Enginees, Construction Engineering Research Laboratory, Champaign, IL, USA, 45-49.

Hart, A.E., (1984),"Experience in the Use of an Inductive System in Knowledge Engineering" in Bramer, M.A., (1984), (editor), "Research and Development in Expert Systems: Proceedings of the 4th Technical Conf. of the British Computer Society Specialist Group on Expert Systems, December 18-20, 1984, University of Warwick,117-126.

Hart, A.E., (1985), "The Role of Induction in Knowledge Elicitation", Expert Systems: The International Journal of Knowledge Engineering, v.2, n.1, 24-28.

Hart, A.E., (1986), "Machine Induction", in Hart, A., (1986), (editor), "Knowledge Acquisition for Expert Systems", Kogan Page, London,109-168.

Hart, A.E., (1985),"Knowledge Elicitation: Issues and Methods", Computer-Aided Design, v.17, n.9, 455-462.

Hart, A.E., (1987), "Role of Induction in Knowledge Elicitation" in Kidd, A.L., (1987), (editor), "Knowledge Acquisition for Expert Systems: A Practical Handbook", Plenum Press, New York, 165-189.

Michalski, R.S., Chilausky,R.L., (1980), "Learning by Being Told and Learning from Examples: An Experimenral Comparison of the Two Methods of Knowledge Acquisition in the Context of Developing an Expert System For Soybean Disease Diagnosis", Int. Journal of Policy Analysis and Information Systems, v.4, n.2, 125-161.

Prerau, D.S., (1987), "Knowledge Acquisition in the Development of a Large Expert System", AI Magazine, Summer 1987, 43-51.

Rasmus, D., (1991), "Knowledge Acquisition: PC Style", PC AI v.5, n.1, 34-39.

Sham, S.H.R., (1991), "The QED Experience: Knowledge Elicitation in Structural Design", Civil Engineering Systems, v.8, 1-9.

Welbank, M., (1983), "A Review of Knowledge Acquisition Techniques for Expert Systems", British Telecommunications, Published by Martlesham Consultancy Services, British Telecom Research Laboratories, Martlesham Heath, Ipswich, England, 1-49.

Zhang, X.J., Yao, J.T.P, (1988), "Automation of Knowledge Organization and Acquisition", Microcomputers in Civil Engineering, v.3, 1-12,

12.0 GRAPHICS

Balachandran, M., Gero, J.S., (1988), "A model for Knowledge-Based Graphical Interfaces", in Gero, J.S., Stanton, R., (1988), (editors), "Artificial Intelligence Developments and Applications", edited selection of papers to the Australian Joint Artificial Intelligence Conference, Sydney, November, 1987, North-Holland, Amsterdam, 147-163.

Kellett, J.M., Winstanley, G., Boardman, J.T., (1989), "A Methodology for Knowledge Engineering Using an Interactive Graphical Tool for Knowledge Modelling", Artificial Intelligence in Engineering, v.4, n.2, 92-102.

Shaw, D.S., (1989) "Knowledge-Based Graphics Dialogues" in "Proceedings: USACERL/ASCE First Joint Conference on Expert Systems June 29-30, 1988", USACERL Conference Proceedings P-89/10, US Army Corps of Enginees, Construction Engineering Research Laboratory, Champaign, IL, USA, 80-84.

Vora, V., "KBES and Interactive Graphics" in Kostem, C.N., Maher, M.L., (1986), "Expert Systems in Civil Engineering", American Society of Civil Engineers, New York.

13.0 STRUCTURAL ENGINEERING

Allen, R.H., (1987), "Expert Systems in Structural Engineering: Works in Progress", Journal of Computing in Civil Engineering, ASCE, v.1, n.4, 312-319.

Allwood, R.J., Cooper,C.N., (1990), "The Use of an Expert System in Selecting Paint Schemes to Protect Structural Steelwork", The Structural Engineer, v.68, n.7, 133-136.

Franck, B.M., Krauthammer, T., (1989), "An Expert System for Field Inspection of Concrete Dams: Part 1, Engineering Knowledge", Engineering with Computers, v.5, 23-38.

Futura, H., Shiraishi, N., Yao, J.T.P., (1986), "An Expert System for Evaluation of Structural Durability", Proceedings of the Fifth International Offshore Mechanics and Arctic Engineering (OMAE) Symposium, ASME, New York, 11-15.

Futura, H., Tu, K.-S., Yao, J.T.P., (1985), "Structural Engineering Applications of Expert Systems", Computer Aided Design, v.17, n.9, 410-419.

Hadripriono, F.C., Larew, R.E., Wong, B., (1988), "A Rule-based weighted rating system for concrete culverts", Civil Engineering Systems, v.5, 99-103.

Hadipriono, F.C., Lim, C.-L., (1987), "Rule-Based Diagnosis of Failure Events" Proceedings of the Fourth International Symposium on Robotics and Artificial Intelligence in Building Construction", Building Research Station-Technion IIT, Haifia, Israel, vol.1, 561-574.

Hadipriono, F.C., Ross, T.J., (1987), "Towards a Rule Based Expert System for Damage Assessment of Protective Structures", Proceedings of Second IFSA, Congress, Japan, v.1, 1-4.

Hadipriono, F.C., Ross, T.J., (1988), "Expert System for Repairability of Protective Structures", Proceedings of NAFIPS'88, 7th Annual Meeting of the North American Fuzzy Information Processing Society, San Francisco State University, 76-80.

Hadipriono, F.C., Ross, T.J., (1988), "Expert System for Repairability of Protective Structures", Proceedings of the 7th Annual Meeting of the North American Fuzzy Information Processing Society, San Francisco, 76-80.

Miyasato, G.H., Dong, W., Levitt, R.E., Boissonnade, A.C., (1986), "Implementation of a Knowledge Based Seismic Risk Evaluation system on Microcomputers", Artificial Intelligence in Engineering, v.1, n.1, 29-35.

Rehek, D.R., Howard, H.C., Sriram, D., (1985), "Architecture of an Integrated Knowledge Based Environment for Structural Engineering Applications", in Gero, J.S. (editor), 1985, "Knowledge Engineering in Computer-Aided Design", Elsevier Science Publishers B.V. (North-Holland), 89-123.

Ross, T.J., Wong, F.S., (1986), "Structural Damage Assessment Using AI Techniques", in Siram, D., Adey, R., (1986), (editors), "Applications of Artificial Intelligence in Engineering Problems: Proceedings of the 1st Int. Conf., Southampton University, April 1986", Computational Mechanics Publications, Southampton, 835-846.

13.1 STRUCTURAL ANALYSIS

Fruchter, R., Gluck, J., (1989), " Adaptive Connections of Multitype Structural Descriptors in a TLS for Structural Analysis", in Topping, B.H.V., (Editor), (1989), "Artificial Intelligence Techniques and Applications for Civil and Structural Engineers", Civil-Comp Press, Edinburgh, 141-148.

Fruchter, R., Gluck, J., (1989), "A Model for Parallel Training and Learning in Dynamic Structural Response Analysis", in Topping, B.H.V., (Editor), (1989), "Artificial Intelligence Techniques and Applications for Civil and Structural Engineers", Civil-Comp Press, Edinburgh, 157-164.

Fruchter, R., Gluck, J., (1987), "Application of AI Programming Techniques to the Analysis of Structures", in Topping, B.H.V. (1987), (editor), "The Application of Artificial Intelligence Techniques to Civil and Structural Engineering", Civil-Comp Press, Edinburgh.

Fruchter, R., Gluck, J., (1989), "Adaptive Multilevel Connections in Training Learning System for Structural Analysis", Gero, J.S., (1989), (editor), "Artificial Intelligence in Design", Computational Mechanics Publications, Southampton, 209-227.

Fruchter, R., Gluck, J., Gold, Y.I., (1988), "Application of AI Programming Techniques to the Analysis of Structures", Computers and Structures v.30, n.3, 747-753.

Fruchter, R., Iwasaki,Y., Law, K.H., (1991), "Generating Qualitative Models for Structural Engineering Analysis and Design", in Cohen, L.F., Rasdorf, W., (1991),, (editors), "Proceedings of the Seventh ASCE Conf. Computing in Civil Engineering and Symposium on Data Bases, Held in Conjunction with A/E/C Systems '91", 268-277.

Fouet, J.-M., (1986), "Artificial Intelligence in Structural Analysis", in Niku-Lari, A., "Structural Analysis Systems", Pergamon Press, Oxford, 3-9.

Lakmazaheri, S., Rasdorf, W.J., (1989), "Constraint Logic Programming for the Analysis and Partial of Truss Structures", AIEDAM V.3, n.3, 157-173.

Miller, G.R., (1989), "An Object Oriented Approach to Structural Analysis and Design", in Topping, B.H.V., (Editor), (1989), "Artificial Intelligence Techniques and Applications for Civil and Structural Engineers", Civil-Comp Press, Edinburgh, 149-156.

Shephard, M.S., Collar, R.R., Baehmann, P.L., Wentorf, R., (1991), "Combining Knowledge-Based and Numerical Techniques for Reliable Engineering Analysis", ICES 91.

Shephard, M.S., Korngold, E.V., Wentorf, R., (1990), "Design Systems Supporting Engineering Idealizations", Geometric Modelling for Product Engineering, Wozny, M. J., Turner, J. U. and Priess, K., editors, North-Holland, Amsterdam, 279-300.

Shephard, M.S., (1990) "Idealization in Engineering Modelling and Design", Research on Engineering Design", Vol 1, 229-238.

Shephard, M.S., Baehmann, P.L., Georges, M.K., Korngold, E.V., (1990), "Framework for the Reliable Generation and Control of Analysis Idealizations, "CMAME", Vol 82, pp 257-280. ,

Shephard, M.S., Korngold, E.V., Collar, R.R., Baehmann, P.L., (1990), "A Modelling Framework for Controlling Structural Idealizations in Engineering Design", Computers and Structures, Vol 37, No 2, pp 181-191.

Wentorf, R., Collar, R.R., Baehmann, P.L., Shephard, M.S., Budhiraja, A., (1991), "Two Prototypes Testing the Use of an Expert System in the Control of Structural Analysis Idealizations", Scientific Computation Research Center, RPI, Troy, NY, SCORECF Report #20-1990, also to appear in Conference Proceedings: Intelligent Scientific Systems, North-Holland.

13.2 CREATIVE DESIGN

Cagan, J., Agogino, A.M., (1987), "Innovative Design of Mechanical Structures from First Principles", AIEDAM, v.1, n.3, 169-189.

Cagan, J., Agogino, A.M., (1989), "Why AI-Design Researchers Should Distinguish Between Creative, Innovative and Routine Levels of Design", in Gero, J. S., (1989), (editor), "WAID: Workshop on research directions for artificial intelligence in design, Stanford University, March 1989" , The Department of Architectural and Design Science, University of Sydney, Australia, 43-51.

Coyne, R.D., (1989), "Logic Models of Design", Pitman, London.

Coyne, R.D., (1990), "Design Reasoning Without Explanations", AI Magazine, Winter 1990, v.11, n.4, 72-80.

Coyne, R.D., Rosenman, M.A., Radford, A.D., Balachandran, M., Gero, J.S., (1990), "Learning and Creativity in Design Systems, Chapter 8", in Coyne, R.D., Rosenman, M.A., Radford, A.D., Balachandran, M., Gero, J.S., (1990), (editors), "Knowledge-Based Design Systems", Addison-Wesley Publishing Company, Reading, Massachusetts, USA, 459-515.

McLaughlin, S., Gero, J.S., (1989), "Requirements of a Reasoning System that Supports Creative and Innovative Design Activity", Knowldge-Based Systems, 62-71.

Rickards, T., (1980), "Designing for Creativity: A State of the Art Review", Design Studies, v.1, n.5, 262-272.

13.3 DESIGN

Abdalla, J.A.,An-Nashif, H., (1991), "An Object-Oriented Data Base for a Structural Engineering System", The Fourth International Conference on Computing in Civil and Building Engineering, Tokyo, Japan, July 29-31.

Abdalla, J.A., Phan, D.H.D., Howard, H.C., Teicholz, P., (1991), "The Primitive-Composite Approach to Data Exchange in Structural Engineering", The Fourth International Conference on Computing in Civil and Building Engineering, Tokyo, Japan.

Abdalla, J.A., Powell, G.H., (1991), "Version Management Needs for Computer Integrated Structural Design", In the Journal of Engineering with Computers, an International Journal for Computer-aided Mechanical and Structural Engineering, Vol 7, No 3.

Adeli, H., (1984), "Artificial Intelligence in Computer-Aided Design of Structures", Proc. Fifth ASCE Engineering Mechanics Division Speciality Conference, University of Wyoming, Laramie, Wyoming, August, 1-4.

Adeli, H., (1984), "Knowledge Based Systems in Structural Engineering", in Topping, B.H.V. (1987), (editor), "The Application of Artificial Intelligence Techniques to Civil and Structural Engineering", Civil-Comp Press, Edinburgh.

Adeli, H., (1989), "Coupled Expert Systems for Engineering Design",
Expert Systems in Civil Engineering: Proceedings of the 1989 Int. Association of Bridge and Structural Engineers (IABSE), Italy, v.58, 231-238.

Adeli, H., Al-Rijeleh, M.M., (1987), "Roof Expert", PC AI, v.1, n.2, 30-34.

Adeli, H., Al-Rijeleh, M.M., (1987), "A Knowledge-Based Expert System for Design of Roof Trusses", Microcomputers in Civil Engineering, v.2, n.3, 179-195.

Adeli, H., Balasubramanyam, K.V., (1988), "A Novel Approach to Expert Systems for Design of Large Structures", AI Magazine, winter, 54-63.

Adeli, H., Chen, Y.S., (1989), "Structuring Knowledge and Data Bases in Expert Systems for Integrated Structural Design", Microcomputers in Civil Engineering, v.4, n.3, 175-204.

Adeli, H., Mak, K.Y., (1988), "Architecture of a Coupled Expert System for Optimum Design of Plate Girder Bridges", Engineering Application of Artificial Intelligence, v.1, n.4, 277-285.

Adeli, H., Paek, Y.T., (1986), "Computer-Aided Analysis of Structures in Interlisp Environment", Computers and Structures, V.23, n.3, 393-407.

Adeli, H., Paek, Y.T., (1986), "Computer-Aided Design of Structures Using Lisp", Computers and Structures, V.22, N.6, 939-956.

Akagi, S., Fujita, K., (1987), "Building an Expert System for the Preliminary Design of Ships", AIEDAM, v.1, n.3, 191-205.

Allen, R.H., Bose, A., (1990), "A Microcomputer-Based Expert System for the Design of Composite Sublaminates", Int. J. Appl. Engng Ed., v.6, n.2, 217-226.

Alkass, S., Abdou, A., (1991), "An Integrated System for Constructibility Assessment of Design Detail", in Rzevski, G., Adey, R.A., (editors), "Applications of Artificial Intelligence in Engineering VI", Computational Mechanics Publications, Southampton, 111-124.

Arciszewski, T., Ziarko, W., (1987), "Adaptive Expert System for Preliminary Engineering Design", Revue Internationale D.E. CFAO et D'intographie, January 1987, v.2, n.1, 27-39.

Arciszewski, T., Ziarko, W., (1987), "Adaptive Expert System for Preliminary Design of Wind Bracing in Steel Skeleton Structures", in Beedle, L.S., (1987), (editor), "Second Century of Skyscrapers: Council on Tall Buildings and Urban Habitat" Van Nostrand Reinhold Co., New York, USA, 847-855.

Balachandran, M., Gero, J.S., (1990), "The Role of Prototypes in Integration of Expert Systems and CAD Systems", Artificial Intelligence in Engineering Design, J. S. Gero, (ed), Computational Mechanics Publications, Southampton, 195-211.

Bedard, C., (1989), "Research Directions for AI in Design: A Building Engineer's View of Design Research", in Gero, J. S., (1989), (editor), "WAID: Workshop on research directions for artificial intelligence in design, Stanford University, March 1989" , The Department of Architectural and Design Science, University of Sydney, Australia, 39-42.

Bedard, C., Gowri, K., (1990), "Automating Building Design Process with KBES", Journal of Computing in Civil Engineering, ASCE, v.4, n.2, 69-83.

Bedard, C., Gowri, K., (1990), "KBS Contribution and Tools in CABD", Int. J. Appl. Engng Ed., v.6, n.2, 155-162.

Bedard, C., Ravi, M., (1990), "Analysis of Building Structural Systems in a Knowledge-Based Design Environment", Proceedings Annual Conf. and 1st Biennial Enviromental Speciality Conf., Materials Computer Applications, May 16-18, Hamilton, Ontario, Canada, Canadian Society for Civil Engineering, v.11-2, 991-1001.

Bedard, C., Ravi, M., (1989), "Preliminary Planning and Design of Multitstory Buildings", Computer Aided Optimum Design of Structures: Applications, Proceedings 1st Int. Conf., June 89, Computational Mechanics, Springer-Verlag, 201-210.

Bedard, C., Ravi, M., (1989), "Planning of Building Layouts for the Preliminary Structural Design of Multistory Buildings", in Barnwell,Jr. T.O., (1989), (editor), "Computing in Civil Engineering, Computers in Engineering Practice, Proceedings of the Sixth Conf., Sponsored by the Technical Council on Computer Practices of the ASCE, Sept. 11-13, 1989, 924-932.

Bedard, C., Ravi, M., (1991), "Knowledge-Based Design of Structural Systems for Multistory Office Buildings", Fourth Int. Conf. on Computing in Civil and Building Engineering, July 1991, Tokoyo, Japan, 1-4.

Bento, J., Feijo, B., (1988), "On the interaction between procedural and declarative programming styles for expert systems in structural engineering design", ESL Technical Report 4/88, Department of Civil Engineering, Imperial College of Science, Technology and Medicine, London.

Bento, J.P., Feijo, B., Dowling, P.J., (1989), "The Knowledge Based Design of Steel Portal Frames for Agricultural Buildings", Expert Systems in Civil Engineering: Proceedings of the 1989 Int. Association of Bridge and Structural Engineers (IABSE), 1989, Italy, 281-290.

Brown, D.C., Chandrasekaran, B., (1982), "An Approach to Expert Systems for Mechanical Design", Proceedings- Trends and Applications 1982, National Bureau of Standards, May 1982, Gaithersberg, Md., 173-180.

Burgoyne, C.J., (1986), "Expert Systems for Structural Design", presented at Construction Industry Computer Conference, London.

Chandrasekaran, B., (1990), "Design Problem Solving: A Task Analysis",
AI Magazine, Winter 1990, v.11, n.4, 59-71.

Chandrasekaran, B., (1989), "A Framework for Design Problem-Solving", in Gero, J. S., (1989), (editor), "WAID: Workshop on research directions for artificial intelligence in design, Stanford University, March 1989" , The Department of Architectural and Design Science, University of Sydney, Australia, 1-19.

Chen, R., Liu, X., (1988), "Expert System RAISE-1" in Adeli, H., (1988), (editor), "Microcomputer Knowledge-Based Expert Systems in Civil Engineering", American Society of Civil Engineers, New York.

Choi, C-K., Kim, E-D., (1987), "A Preliminary Model of I-Builds: An Intelligent Building Design System", in Siram, D., Adey, R.A., (1987), (editors), "Knowledge Based Expert Systems in Engineering: Planning and Design", Computational Mechanics Publications, Southampton, 331-343.

Coyne, R.F., Flemming, U., (1990), "Planning in Design Synthesis: Abstraction-Based LOOS" in Gero, J.S., (editor), "Applications of Artificial Intelligence in Engineering V, Vol 1: Design", Computational Mechanics Publications, Southampton, 91-113.

Coyne, R.D., Rosenman, M.A., Radford, A.D., Balachandran, M., Gero, J.S., (1990), (editors), "Knowledge-Based Design Systems", Addison-Wesley Publishing Company, Reading, Massachusetts, USA.

Dixon, J.R, Simmons, M.K., (1983), "Computers That Design: Expert Systems for Mechanical Engineers", Computers in Mechanical Engineering, 10-18.

Dym, C.L., (1987), "Issues in the Design and Implemetation of Expert Systems", AIEDAM, v.1, n.1, 37-46.

Dym, C.L., (1984), "Expert Systems: New Approaches to Computer-Aided Engineering", Proceedings of the 25th AIAA/ASME/ASCE/AHS Structures, Structural Dynamics and Materials Conf., May 1984, Palm Spring, Ca., USA, 99-115.

Dym, C.L., Salata, S.E., (1989), "Representation of Strategic Choices in Structural Modeling", Proceedings of the ASCE Structures Congress, San Francisco, CA.

Fazio, P., Bedard, C., Gowri, K., (1988), "Design of Reinforced Concrete Walls using PC Plus", Proceedings of the Canadian Society for Civil Engineering Annual Conference, Calgary, Canada, v.1, 31-40.

Feijo, B., Bento, J.P., (1991), "A Framework for Cognitive Models of Design", Relatorio CMEST, Report AI 15/91, 1991, 1-21.

Fenves, S.J., (1989), "Expert Systems: Expectations Versus Realities", Expert Systems in Civil Engineering: Proceedings of the 1989 Int. Association of Bridge and Structural Engineers(IABSE), 1989, Italy, 1-19.

Fenves, S.J., Flemming, U., Hendrickson, C., Maher, M.L., Schmitt, G., (1988), "An Integrated Software Environment for Building Design and Construction", Proceedings of the 5th Conf. on Computing in Civil Engineering-ASCE, March 1988, Alexandria, Va., USA, 21-32.

Fenves, S.J., Flemming, U., Hendrickson, C., Maher, M.L., Schmitt, G., (1988), "An Integrated Software Environment for Building Design and Construction", Report EDRC-12-21-88, 1-12.

Fenves, S.J., Norabhoompipat, T., (1978), "Potentials for Artificial Intelligence Applications in Structural Engineering Design and Detailing", in Latombe, J-C., (1978), (editor), "Artificial Intelligence and Pattern Recognition in Computer Aided Design: Prooceedings of the IFIP Working Conf. Orginized by Working Group 5.2, Computer Aided Design, March 17-19,1978,Grenoble, France, 105-122.

Finger, S., Dixion, J.R., (1989), "A Review of Research in Mechanical Engineering Design", in Gero, J. S., (1989), (editor), "WAID: Workshop on research directions for artificial intelligence in design, Stanford University, March 1989" , The Department of Architectural and Design Science, University of Sydney, Australia, 53-58.

Fox, M.S., (1989), "Issues in Architectures for Design", in Gero, J. S., (1989), (editor), "WAID: Workshop on research directions for artificial intelligence in design, Stanford University, March 1989", The Department of Architectural and Design Science, University of Sydney, Australia, 59-60.

Fukuda, S., (1986), "Development of an Expert System for the Design Support of an Oil Storage Tank", in Siram, D., Adey, R., (1986), (editors), "Applications of Artificial Intelligence in Engineering Problems: Proceedings of the 1st Int. Conf., Southampton University, April 1986", Computational Mechanics Publications, Southampton, 791-796.

Ganeshan, R., Finger, S., Garrett,Jr., J.H., (1991), "Representing and Reasoning with Design Intent", Proceedings of AI in Design, Edinburgh, UK, June 1991, 1-15.

Garcia, A.C.B., and Howard, H.C., (1991), "Building a Model for Augmented Design Documentation", Artificial Intelligence in Design, Butterworth-Heinemann Ltd., Oxford, England, 723-736.

Gero, J.S., (1989), "Research Directions for Artificial Intelligence in Design", in Gero, J. S., (1989), (editor), "WAID: Workshop on research directions for artificial intelligence in design, Stanford University, March 1989" , The Department of Architectural and Design Science, University of Sydney, Australia, 61-64.

Gero, J.S., (1990), "Design Prototypes: A Knowledge Representation Schema for Design", AI Magazine, Winter 1990, v.11, n.4, 26-36.

Gero, J.S., Radford, R., Coyne, R., Akiner, V.T., (1985), "Knowledge-Based Computer-Aided Design Architectural Design", in Gero, J.S., (1985), (editor), "Knowledge Engineering in Computer-Aided Design", Elsevier Science Publishers B.V. (North-Holland), 57-88.

Gero, J.S., Rosenman, M.A., (1990), "A Conceptual Framework for Knowledge-Based Design Research at Sydney University's Design Computing Unit", Artificial Intelligence in Engineering, v.5, n.2, 65-77.

Gero, J.S., Balachandran, M., (1986), "Knowledge and Design Decision Processes", in Siram, D., Adey, R., (1986), (editors), "Applications of Artificial Intelligence in Engineering Problems: Proceedings of the 1st Int. Conf., Southampton University, April 1986", Computational Mechanics Publications, Southampton, 343-352.

Gero, J.S., Maher, M.L., Zhang, W., (1988), "Design Knowledge and Representation", in Gero, J.S., (1988), (editor), "Artificial Intelligence in Engineering: Design", Computational Mechanics Publications, Southampton, 3-21.

Gowri, K., Bedard, C., Fazio, P., (1987), "Knowledge Based System Approach to Reinforced Concrete Design", Proceedings of the Can-Cam Conference, Edmonton, Alberta, v.2, 112-113.

Gowri, K., Bedard, C., Fazio, P., (1991), "Knowledge-Based Assistant for Aluminum Component Design", Computers & Structures, v.38, n.1, 9-20.

Gregory, B.L., Shepherd, M.S., (1986), "Design of Knowledge Based System to convert airframe geometric models to structural models", in Kostem, C.N., Maher, M.L., (1986), "Expert Systems in Civil Engineering", American Society of Civil Engineers, New York.

Grierson, D.E., Cameron, G.E., (1987), "A Knowledge-Based Expert System for Computer Automated Structural Design", in Topping, B.H.V. (1987), (editor), "The Application of Artificial Intelligence Techniques to Civil and Structural Engineering", Civil-Comp Press, Edinburgh.

Grierson, D.E., Cameron, G.E., (1988), "An Expert System for Structural Steel Design", in Gero, J.S., (1988), (editor), "Artificial Intelligence in Engineering: Design", Computational Mechanics Publications, Southampton, 279-293.

Hartmann, D.,(1989),"Knowledge-Based Systems in Civil Engineering (from CAD to KAD)", Expert Systems in Civil Engineering: Proceedings of the 1989 Int. Association of Bridge and Structural Engineers (IABSE), 1989, Italy,v.58,249-259.

Harty, N., (1987), "An Aid to Preliminary Design", in Siram, D., Adey, R.A., (1987), (editors), "Knowledge Based Expert Systems in Engineering: Planning and Design", Computational Mechanics Publications, Southampton, 377-392.

Horner, R., Brown, D.C., (1990), "Knowledge Compilation using Constraint Inheritance", in Gero, J.S., (editor), "Applications of Artificial Intelligence in Engineering V, Vol 1: Design", Computational Mechanics Publications, Southampton, 161-176.

Howard, H.C., Abdalla, J.A., and Phan, D.H.D., (1991), "A Primitive-Composite Approach to Structural Data", The Seventh American Society of Civil Engineers Conference in Computing in Civil Engineering, Washington, D.C., USA, May 6-8, 799-809.

Howard, H,C., Wang, J., Daube, F., Rafiq, T., (1989), "Applying Design-Dependent Knowledge in Structural Engineering Design", AIEDAM v.3, n.2, 111-123.

Howe, A.E., Cohen, P.R., (1986), "Dominic: A Domain-Independent Program for Mechanical Engineering Design", Artificial Intelligence in Engineering", v.1, n.1, 23-28.

Jain, D., Krawinkler, H., Law, K.H., Luth, G.P., (1991), "A Formal Approach to Automating Conceptual Structural Design, Part II: Application to Floor Framing Generation", Engineering with Computers, v.7, 91-107.

Jayachandran, P., Tsapatsaris N., (1988), "A Knowledge-Based System for the Seclection of Structural Systems for Tall Buildings", in Adeli, H., (1988), (editor), "Microcomputer Knowledge-Based Expert Systems in Civil Engineering", American Society of Civil Engineers, New York.

Jones, M.S., Saouma, V.E., (1988), "Prototype Hybrid Expert System for R/C Design", Journal of Computing of Civil Engineeing, ASCE, v.2, n.2, 136-143.

Kalyanaraman, V., and Sakthivel, T.S., (1990), "Computer Aided Engineering of Steel Structures", Kalyanaraman, V., (Editor), Advances in Steel Structures, Tata-McGraw Hill Publications, New Delhi, 71-95.

Krauthammer, T., Kohler, S., (1986), "R.C. Structures under Severe Loads-An Expert Systems Approach", in Kostem, C.N., Maher, M.L., (1986), "Expert Systems in Civil Engineering", American Society of Civil Engineers, New York.

Kumar, B., Chung, P.W.H., Rae, R.H., Topping., B.H.V., (1987), "A Knowledge-Based Approach to Structural Design", in Topping, B.H.V. (1987), (editor), "The Application of Artificial Intelligence Techniques to Civil and Structural Engineering", Civil-Comp Press, Edinburgh.

Kumar, B., Topping, B.H.V., (1988), "An Integerated Rule-Based System for Industrial Building Design" in Adeli, H., (1988), (editor), "Microcomputer Knowledge-Based Expert Systems in Civil Engineering", American Society of Civil Engineers, New York.

Kumar, B., Topping, B.H.V., (1988), "INDEX: An INDustrial Building Design EXpert" Civil Engineering Systems, v.5, n.2, 65-76.

Kumar, B., Topping, B.H.V., (1990), "Non-Monotonic Reasoning for Structural Design", Civil Engineering Systems, v.7, n.4, 209-218.

Kusiak, A., Wang, J., (1991), "Decomposition of Design Activities", in Rzevski, G., Adey, R.A., (editors), "Applications of Artificial Intelligence in Engineering VI", Computational Mechanics Publications, Southampton, 163-178.

Lakmazaheri, S., Rasdorf, W.J. (1990), "A Formal Approach to Structural Design Automation Using Predicate Logic", Proceedings of the Design Theory and Methodology Conf., ASME, Chicago, Il., USA, 259-266.

Law, K.H., Barsalou, T., Wiederhold, G., (1990), "Managemement of Complex Structural Engineering Objects in a Relational Framework", Engineering with Computers, v.6, 81-92.

Law, K.H., Wiederhold, G., Siambela, N., Sujansky, W., Zingmond, D., Singh, H., Barsalou, T., (1991), "Architecture for Managing Design Objects in a Shareable Relational Framework", Int. Journal of Systems Automation: Research and Applications (SARA), v.1, 47-65.

Luth, G.P., Jain, D., Krawinkler, H., Law, K.H., (1991), "A Formal Approach to Automating Conceptual Structural Design, Part I: Methodology", Engineering with Computers, v.7, 79-89.

MacCallum, K.J., (1990), "Does Intelligent CAD Exist", Artificial Intelligence in Engineering, v.5, n.2, 55-64.

MacCallum, K.J., Duffy, A., (1990), "Representing and using numerical empiricism in Design", Gero, J.S., (editor), "Applications of Artificial Intelligence in Engineering V, Vol 1: Design", Computational Mechanics Publications, Southampton, 115-136.

McCarthy, T.J., Nouas, Z., (1989), "An Experimental Expert System for the Design of Column Base Plates", in Topping, B.H.V., (Editor), (1989), "Artificial Intelligence Techniques and Applications for Civil and Structural Engineers", Civil-Comp Press, Edinburgh, 217-220.

Mackenzie, C.A., (1988), "Inducing Relational Grammars from Design Interpretations", in Gero, J.S., Stanton, R., (1988), (editors), "Artificial Intelligence Developments and Applications", Elsevier Science Publishers B.V.(North-Holland), Amsterdam, 315-328.

MacLeod, I.A., Rafiq, M.Y., (1988), "Integrated Computer Aided Structural Design", The Structural Engineer, v.66, n.20, 325-339.

McLaughlin, S., Gero, J.S., (1987), "Acquiring Expert Knowledge from Characterized Designs", AIEDAM, v.1, n.2, 73-87.

Maher, M.L., (1985), "HI-RISE and Beyond: Directions for Expert Systems in Design", Computer Aided Design, v.17, n.9, 420-427.

Maher, M.L., (1987), "Expert Systems in Structural Engineering", in Maher, M.L., (1987), (editor), "Expert Systems for Civil Engineers: Technology and Application", American Society of Civil Engineers, New York.

Maher, M.L., (1987), "Expert Systems for Structural Design", Journal of Computing in Civil Engineering, ASCE, v.1, n.4, 270-283.

Maher, M.L., (1990), "Process Models for Design Synthesis", AI Magazine, Winter 1990, v.11, n.4, 49-58.

Maher, M.L., (1989), "Synthesis of Structural Systems", Expert Systems in Civil Engineering: Proceedings of the 1989 Int. Association of Bridge and Structural Engineers(IABSE), 1989, Italy, v.58, 261-269.

Maher, M.L., (1987), "Engineering Design Synthesis: A Domain Independent Representation", AIEDAM, v.1, n.3, 207-213.

Maher, M.L., (1987), "Expert Systems for Structural Design", Journal of Computing in Civil Engineering, ASCE, v.1, n.4, 270-283.

Maher, M.L., (1989), "Research Directions: AI in Design", in Gero, J. S., (1989), (editor), "WAID: Workshop on research directions for artificial intelligence in design, Stanford University, March 1989" , The Department of Architectural and Design Science, University of Sydney, Australia, 65-68.

Maher, M.L., Fenves, S.J., (1985), "HI-RISE: A Knowledge-Based Expert System for the Preliminary Design of High Rise Buildings", Report No. R-85-146, Department of Civil Engineering, Carnegie-Mellon University, U.S.A.

Maher, M.L., Fenves, S.J., (1985), "Hi-Rise: An Expert System for the Preliminary Design of High Rise Buildings", in Gero, J.S., (1985), (editor), "Knowledge Engineering in Computer-Aided Design", Elsevier Science Publishers B.V.(North Holland), 125-140.

Maher, M.L., Fenves, S.J., Garrett, J.H., (1988), "Expert Systems for Structural Design", in Adeli, H., (1988), (editor), "Expert Systems in Construction and Structural Engineering", Chapman and Hall, London.

Miles, J.C., Moore, C.J., An Expert System for the Conceptual Design of Bridges", in Topping, B.H.V., (Editor), (1989), "Artificial Intelligence Techniques and Applications for Civil and Structural Engineers", Civil-Comp Press, Edinburgh, 171-176.

Mittal, S., Dym, C.L., (1985), "Knowledge Acquistion from Multiple Experts", AI Magazine, v.6, n.2, 32-36.

Mostow, J., (1985), "Toward Better Models of the Design Process", AI Magazine, v.6, n.1, 44-57.

Murlidharan, T.L., Aravind, H.B., Suryakumar, Raman, N.V., (1991), "Expert Tower Analysis and Design System. I: Architecture and Heuristics", Journal of Computing in Civil Engineering, ASCE, v.5, n.2, 175-192.

Murlidharan, T.L., Aravind, H.B., Suryakumar, Raman, N.V., (1991), "Expert Tower Analysis and Design System. II: Search Strategies and Learning", Journal of Computing in Civil Engineering, ASCE, v.5, n.2, 193-210.

Murphy, J., O'Flaherty, B., Lewis, A.W., (1991), "Design of Breakwaters using Knowledge-Based Systems", in Rzevski, G., Adey, R.A., (editors), "Applications of Artificial Intelligence in Engineering VI", Computational Mechanics Publications, Southampton, 199-218.

Murthy, S.S., Addanki, S., (1987), "PROMPT: An Innovative Design Tool", AAAI 87: Sixth National Conf. on Artificial Intelligence, v.2, 637-642.

Naeim, F., Martin, Jr.J.A., (1986), "Application of Artificial Intelligence in Preliminary Structural Design", in Will, K.M., (1986), (editors), "Electronic Computation: Proceedings of the 9th conf." 53-64.

Navinchandra, D., (1989), "Research Directions for AI in Design", in Gero, J. S., (1989), (editor), "WAID: Workshop on research directions for artificial intelligence in design, Stanford University, March 1989", The Department of Architectural and Design Science, University of Sydney, Australia, 69-76.

Nguyen, G.T., Rieu, D., (1987), "Expert Database Concepts for Engineering Design", AIEDAM, v.1, n.2, 89-101.

Nishido, T., Maeda, K., Nomura, K., (1989), "Practical System for Type Selection of Bridge Crossing River", Expert Systems in Civil Engineering: Proceedings of the 1989 Int. Association of Bridge and Structural Engineers(IABSE), 1989, Italy, v.58, 311-319.

Oxman, R., (1990), "Design Shells: A Formalism for Prototype Refinement in Knowledge-Based Design Systems", Artificial Intelligence in Engineering, v.5, n.1, 2-8.

Paek, Y., Adeli, H., (1988), "SDL: An Environment for Building Integrated Structural Design Expert Systems", in Adeli, H., (1988), (editor), "Microcomputer Knowledge-Based Expert Systems in Civil Engineering", American Society of Civil Engineers, New York.

Paek, Y., Adeli, H., (1988), "STEELEX: A coupled expert system for integrated design of steel structures", Engineering Applications of AI, v.3, 170-180.

Paek, Y., Adeli, H., (1988), "Representation of Structural Design Knowledge in Symbolic Language", Journal of Computing in Civil Engineering, ASCE, v.2, n.4, 346-364.

Payne, P.N., (1989),Structural Design using Intelligent Objects", in Topping, B.H.V., (Editor), (1989), "Artificial Intelligence Techniques and Applications for Civil and Structural Engineers", Civil-Comp Press, Edinburgh, 177-186.

Powell, G.H. and Abdalla, J.A., (1989), "Object Management in an Integrated Design System", The Sixth American Society of Civil Engineers Conference in Computing in Civil Engineering, Atlanta, Ga, USA, 556-563.

Powell, G.H., Abdalla, J.A. and Fillipou, F.C., (1988), "An Object-Oriented Approach to Computer-Aided R/C Design", In the Proceedings of the Third International Conference in Computing in Civil Engineering, Vancouver, Canada, 119-126.

Powell, G.H., Abdalla, J.A. and Sause, R., (1989), "Object-Oriented Knowledge Representations: Cute things and Caveats", The Sixth American Society of Civil Engineers Conference in Computing in Civil Engineering, Atlanta, Ga, USA, 1-9.

Radford, A.D., Hung, P., Gero, J.S., (1984), "New Rules of Thumb from Computer-Aided Structural Design: Aquiring Knowledge for Expert Systems", in J. Wexler, (1984), (editor), "CAD'84", Butterworths, 558-566.

Rasdorf, W.J., (1985), "Perspective on Knowlege in Engineering Design", Proceedings of the Int. Computers in Engineering Conf., ASME, Boston, MA., USA, V.2, 249-253.

Rasdorf, W.J., Parks,L.M., (1986), "Expert Systems and Engineering Design Knowledge", in Will, K.M., (1986), (editors), "Electronic Computation: Proceedings of the 9th conf.", 28-42.

Ray, M.H., (1986), "The Design of Reinforced Concrete Members Using an Expert System", in Siram, D., Adey, R., (1986), (editors), "Applications of Artificial Intelligence in Engineering Problems: Proceedings of the 1st Int. Conf., Southampton University, April 1986", Computational Mechanics Publications, Southampton, 771-780.

Roddis, W.M.K., Connor,J., (1988), "Using Prolog on a Macintosh to Build an Engineering Expert System" in Adeli, H., (1988), (editor), "Microcomputer Knowledge-Based Expert Systems in Civil Engineering", American Society of Civil Engineers, New York.

Rogers, J.L., (1990), "Knowledge-Based Tool for Decomposing Complex Design Problems", Journal of Computing in Civil Engineering, ASCE, v.4, n.4, 298-312 also in NASA Technical Paper 2903, Langley Research Center, Hampton, 1989.

Rogers, J.L., (1989), "The Potential Application of the Blackboard Model of Problem to Multidisciplinary Design", NASA Technical Memorandum 101633, Langley Research Center, Hampton.

Rogers, J., Feyock, S., Sobieszczanski-Sobieski, J., (1988), "STRUTEX-a prototype knowledge based system for intially configuring a structure to support point loads in two dimensions", in Gero, J.S., (1988), (editor), "Artificial Intelligence in Engineering: Design", Computational Mechanics Publications, Southampton, 315-335 also in NASA Technical Memorandum 100613, Langley Research Center, Hampton.

Rooney, M.F., Smith, S.E., (1982), "Artificial Intelligence in Simple Beam Design", Journal of Structural Division, ASCE, v.108, n.ST10, 2344-2348.

La Rota, J.L., Biswas, G., Basu, P.K., (1990), "A Model-Based Approach to Structural Design" in Gero, J.S., (editor), "Applications of Artificial Intelligence in Engineering V, Vol 1: Design", Computational Mechanics Publications, Southampton, 3-22.

Russell, S.O.D., (1991), "The Design Process: Insights from expert system development", Civil Engineering Systems, v.8, n.1, 9-12.

Rychener, M.D., (1985), "Expert Systems for Engineering Design", Expert Systems: The International Journal of Knowledge Engineering, v.2, n.1, 30-44.

Sacks, R., Buyukozturk, O., (1986), "Expert Interactive Design of R/C Columns Under Biaxial Bending", Journal of Computing in Civil Engineering, ASCE, v.1, n.2, 69-81.

Sakr, K.M., Hosain, M.U., (1989), "Applications of Expert System Tools in Structural Design", Proceedings 6th Conf. on Computing in Civil Engineering, September 1989, Atlanta Ga., USA, 434-441.

Saleh, H.S.A., (1986), "Artificial Intelligence and Computer Aided Design in Civil Engineering", in Siram, D., Adey, R., (1986), (editors), "Applications of Artificial Intelligence in Engineering Problems: Proceedings of the 1st Int. Conf., Southampton University, April 1986", Computational Mechanics Publications, Southampton, 781-789.

Saleh, H.S., (1987), "Artificial Intelligence Based Computer Aided Design of Shell Structures", in Siram, D., Adey, R.A., (1987), (editors), "Knowledge Based Expert Systems in Engineering: Planning and Design", Computational Mechanics Publications, Southampton, 421-427.

Sause,R., Powell, G., (1989), "Issues in Research Directions for Artificial Intelligence in Design", in Gero, J. S., (1989), (editor), "WAID: Workshop on research directions for artificial intelligence in design, Stanford University, March 1989", The Department of Architectural and Design Science, University of Sydney, Australia, 77-81.

Simmonds, S.H., Elezaby, Y.K., (1990), "Some Thoughts on KBES for Structural Design", in Topping, B.H.V., (1990), (editor), "Developments in Structural Engineering: Proceedings of the Forth Rail Bridge Centenary Conf., Volume Two; Analysis, Design and Assessments of Structures", Chapman Hall, London, v.2 1338-1347.

Slater, J.H., (1986), "Qualitative Physics and the Prediction of Structural Behaviour", in Kostem, C.N., Maher, M.L., (1986), "Expert Systems in Civil Engineering", American Society of Civil Engineers, New York.

Smith, S.E., (1983), "Artificial Intelligence Applied to the Preliminary Design of Simply Supported Beams", a thesis in partial fulfillment of the requirement for the degree of Master of Engineering, Rensselaer Polytechnic Institute, Troy, New York.

Smithers, T., Conkie, A., Doheny, J., Logan, B., Millington, K., Xi Tang, M., (1990), "Design as Intelligent Behaviour: An AI Design Research Programme", Artificial Intelligence in Engineering, v.5, n.2, 78-109.

Soh, C-K. Soh, A-K., (1987), "Example of Intelligent Structural Design System", Journal of Computing in Civil Engineering, ASCE, v.2, n.2, 329-345.

Spillers, W.R., (1966), "Artificial Intelligence and Structural Design", Journal of the Strucutral Division, ASCE, v.92, 491-497.

Srinivasan, S., Allen, R.H., (1989), "Patitioning and Guided Search: A Generalized Approach to Problem Solving in Preliminary Design", in Design Theory and Methodology DTM89, (W.P. Seering, editor), ASME, DE-v. 17, 37-43.

Sriram, D., (1986), "Knowledge-Based Approaches for Structural Design", a thesis submitted in partial fulfillment of the requirements for the degree of Doctor of Philosophy, Department of Civil Engineering, Carnegie-Mellon University, Pittburgh, U.S.A.

Sriram, D., (1986), "DESTINY: A Model for Integrated Structural Design",in Smith, A., (editor), "CAD86", Butterworths, London, 226-235.

Sriram, D., (1991), "Computer-Aided Collaborative Product Development", Research Report R91-14, Order No. IESL91-05, May, MIT, Intelligent Engineering Systems Lab., Dept. of Civil Engineering, 1-54.

Sriram, D., (1987), "Knowledge-Based Approaches for Structural Design, CM Publications, Southampton.

Sriram, D., Logcher, R., Wong, A., and Ahmed, S., (1991), "Computer-Aided Cooperative Product Development: A Case Study", of Systems Automation: Research & Applications, Vol 1, No 1, pp 91-114.

Sriram, D., Logcher, R., Wong, A., and Ahmed, S., (1991), "An Object-Oriented Environment for Computer-Aided Collaborative Engineering Design", In Computer-Aided Cooperative Product Development, Edited by Sriram, D., Logcher, R., and Fukuda, S., Springer Verlag Inc.

Sriram, D., Logcher, R., Wong, A., Ahmed, S., (1991), "Computer-Aided Cooperative Product Development: A Case Study", Int. J. of Systems Automation: Research and Applications (SARA), v.1, 89-112.

Sriram, D., Maher, M.L., Fenves, S.J., (1985), "Knowledge-Based Expert Systems in Structural Design", Computers and Structures, v.20, n.1-3, 1-9.

Sriram, D., Stephanopoulos, G., Logcher, R., Gossard, D., Groleau, N., Serrano, D., Navinchandra, D., (1989), "Knowledge-Based System Applications in Engineering Design: Research at MIT", AAAI, 79-96.

Steier, D., (1990), "Creating a Scientific Community at the Interface Between Engineering Design and AI", AI Magazine, Winter 1990, v.11, n.4, 18-22.

Thompson, D.R., (1990), "A proposed model of the structural design process with special reference to non-domestic timber roof structures", Civil Engineering Systems, v.7, n.1, 5-10.

Thompson, R., Tomski, T., Ellacott, S.W., Kuczora, P., (1989), "An Expert System for the Preliminary Design of Timber Roofs", in Topping, B.H.V., (Editor), (1989), "Artificial Intelligence Techniques and Applications for Civil and Structural Engineers", Civil-Comp Press, Edinburgh, 187-196.

Turkiyyah, G.M. and Fenves, S.J., (1987), "Knowledge-Based Analysis of Structural Systems", in Siram, D., Adey, R.A., (1987), (editors), "Knowledge Based Expert Systems in Engineering: Planning and Design", Computational Mechanics Publications, Southampton, 273-284.

Tweed, C., Bijl, A., (1988), "MOLE: A Reasonable Logic for Design", in ten-Hagen, P.J.W., Tomiyama, T., Akman, V., (1988),, (editors), "Published in Intelligent CAD System 2: Implementation Issues, Springer-Verlag, Berlin, 1988, 1-23.

Wang, J., Howard, H.C., (1988), "Design-Dependent Knowledge for structural engineering design", in Gero, J.S., (1988), (editor), "Artificial Intelligence in Engineering: Design", Computational Mechanics Publications, Southampton, 267-278.

Wang, J., and Howard, H.C., (1991), "A Design-dependent Approach to Integrated Structural Design", Artificial Intelligence in Design, Butterworth-Heinemann Ltd., Oxford, England, 151-171.

Yang, B., Datta, U., Datseris, P., Wu, Y., (1989), "An Integrated System for Design of Mechanisms by an Expert System-Domes", AIEDAM, v.3, n.1, 53-70.

Yehia, N.A.B., Bechara, R.M., (1988), "A Simple Knowledge-Based Design Approach", Concrete International, ACI, 50-55.

Yehia, N.A.B., Sabbah, A.A.L., Mosallam, A.M., (1989), "A Knowledge-Based System for the Design of Reinforced Concrete Halls", in Topping, B.H.V., (Editor), (1989), "Artificial Intelligence Techniques and Applications for Civil and Structural Engineers", Civil-Comp Press, Edinburgh, 201-210.

Zhang, X.J., Yao., J.T.P., (1988), "Expert systems for condition evaluation of existing structures", in Adeli, H., (1988), (editor), "Expert Systems in Construction and Structural Engineering", Chapman and Hall, London.

Zhang, X.J., Yao, J.T.P., (1988), "Knowledge Acquisition and Machine Learning in an Expert System" in Adeli, H., (1988), (editor), "Microcomputer Knowledge-Based Expert Systems in Civil Engineering", American Society of Civil Engineers, New York.

13.4 CONSTRAINTS, CODES, RULES AND DESIGN

Atabakhsh, H., (1991), "A Survey of Constraint Based Scheduling Systems Using an Artificial Intelligence Approach", Artificial Intelligence in Engineering, v.6, n.2, 58-73.

Adeli, H., Chen, Y.S., (1989), "Structuring Knowledge and Data Bases in a Coupled Expert System for Integrated Structural Design", Microcomputers in Civil Engineering, v.4, n.3.

Akman, V., Hager, P., Rogier, J., Veerkamp, P., (1988), "Knowledge Engineering in Design", Knowledge Based Systems, v.1, n.2, 67-77.

Arciszewski, T., (1989), "Model-Based Expert System for Engineering Design", in "Proceedings: USACERL/ASCE First Joint Conference on Expert Systems June 29-30, 1988", USACERL Conference Proceedings P-89/10, US Army Corps of Enginees, Construction Engineering Research Laboratory, Champaign, IL, USA, 12-16.

Balachandran, M., Gero, J.S., (1987), "A Knowledge-Based Graphical Interface for Structural Design", in Siram, D., Adey, R.A., (1987), (editors), "Knowledge Based Expert Systems in Engineering: Tools and Techniques", Computational Mechanics Publications, Southampton, 335-346.

Barbuceanu, M., (1985), An Object-Centred Framework for Expert Systems in Computer Aided Design", in Gero, J.S., (1985), (editor), "Knowledge Engineering in Computer-Aided Design", Elsevier Science Publishers B.V.(North Holland), 223-258.

Bedard, C., (1989), "Research directions for AI in design: a building engineers view of design research", in Gero, J.S., (1989), (editor), "WAID: Workshop on research directions for artificial intelligence in design, Stanford University, March, 1989", The Department of Architectural and Design Science, University of Sydney, Australia, 39-42.

Bedard, C., (1989), "CODES Prcessing and Preliminary Design with Expert Systems", Proceedings 1989 CSCE/CPCA Structural Conf. March 20-21, 1989, cosponsored by: Construction Control Ltd, Carleton Univesity, McGill University Carruthers and Wallace Ltd. Ontario Ministry of Trans. & Comm., CPCI, 177-193.

Brotchie, J.F., Sharpe, R., Marksjo, B., (1986), "Introducing Intelligent and Knowledge into CAD", in Siram, D., Adey, R., (1986), (editors), "Applications of Artificial Intelligence in Engineering Problems: Proceedings of the 1st Int. Conf., Southampton University, April 1986", Computational Mechanics Publications, Southampton, 797-810.

Brown, D.C., Breau, R., (1986), "Types of Constraints in Routine Design Problem-Solving", in Siram, D., Adey, R., (1986), (editors), "Applications of Artificial Intelligence in Engineering Problems: Proceedings of the 1st Int. Conf., Southampton University, April 1986", Computational Mechanics Publications, Southampton, 383-390.

Cameron, G.E., Grierson, D.E., (1989),"Developing an Expert System for Structural Steel Design: Issues and Items", in Gero, J.S., (1989), (editor), "Artificial Intelligence in Design", Computational Mechanics Publications, Southampton, 15-40.

Cagan, J., Agogino, A.M., (1989), "Why AI-Design researchers should distinguish between creative, innovative and routine levels of design", in Gero, J.S., (1989), (editor), "WAID: Workshop on research directions for artificial intelligence in design, Stanford University, March, 1989", The Department of Architectural and Design Science, University of Sydney, Australia, 43-51.

Case, M., Quek, L.H., (1989), "Using CAD for Knowledge Based Systems" in "Proceedings: USACERL/ASCE First Joint Conference on Expert Systems June 29-30, 1988", USACERL Conference Proceedings P-89/10, US Army Corps of Enginees, Construction Engineering Research Laboratory, Champaign, IL, USA, 17-21.

Chan, W, Paulson, B.C., (1987), "Exploratory Design using Constraints", AIEDAM, v.1, n.1, 59-71.

Chan, W., Paulson, B.C., (1988), "Logic Programming to Manage Constraint-Based Design", in Adeli, H., (1988), (editor), "Microcomputer Knowledge-Based Expert Systems in Civil Engineering", American Society of Civil Engineers, New York.

Chandresekaran, B., (1989), "A framework for design problem-solving", in Gero, J.S., (1989), (editor), "WAID: Workshop on research directions for artificial intelligence in design, Stanford University, March, 1989", The Department of Architectural and Design Science, University of Sydney, Australia,1-19.

Cronembold, J.R., Law, K.H., (1988), "Automated Processing of Design Standards", Journal of Computing in Civil Engineering, ASCE, v.2, n.3, 255-273.

Das, M.L., (1986), "Knowledge-Base for Structural Design", in Siram, D., Adey, R., (1986), (editors), "Applications of Artificial Intelligence in Engineering Problems: Proceedings of the 1st Int. Conf., Southampton University, April 1986", Computational Mechanics Publications, Southampton, 33-44.

Fazio, P., Bedard, C., Gowri, K., (1988), "Knowledge-Based System Development Tools for Processing Design Specifications", Microcomputers in Civil Engineering", Mircocomputers in Civil Engineering, 3, 333-344.

Fazio, P., Bedard, C., Gowri, K., (1988), "A Rule-Based System for Estimating Snow Loads on Roofs", Proceedings of Knowledge-Based Expert Systems in Civil Engineering, ASCE, May 1988, Nashville, TN., USA, 73-78.

Fenves, S.J., Garrett, Jr., J.H., (1985), "Standards Representation and Processing", Steel in Buildings, IABSE, 107-114.

Fenves, S.J., Garrett, J.H., (1986), "Knowledge Based Standards Processing", Artificial Intelligence in Engineering, v.1, n.1, 3-14.

Fenves, S.J., Liadis, S.A., (1976), "A data structure for computer-aided design of buildings", APEC Journal, 14-17, Fall.

Fenves, S.J., Rasdorf, W.J., (1982), "Treatment of Engineering Design Constraints in a Relational Database", Carnegie-Mellon University, Report DRC-12-14-82.

Finger, S., Dixon, J.R., (1989), "A review of resesrch in Mechanical Engineering Design", in Gero, J.S., (1989), (editor), "WAID: Workshop on research directions for artificial intelligence in design, Stanford University, March, 1989", The Department of Architectural and Design Science, University of Sydney, Australia, 53-58.

Fox, M.S., (1989), "Issues in architecture for desihn" in Gero, J.S., (1989), (editor), "WAID: Workshop on research directions for artificial intelligence in design, Stanford University, March, 1989", The Department of Architectural and Design Science, University of Sydney, Australia, 59-60.

Fukuda, S., (1988), "Codes and rules and their roles as constraints in expert systems for structural design", in Adeli, H., (1988), (editor), "Expert Systems in Construction and Structural Engineering", Chapman and Hall, London.

Garrett, Jr., J.H., (Oct 1989), "An Object-Oriented Representation of Design Standards", in the Proceedings of the International Association of Bridge and Structural Engineers (IABSE) Colloquium on Expert Systems in Civil Engineering, Bergamo, Italy, 373-382.

Garrett J.H., Fenves, S.J., (1986), "A Knowledge-Based Standards Processor for Structural Component Design", Report No. R-86-157, Department of Civil Engineering, Carnegie-Mellon University, U.S.A.

Garrett, Jr., J.H., Fenves, S.J., (1987), "A Knowledge-Based Standards Processor for Stuctural Component Design", Engineering with Computers, v.2, 219-238.

Garrett, Jr., J.H., Fenves, S.J., (1989), "Knowledge-Based Standard-Independent Member Design", Journal of Structural Engineering, ASCE, v.115, n.6, 1396-1411.

Gero, J.S., (1989), "Research directions for artificial intelligence in design", in Gero, J.S., (1989), (editor), "WAID: Workshop on research directions for artificial intelligence in design, Stanford University, March, 1989", The Department of Architectural and Design Science, University of Sydney, Australia, 61-64.

Gero, J.S., Balachandran, B., (1986), "Knowledge and Design Decision Processes" in Siram, D., Adey, R., (1986), (editors), "Applications of Artificial Intelligence in Engineering Problems: Proceedings of the 1st Int. Conf., Southampton University, April 1986", Computational Mechanics Publications, Southampton, 343-352.

Gero, J.S., Coyne, R.D., (1986), "Developments in Expert Systems for Design Synthesis" in Kostem, C.N., Maher, M.L., (1986), "Expert Systems in Civil Engineering", American Society of Civil Engineers, New York.

Jain, D., Maher, M.L., (1988), "Combining Expert Systems and CAD Techniques" in Gero, J.S., Stanton,R., (1988), (editors), "Artificial Intelligence Developments and Applications", edited selection of papers to the Australian Joint Artificial Intelligence Conference, Sydney, November, 1987, North-Holland, Amsterdam, 65-81.

Jain, D., Law, K.H., Krawinkler, H., (1989), "On Processing Standards with Predicate Calculus", in Barnwell,Jr. T.O., (1989), (editor), "Computing in Civil Engineering, Computers in Engineering Practice, Proceedings of the Sixth Conf., Sponsored by the Technical Council on Computer Practices of the ASCE, Sept. 11-13, 1989, 259-266.

Harris, J.R., Wright, R.N., (1980), "Organization of Building Standards: Systematic Techniques for Scope and Arrangement", Building Science Series, NBS BSS 136, National Bureau of Standards, Washington, D.C., U.S.A., Dec.

Holtz, N.M., Fenves, S.J., (1982), "Symbolic Manipulation of Design Constraints", Technical Report # DRC-02-12-82, Design Research Center, Carnegie Mellon University, Pittburgh, USA.

Kumar, B., Topping, B.H.V., (1988), "Issues in the development of a Knowledge-Based system for the detailed design of structures",in Gero, J.S., (1988), (editor), "Artificial Intelligence in Engineering: Design", Computational Mechanics Publications, Southampton, 295-314.

Kumar, B., Topping, B.H.V., (1989), "A Prolog-Based Representation of Standards for Structural Design", in Topping, B.H.V., (Editor), (1989), "Artificial Intelligence Techniques and Applications for Civil and Structural Engineers", Civil-Comp Press, Edinburgh, 165-170.

Landsdown, J., (1989), "The Designer's information environment some tools for design knowledge manipulation", Civil Engineering Systems, v.6, n1/2, 5-10.

Law, K.H., Jouaneh, M.K., (1986), "Data Modelling for Building Design", in Lenocker, W.T., (editor), "Computing in Civil Engineering", Proceedings of the Fourth Conference, American Society of Civil Engineers, New York, U.S.A., 21-36.

Law, K.H., Jouaneh, M.K., Spooner, D.L., (1987), "Abstraction Database Concept for Engineering Modeling", Engineering with Computers, v.2, 79-94,

MacCallum, K.J., (1982), "Creative Ship Design by Computer", in Rogers,D.F., Nehrling, B.C., Kuo, C., (1982), "Computer Applications in the Automation of Shipyard Operation and Ship Design IV", North Holland Publishing, 55-62.

MacCallum, K.J., (1982), "Understanding relationships in marine systems design", Proc. First Int. Marine Systems Design Conf., Theory and Prctice of Marine Design, Royal Institute of Naval Architects, April, 1-9.

Mackenzie, C.A., Gero, J.S., (1987), "Learning design rules from decisions and performances", Artificial Intelligence in Engineering, v.2, n.1, 2-10.

McLaughlin, S., Gero, J.S., (1987), "Learning from Characterized Designs", in
Siram, D., Adey, R.A., (1987), (editors), "Knowledge Based Expert Systems in Engineering: Tools and Techniques", Computational Mechanics Publications, Southampton, 347-359.

McLaughlin, S., Gero, J.S., (1989), "Requirements of a Reasoning System that supports creative and innovative design activity", Knowledge Based Systems, v.2, n.1, 62-71.

Maher, M.L., (1989), "Synthesis and Evaluation of Preliminary design", in Gero, J.S., (1989), (editor), "Artificial Intelligence in Design", Computational Mechanics Publications, Southampton, 3-14.

Maher, M.L., (1989), "Research directions: AI in design", in Gero, J.S., (1989), (editor), "WAID: Workshop on research directions for artificial intelligence in design, Stanford University, March, 1989", The Department of Architectural and Design Science, University of Sydney, Australia, 65-68.

Mostow, J., (1989), "Towards knowledge compilation as an approach to computer aided design" in Gero, J.S., (1989), (editor), "WAID: Workshop on research directions for artificial intelligence in design, Stanford University, March, 1989", The Department of Architectural and Design Science, University of Sydney, Australia, 21-37.

Navinchandra, D., (1989), "Research directions for AI in design", in Gero, J.S., (1989), (editor), "WAID: Workshop on research directions for artificial intelligence in design, Stanford University, March, 1989", The Department of Architectural and Design Science, University of Sydney, Australia, 69-76.

Neveu, B., Haren, P., (1986), "Smeci: An Expert System for Civil Engineering Design", in Siram, D., Adey, R., (1986), (editors), "Applications of Artificial Intelligence in Engineering Problems: Proceedings of the 1st Int. Conf., Southampton University, April 1986", Computational Mechanics Publications, Southampton, 317-326.

Pettit, J.H., (1990), "An Expert System for the Compilation of Engineering Specification documents", Eng. Appli. of AI., V.3, 288-297.

Rasdorf, W.J., Fenves, S.J., (1986), "Constraint Enforcement in Structural Design Data Base", Journal of Structural Engineering, v.112, n.12, 2565-2577.

Rasdorf, W.J., Lakmazaheri, S., (1991), "A Logic Based Approach for Processing Design Standards", AIEDAM v.4, n.3, 179-192.

Rasdorf, W.J., Lakmazaheri, S., (1990), "A Logic-Based Approach For Modeling, the Organization of Design Standards", Journal of Computing in Civil Engineering, ASCE, v.4, n.2, 102-123.

Rasdorf, W.J., Parks, L.M., (1986), "Natural Language Prototypes for Analysisng Design Standards",in Siram, D., Adey, R.A., (1987), (editors), "Knowledge Based Expert Systems in Engineering: Tools and Techniques", Computational Mechanics Publications, Southampton, 147-160.

Rasdorf, W.J., Ulberg, K.J., Baugh, J.W., (1987), "A Structure-Based Model of Semantic Integrity Constraints for Relational Databases", Engineering with Computers, v.2, n.1, 31-39.

Rasdorf, W.J., Wang, T.E., (1987), "SPIKE: A Generic Design Standards Processing Expert System" Siram, D., Adey, R.A., (1987), (editors), "Knowledge Based Expert Systems in Engineering: Planning and Design", Computational Mechanics Publications, Southampton, 241-257.

Rasdorf, W.J., Wang, T.J.E., (1988), "Generic Design Standards Processing in an Expert System Environment", Journal of Computing in Civil Engineering, ASCE, v.2, 68-87.

Rooney, M.F., Smith,S.E., (1983), "Artificial Intelligence in Engineering Design", Computers and Structures, v.16, 279-288.

Rosenman, M.A., Gero, J.S., (1985), "Design Codes as Expert Systems", Computer Aided Design, v.17, n.9, 399-409.

Rosenmann, M.A., Gero, J.S., Hutchinson, P.J., and Oxman, R., (1986), in "Expert Systems Applications in Computer-Aided Design", in Smith, A., (editor), "CAD86", Butterworths, London, 218-225.

Rosenman, M.A., Gero, J.S., Oxman, R., (1986), "An Expert System for Design Codes and Design Rules", in Siram, D., Adey, R., (1986), (editors), "Applications of Artificial Intelligence in Engineering Problems: Proceedings of the 1st Int. Conf., Southampton University, April 1986", Computational Mechanics Publications, Southampton, 745-758.

Ross, T.J., Wong, F.S., (1989), "Integrated Knowledge-Based System for Structural Design", in "Proceedings: USACERL/ASCE First Joint Conference on Expert Systems June 29-30, 1988", USACERL Conference Proceedings P-89/10, US Army Corps of Enginees, Construction Engineering Research Laboratory, Champaign, IL, USA, 72-79.

Sause, R., Powell, G., (1989), "Issues in research directions for artificial intelligence in design", in Gero, J.S., (1989), (editor), "WAID: Workshop on research directions for artificial intelligence in design, Stanford University, March, 1989", The Department of Architectural and Design Science, University of Sydney, Australia, 77-82.

Simmons, M.K., (1984), "Artificial Intelligence for Engineering Design", Computer Aided Engineering Journal, 75-83.

Smithers, T., (1985), "The Alvey Large Scale Demonstrator Project "Design to Product", 3rd Int. Conf. on Advanced IT: AI in Manufacturing - Key to Integration, Zurich, Switzerland, Nov.

Smithers, T., (1989), "Intelligent control in AI-based design support systems", in Gero, J.S., (1989), (editor), "WAID: Workshop on research directions for artificial intelligence in design, Stanford University, March, 1989", The Department of Architectural and Design Science, University of Sydney, Australia, 83-90.

Oey, K.H., Passchier, E., (1987), "An Expert System Approach for Codes of Practice" KM 87/4, Instituut TNO voor Bouwmaterialen en Bouwconstructies, TNO, Delft.

Rosenman, M.A., Gero, J.S., Oxman, R., (1986), "An Expert System for Design Codes and Design Rules", in Siram, D., Adey, R., (1986), (editors), "Applications of Artificial Intelligence in Engineering Problems: Proceedings of the 1st Int. Conf., Southampton University, April 1986", Computational Mechanics Publications, Southampton.

Sriram, D., (1986), "DESTINY: A Model for Integrated Structural Design", Artificial Intelligence in Engineering, v.1, n.2, 109-116.

Sriram, D., (1987), "ALL-RISE: A Case Study in Constraint-Based Design", Artificial Intelligence in Engineeering, v.2, n.4, 186-203.

Sriram, D., Mather, M.L., (1986), "The Representation and Use of Constraints in Structural Design", in Siram, D., Adey, R., (1986), (editors), "Applications of Artificial Intelligence in Engineering Problems: Proceedings of the 1st Int. Conf., Southampton University, April 1986", Computational Mechanics Publications, Southampton, 355-386.

Taylor, N.K., Corlett, E.N., (1987), "An Expert System which Constrains Designs", Artificial Intelligence in Engineering, v.2, n.2, 72-75.

Tong, C., (1987), "Toward an Engineering Science of Knowledge-Based Design", Artificial Intelligence in Engineering, v.2, n.3, 133-166.

Tong, C., (1989), "Open research areas in knowledge-based design", in Gero, J.S., (1989), (editor), "WAID: Workshop on research directions for artificial intelligence in design, Stanford University, March, 1989", The Department of Architectural and Design Science, University of Sydney, Australia, 91-107.

Topping, B.H.V., Kumar, B., (1989), "Knowledge Representation and processing for structural engineering design codes", Eng. Appl. of A.I., v.2, n.3, 214-228.

Twelt, C., Moskowitz, D., (1991), "Automated Text Generation for Building Standards", Journal of Computing in Civil Engineering, ASCE., v.4, n.1, 20-36.

Vaish, A.K., Kamil, H., (1988), "An Expert System for Structural Analysis and Design", AIAA/ASME/ASCE/AHS, 29th Structures, Structural Dynamics and Materials Conf., Part1, A Collection of Technical Papers, April 18-20, 1988, Williamsburg, Virginia, 1333-1348.

Wentorf, R., Law, K.H., (1986), "Natural Language Processing and Computer Graphics for Engineering Design", in Lenocker, W.T., (editor), "Computing in Civil Engineering", Proceedings of the Fourth Conference, American Society of Civil Engineers, New York, U.S.A., 37-52.

Wentof, R., Law, K.H., Ackroyd, M.H., (1986), "A Natural Language Interepreter for Plate Girder Design", Engineering with Computers, v.1, 127-146.

Wolverton, M., Hayes-Roth, B., (1989), "Using analogy to creatively design", in Gero, J.S., (1989), (editor), "WAID: Workshop on research directions for artificial intelligence in design, Stanford University, March, 1989", The Department of Architectural and Design Science, University of Sydney, Australia, 109-114.

13.5 DATABASE SYSTEMS

Ahmed, S., Sriram, D., and Logcher, R., (1991), "A Comparative Analysis of Object Oriented Databases for Collaborative Engineering Work", Seventh ASCE Conference on Computing in Civil Engineering, Washington, D.C.

Ahmed, S., Sriram, D., and Logcher, R., (1991), "Transaction and Version Management in Object-Oriented Databases", Seventh ASCE Conference on Computing in Civil Engineering, Washington, D.C. ,

Asgari, D., Modesitt, K.L., (1986), "Space Shuttle Main Engine Test Analysis: A Case Study for Inductive Knowledge-Based Systems Involving Very Large Data Bases", Proceedings of the 10 Annual Conf., IEEE Computer Society COMPSAC '86, October 1986, Chicago, Ill., USA, 65-71.

Ghosh, D.K., Kalyanaraman, V., (1990), "A Relational Database Interface for Design Expert Systems", Microcomputers in Civil Engineering, V 5, 151-156.

Howard, H.C., (1988), "KADBASE: An Expert System/Database Interface", The Fifth American Society of Civil Engineers Conference in Computing in Civil Engineering, Alexandria, VA, USA, March 29-31, 11-20.

Howard, H.C., Rehak, D.R., (1989), "KADBASE Interfacing Expert with Databases", IEEE Expert, 65-75.

Howard, H.C., Rehak, D.R., (1986), "Expert Systems and CAD Databases", in Smith, A., (editor), "CAD86", Butterworths, London, 236-248.

Howard, H.C., and Rehak, D.R., (1989), "KADBASE - A Prototype Expert System-Database Interface for Integrated CAE Environments", IEEE Expert, 804-808.

Lafue, G.M.E., Mitchell, T., (1983), "Database Management Systems and Expert Systems for CAD", in Bo, K., Lillehagen, F.M., (1983), (editors), "CAD Systems Framework", North-Holland Publishing Company, 313-323.

Powell, G.H., Bhateja, R., Abdalla, J.A., An-Nashif, H., Sause, R. and Martini, K., (1988), "A Database Concept for Computer Integrated Structural Engineering Design", The Fifth American Society of Civil Engineers Conference in Computing in Civil Engineering, Alexandria, VA, USA, 521-529.

Rasdorf, W.J., Salley, G.C., (1985), "Generative Engineering Databases- Toward Expert Systems", Computers and Structures, v.20, n.1-3, 11-15.

Rasdorf, W.J., Wang, T.E., (1986), "Expert System Integrity Maintenance for the Use of Engineering Data", Proceedings of the Fourth Conference on Computing in Civil Engineering, ASCE, Boston, Ma, 654-668.

Rasdorf, W.J., Wang, T.E., (1986), "CDIS: An Engineering Constraint Definition and Integrity Enforcement System for Relational Databases", Proceedings of the Int. Computers in Engineering Conf., ASME, Chicago, Il, USA, V.2, 273-280.

Rasdorf, W.J., Watson, B.R., (1986), "ADI: An Adaptive Database Interface for Dynamic Databases", Proceedings of the 1986 ASME Symposium on Knowledge-Based Expert Systems for Manufacturing, ASME, Production Engineering Division, Anaheim, Ca, 119-130.

Rasdorf, W.J., Watson., B.R., (1988), "A Knowledge-based approach to engineering information retrieval and management", in Adeli, H., (1988), (editor), "Expert Systems in Construction and Structural Engineering", Chapman and Hall, London.

Watson, A.S., Chan, S.H., (1989), "A Prolog Based Object Oriented Engineering DBMS", in Topping, B.H.V., (Editor), (1989), "Artificial Intelligence Techniques and Applications for Civil and Structural Engineers", Civil-Comp Press, Edinburgh, 37-48.

13.6 BRIDGE ENGINEERING

Adeli, H., Balasubramanyam, K.V., (1987), "Heuristic Analysi of Bridge Trusses under AASHTO Live Loads", Microcomputers in Civil Engineerin, v.2, n.2 147-160.

Adeli, H., Balasubramanyam, K.V., (1988), "A Knowledge-Based System for Design of Bridge Trusses", Journal of Computing in Civil Engineering, ASCE, v.2, n.1, 1-20.

Adeli, H., Mak, K.Y., (1988), "Architecture of a coupled expert system for optimum design of plate girder bridges", Engineering Applications of Artificial Intelligence, v.1, 277-285.

Adeli, H., Mak, K.Y., (1989), "Application of a Coupled System for Optimum Design of Plate Girder Bridges", Engineering Applications of Artificial Intelligence, v.2, nl.

Burgoyne, C.J., Sham, S.H.R., (1987), "Application of Expert Systems to Prestressed Concrete Bridge Design", Civil Engineering Systems, Vol. 4, n.1., 14-19.

Futura, H., Shiraishi, N., (1988), "Safety Assessment of Bridges during Constructiion", in Iri, M., Yajima, K., (1988), "System Modelling and Optimization", Springer Verlag, Berlin, 498-507.

Futura, H., Shiraishi, N., "Applications of Expert Systems to Structural Engineering", in Choi, C.-K., Shiraishi, N., (1988), Proceedings of Korea-Japan Joint Seminar on "Emerging Technologies in Structural Engineering and Mechanics", 1-10.

Kostem, C.N., (1986), "Design of an Expert System for the Rating of Multibeam Highway Bridges", in Kostem, C.N., Maher, M.L., (1986), "Expert Systems in Civil Engineering", American Society of Civil Engineers, New York.

Sham. S.H.R., (1986), "Expert Systems in Bridge Design: A Research Proposal", ESL Technical Report 2/86, Department of Civil Engineerring, Imperial College of Science, Technology and Science, London.

Sham, S.H.R., (1989), "An Overview of the Computational Aspects of the QED Project" ESL Technical Report 8/89, Department of Civil Engineering, Imperial College of Science, Technology and Medicine, London.

Shiraishi, N., Futura,H., Umano, M., Kawakami, K., (1987), "An Expert System for Damage Assessment of Reinforced Concrete Brideg Deck", Proceedings of Second IFSA Congress, Tokyo, 160-163.

Spencer, W.J., Atkins, R.M., Podlaha, P., (1989), "The Development of an Expert System for the preliminary design of bridges", Civil Engineering Systems, v.6, n1/2, 51-57.

13.7 STRUCTURAL LOADING

Fazio, P., Bedard, C., Gowri, K., (1988), "A Rule Based System for Estimating Snow Loads on Roofs" in Adeli, H., (1988), (editor), "Microcomputer Knowledge-Based Expert Systems in Civil Engineering", American Society of Civil Engineers, New York.

Marksjo, B., Sharpe, R., Holmes, J., Fitchett, P., Ho, F., (1989), "Windloader KBS- from prototype to the market", Civil Engineering Systems, v.6, n.1/2, 36-43.

Miyasato, G.H., Wong, W.M., Levitt, R.E., Boissonnade, A.C., Shah, H.C., (1986), "Seismic Risk Analysis Systems", in Kostem, C.N., Maher, M.L., (1986), "Expert Systems in Civil Engineering", American Society of Civil Engineers, New York.

13.8 OFFSHORE ENGINEERING

Ahmad, K., Langdon, A., Frieze, P.A., (1991), "An Expert System for Offshore Structure Inspection and Maintenance", Computers & Structures, v.40, n.1, 143-159.

Brysland, C., MacCallum, K.J., (1988), "A Knowledge based assistant for the scheduling of subsea maintenance", in Gero, J.S., (1988), (editor), "Artificial Intelligence in Engineering: Robotics and Processes", Computational Mechanics Publications, Southampton, 311-329.

Pek, P.-K., (1990), "Development of a PC-Based Expert System for the Design of Offshore Topside Structures", Engineering Application of Artificial Intelligence, v.3, n.2, 91-100.

Gupta, A.,(1988),"Knowledge-Based Expert Systems in Offshore Structural Design", Computers and Structures,v.28,n.6,745-748.

Gupta, A., Jain, A.K., (1986), "Application of Artificial Intelligence in Offshore Structures", in Siram, D., Adey, R., (1986), (editors), "Applications of Artificial Intelligence in Engineering Problems: Proceedings of the 1st Int. Conf., Southampton University, April 1986", Computational Mechanics Publications, Southampton, 669-706.

Langdon, A., Ahmad, K., Frieze, P.A., (1989), "An Expert System for Offshore Structure Inspection and Maintenance", in Topping, B.H.V., (Editor), (1989), "Artificial Intelligence Techniques and Applications for Civil and Structural Engineers", Civil-Comp Press, Edinburgh, 221-232.

Reddy, D.V., Arockiasamy, M., Sriram, D., and Connor, J., (1988), "Knowledge-Based Systems for Design and Analysis of Offshore Structures", In ASME Conference on Computers in Artic Regions.

Soh, C.K., Miles, J.C., (1989), " The Design of Steel Offshore Structures using an Expert System, in Topping, B.H.V., (Editor), (1989), "Artificial Intelligence Techniques and Applications for Civil and Structural Engineers", Civil-Comp Press, Edinburgh, 197-200.

Soh, C-K., Soh, A-K., (1989), "Using meta-level knowledge in the design of offshore structures", Eng. Appl. of A.I., v.2, n.3, 228-237.

13.9 DYNAMICS AND EARTHQUAKE ENGINEERING

Adeli, H., Hung, S.L., (1989), "A Knowledge-Based Assistant for Earthquake Resistant Design", in Topping, B.H.V., (Editor), (1989), "Artificial Intelligence Techniques and Applications for Civil and Structural Engineers", Civil-Comp Press, Edinburgh, 241-248.

Alim, S., (1987), "A prolog-based expert system for encoding seismic design provisions", Civil Engineering Systems, v.4, n.1, 39-44.

Baker, D.J., Manolis, G.D., Mohan, S., (1989), "HIBIC: Expert System for Highway Bridge Dynamics", Journal of Computing in Civil Engineering, ASCE, v.3, n.4, 348-366.

Dong, W-M., Shah, H.C., Wong, F.S., (1988), "Condensation of the Knowledge Base in Expert Systems with Applications to seismic risk evaluation" in Adeli, H., (1988), (editor), "Expert Systems in Construction and Structural Engineering", Chapman and Hall, London.

Dussom, K.B., Hadj-Hamou, T., Bakeer, R.M., (1989), "Quake: An Expert System for the Selection of Design Earthquake Accelerogram", in Topping, B.H.V., (Editor), (1989), "Artificial Intelligence Techniques and Applications for Civil and Structural Engineers", Civil-Comp Press, Edinburgh, 233-240.

Ganguly, J., Kausel, E., and Sriram, D., (1990), "QSEIS: A Knowledge-Based System for Deriving Qualitative Seismic Behaviour of Structural Systems from Structural Descriptions", In Artificial Intelligence: Applications in Computational Engineering, Edited by Kleiber, M., Published by Ellis Horwood.

Ganguly, J., Kausel, E., Sriram, D., (1991), "An Integrated Knowledge-Based Framework for Deriving (Seismic) Behavior Qualitatively from Structural Descriptions", Journal of Computing in Civil Engineering, ASCE, 1-37.

Hosking, J.G., Lomas, S., Mugridge, W.B., Cranston, A.J., (1989), "The development of an expert system for seismic loading", Civil Engineering Systems, v.6, n.1/2, 27-35.

Kim, J., Dong, W., Wong, F., Shah, H.C., (1988), "Expert Systems for the Earthquake Related Industry", in Adeli, H., (1988), (editor), "Microcomputer Knowledge-Based Expert Systems in Civil Engineering", American Society of Civil Engineers, New York.

Lui, X., Wang, P., (1988), "An Expert System for Earthquake intensity evaluation", in Adeli, H., (1988), (editor), "Expert Systems in Construction and Structural Engineering", Chapman and Hall, London.

Subramani, M., Gergely, P., Conely, C.H., Abel, J.F., Zaghw, A., (1989), "A Knowledge-Based Approach to Structural Design of Earthquake-Resistant Buildings", Technical Report NCEER-89-0006, National Center for Earthquake Engineering Research, State University of New York at Buffalo, USA.

Subramani, M., Zaghw, A., Conley, C.H., Gergely, P., Abel, J.F., (1989), "A KBES for Seismic Design of Buildings", in Nelson, J.K., (1989), "Computer Utilization in Structural Engineering", American Society of Civil Engineers, 342-351.

Thurston, H.M., (1990), "Seismic Damage Evaluation of Tilt-Up Buildings", Journal of Computing in Civil Engineering, ASCE, v.4, n.4, 349-369.

13.10 STRUCTURAL OPTIMIZATION

Adeli, H., Balasubramanyam, K.V., (1987), "Interactive Layout Optimization of Trusses", Journal of Computing in Civil Engineering, ASCE, v.1, n.3, 183-196.

Arciszewski, T., Ziarko, W., (1991), "Structural Optimization: Case-Based Approach", Journal of Computing in Civil Engineering, ASCE, v.5, n.2, 159-174.

Arora, J.S., Baenziger, G., (1986), "Uses of Artificial Intelligence in Design Optimization", Computer-Methods in Applied Mechanics and Engineering, v.54, 303-323.

Azarm, S., Pecht, M., (1986), "An Approach for Building a Rule-Based System for Design Optimization", in Niku-Lari,A., (1986), (editor), "Structural Analysis Systems", Pergamon Press, Oxford, 363-383.

Baier, H.J., (1986), "Improved Elements for Computer Generated Design of Dynamically Exiited Structures based on Expert Systems and Quasi-Optimization Routine", in Niku-Lari,A., "Structural Analysis Systems", Pergamon Press, Oxford, 11-25.

Balachandran, M., (1991), "Knowledge Based Optimium Design", Computational Mechanics Publications, Southampton.

Balachandran, M., Gero, J.S., (1988), "Development of a Knowledge-Based System for Structural Optimization", Structural Optimization, G. I. N., Rozvany, and B. L. Karihaloo, (eds), Kluwer Academic Publishers, 17-24.

Balachandran, M., Gero, J.S., (1987), "A Knowledge-based approach to mathematical design modeling and optimization", Engineering Optimization, v.12, n.2, 91-115.

Balachandran, M., Gero, J.S., (1987), "Use of Knowledge in selection and control of optimization algorithms", Engineering Optimization, v.12, n.2, 163-173.

Balachandran, M., Gero, J.S., (1990), "Knowledge Engineering and Multicriteria Optimization", Multicriterion Design Optimization, H. Eschenauer, J. Koski and A. Osyczka, (eds), Springer-Verlag, Berlin, pp 115-147.

Hartmann, D., (1986), "Selection and Evaluation of Structural Optimization Strategies by means of Expert Systems", in Niku-Lari,A., (1986), (editor), "Structural Analysis Systems", Pergamon Press, Oxford, 39-56.

Jozwiak, S.F., (1986), "Application of Artificial Intelligence Notions in Structural Optimization Programs", Computers and Structures, v.24, n.6, 1009-1013.

Jozwiak, S.F., (1986), "Application of Artificial Intelligence in Structural Optimization", in Siram, D., Adey, R., (1986), (editors), "Applications of Artificial Intelligence in Engineering Problems: Proceedings of the 1st Int. Conf., Southampton University, April 1986", Computational Mechanics Publications, Southampton, 823-831.

Jozwiak, S., (1986), "Technical Note: Application of Artificial Intelligence Notations in Structural Optimization Programs", Computers and Structures v.24, n.6, 1009-1013.

Jozwiak, S., (1987), "Improving Structural Optimization Programs Using Artificial Intelligence Concepts", Engineering Optimization, v.12, 155-162.

Jozwiak, S., (1986), "Application of Artificial Intelligence in Structural Optimization", in Adey, R., Sriram, D., (1986), (editors), "Applications of Artificial Intelligence in Engineering Problems, CMP, Southampton, 823-831.

Jozwiak, S., (1986), "Artificial Intelligence: The Means for Developing More Effective Structural Optimization Programs", Proceedings Int. Conf. on Comp. Mech.(ICCM 86), May 1986, Tokyo, Japan, 103-108.

Jozwiak, S.F., (1989), "A Knowledge-Based Interface for Structural Design and Optimization Programs", in Gero, J.S., (1989), (editor), "Artificial Intelligence in Design", Computational Mechanics Publications, Southampton, 197-208.

Lehner, K., Hartmann, D., (1991), "Knowledge-Based Software Components in Structural Optimization", Conf. on Computer Aided Optimum Design of Structures", June 25-27, Boston, Ma., USA.

Mackerle, J., Orsborn, K., (1988), "Expert systems for Finite element analysis and design optimization-a review", Engineering Computations, v.5, 90-102.

Morris, A.J., (1987), "Potential of the AI Methods in Optimisation", in Mota Soares, C.A., (editor), (1987), "Computer Aided Optimal Design: Structural and Mechanical Systems", Springer Verlag, Berlin Heidelberg, 1987.

Papalambros, P., (1987), "Knowledge-Based Systems in Optimal Design", in Mota Soares, C.A., (editor), (1987), "Computer Aided Optimal Design: Structural and Mechanical Systems", Springer-Verlag, Berlin Heidelberg.

Rajeev, S., Kirshnamoorthy, C.S., (1987), "Expert System for Optimal Design of Structural Steel Memebers", in Siram, D., Adey, R.A., (1987), (editors), "Knowledge Based Expert Systems in Engineering: Planning and Design", Computational Mechanics Publications, Southampton, 405-419.

Rodriguez-Velazquez, J., Seireg, A.A., (1985), "Optimimizing the Shapes of Structures via a Rule-Based Computer Program", Computers in Mechanical Engineering, July, 20-28.

Rogers, J.L., Barthelemy, J-F. M., (1985), "An Expert System for choosing the best combination of options in a general-purpose program for automated design synthesis", NASA Technical Memorandum 86329, Langley Research Center, Hampton.

13.11 FINITE ELEMENT ANALYSIS, MODELLING AND IDEALISATION

Baugh, J.W., Rehak, D.R., (1987), "FOOBAR: An Artificial Intelligence Based Finite Element System", in Siram, D., Adey, R.A., (1987), (editors), "Knowledge Based Expert Systems in Engineering: Classification, Education and Control", Computational Mechanics Publications, Southampton, 209-220.

Forde, B.W.R., Stiemer, S.F., (1988), "ESA: Expert Structural Analysis for Engineers", Computers and Structures, v.29, n.1, 171-174.

Gregory, B.L., Shepard, M.S., (1987), "The Genaration of Airframe Finite Element Models using an Expert System", Engineering with Computers, 2, 65-77.

Kissil, A., Kamel, H.A., (1986), "An Expert System Finite Element Modeler", in Siram, D., Adey, R., (1986), (editors), "Applications of Artificial Intelligence in Engineering Problems: Proceedings of the 1st Int. Conf., Southampton University, April 1986", Computational Mechanics Publications, Southampton, 1179-1186.

Mackerle, J., Orsborn, K., (1988), "Expert Systems for Finite Element Analysis and Design Optimization-A Review", Engineering Computations, v.5, 90-102.

Rank, E., Babuska, I., (1987), "An Expert System for the Optimal Mesh Design in the hp-version of the Finite Element Method, Int. J. for Numerical Methods in Eng., v.24, 2087-2106.

Shephard, M.S., Korngold, E.V., Wentorf, R., (no date), "Design Systems Supporting Engineering Idealisations", Technical Repaort, TR-89004, Rensselaer Design Research Center, Rensselaer Polytechnic Institute, Troy, New York.

Shephard, M.S., Korngold, E.V., Wentorf, R., (1989), "Idealised Models in Engineering Analysis", in Nelson, Jr., J.K., "Computer Utilization in Structural Engineering" American Society of Civil Engineers, New York, 81-90

Taig, I.C., (1986), "Expert Aids to Finite Element System Applications" in Siram, D., Adey, R., (1986), (editors), "Applications of Artificial Intelligence in Engineering Problems: Proceedings of the 1st Int. Conf., Southampton University, April 1986", Computational Mechanics Publications, Southampton, 759-769.

13.12 STRUCTURAL ASSESSMENT

Hadipriono, F.C. and Ross, J.T., (in press), "A Rule-Based Fuzzy Logic Deduction Technique for Damage Assessment of Protective Structures", in a Special Issue of Journal of Fuzzy Set and Fuzzy System.

Hadipriono, F.C., and Ross, T.J., (1987), "Towards a Rule-Based Expert System for Damage Assessment of Protective Structures", Proceedings of the Second International Fuzzy Set Association Congress, Tokyo, Japan, 156-159.

Ross, T.J., and Hadipriono, F.C., (1987), "Fuzzy Production Rules for a Damage Assessment Expert System", Proceedings of the International Symposium on Fuzzy Systems and Knowledge Engineering, Guangzhou, China, 249-256. ,

Ross, T.J., (1988), "Approximate Reasoning in Structural Damage Assessment", in Adeli, H., (1988), (editor), "Expert Systems in Construction and Structural Engineering", Chapman and Hall, London.

Ross, T.J., Hadipriono, F.C., (1987), "Fuzzy Production Rules for Damage Assessment Expert System", Proceedings of the International Symposium on Fuzzy Systems and Knowledge Engineering, Guangdong, China,

Ross, T.J., Sorensen, H.C., Savage, S.J., Carson, J.M., (1990), "DAPS: Expert System for Structural Damage Assessment", Journal of Computing in Civil Engineering, ASCE, v.4, n.4, 327-348.

Ross, T.J., Wong, F.S., Savage, S.J., Sorensen, H.C., (1986), "DAPS An Expert System for Damage Assessment of Protective Structures", in Kostem, C.N., Maher, M.L., (1986), "Expert Systems in Civil Engineering", American Society of Civil Engineers, New York.

Shiraishi, N., Furuta, H., Umano, M., Kawakami, K., (1989), "Knowledge-Based Expert System for Damage Assessment Based on Fuzzy Reasoning", in Topping, B.H.V., (Editor), (1989), "Artificial Intelligence Techniques and Applications for Civil and Structural Engineers", Civil-Comp Press, Edinburgh, 211-216.

Tee, A.B., Bowman, M.D., Sinha, K.C., (1988), "A Fuzzy Mathematical Approach for Bridge Condition Evaluation", Civil Engineering Systems, v.5, 17-24.

14.0 EDUCATION: Structural Engineering

Alkass, S., Harris, F.C., (1989), "The Educational Value of Expert Systems", in Topping, B.H.V., (Editor), (1989), "Artificial Intelligence Techniques and Applications for Civil and Structural Engineers", Civil-Comp Press, Edinburgh, 275-284.

Arciszewski, T., (1989), "Teaching Expert Systems Techniques at Wayne State University" in Mohan, S., Maher, M.L., (1989), (editors), "Expert Systems for Civil Engineers: Education", American Society of Civil Engineers, New York.

Argialas, D.P., (1989), "Teaching Expert Systems Techniques at Louisiana State University" in Mohan, S., Maher, M.L., (1989), (editors), "Expert Systems for Civil Engineers: Education", American Society of Civil Engineers, New York.

Barker, P., (1988), "Expert Systems in Engineering Education", Eng. Appli of AI, v.1, 47-58.

Clancey, W.J., Letsinger, R., (1987), "NEOMYCIN: Reconfiguring a Rulebased Expert System for Application to Teaching", Artificial Intelligence: Proceedings of the 7th IJCAI, August 1987, Vancouver, Canada, 829-836.

Durkin, J., (1990), "Introducing Students to Expert Systems", Expert Systems: The International Journal of Knowledge Engineering, v.7, n.2, 70-80.

Gunaratnam, D.J., (1987), "A Knowledge-Based Expert System for Tutoring in Structural Engineering", in Topping, B.H.V. (1987), (editor), "The Application of Artificial Intelligence Techniques to Civil and Structural Engineering", Civil-Comp Press, Edinburgh.

Ludvigsen, P., Maher, M.L., (1989), "Hardware and Software for Expert Systems Courses", in Mohan, S., Maher, M.L., (1989), (editors), "Expert Systems for Civil Engineers: Education", American Society of Civil Engineers, New York.

MacFarlane, A.G.J., (1990), "Interactive Computing: A revolutionary medium for teaching and design", Computer and Control Engineering Journal, July, 149-158.

MacLeod, I.M., Lun, V., (1990), "The Development of a Rule-Based Expert-System Tool for Engineering Teaching and Research", Engineering Application of Artificial Intelligence, v.3, n.2, 145-156.

Maher, M.L., (1989), "The ASCE Short Course on Expert Systems Techniques", in Mohan, S., Maher, M.L., (1989), (editors), "Expert Systems for Civil Engineers: Education", American Society of Civil Engineers, New York.

Niedzwecki, J.M., (1989), "Teaching Experts Systems Techniques at Texas A & M University", in Mohan, S., Maher, M.L., (1989), (editors), "Expert Systems for Civil Engineers: Education", American Society of Civil Engineers, New York.

Niedzwecki, J.M., Earles, J.A., (1990), "Strategy for Design of Intelligent Tutoring Systems", Journal of Computing in Civil Engineering, ASCE, v.5, n.2, 211-229.

Palmer, R.N., (1989), "Teaching Expert Systems Short Course, A Personal Perspective", in Mohan, S., Maher, M.L., (1989), (editors), "Expert Systems for Civil Engineers: Education", American Society of Civil Engineers, New York.

Rasdorf, W.J., Storaasli, O.O., (1987), "Educational Fundamentals of Computer-Aided Engineering", International Journal of Applied Engineering Education, v.3, n.3, 247-254.

Roy, G.G., (1989), "Force-Solver: A Knowledge-based approach to teaching statics", Civil Engineering Systems, v.6, n.1/2, 44-50.

Sriram, D., (1989), "Teaching Expert Systems Techniques at M.I.T." in Mohan, S., Maher, M.L., (1989), (editors), "Expert Systems for Civil Engineers: Education", American Society of Civil Engineers, New York.

Whiting, J., (1988), "Development of intelligent tutoring systems for open and distance learning", Eng. Appli. of AI, v.1, 126-137.

Wright, J.R., (1989), "Teaching Expert Systems Techniques at Purdue", in Mohan, S., Maher, M.L., (1989), (editors), "Expert Systems for Civil Engineers: Education", American Society of Civil Engineers, New York.

15.0 Acknowledgements

We should like to thank the many authors who sent us copies of papers, reports and books.